Sustainable Crop Protection under Protected Cultivation

P. Parvatha Reddy

Sustainable Crop Protection under Protected Cultivation

Springer

P. Parvatha Reddy
Indian Institute of Horticultural Research
Bangalore, Karnataka, India

ISBN 978-981-13-5733-6 ISBN 978-981-287-952-3 (eBook)
DOI 10.1007/978-981-287-952-3

Springer Singapore Heidelberg New York Dordrecht London

Softcover re-print of the Hardcover 1st edition 2016

Printed on acid-free paper

Springer Science+Business Media Singapore Pte Ltd. is part of Springer Science+Business Media (www.springer.com)

Foreword

भारतीय कृषि अनुसंधान परिषद
कृषि अनुसंधान भवन-II
पूसा, नई दिल्ली -110 012

डॉ एन. के. कृष्ण कुमार
उप महानिदेशक (बाग. वि.)
Dr. N. K. Krishna Kumar
DEPUTY DIRECTOR GENERAL (HORTICULTURAL SCIENCE)

INDIAN COUNCIL OF AGRICULTURAL RESEARCH
KRISHI ANUSANDHAN BHAWAN-II
PUSA, NEW DELLHI-110 012
June 10, 2015

Indian agriculture is passing through transition in this century under the influence of globalized trade and liberalization of Indian economy during early 1990s. Horticulture in particular has attained a pride place contributing 28.65 % to GDP from merely 8.5 % of area and has a high potential for further value addition and foreign exchange. In view of the competitive world environment in the context of the WTO agreement, renewed concentration on productivity, quality, and value addition has become imperative. To keep pace with the growing population, we need to increase horticulture production to 540 million tons by 2050 from the current level of 277.35 million tons (NHB 2014). The prospect of enhancing horticultural production by increasing area is limited. Thus, it becomes absolutely essential to increase the productivity.

Adoption of modern technologies in horticulture (protected cultivation, micro-irrigation, fertigation, micro-propagation, high-tech postharvest management including cold chain, etc.) has become essential to enhance productivity. Protected cultivation has a tremendous potential to increase productivity, especially under peri-urban horticulture.

Reducing the cost of production, a cropping pattern to minimize pest damage and reduced application of pesticides will significantly contribute to sustainable production, profitability, and food safety of vegetables produced in protected structures. The concept of sustainability should address holistically economic sustainability, soil health, and food safety.

In this context, the book on *Sustainable Crop Protection Under Protected Cultivation* by Dr. P. Parvatha Reddy provides valuable information. Such a commendable work comprehensively deals with important basic and applied aspects on integrated pest management in greenhouse crops using eco-friendly components.

I am sure that the book will be of immense value to policy makers and the scientific community involved in teaching, research, and extension activities on protected horticulture. It would also be useful for teaching at postgraduate level as well as for growers engaged in greenhouse cultivation in the country.

(N.K. Krishna Kumar)

Preface

The prospects of increasing production of horticultural crops by increasing land under cultivation are very much limited. The average productivity of most horticultural crops in India is low. There is a wide gap between yields obtained and potential yields with improved varieties and technologies. The productivity enhancing technologies include the use of high-yielding varieties/F1 hybrids, high-density planting, integrated nutrient and pest management, micro-irrigation and fertigation, use of bioregulators, genetically modified crops, and protected cultivation. Among the productivity enhancing technologies, protected cultivation has a tremendous potential to increase the yield of vegetables and flower crops by several-folds. Hence, it is essential to increase productivity of horticultural crops through protected cultivation in order to meet the future demand and fulfillment of commitment to the sustainable horticultural development.

High-tech production of high-value horticultural crops under protected conditions adopting modern technology is the recent development in this field. Greenhouse crop production is now a growing reality throughout the world with an estimated 405,000 ha of greenhouses spread over in more than 55 countries in the world where cultivation of crops is undertaken on a commercial scale under cover and it is continuously growing at a fast rate internationally.

Tomato, capsicum, cucumber, lettuce, rose, carnation, gerbera, orchids, and anthuriums are the most extensively grown vegetable and ornamental crops under greenhouses to achieve higher returns. The purpose of growing crops under greenhouse conditions is to extend their cropping season and to protect them from adverse environmental conditions, such as extreme temperatures and precipitation, and from diseases and pests. Quality is a high priority for greenhouse crops, requiring much care in pest and disease management, not only to secure yields but also to obtain a high cosmetic standard. Historically, not enough attention has been paid to exploiting and amending production technology for the control of pests and diseases. This makes the control of pests and diseases in protected crops even more challenging, with many important problems being unresolved and new ones arising as the industry undergoes more changes in production systems.

Pests and diseases are one of the major challenges to protected cultivation (glasshouses and plastic houses and tunnels). Year-round warm temperatures and relatively high humidity and abundant food make the protected environment of greenhouses highly favorable to pests and diseases. While much is known about disease epidemiology and insect behavior, scant attention, however, has been paid to the manipulation of greenhouse environments expressly to avoid disease epidemics and insect infestations, which together can easily account for 30 % crop losses. This is a significant factor in a grower's balance sheet which is often overlooked and usually dealt with simplistically by indiscriminate pesticide applications. Hence, integrated management must be incorporated into the present production systems, and appropriate research and experimentation programs must be developed in order to generate a pest and disease control technology adapted to the ecological conditions and predominant species.

The information on pest and disease management under protected cultivation of horticultural crops is very much scattered. There is no book at present which comprehensively and exclusively deals with the above aspects. This book deals with pests (insect and mite) and diseases (fungal, bacterial, viral, and nematode) in protected horticultural crops (fruits, vegetables, and ornamentals) in detail using physical, cultural, chemical, biological, host resistance, and integrated methods. This book is divided into 35 chapters which comprehensively cover all the aspects of pest management in horticultural crops in greenhouse.

The book opens with initial chapters describing the scenario where integrated pest and disease control operates, that is, the greenhouse and its environment. Ensuing chapters provide the basic strategies and tactics of different control methods including integrated control, with special reference to greenhouse crops. Further chapters include the different facets of biological pest and disease control – its scientific bases, its development in practice, and its commercialization and quality control. The preeminence of biological control in the book is not surprising since without doubt it is the cornerstone of integrated insect pest control and is also becoming increasingly more important in disease control. The concluding chapters of the book show us the present situation of integrated pest and disease control in the most important greenhouse crops (fruits, vegetables, and flower crops) worldwide. The final chapter deals with the future challenges assigned to researchers to identify non-pesticide methods and integrate sustainable pest management technologies that can contribute to increased productivity, such as breeding for durable resistance, biological control, and cultural methods, and which will have minimal adverse environmental and social impacts.

This is a reference book which attempts to provide postgraduate and professional readers already familiar with the subject with a means to acquire deeper knowledge on integrated control of pests and diseases in greenhouse crops and furthermore suggest possible roads to take in future tasks. Another decisive stimulant for this endeavor was the realization of the growing need to incorporate integrated systems of protection from arthropod pests and diseases for the thousands of hectares of protected crops in the world. The fruit, vegetable, and ornamental plant markets and the technical and economic efficiency of crop protection require these integrated control systems. The book adopts multidisciplinary approach (involving entomology, plant pathology, nematology, and horticulture) in addressing both basic and applied aspects of integrated pest management in greenhouse crops.

This book will be of immense value to the scientific community as a whole involved in teaching, research, and extension activities. The material can also be used for teaching postgraduate courses. The book can serve as a very useful reference for policy makers and practicing farmers. Suggestions to improve the contents of the book are most welcome (e-mail: reddy_parvatha@yahoo.com). The publisher, Springer (India) Pvt Ltd, New Delhi, deserves commendation for their professional contribution.

Bangalore, India
June 25, 2015

P. Parvatha Reddy

Contents

About the Author

Dr. P. Parvatha Reddy obtained his Ph.D. degree jointly from the University of Florida, USA, and the University of Agricultural Sciences, Bangalore.

Dr. Reddy served as the Director of the prestigious Indian Institute of Horticultural Research (IIHR) at Bangalore from 1999 to 2002 during which period the Institute was honored with "ICAR Best Institution Award". He also served as the Head, Division of Entomology and Nematology at IIHR and gave tremendous impetus and direction to research, extension and education in developing bio-intensive integrated pest management strategies in horticultural crops. These technologies are being practiced widely by the farmers across the country since they are effective, economical, eco-friendly and residue-free. Dr. Reddy has about 34 years of experience working with horticultural crops and involved in developing an F1 tomato hybrid "Arka Varadan" resistant to root-knot nematodes.

Dr. Reddy has over 250 scientific publications to his credit, which also include 30 books. He has also guided two Ph.D. students at the University of Agricultural Sciences, Bangalore.

Dr. Reddy served as Chairman, Research Advisory Committee, Indian Institute of Vegetable Research, Varanasi; Member, RAC of National Research Centre for Integrated Pest Management, New Delhi; National Research Centre for Citrus, Nagpur and the Project Directorate of Biological Control, Bangalore. He served as a Member, QRT to review the progress of AICRP on Nematodes; AINRP on Betelvine; Central Tuber Crops Research Institute, Trivandrum and AICRP on Tuber Crops. He also served as a Member of the Expert Panel for monitoring the research program of National Initiative on Climate Resilient Agriculture (NICRA) in the theme of Horticulture including Pest Dynamics and Pollinators. He is the Honorary Fellow of the Society for Plant Protection Sciences, New Delhi; Fellow of the Indian Phytopathological Society, New Delhi and Founder President of the Association for Advancement of Pest Management in Horticultural Ecosystems (AAPMHE), Bangalore.

Dr. Reddy has been awarded with the prestigious "Association for Advancement Pest Management in Horticultural Ecosystems Award", "Dr. G.I. D'souza Memorial Lecture Award", "Prof. H.M. Shah Memorial Award" and " Hexamar Agricultural Research and Development Foundation Award" for his unstinted efforts in developing sustainable, bio-intensive and eco-friendly integrated pest management strategies in horticultural crops.

Dr. Reddy has organized "Fourth International Workshop on Biological Control and Management of Chromolaena odorata", "National Seminar on Hitech Horticulture", "First National Symposium on Pest Management in Horticultural Crops: Environmental Implications and Thrusts" and "Second National Symposium on Pest Management in Horticultural Crops: New Molecules and Biopesticides".

1 Protected Cultivation

Abstract

Among the productivity-enhancing technologies, protected cultivation has a tremendous potential to increase the yield of vegetables and flower crops by several folds. Greenhouse crop production is now a growing reality throughout the world with an estimated 405,000 ha of greenhouses spread over all the continents. There are more than 55 countries now in the world where cultivation of crops is undertaken on a commercial scale under cover, and it is continuously growing at a fast rate internationally. The need for protected cultivation, present status, and different vegetable and flower crops grown in greenhouse are discussed.

Keywords

Protected cultivation • Flowers • Vegetables • Advantages • Limitations

1.1 Introduction

Horticulture in India is fast emerging as a major commercial venture, because of higher remuneration per unit area. Through the "Green Revolution" in the late 1960s, India achieved self-sufficiency in food production, which was hailed as a breakthrough on the farm front by international agricultural experts. But still the country has not achieved self-sufficiency in production of horticultural crops. There is an urgent need to increase horticultural production in order to provide nutritional security to the fast-growing population of the country (increasing at 1.6 % annually) which is expected to reach 1331 million by 2020. India plans to increase the production of horticultural crops to 200 million tons by 2020 from the current level of 125 million tons (NHB 2008).

The average productivity of most horticultural crops in India is low. There is a wide gap between yields obtained and potential yields with improved varieties and technologies. The prospects of increasing production of horticultural crops by increasing land under cultivation are very much limited. Hence, it is essential to increase productivity of horticultural crops in order to meet the future demand and fulfillment of the country's commitment to the sustainable horticultural development.

P.P. Reddy, *Sustainable Crop Protection under Protected Cultivation*,
DOI 10.1007/978-981-287-952-3_1

The productivity-enhancing technologies include use of high-yielding varieties/F1 hybrids, high-density planting, integrated nutrient and pest management, micro-irrigation and fertigation, use of bioregulators, GM crops, and protected cultivation. Among the productivity-enhancing technologies, protected cultivation has a tremendous potential to increase the yield of vegetables and flower crops by several folds.

Indian agriculture is passing through a transitional era in this century due to globalization of trade and liberalization of the Indian economy during the early 1990s. Horticulture in particular has attained pride status in the field of modern agriculture as horticulture crops contribute about 28.65 % to GDP from merely 8.5 % of area and have a high potential for value addition and foreign exchange earnings. High-tech production of high-value horticultural crops under cover as well as in open conditions adopting modern technology is the recent development in this field.

In view of the competitive world environment in the context of the WTO agreement, renewed concentration on productivity, quality, and value addition is one of the concerns of the farmers of our country. Hence, adoption of modern technologies in horticulture (greenhouse technology, micro-irrigation, fertigation, micro-propagation, nursery technology, high-tech postharvest technologies including cold chain, etc.) has become necessary to increase the productivity of horticultural crops.

Protected cultivation in India has started during the early 1990s and Karnataka is the leading state in adoption. Earlier it was started around Bangalore and surrounding districts, and recently progressive farmers in North Karnataka around Belgaum, Dharwad, Hubli, Bagalkot, Sirsi, etc. have started protected cultivation of flowers and vegetables, besides high-tech production of high-value horticulture crops under open field. Day by day, many progressive farmers, youngsters, and educated unemployed youth are taking keen interest in this field.

1.2 Protected Cultivation

Protected cultivation practices can be defined as a cropping technique wherein the microenvironment surrounding the plant body is controlled partially/fully as per the plants' need during their period of growth to maximize the yield and resource saving. The greenhouse is the most practical method of achieving the objectives of protected horticulture, where natural environment is modified by the use of sound engineering principles to achieve optimum plant growth and yield (more produce per unit area) with increased input use efficiency (Nagarajan et al. 2002).

The greenhouse is generally covered by a transparent or translucent material such as glass or plastic. The greenhouse covered with simple plastic sheet is termed as a polyhouse. The greenhouse generally reflects back about 43 % of the net solar radiation incident upon it allowing the transmittance of the "photosynthetically active solar radiation" in the range of 400–700-nm wavelength. The sunlight admitted to the protected environment is absorbed by the crops, floor, and other objects. These objects in turn emit long-wave thermal radiation in the infrared region for which the glazing material has lower transparency. As a result, the solar energy remains trapped in the protected environment, thus raising its temperature. This phenomenon is called the "greenhouse effect."

Tomato, bell pepper, cucumber, rose, carnation, and gerbera are the most extensively grown vegetable and ornamental crops under greenhouses to achieve higher returns (Chandra et al. 2000). Growing of crops using cost-effective plastic greenhouses provides an alternative for raising crop in the period of scarcity. This also ensures to meet year-round supply of fresh produce with more efficient resource utilization (Sharma et al. 2009). New features added to these structures have cut down the requirement of water and energy in such cultivation through novel means like micro-irrigation-cum-fertilization (fertigation) and rainwater harvesting.

1.2.1 Need for Protected Cultivation

Talukdar et al. (2003) reported the following major factors (Fig. 1.1) in protected cultivation:

- Productivity in the greenhouse is increased manifold in comparison to open field (Tables 1.1 and 1.2).
- Better-quality produce – protection from pests and diseases.
- Early maturity (Table 1.3).
- Round-the-year cultivation.
- Cultivation is possible in hostile climates.
- Create microclimate for optimum plant performance – controlled temperature, humidity, and light as per plant requirement.
- Cultivation of crops during off-season when it is not possible to grow in open fields.
- Tropical vegetables like cucurbits, capsicum, brinjal, okra, etc. are rare in hilly regions, but these crops can be grown in the greenhouse.
- The greenhouse provides an excellent opportunity to produce quality crops for export.
- Raising nursery earlier and advancing the availability are also possible by use of greenhouse.
- Conservation of valuable planting materials and its cultivation are also possible under the greenhouse.
- The greenhouse can also be used for growing of flower plants, strawberries, and grapes and propagation of quality fruit plants.
- Pest and weed management is easier in the greenhouse than in open field.
- The greenhouse conserves the moisture. Hence, frequency of watering of plants is reduced.
- The greenhouse is ideally suited for Indian farmers having small holdings.
- Productivity per unit area and time can be increased by adopting suitable crop sequences.

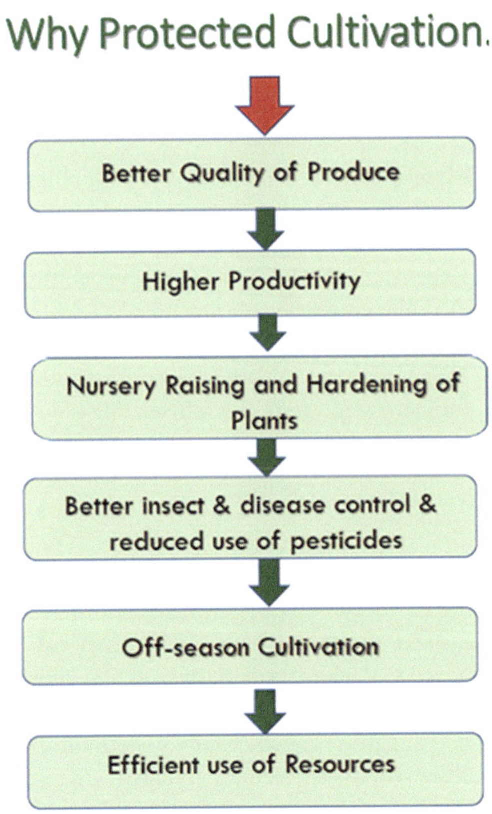

Fig. 1.1 The need for protected cultivation

Table 1.1 Comparative performance of Solanaceous vegetable crops in greenhouse

Crop	Fruit yield (kg/m^2)		
	Fiberglass-reinforced panel	Trench	Open
Capsicum	1.82	0.83	0.54
Brinjal cv. PH-5	1.25	1.09	0.63
Chili cv. BSS 344	1.35	1.24	0.13
Tomato cv. BSS 347	12.69	–	7.25

Table 1.2 Yield of capsicum and tomato under greenhouse at different locations in India

Location	Capsicum (t/ha)	Tomato (t/ha)	Cucumber (t/ha)
Pune	203	124	–
Coimbatore	148	186	–
Bangalore	147	152	–
Solan	79	95	–
Hissar	302	211	151

Table 1.3 Effect of greenhouse in achieving earliness in some Solanaceous vegetables

	Days for first picking		Harvesting period (days)	
Crop	Fiberglass-reinforced panel	Open	Polycarbonate	Open
Tomato	70	105	100	50
Capsicum	90	127	90	45
Brinjal	92	128	–	–

1.2.2 Advantages and Limitations

1.2.2.1 Advantages

Protected production can reduce the amount of water and chemicals used in production of high-value crops compared to open-field conditions. The comparative advantages are:

- Year-round production of crops.
- Adverse climate for crop production can be overcome by different systems of protected cultivation.
- Multiple cropping on the same piece of land is possible.
- Off-season crop production to get better return to growers.
- Production of high-quality and healthy seedlings for transplanting in open field supporting early crop and strong and resistant crop stands.
- Use of protected cultivation can increase production as well as productivity per unit of land, water, energy, and labor. It supports the production of high-quality and clean products.
- It makes cultivation of crops possible in areas where it is not possible in open conditions such as high-altitude deserts.
- It makes vertical cultivation of crops possible using technologies like hydroponics, aeroponics, etc. and by use of vertical beds for production.
- Disease-free seed production of high-tech crops becomes easy under protected structures.
- The potential of polyhouse production technology to meet the demand of producing good nutrition and healthy foods and quality vegetables free from pesticides can be fully exploited.
- Controlled environmental conditions are used for early raising of nurseries, off-season production of vegetables, seed production, and protecting the valuable germplasm.
- Crops can be grown under adverse weather conditions round the year and off-season.
- Management and control of insect pests, diseases, and weeds is easier.
- Maintenance of stock plants and cultivation of grafted plantlets and micro-propagated plants are easy.

1.2.2.2 Limitations

- Manual or hand pollination in cross-pollinated vegetables like cucurbits or development of their parthenocarpic hybrids/varieties.
- Expensive, short life, and nonavailability of cladding materials.
- Lack of appropriate tools and machinery.
- Structure cost initially looks unaffordable. Farmers with zero risk affordability do not come forward to adopt it.

1.2.3 Increase in Productivity

The yield under polyhouse cultivation can be achieved to the level of 5–8 times as compared to the open crop cultivation. Various trials conducted at agro-research centers in northern India indicate that capsicum (planted in mid-September), cucumber (planting in mid-October), and tomato (planted in November) under the polyhouse produced 1060, 1460, and 1530 kg/100 m^2. The duration of these crops was 4–9 months, and more than 90 % of total yield were obtained during off-season (during winter before the start of summer) which fetches a significantly higher market price (2–4 times than normal season). Further, the crop duration can be extended up to July–August with the application of micro-irrigation and fertigation and yield can be achieved to the level of 20–25 kg/m^2. Therefore, it is possible to harvest a single crop round the year with minimum additional inputs and higher income can be generated. Further cut flowers like carnations, gerbera, lily, rose, orchids, anthurium, etc. can be grown under polyhouses/net houses giving high returns and top-quality produce. The potential of floriculture under protected cultivation is huge for Indian and global markets.

1.3 Present Status

1.3.1 Global Scenario

The idea of growing plants in environmentally controlled areas had existed since the Roman times. The Roman Emperor Tiberius (42–37 BC) daily ate cucumber grown through artificial methods (similar to the greenhouse system). The gardeners planted cucumbers in carts that were wheeled into the sunlight by day and brought indoors by night to protect them from the elements. This was an innovative form of the greenhouse. In the sixteenth century, European explorers brought back exotic plants acquired in the course of their travels. Many were tropical plants that could not survive the cold European climates. The result was the creation of greenhouses that would progress from the Italian "botanical gardens" to wooden structures to the more stable cast-iron frames and eventually to glasshouses. Later, with the advent of plastics during World War II, a new phase in the greenhouse technology emerged. At present, nearly 90 % of the new greenhouses are being constructed by utilizing ultraviolet (UV)-stabilized polyethylene sheets as the glazing material.

Greenhouse crop production is now a growing reality throughout the world with an estimated 405,000 ha of greenhouses spread over all the continents (FAO 2013). There are more than 55 countries now in the world where cultivation of crops is undertaken on a commercial scale under cover and it is continuously growing at a fast rate internationally. China is the largest user of greenhouses. The development of greenhouse technology in China has been faster than in any other country in the world. With a modest beginning in the late 1970s, the area under greenhouses in China has increased tremendously. The world scenario of greenhouse production is given in Table 1.4.

In Europe, Spain is leading in protected agriculture with 52,170 ha mostly under low-cost polyhouses. In Asia, China has the largest area under protected cultivation, 2.76 million ha under polyhouse/greenhouse. Protected vegetable production is an important component of protected agriculture. Protected vegetable production is practiced throughout the world irrespective of the altitude of the place since several hundred years ago.

1.3.2 Indian Scenario

Protected cultivation technology is a relatively new technology for India. The total area covered under protected cultivation is approxi-

Table 1.4 Worldwide total area in major greenhouse production (Kacira 2011)

Countries	Greenhouse area (ha)
China	2,760,000
Korea	57,444
Spain	52,170
Japan	49,049
Turkey	33,515
India	30,000
Italy	26,500
Mexico	11,759
Netherlands	10,370
France	9,620
USA	8,425

mately 30,000 ha. There has been a very good development in this area during the last 5 years. The leading states in the area of protected cultivation are Maharashtra, Karnataka, Himachal Pradesh, northeastern states, Uttarakhand, Tamil Nadu, and Punjab. The major crops grown in the protected cultivation are tomato, bell pepper, cucumber, melons, rose, gerbera, carnation, and chrysanthemum. Nursery grown in the protected cultivation is becoming a very popular venture for income and employment generation.

India's first exposure to truly high-tech protected farming of vegetables and other high-value horticultural produce came through the Indo-Israel project on greenhouse cultivation, initiated at the New Delhi-based Indian Agricultural Research Institute (IARI) in 1998, shortly after the establishment of diplomatic ties with that country. However, the Israeli experts left India in 2003 at the end of the 5-year project; IARI continued to maintain the facility, calling it the Centre for Protected Cultivation Technology (CPCT). It has, in the past 10 years, managed to refine and upscale the system to reduce costs, besides designing greenhouse structures to suit local conditions.

1.4 Protected Cultivation of Horticultural Crops

Greenhouses are often used for growing flowers (rose, carnation, gerbera, chrysanthemum, orchids), vegetables (tomato, bell pepper, cucumber, cabbage), fruits (strawberry), and transplants (Figs. 1.2 and 1.3). Special greenhouse varieties of certain crops are generally used for commercial production. Many vegetables and flowers can be grown in greenhouses in late winter and early spring.

1.4.1 Flower Crops

1.4.1.1 Rose

In an evaluation of rose varieties for commercial cultivation under polyhouse conditions, red variety Tropical Amazon produced higher productivity of 127.00 flowers/m^2 and was found suitable for polyhouse cultivation.

1.4.1.2 Carnation

In an evaluation trial on carnation varieties under a greenhouse, it was found that variety Yellow Firato produced maximum number of flowers (8.45/plant and 280.60/m^2) followed by Firato (8.37/plant and 277.58/m^2).

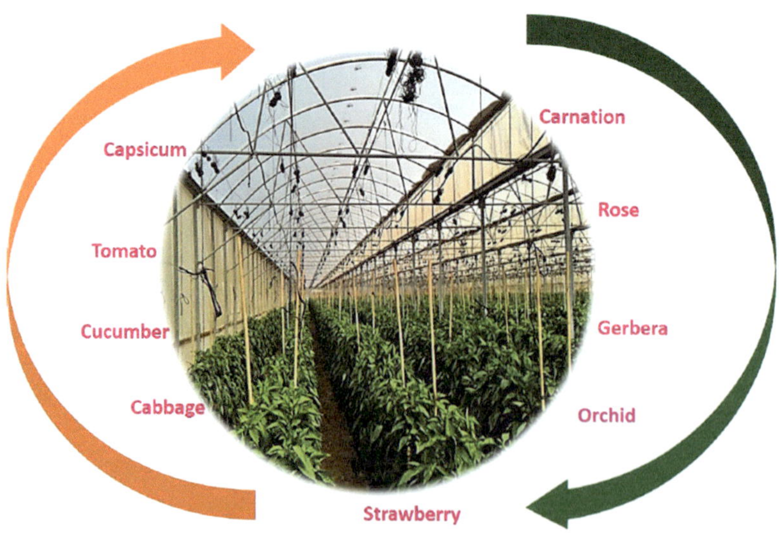

Fig. 1.2 Crops under protected cultivation

Fig. 1.3 Crops under shade net house

1.4.1.3 Gerbera

In an evaluation of gerbera varieties for commercial cultivation, the red variety Savannah produced higher productivity of 235.20 flowers/m^2 and was found profitable for polyhouse cultivation.

1.4.1.4 Alpinia

In a nutritional study on alpinia grown under a shade house (Fig. 1.4), it was observed that a higher level of NPK (15 g/plant) showed higher vegetative growth and produced maximum flower yield of 27.50 stems/m^2 with superior-quality flowers having maximum stem and flower length (132 and 26 cm, respectively).

1.4.1.5 Bird-of-Paradise

In a trial on crop geometry for cultivation of bird-of-paradise (BOP) under a shade house (Fig. 1.5), it was noticed that closer spacing of 75 × 60 cm produced maximum mean flower stem yield of 11.66/m^2 with a higher benefit-to-cost ratio of 2.75.

1.4.1.6 Orchids

In a varietal evaluation of orchids for commercial cultivation under a greenhouse, purple variety Sonia-17 produced higher productivity of 88.50 flowers/m^2.

The study on the influence of growing media and system for cultivation of orchids (variety Sonia-17) under a greenhouse revealed that a pot system having wider openings and growing media as brick pieces + charcoal was best for higher productivity of 74.44 flower spikes/m^2 with superior-quality flowers.

1.4.1.7 Anthurium

In an evaluation of anthurium varieties for commercial cultivation under a greenhouse, variety Esmeralda produced higher productivity of 83.36 flowers/m^2 and was found to be a highly profitable variety for polyhouse cultivation.

The standardization trial for cultivation of anthuriums under greenhouse conditions revealed that a growing system of 90 % shading net as medium holding material was best for high productivity of 126.50 flowers/m^2 with superior-quality flowers.

1.4.2 Vegetable Crops

1.4.2.1 Tomato

In a study on standardization of training systems for tomato under a shade house, the results revealed that, among the two training systems, the double-stem method recorded maximum yield of 160.55 t/ha.

Fig. 1.4 Alpinia under protected cultivation

Fig. 1.5 Bird-of-paradise under protected cultivation

1.4.2.2 Bell Pepper

In an evaluation trial of bell pepper hybrids for commercial cultivation under protected conditions, green hybrid Indra produced a higher yield of 85.69 t/ha with a benefit-to-cost ratio of 2.75, but colored hybrids red Bomby and yellow Orobelle produced an average yield of 75.83 t/ha with a higher benefit-to-cost ratio of 3.25.

In a planting geometry study for cultivation of bell pepper hybrids under different growing conditions, planting at 45×30 cm under polyhouse conditions produced a higher yield of 95.72 t/ha with superior-quality produce, while, under shade house conditions, planting at 45×45 cm produced a higher yield of 81.70 t/ha with superior-quality produce.

In a standardization trial of training of plants for cultivation of bell pepper hybrids under different growing conditions, training of four stems per plant for bell pepper was found best for a higher yield of 77.25 t/ha with superior-quality produce under all different growing conditions and hybrids tried.

1.4.2.3 Lettuce

In a study on the influence of spacing on lettuce grown under open and protected conditions, lettuce leaf yield per hectare was maximum under shade house (24.0 t/ha) followed by polyhouse (19.60 t/ha) and least under open field (13.45 t/ha). With respect to planting geometry, leaf yield per hectare was maximum in 30×30-cm spacing (24.30 t/ha).

1.4.2.4 Chinese Cabbage

In the spacing trial of Chinese cabbage grown under open and protected conditions (Fig. 1.6), Chinese cabbage head yield per hectare was maximum under shade house (63.65 t/ha) followed by polyhouse (52.20 t/ha) and least under open field (24.35 t/ha). With respect to planting geometry, head yield per hectare was maximum in 45 × 45-cm (51.76 t/ha) spacing.

1.4.2.5 Broccoli

The study on influence of spacing on broccoli grown under open and protected conditions revealed that broccoli flower head yield per hectare was maximum under shade house (15.05 t/ha) followed by polyhouse (13.10 t/ha) and least under open field (6.7 t/ha). With respect to planting geometry, flower head yield per hectare was maximum in 45 × 45-cm (14.5 t/ha) spacing.

1.4.2.6 Celery

In a trial on influence of spacing on celery grown under open and protected conditions (Fig. 1.7), the celery leafstalk yield per hectare was maximum under shade house (23.15 t/ha) followed by polyhouse (20.02 t/ha) and least under open field (14.10 t/ha). With respect to planting geometry, leafstalk yield per hectare was maximum in 30 × 15-cm (30.0 t/ha) spacing.

1.4.2.7 Red Cabbage

In a trial on influence of spacing on red cabbage under open and protected conditions (Fig. 1.8), head yield was maximum under shade house (35.75 t/ha) followed by polyhouse (32.85 t/ha) and least under open field (16.10 t/ha). With respect to planting geometry, head yield was maximum in 45 × 30-cm (34.0 t/ha) spacing.

1.4.2.8 Zucchini

In a trial on influence of spacing on zucchini grown under open and protected conditions (Fig. 1.9), the zucchini fruit yield was maximum under shade house (23.45 t/ha) followed by polyhouse (19.00 t/ha) and least (8.40 t/ha) under open field. With respect to planting geometry, fruit yield per hectare was maximum in 100 × 90-cm spacing (22.15 t/ha).

Fig. 1.6 Chinese cabbage under protected cultivation

Fig. 1.7 Celery under protected cultivation

Fig. 1.8 Red cabbage under protected cultivation

Fig. 1.9 Zucchini under protected cultivation

1.4.2.9 Leek

In a trial on influence of spacing on leek under protected conditions (Fig. 1.10), the yield per hectare was maximum under shade house (39.41 t/ha) followed by polyhouse (33.75 t/ha) and least (19.55 t/ha) under open field. With respect to planting geometry, yield per hectare was maximum in 20 × 15-cm spacing (30.5 t/ha).

1.5 Conclusions

The protected cultivation of high-value crops has become irreplaceable both from economic and environmental points of view. It offers several advantages to grow high-value crops with improved quality even under unfavorable and marginal environments. However, high training needs of the greenhouse growers and some poor-quality produce with pesticide residues have been a matter of great concern. These issues can easily be addressed by integrating various production and protection practices including location-specific designing and construction of the polyhouses for efficient input use. Creating awareness among the greenhouse growers for judicious use of pesticides for safe production can be instrumental in providing quality products without polluting the environment.

Fig. 1.10 Leek under protected cultivation

References

Chandra P, Sirohi PS, Behera TK, Singh AK (2000) Cultivating vegetables in polyhouse. Indian Hort 45:17–25

FAO (2013) Good agricultural practices for greenhouse vegetable Crops. FAO plant production and protection paper 217. Food and Agriculture Organization of the United Nations, Rome

Kacira M (2011) Greenhouse production in US: status, challenges, and opportunities. In: Presented at CIGR 2011 conference on sustainable bioproduction WEF 2011, Tower Hall Funabori, Tokyo

Nagarajan M, Senthilvel S, Palanysamy D (2002) Material substitution in greenhouse construction. Kisan World 11:57–58

National Horticulture Board (2008) National Horticulture Board data base. (http://nhb.gov.in/)

Sharma M, Negi S, Kumari S (2009) Effect of different growing media and fertigation levels on production of cucumber (*Cucumis sativus* L.) under protected conditions in the hills. Indian J Agric Sci 79(10):853–856

Talukdar MC, Sarma B, Das S, Mahanta S (2003) Evaluation of spray chrysanthemum cultivars under open and polyhouse conditions. In: National symposium on recent advances in Indian floriculture, Kerala Agricultural University, Kerala

2 Greenhouse Technology

Abstract

The purpose of growing crops under greenhouse conditions is to extend their cropping season and to protect them from adverse environmental conditions, such as extreme temperatures and precipitation, and from diseases and pests. Greenhouse production normally requires a high level of technology to obtain adequate economic returns on investments. Quality is a high priority for greenhouse crops, requiring much care in pest and disease management, not only to secure yields but also to obtain a high cosmetic standard. Agroclimate, types of glazing materials, types of greenhouses, and greenhouse environmental control are some of the topics discussed in this chapter.

Keywords

Greenhouse production • Agroclimate • Glazing materials • Greenhouse • Environmental control

2.1 Introduction

Greenhouse technology implies production of plants for economic use in a covered structure that allows rapid harvesting of solar radiation and modification of agroclimatic conditions conducive for plant growth and development. The technology embraces infrastructure modeling, selection of plants for adaptation, production economics, agronomic management and commercial potential, etc.

"A greenhouse is a framed or an inflated structure covered with a transparent or translucent material which permits at least partial control of plant environment and which are large enough to permit a person to carry out cultural operations" (Chandra and Panwar 1987). While full advantage is taken of the available sunshine for crop production by way of selecting proper covering materials, the enclosure provides an opportunity to control the other environmental parameters. As a result, greenhouse crop productivity is

P.P. Reddy, *Sustainable Crop Protection under Protected Cultivation*,
DOI 10.1007/978-981-287-952-3_2

largely independent on outdoor environmental conditions.

The purpose of growing crops under greenhouse conditions is to extend their cropping season and to protect them from adverse environmental conditions, such as extreme temperatures and precipitation, and from diseases and pests (Hanan et al. 1978). Greenhouse structures are essentially light scaffolding covered by sheet glass, fiberglass, or plastic. Such materials have a range of energy-capturing characteristics, all designed to maximize light transmission and heat retention. Crops may be grown in ground bed soil, usually amended with peat or farmyard manure, in benches, in pots containing soil or soil mixtures or soil substitutes, and in hydroponic systems, such as sand or rock wool cultures and flowing nutrient systems, without a matrix for the roots.

Modern technology has given the grower some powerful management tools for production. Generally, added-value crops are grown under protection. Most of them are labor intensive and energy demanding during cold weather. Greenhouse production therefore normally requires a high level of technology to obtain adequate economic returns on investments. Quality is a high priority for greenhouse crops, requiring much care in pest and disease management, not only to secure yields but also to obtain a high cosmetic standard. Although technological changes are ultimately intended to reduce production costs and maximize profits, precise environmental and nutritional control push plants to new limits of growth and productivity. This can generate chronic stress conditions, which are difficult to measure but apparently conducive to some pests and diseases. Historically, not enough attention has been paid to exploiting and amending production technology for the control of pests and diseases. This makes the control of pests and diseases in protected crops even more challenging, with many important problems being unresolved and new ones arising as the industry undergoes more changes in production systems.

Growing of vegetable and floriculture crops under protected cultivation is receiving utmost attention and gaining popularity among farming community across India. The government of India launched the Horticulture Technology Mission Project, and under this program, a large number of polyhouses in the states of Kerala, Himachal Pradesh, Karnataka, Maharashtra, the northeastern hill region, Haryana, Gujarat, Tamil Nadu, Andhra Pradesh, and the NCR region have come up in a big way. The protected cultivation has shown promise in respect of higher crop productivity both in terms of quality and quantity of the produce.

Haryana Agricultural University, Hisar, has achieved the productivity of 302 t/ha in tomato, 211 t/ha in bell pepper, and 151 t/ha in cucumber per crop cycle.

2.2 Agroclimate

There are several climatic factors that need care in the management of a greenhouse. The following factors are important for greenhouse cultivation.

2.2.1 Sunlight

The direction of the greenhouse should be such that it receives sunlight for maximum hours. In hilly regions, greenhouse should face the southeast direction and its location should not be under shade. However, under the tropical condition, temperature retention can be reduced by choosing a less sunny site. Sunlight is essential for augmentation of the process of photosynthesis and minimum sun requirement of crop should be kept in mind. In addition, sunlight also provides requisite thermal environment for various other physiological processes including photosynthesis. Hilly regions receive clear sunny days in winter so sunlight as such may not be a limiting factor, but its tapping and conservation could be a major concern for further investigation.

2.2.2 Temperature

Sunlight is the major rather than the sole source of temperature. Examples of providing external heat source other than sunlight are rare in hilly regions. Temperature during night hours may drop down to below 0 °C and heat obtained from trapped solar radiation is not properly conserved. The following steps are advantageous to maintain a thermal environment of the greenhouse:

- Blackening of internal surface of brick wall and soil bed surface
- Providing insulation by doubling the brick wall
- Covering polyethylene or glass external surface with multilayer covering during night hours
- Mud plastering in the case of stone walls
- Placing of black-colored stones in the greenhouses
- Keeping containers filled with water

2.2.3 Relative Humidity

The minimum limit of relative humidity for normal physiological functioning of the plants is 50 %. Extreme xeric conditions hardly allow relative humidity to rise to this level in the outer environment, but inside the greenhouse, sufficient water application and restricted air circulation gear up and maintain relative humidity at higher level. It may be increased by keeping pan evaporation. A mechanical device automatically controls relative humidity by periodical sprinkling of water. Cooling pads are usually installed in low-humidity and tropical areas.

2.2.4 Carbon Dioxide Enrichment

Carbon dioxide (CO_2) is the basic ingredient required for manufacturing of organic matter by process of photosynthesis. In cold arid regions, CO_2 concentration is below normal (NAEDB 1992), and in the closed structures like greenhouses, it may drop down further because of its consumption by the plants. At relatively high sunlight intensities and temperature, photosynthetic rate is directly proportional to CO_2 concentration up to 2400 ppm. Hence supplementary application of CO_2 will boost plant productivity inside the greenhouse. Under the hilly region situation, appropriate air circulation is a better and economic means to prevent undesirable lowering of CO_2, while for conservation of CO_2 produced at night, a tightly closed greenhouse is a prerequisite.

2.2.5 Soil

Soil serves as growth medium for plants. Sandy loam soil is the best choice. It should be supplemented with well-rotten FYM, compost, or biofertilizers. Inorganic fertilizers need to be used as per recommendations. Continuous cultivation inside the greenhouse may give rise to buildup of pathogens. To destroy pathogens, pasteurization of soil is recommended (steam aeration) before every crop and soil replacement is recommenced after every 3 years. Steam aeration temperature of 60 °C for 30 min is good enough to accomplish the task. Chemical fumigation can also be used but steam is the best as it is not selective thus attacking all sort of pathogens.

2.3 Types of Glazing Materials

2.3.1 Glass

Glass has been the preferred covering material for greenhouses worldwide because of its light transmissivity characteristics. Transmissivity of 40–50-year-old glass differs a little from that of new glass. Temperature retention in a glasshouse is pretty good. However, high installation cost is a major limitation. Moreover, due to transportation of materials and frequent damage of glasses due to high wind velocity, a glasshouse is discouraged.

2.3.2 Polyethylene

Plastic polyethylene is the most widely used greenhouse film around the globe. It is produced by mixing homopolymers of ethylene with or without an ultraviolet (UV) inhibitor package, but only a UV-stabilized polyethylene sheet is recommended because non-UV inhibitor package-added polyethylene will break down after 3–5 months due to photochemical reactions. Polyethylene film is tough, flexible, and relatively inexpensive. It can withstand as low as −50 °C but 80–90 °C temperature will cause it to melt. Above 60 °C, it loses much of its strength and may stretch markedly. The life of UV-stabilized polyethylene film is 1–3 years. However, Rigidex polyethylene sheet may be used for 5–6 years on a trench.

2.3.3 Fiberglass-Reinforced Panel (FRP)

These panels consist of fiberglass-reinforced polyester. The panels have been very popular in areas of high light intensity such as Southern USA, but about 10–15 % heat loss due to increased exposed area makes it unfit for most of the Indian region. Further, with its age under use, the transparency gets reduced due to yellowing.

2.3.4 Polycarbonate Panels

The panels are available in double-skinned sheets. SDP polycarbonate sheets are similar in physical dimension to the SDP acrylite but are stronger and have a lower light transmissivity. Its price is approximately 25 % higher than SDP acrylite.

2.4 Types of Greenhouses

Various types of greenhouses are also seen in India. The brief description of each type is being discussed below.

2.4.1 Glasshouse

As its name indicates, glass is used as glazing material (Fig. 2.1) in this greenhouse. Perhaps this is the first and oldest among all types of greenhouse structures. Glass panels are fitted with the help of a wooden or metal frame. It can be of any shape and size and it is pretty effective for winter cultivation, but due to increase in day temperature in summer, it becomes unfit for cultivation during summer. High initial cost, difficulty in construction, and frequent damage of glass panels by strong winds are other discouraging factors. In hilly regions, only a few of such structures are found for research.

2.4.2 Polyhouse

Ideal features of polyethylene have increased the use of polyhouses (Fig. 2.2) in place of the glasshouse throughout the globe. It has not only reduced the initial cost but also increased the popularity of the greenhouse by simplifying the installation technology. Generally, there are two types of polyhouses in the hilly region.

2.4.2.1 Prefabricated Polyhouses

A prefabricated metal structure of convenient size is installed and a polyethylene film is covered over it. This structure is not recommended at all for the hilly region because of poor temperature retention, low crop yield, and high installation cost (Singh et al. 1998a).

2.4.2.2 Ladakh Polyhouse

This is one of the innovative and low-cost greenhouses of the Ladakh region. It is similar to normal, but the only difference is its surrounding mud brick wall in place of polyethylene sheets which not only cuts down the installation cost but also reduces the adverse effects of strong winds and also increases temperature retention and ultimately increases net profit. This is generally the lean type and has three sides made of mud bricks. The back wall is 7 ft in height, while the front has no wall. The length is 32 ft with a width of 16 ft

Fig. 2.1 Glasshouse

Fig. 2.2 Polyhouse

(Singh et al. 2000). The polyethylene is supported on wooden poles and sidewalls. The two sidewalls are descending toward the front. Jammu and Kashmir State Department of Agriculture/Horticulture provides cash subsidy besides the 32 × 16-ft polyethylene sheet. This has brought a boost to the greenhouse revolution in Ladakh.

2.4.3 Trench (Underground Greenhouse)

This is a unique, innovative, very simple, cheap, and useful underground greenhouse structure for the hilly region and thus has unlimited potential in the region. This may be of any convenient dimension. However, a trench of 30 × 10 × 3-ft size is ideal. In this pit type of structure, wooden poles are used to hold UV-stabilized polyethylene film. The polyethylene is also covered by an additional or woolen or cotton sheet polyethylene film during night to reduce the heat loss during extreme winter. The damage of blowing off the polyethylene film by strong winds is minimized by putting stones along the sides. Cultural practices and other operations are done by removing the polyethylene sheet from the top of the trench. The structure does not require much skill in its construction and management. Its cost is lowest among all other greenhouses, and being an underground structure, heat loss is minimal and temperature retention is high (Singh and Dhaulakhandi 1998) and thus yields good crop. Strong winds do not affect polyethylene cover much and hence it is long lasting. This structure is therefore being recommended as the most suitable greenhouse for the hilly region (Singh et al. 1998, 2000).

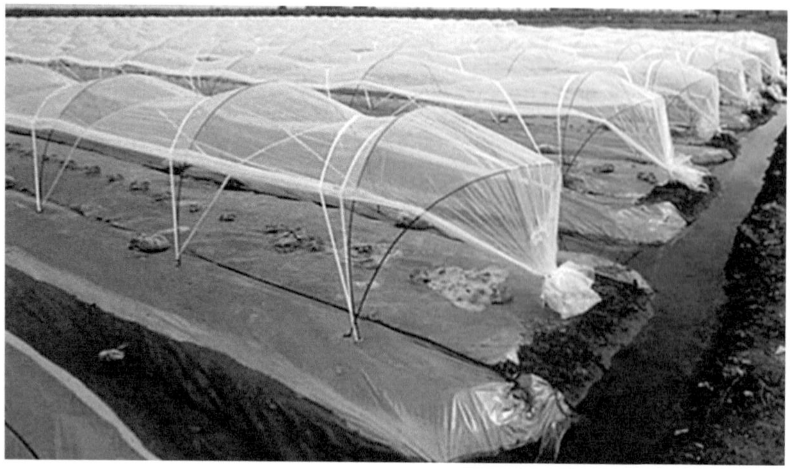

Fig. 2.3 Nylon netting single-bed tunnels

2.4.4 Plastic Low Tunnel

It is a small semi-spherical structure frame made of metal, wood, or plastic and covered with polyethylene or fiber-reinforced plastics to create a protected environment (Fig. 2.3). Plastic low tunnels are flexible transparent coverings that are installed over rows of individual beds of transplanted vegetables to enhance the plant growth by warming the air around the plants (microclimate) in open field during winters. These help in warming the soil and protect the crop against the hails and snow and advance the crop by 30–60 days than their normal season. This low-cost technology is very simple and highly profitable for off-season cultivation and catching the early market. Generally galvanized iron arches are fixed manually 1.5–2.0 m apart to support the plastic tunnel. Width and height is kept about 45–60 cm to cover the plant as per the crop requirement. Transparent non-perforated plastic of 30–50 thickness is sufficient. It partly reflects infrared radiation to keep the temperature of the low tunnel higher than outside. Small vents can be made at the side of the tunnel to facilitate proper aeration in tunnel. These structures serve as a mini greenhouse. This is a temporary structure which can easily be shifted to any place. However, a major problem in tunnels is faced in watering, weeding, and harvesting which involves removal of cover and again putting them back.

2.4.5 Double-Wall Polyench

This structure is designed by the Forest Research Laboratory (FRL), Leh, to harness the soil and solar heat for growing the vegetables especially during winter months. The polyench refers to a polyhouse erected over a trench where back walls and sidewalls are made up of mud bricks. The sidewalls are provided with double walls filled with insulating material available locally, i.e., sawdust for better retention of heat inside the polyench. The inner walls are painted black to absorb more solar radiation during morning hours. The structure is also like gable uneven span with modified roof having a polythene sheet toward the sun-facing side and grass thatch support over a wooden frame opposite to the sun-facing side with provision of ventilators in the roof. Since locally available material has been used in this greenhouse, the cost is reduced drastically compared to a normal greenhouse with better efficiency of temperature retention than the trench and polyhouse. Considering the wear and tear that occurred due to high wind velocity in the hilly regions, the polythene sheet has been replaced with FRL sheets to increase the life span of the structure. The structure has been accepted by the Ladakh Autonomous Hill Development Council (LAHDC), and the state government is providing a subsidy of 50% to the farmers for promoting the protected cultivation in the Ladakh region.

Fig. 2.4 Net houses

2.4.6 Multipurpose Nets

Shade nets are used to reduce the adverse effect of scorching sun and heavy rain (Fig. 2.4). Shade houses are becoming popular for growing crops and nursery during summer season. Net houses are used for raising vegetables/fruits/flowers/medicinal plants in high-rainfall regions. The roof of the structure is covered with suitable cladding material, mostly HDPE, which does not absorb moisture. Slides are made of wire mesh of different gauges 25–90% shade depending upon requirements. Such structures are popular in the northeastern region of the country. Similarly weed nets are made of HDPE and covered on the ground to control the weeds by reducing light on the soil. Insect-proof nets are effective to reduce the incidence of a number of pests and viral diseases in crops. These nets are used like a mosquito net around the crops, having 40–50-mesh size. Other similar nets include bird protection nets to control the bird damage of plants/crops, hail protection nets, reflector nets, etc.

2.5 Greenhouse Environmental Control

A greenhouse is essentially meant to permit at least a partial control of microclimate within it. The control of greenhouse environment means the control of temperature, light, air composition, and nature of the root medium. Obviously, a control over all these parameters makes a greenhouse a completely controlled structure.

2.5.1 Ventilation

A greenhouse is ventilated for either reducing the very high greenhouse air temperatures or for replenishing carbon dioxide supply or for moderating the relative humidity in the greenhouse. It is quite possible to bring greenhouse air temperature down during spring and autumn seasons by providing adequate ventilation for the greenhouse. The ventilation in a greenhouse could either be natural or forced. In the case of a small greenhouse (less than 6 m wide), natural ventilation could be quite effective during spring and autumn seasons. However, fan ventilation is essential to have precise control over air temperatures, humidity, and carbon dioxide levels. Orientation of the greenhouse is another important factor. An east–west-oriented freestanding greenhouse maintains better winter light level as compared to a north–south-oriented greenhouse. Therefore, in northern India, a greenhouse should be oriented in the east–west direction. Gutter-connected greenhouse should be oriented north–south to avoid continuous shading of certain portions of the greenhouse due to structural members.

A greenhouse structure has three distinct segments, i.e., frame, glazing material, and control/monitoring equipment. All the three components have different designed life periods. Whereas a metallic greenhouse frame is designed for a service period of 15–25 years, glazing materials have a life span of 2–20 years. Control and monitoring equipment normally wear out in 5–10 years. In the prevailing economic conditions, where capital is a scarce input, the choice often favors a low-initial-investment greenhouse. Galvanized mild steel pipe as a structural member in association with wide-width UV-stabilized polyethylene film is a common option selected by greenhouse designers.

A 600–800-gauge-thick polyethylene film can safely withstand normal wind loads prevailing in most parts of the country. A single-piece polyethylene film to cover a greenhouse is preferred due to material economy, easy handling, and improved environmental control. An 800-gauge-thick polyethylene film costs approximately Rs. 60–80/m^2 and has a service span of 2–5 years.

The selection of greenhouse equipment depends on local climate conditions and the crops to be grown. A heating unit is a must in cold regions and a cooling unit is required in almost all climates in India.

2.5.2 Cooling Systems

While ventilation may be used for cooling during autumn and spring seasons, other methods have to be employed for cooling during summers. Roof shading helps in cooling the greenhouses. The amount of solar radiant energy entering the greenhouse can be reduced by applying opaque coatings directly to the glazing or by placing wood or aluminum over the glazing. Commercial shading compounds or mixtures prepared with paint pigments are preferred for this purpose. White compounds are preferred for they reflect a maximum amount of sunlight, 83 % versus 43 % for green and 25 % for blue or purple.

2.5.2.1 Water Film on the Greenhouse Cover

To absorb infrared radiation, a water layer must be at least 1.0 cm thick. But on a sloping greenhouse roof, it is limited to about 0.05 cm which is not thick enough. Cooling is most effective when cold water is used in the water film.

2.5.2.2 Evaporative Cooling (EC)

The degree of cooling obtained from an evaporative system is directly related to the wet-bulb depression that occurs with a given set of climate conditions. EC systems are most effective in areas where a consistently low relative humidity exists.

2.5.2.3 Fan and Pad System (F & P)

It is adaptable to both large and small greenhouses. In this system, low-velocity and large-volume fans draw air through wet fibrous pads mounted on the opposite side or end wall of the greenhouse. The outside air is cooled by evaporation to 20 °C of the wet-bulb temperature. Either vertical or horizontal pads can be used in the F & P systems. However, vertical pads accumulate salts and sag rand; thus, they create openings that allow hot air to enter the greenhouse. Various materials, viz., gravel, pine bark, straw, burlap, aspen wood fiber (shredded *Populus tremuloides* mats), honeycomb paper, etc., can be used for the pad. However, pumice and volcanic rock (1–4 cm in diameter) are reported to function very satisfactorily.

2.5.2.4 High-Pressure Mist System

Water is sprayed into the air above the plants at pressures of 35–70 kg/cm^2 from low-capacity nozzles (1.8–2.8 L/h). Although most of the mist evaporates before reaching the plant level, some of the water settles on the foliage where it reduces leaf temperatures.

2.5.2.5 Low-Pressure Mist System

Misting with water pressure at less than 7 kg/cm^2 has achieved an air temperature that is 5 °C cooler in a greenhouse compared to natural ventilation. The water droplets from a low-pressure misting system are quite large and do not evaporate quickly. Leaching of nutrients from the foliage and the soil is a serious drawback of using this technique.

2.5.3 Humidity Control

For most crops, the acceptable range of relative humidity is between 50 % and 80 %. However, for plant propagation work, relative humidity up to 90 % may be desirable. Humidification in summers can be achieved in conjunction with greenhouse cooling by employing appropriate evaporative cooling methods, such as fan–pad and fogging systems. Sometimes during winters, when sensible heat is being added to raise the greenhouse air temperature during nights, the relative humidity level might fall below the acceptable limit.

2.5.4 Greenhouse Heating

There are essentially three main categories of efforts needed to maintain desirable greenhouse temperatures during winter:

- Design of energy-efficient greenhouse with passive solar heating components.
- Design of active heating systems based on renewable energy sources such as solar and biogas.
- Design of an active heating system based on conventional fuels.
- While the conventional fuel-based heating systems are many and dependable, the other two categories of efforts are still evolving.

2.5.4.1 Heating Systems Based on Conventional Fuels

Traditionally, glass greenhouses have been heated by hot water systems. Most of the larger commercial greenhouses are heated with some type of boiler system. Gas-fired unit heaters for greenhouse heating deliver heat at approximately half of the cost of the steam/hot water systems. The hot air is distributed through a perforated polytube running along the greenhouse length. Electric heaters, both radiative and convective, are simple and convenient to control. But, the nonavailability of electricity and its high cost limit its use to small and/or experimental greenhouse. The heating systems should not only raise the greenhouse temperatures but should also achieve uniformity of temperature distribution. Therefore, placement of heating units and the type of distribution system to be selected are important.

2.5.4.2 Heating Systems Based on Nonconventional Energy Sources

A number of active solar heating have been developed for the greenhouse during the last two decades in order to reduce the dependence of the greenhouse industry on conventional fuels. The normal components of a solar heating system are solar collector, heat transfer medium, heat exchanger, and heat storage. Biogas, generated from agricultural wastes/residues, can also be used for greenhouse heating.

2.5.4.3 Design of Passive Solar Greenhouse

A passive solar greenhouse is one which not only attempts to capture maximum solar energy but also minimizes the unwanted thermal exchange between the greenhouse and the surroundings in order to maintain desirable temperatures. The following points summarize the useful results so far:

- A greenhouse should be oriented east–west.
- The north side of a greenhouse structure should be thermally insulated.
- The north side of a greenhouse facing the crop should be covered with a reflective surface so that the sunlight incident on it from the south side is reflected on to the crops.

2.6 Future Thrusts

Greenhouse technology development has made good progress in India during the last two decades, and steps necessary for promoting greenhouse cultivation of flower and vegetable crops have been started. Energy-efficient greenhouse cultivation continues to be an area of active research and development, and this is sought to be achieved through precision equipment and protocols. While the gap between the demand and supply of most horticultural crops remains wide and the country plans to double the production of horticultural crops by 2012, the protected cultivation technology holds the key to meet the targets. It clearly emerges that greenhouse technology has multipurpose application for sustainable development of a hilly zone. During harsh winter which otherwise threatens the survival of human beings in these remote hilly areas, popularization and further improvements in greenhouse management could provide a sign of relief in solitude of white snow cover. Thrust areas for future research are enumerated below:

- Conservation of more thermal radiation in winter.
- Durable infrastructure modeling.
- Development of successful commercial models.
- Modernization of polygreenhouse to permit agroclimatic manipulations of higher magnitude.
- Identification of new crops/varieties suitable for cultivation inside the greenhouse.
- Socioeconomic impacts of polygreenhouse technology in hilly region/cold arid zone.
- Plastic films of suitable grades need to be made available in adequate quantity.
- Development of area-specific suitable greenhouse designs.
- Standardization of package of practices for growing various crops in greenhouse for specific regions.
- Varieties suitable for greenhouse cultivation need to be made available to enhance the productivity.
- Vocational training to the unemployed youth and greenhouse practitioners in remote areas to update their skills in this technology.
- Financial assistance such as low-interest loans should be made available for poor farmers to adopt this technology.
- Insurance of greenhouse structure and crop to protect the farmers against the loss caused by natural calamities.
- Rigorous extension program to disseminate the technology information to the poorest of the society.

References

Chandra P, Panwar J (1987) Greenhouse technology and its scope in India. In: Proceedings of the national symposium on use of plastics in agriculture, New Delhi, pp 62–66

Hanan JJ, Honey WD, Goldsberry KL (1978) Greenhose management. Springer, New York

NAEDB (1992) Action plans on cold deserts: an integrated approach for sustainable development. NAEDB, UHF, Solan – 173230 (HP)

Singh B, Dhaulakhandi AB (1998) Application of solar greenhouse for vegetable production in cold desert. In: Proceedings of world renewable energy congress, Florence, V:2311–2314

Singh B, Dwivedi SK, Paljor Eli (1998a) Studies on suitability of various structures for winter vegetable production at subzero temperatures (Abstr). In: 25th Indian horticulture congress, Belgium, p 290

Singh B, Dwivedi SK, Singh N, Paljor Eli (1998b) Sustainable horticultural practices for cold arid areas. In: International symposium on sustainable agriculture in Hill areas (Abstr). Himachal Pradesh Krishi Vidyapeeth, Palampur, pp 35–36

Singh B, Dwivedi SK, Sharma JP (2000) Greenhouse technology and winter vegetable production in cold arid zone. In: Sharma JP, Mir AA (eds) Dynamics of cold and agriculture. Kalyani Publishers, New Delhi, pp 279–293

3 Crop Protection

Abstract

Year-round warm temperatures and relatively high humidity and abundant food make the protected environment of greenhouses highly favorable to pests and diseases. While much is known about disease epidemiology and insect behavior, scant attention, however, has been paid to the manipulation of greenhouse environments expressly to avoid disease epidemics and insect infestations, which together can easily account for 30 % crop losses. This makes the control of pests and diseases in protected crops even more challenging, with many important problems being unresolved and new ones arising as the industry undergoes more changes in production systems. The topics discussed in this chapter include location and orientation of greenhouse, structures and equipment, factors stimulating sustainable crop protection, managing crop pests and diseases, managing the crop environment, cultural techniques, and factors favorable to pest and disease development.

Keywords

Pests • Diseases • Structures • Equipment • Pest management

3.1 Introduction

The purpose of growing crops under greenhouse conditions is to extend their cropping season and to protect them from adverse environmental conditions, such as extreme temperatures and precipitation, and from diseases and pests (Hanan et al. 1978). Quality is a high priority for greenhouse crops, requiring much care in pest and disease management, not only to secure yields but also to obtain a high cosmetic standard. Although technological changes are ultimately intended to reduce production costs and maximize profits, precise environmental and nutritional control pushes plants to new limits of growth and productivity. This can generate chronic stress conditions, which are difficult to measure but apparently conducive to some pests and diseases. Historically, not enough attention has been paid to exploiting and amending production technology for the control of pests and diseases. This makes the control of pests and diseases in protected

P.P. Reddy, *Sustainable Crop Protection under Protected Cultivation*,
DOI 10.1007/978-981-287-952-3_3

crops even more challenging, with many important problems being unresolved and new ones arising as the industry undergoes more changes in production systems.

Greenhouses were initially built in areas with long, cold seasons to produce off-season vegetables, flowers, and ornamental plants. Northern Europe is the paradigm of pioneering areas of greenhouse cultivation. The development of international exchanges of agricultural products and the availability of a variety of cheap plastic materials for covering simple structures have led to a spectacular increase in the area of protected crops in warmer regions like the Mediterranean basin and East and Southeast Asia (Wittwer and Castilla 1995). These new regions are commonly characterized by low or irregular annual precipitation and poor vegetation development. The insertion of greenhouse patches leads to drastic changes in the structure and ecology of the landscape. In early stages of greenhouse cultivation in a new area, greenhouses are isolated spots, like oases, where some phytophagous insects find good seasonal conditions for rapid increases in density. But optimal weather and host plant conditions rarely last throughout the year and for a few months – usually the hottest – the increase in the herbivore population is interrupted. When greenhouses become more common in the area, the mosaic pattern may evolve to a large area of protected crops, with a succession of crops throughout most of the year and with polyphagous pests. These pests are able to feed on many agricultural plants and migrate between greenhouses. Additionally, field crops may be excellent refuges for pests in hot seasons, when the temperature is too high for greenhouse cultivation. This has several consequences, as the immigration of pests into the greenhouse causes sudden and largely unpredictable pest density increases.

Exotic pests quickly become established, especially if ornamental plants are cultivated. Polyphagous pests (like whiteflies; spider mites; thrips; leaf miners; several aphid species, especially *Aphis gossypii*; leaf-eating caterpillars; and soil worms), which may exploit several crops successively, become prevalent. As pest densities increase, crops are increasingly sprayed with insecticides, native natural enemies become very rare, and natural control loses effectiveness. Unexpected and high pest pressure from the outside makes biological control very difficult. Under such conditions, a more holistic approach would consider the fields outside the greenhouse and the crop inside the greenhouse as a single entity for applying integrated strategies against pests and diseases. Programs for conserving native or introduced natural enemies in the area should both lower pest pressure on greenhouse crops and incorporate beneficial fauna into the outside–inside greenhouse cycle of the pest–natural enemy complex.

Additionally, the international trade in ornamental and flower plants facilitates the spread of pests and diseases around the world and their establishment in new areas. In Europe, for example, at least 40 new pests have been recorded in protected crops in the last 25 years. The increasing complexity of pest and disease problems and the high cosmetic standards of vegetable, ornamental, and flower products have led growers to apply intensive preventive chemical programs, which result in pests and pathogens becoming resistant to the most frequently used pesticides in a few years, which, in turn, increases control costs. In southern Spain, the average cost of pesticide application in 1992 in protected vegetables was estimated as US\$ 0.14/m^2 (16.5 % of the total production cost) (Cabello and Cañero 1994), and several whitefly, thrips, aphid, and fungus species are suspected to be resistant to several active ingredients. A similar figure is valid for Italy, where the most sophisticated structures are located in the northern part of the country: pesticides are widely applied and pest and disease resistance is quite widespread (Gullino 1992). In the Netherlands, pest and disease control costs for vegetables are still limited and are normally below 3 % of the total costs to produce a crop (van Lenteren 1995).

As control costs increase, pesticide resistance spreads, and consumers become aware of the risks of pesticide residues in fresh vegetables, a strong demand for nonchemical control methods is emerging in many countries. Integrated systems for greenhouse pest and disease control have been developed and implemented in northern Europe and Canada, but implementation is still cumbersome in other parts of the world.

3.2 Location and Orientation of Greenhouse

On a worldwide basis, commercial greenhouse production is concentrated in regions between 25° and 65° latitude where the climate is moderate and local weather patterns are favorable. At high latitudes solar irradiance is low, day length is short, and temperatures are low during the winter months resulting in poor growth and increased susceptibility to disease. Under such conditions, diapauses of predatory insects may make biological control difficult. At low latitudes, high solar irradiance stresses crops making them more susceptible to disease. More outside ventilation air is required which brings with it more pathogen propagules and insect pests.

Within any given region, the location of a particular greenhouse operation makes a significant difference in the management of disease and insect problems. Field crops and natural vegetation growing in close proximity to a greenhouse create disease and insect pressure, especially if those crops and the vegetation are susceptible to the same disease and insect pests as the greenhouse crop. This pressure is intensified when pathogen propagules are stirred up by field operations or when the outdoor crop is harvested or senesces and insects are forced to find a new host. Low temperatures force insects to seek out warmer climates indoors. On the other hand, freezing outdoor temperatures reduce pest pressures by inactivating pathogens and arthropod pests. Insects and pathogen propagules are carried into greenhouses through vents and doors by wind. By locating a greenhouse away from and/or upwind of outdoor crops, many pest problems can be reduced to manageable levels.

Out of concern for maximizing productivity and crop uniformity, greenhouses are oriented toward east–west for maximum light penetration. Achieving good lighting uniformity over the course of a day is important for IPM because insects and diseases proliferate in shaded areas and on stunted plants.

The environs of the greenhouse may be reservoirs of pathogens and pests. Greenhouses are often in an arable area, with trash piles, weeds, and crops botanically related to the crop being grown in the greenhouse to provide ample inoculums and infestations of pathogen vectors (Jarvis 1992). Entry into the greenhouse can be rapid and on a massive scale: windblown dust carries spores and bacteria, air currents with or without forced ventilation carry spores and viruliferous insects from trash piles and weeds, water runoff into the greenhouse can carry soilborne pathogens such as *Pythium* and *Phytophthora* species and chytrid vectors of viruses, and dirt on feet and machinery carries pathogens. A foot bath containing a disinfectant reduces this latter risk when placed at the doorway. To surround greenhouses by a 10-m band of weed-free lawn and to eliminate trash piles may prevent or delay pest and pathogen inoculum entrance into greenhouses. Though whitefly-proof screens can keep out most insects (and keep in pollinator insects), fungal spores and bacteria cannot be excluded. Diseases of tomato such as *Verticillium* wilt, *Fusarium* crown and root rot, and bacterial canker are often first noticed directly beneath root vents or just inside doorways, as is the *Diabrotica-borne* bacterial wilt of cucumber (*Erwinia tracheiphila*).

Overlapping of cropping, i.e., raising seedlings and transplants alongside production crops, is unsound hygiene, inviting infection and infestation of the new crop from large reservoirs in the old crop.

3.3 Structures and Equipment

The low-cost, low-height, plastic film-covered structures that are often first built by growers provide some protection from outdoor weather and pests, but without any means for climate control, conditions inside are often more favorable for diseases and pests than outside. Higher structures with more substantial framing members are required to accommodate climate control equipment. Increased gutter heights have also been beneficial for IPM because they increase the height that insects and pathogen propagules must be transported by wind to find their way into greenhouses through vents. With larger complexes and the economies of scale they provide, it is feasible to incorporate features in a greenhouse design that favor IPM.

Separate foot baths, refuse handling facilities, concrete floors, mat, etc. which reduce the transport of insects and pathogen propagules into the growing areas can be justified. The costs of pressure washing equipment and specialized potting and growing medium sterilizing equipment are easier to justify. Also, for large-scale operations, it is feasible to have separate propagation facilities specially designed for the production of disease-free transplants. On the other hand, because of the increased number of nooks and crannies, it is more difficult to eradicate insects and disease propagules from large complexes once they have gained a foothold.

3.3.1 Greenhouse Design

The greenhouse design (particularly its height, shape, opening systems, and cladding material) strongly influences climatic conditions inside, thus having a profound impact on pest and disease development. Plastic houses almost always have a more humid climate and large diurnal temperature variation and are more difficult to ventilate. Typically, they result in more problems with high humidity-dependent diseases, such as gray mold, downy mildews, and rusts (Jarvis 1992). Regulating the atmosphere throughout the day and night is important for disease control and for reducing the total amount of chemicals sprayed. This has been demonstrated in the case of gray mold (*Botrytis cinerea*) in tomato (Gullino et al. 1991) and cucumber (Yunis et al. 1994) and of downy mildew (*Bremia lactucae*) in lettuce (Morgan 1984).

The influence of greenhouse structures and covers on greenhouse climatic regimes may have strong consequences for pests and their natural enemies, as they have for diseases. A typical case of climate influence on pests and natural enemies concerns the spider mite and its predator *Phytoseiulus persimilis*: low humidity regimes may constrain effective use of *P. persimilis* (Stenseth 1979). In high-tech greenhouses, regulation of temperature and water pressure deficit enables the creation of conditions less favorable to pathogens and, in some cases, more favorable to biocontrol agents. The use of heating to limit development of a number of pathogens is well known (Jarvis 1992): however, heating is not economically feasible in all greenhouse systems. Recently, with the development of soilless systems, the effect of managing the temperature of the circulating solution has been studied and has proven to be effective against certain pathogens. The use of high root temperatures in winter-grown tomatoes in rock wool offers a nonchemical method of controlling root rot caused by *Phytophthora cryptogea*. The high temperature was shown to enhance root growth while simultaneously suppressing inoculum potential and infection and, consequently, reducing or preventing aerial symptoms (Kennedy and Pegg 1990). Careful control of the temperature also proved important in the case of hydroponically grown spinach and lettuce, in which it prevented or reduced attack by both *Pythium dissotocum* and *P. aphanidermatum* (Bates and Stanghellini 1984). Recently, attacks of *P. aphanidermatum* on nutrient film technique (NFT)-grown lettuce in Italy were related to the high temperature (>29 °C) of the nutrient solution. Root rot was inhibited by reducing the temperature below 24 °C (Carrai 1993).

3.3.2 Cladding Material/Covers

With respect to the cladding material used, in some cases a possible effect on diseases has been reported, mostly through the direct influence of radiation on sporulation (Jarvis 1992). Certain UV-absorbing plastic coverings for greenhouses that absorb light at 340 nm have been exploited to inhibit the sporulation of *Sclerotinia sclerotiorum* (Honda and Yunoki 1977) and species of *Alternaria* and *Botrytis squamosa* (Sasaki et al.

1985). Reuveni et al. (1989) observed a reduction in the number of infection sites of *B. cinerea* on tomato and cucumber when a UV-absorbing material was added to polyethylene film to increase the ratio of blue light to transmitted UV light. Recently, blue photoselective polyethylene sheets have been suggested for their ability to reduce gray mold on tomato (Reuveni and Raviv 1992) and downy mildew on cucumber (Reuveni and Raviv 1997). Green-pigmented polyethylene reduced the conidial load and gray mold in commercial tomato and cucumber greenhouses by 35–75 %. *Sclerotinia sclerotiorum* on cucumber, *Fulvia fulva* (= *Cladosporium fulvum*) on tomato, and cucumber powdery mildew were also reduced (Elad 1997).

Ideally, a cover material should have a low ultraviolet (UV) transmission to inhibit sporulation of fungi. Depending on latitude and local climate, some cover materials have been found better than others for IPM. Glass, as a covering material, does transmit the UV radiation necessary for the sporulation of fungi and has relatively high air leakage which can lead to very low humidity during cold periods with high heat demand. During these periods, it is necessary to humidify glass greenhouses to ensure the continued activity of biological control agents. Polyethylene is the preferred greenhouse cover material at lower latitudes where high photosynthetically active radiation (PAR) transmission is not as critical and where retention of humidity for IPM is important. Some manufacturers include admixtures in their polyethylene films to block the UV wavelengths necessary for sporulation of fungi. During cool wet periods, high humidity and condensation on the underside of polyethylene films is a problem that can lead to indiscriminate dripping and spread of diseases in the crop.

3.3.3 Heating Systems

A carefully designed heating system to maintain air and root zone temperatures close to recommended levels is essential for an effective IPM program in greenhouses. During very cold weather, operation of additional heating pipes around the perimeter and under gutters in hot water and steam-heated greenhouses is required to prevent cold spots where diseases are prone to develop.

3.3.4 Misting Systems

A common reason for failure of biological disease and insect controls early in the greenhouse growing season, and later on when outdoor conditions become hot and dry, is very low humidity levels in the greenhouse air. Under these conditions, transpiration of the crop is not adequate to maintain humidity levels in the optimum range for biological controls, and it is necessary to add humidity to the air. Under hot and dry conditions, addition of humidity to the greenhouse has the added benefit of evaporative cooling the greenhouse air.

3.3.5 Ventilation Systems

In actively and passively ventilated greenhouses, the pressure differential between inside and outside is usually negative, and it is easy for airborne pathogens and insects to enter the greenhouse, particularly if doors and ventilators are left open in hot weather. In special circumstances where it is essential to exclude pests and disease propagules, it may be necessary to maintain a positive pressure differential. With such a ventilation system, air can be filtered as it is drawn into the greenhouse to remove insects, but removing airborne fungal spores and bacteria is impracticable. With a positive pressure differential, there are fewer tendencies for infiltration of insects and disease propagules from outside through cracks in the greenhouse cover.

3.3.6 Thermal/Shade Curtains

Thermal curtains and shade curtains are generally beneficial for IPM because they reduce the extremes in climate that stress the crop and biological controls. Moveable shading systems

(Jewett and Short 1992) are also useful for acclimatizing crops and biological controls to rapidly changing solar radiation conditions.

Currently available greenhouse control software enables greenhouse operators to schedule climate set points for the conditions that they believe are best for production and IPM.

3.3.7 Insect Screening

In the Mediterranean basin, protecting crops from arthropods is regarded as more important than protecting them from the weather, so the physical exclusion of insects from the greenhouse should help in reducing the incidence of direct crop damage and also of insect-transmitted virus diseases; theoretically, this exclusion can be done by fitting fabric screens of mesh aperture smaller than the insects' body width over ventilators and doorways or by insect-repellent fabrics, but in practice there still can be significant insect penetration. Moreover, screens impede ventilation and reduce light transmission, so compromises in the management of light, temperature, and humidity are necessary to avoid adverse effects on crops and their susceptibility to diseases.

Screens do not suppress or eradicate pests; they merely exclude most of them. Therefore, they must be installed prior to their appearance, and supplementary pest control measures, such as biocontrol, are still required (Berlinger et al. 1988). Insect parasitoids and predators that are smaller than their prey can still immigrate through pest screens into the greenhouse, but larger ones have to be introduced. Since they offer an economical method of biological control of pests, they must be preserved, and destructive insecticides should be avoided. Screens impede ventilation (Baker and Shearin 1994), resulting in overheating and increased humidity. Increased humidity necessitates more frequent fungicide sprays than were required previously in an unscreened greenhouse. In Israel, 5–6 sprays per season (as opposed to 2–3 previously) are required in screened greenhouses. To minimize these harmful effects, growers add forced ventilation, but this only helps to pull whiteflies through the screen, while exhausting air from the screenhouse increases the intake of small insects.

Application of positive air pressure, pushing air into the structure through an insect-proof filter, reduces whitefly influx (Berlinger and Lebiush-Mordechi 1995).

Thus, while screens can reduce immigrant populations of pests, they also reduce the immigration of beneficial arthropods. In neither case exclusion is total. Screens are disadvantageous in that temperatures and humidities tend to rise, promoting plant stress and susceptibility to diseases, and they also reduce light. Access to the greenhouse by workers and machinery is more difficult.

Various types of screens and plastic covers have been developed to protect crops from insects; the challenge for the grower is to match the proper type of screen to local insect populations.

3.3.7.1 Woven Screens

The conventional woven screens are made from plain woven plastic yarns. Weaving leaves gaps (slots) between the yarns both in the warp and in the weft. In commercial screens the slot is rectangular whose width must be smaller than the whitefly's body size, about 0.2 nm, but it must allow maximum air and light transmission. Elongating the slot to improve ventilation is not feasible, since the threads slide apart, allowing insect penetration.

Bethke and Pain (1991) found that screens designed to exclude *Bemisia tabaci* still permitted some to penetrate and they failed to exclude *Frankliniella occidentalis*. They did, however, exclude larger insects such as moths, beetles, leaf miners, aphids, and leafhoppers, and they retained bumble bee pollinators.

3.3.7.2 Unwoven Sheets

These are made of porous, unwoven polyester and polypropylene or of clear, micro-perforated, polyethylene fabric. All are very light materials which can be applied loosely and directly over transplants or seeded soil, without the need of mechanical support. They have been used primarily in the open field, in early spring, and as spun-bonded row covers, to enhance plant growth and to increase yield. At the same time, they also proved to protect plants from insects. A polypropylene perforated sheet protected tomatoes from tomato yellow leaf curl virus (TYLCV) transmission by *B. tabaci* (Berlinger et al. 1988).

3.3.7.3 Knitted Screens

Because of irregularity in the shape of the holes, whiteflies are not excluded. Reducing slot size to block whiteflies reduced ventilation to an impractical level. However, knitted screens can exclude larger insects.

3.3.7.4 Knitted–Woven Screen

This plastic screen is produced by a technique that combines knitting and weaving. The slot is almost three times longer than in the commercial woven screen, while the width is smaller than the whitefly body size. The insect cannot pass, but ventilation is improved. A laboratory test confirmed the screen's high blockage capacity for whiteflies, which was similar to that of a conventional screen (0.1 % vs. 0.5 % penetration, respectively).

3.3.7.5 UV-Absorbing Plastic Sheets

The unconfirmed reports claim that UV-absorbing plastic sheets protect crops from insect pests and virus diseases vectored by insects, by modifying insect behavior (Antignus et al. 1996). Nevertheless, these UV-absorbing plastic sheets have become available for commercial use.

3.3.8 Whitefly Exclusion

The sweet potato whitefly (*B. tabaci*) is a small insect, about 0.2 mm wide, which transmits TYLCV, and has become the limiting factor in vegetable and flower production in Israel (Cohen and Berlinger 1986). Its physical exclusion from greenhouses is crucial, and accordingly whitefly-proof screens were developed (Berlinger et al. 1991). While the rate of whitefly exclusion is generally proportional to the screen's mesh ($R^2=0.85$) (Berlinger and Lebiush-Mordechi 1995), the insect's ability to pass through any barrier could not be predicted solely from thoracic width and mesh size (Bethke and Pain 1991). There is an unexpectedly high rate of whitefly penetration resulting from a great variability among the samples of the same screen resulting from uneven and slipping weave.

3.3.9 Thrips Exclusion

Whitefly-proof (50-mesh) woven screens are by far the most widely used covers for the exclusion of whiteflies and bigger insects. In laboratory tests, thrips, with a body width of only 245 μm, moved freely through this screen. However, in the field, a high proportion (50 %) is excluded, possibly because of the optical features of the plastic (Berlinger et al. 1993).

Western flower thrips are strongly affected by color. Loose shading net of aluminum color, through which even whiteflies penetrated freely in the laboratory test, was tested in the field and in a walk-in tunnel. The aluminum screen reduced thrips penetration by 55 % over an identically shading net but white in color (Berlinger et al. 1993). The closer aluminum fabric is placed around the entrance, the more effectively it works (Mcintyre et al. 1996).

3.3.10 Operation and Maintenance of Equipment

Proper operation and maintenance of climate control equipment is essential for healthy crops and avoidance of disease and insect problems in greenhouses. Mistakes in climate control settings or failures of key pieces of equipment can lead to devastating losses in a matter of minutes. Even if such events do not cause immediate crop losses, physiological, disease, and insect problems often show up some time later. The key to avoiding such problems is skilled operators and preventive maintenance programs. Insect screens have to be repaired if damaged. Also, insect screens have to be cleaned periodically to prevent blockages of light and airflow.

3.3.11 Worker Education

For an effective IPM program, greenhouse workers have to be trained to recognize nutrient deficiencies and disease and insect problems and to take appropriate action. Personal protective gear, disinfectants, disposal bins, markers, etc. have to be made available to workers so that they can play their part in an IPM program. In large operations, it is necessary to have a large site map of the greenhouses and a good record keeping system so that disease and pest outbreaks as well as control actions that have been taken can be noted for the information of all greenhouse staff. New decision support software programs (Clarke et al. 1994) offer great potential for education of workers and record keeping of all greenhouse activities, including IPM.

3.4 Factors Favorable to Pest and Disease Development

Well-grown and productive crops are generally less susceptible to diseases, but in many cases compromises have to be made between optimum conditions for economic productivity and conditions for disease and pest prevention. Well-fertilized and irrigated crops are, however, often more sensitive to pests, like aphids, whiteflies, and leaf miners.

Ground bed crops are rarely rotated, so soilborne pathogens and pests pupating in the soil accumulate if the soil is not disinfested. Soil disinfestations, although effective, create a "biological vacuum" (Katan 1984). Major changes in cultural techniques include the use of hydroponic and soilless cultures and artificial substrates controlled by computerized systems. Although these changes are ultimately intended to reduce production costs and maximize profits, precise environmental and nutritional control that pushes plants to new limits of growth and productivity can generate chronic stress conditions, which are difficult to measure but are apparently conducive to diseases caused by pathogens such as *Penicillium* spp. or *Pythium* spp. (Jarvis 1989). Some soil substitutes and soilless systems do not always provide sufficient competition for pathogens, due to their limited microflora.

High host plant densities and the resulting microclimate are favorable to disease spread. Air exchange with the outside is restricted, so water vapor transpired by the plants and evaporated from warm soil tends to accumulate, creating a low vapor pressure deficit (high humidity). Therefore, the environment is generally warm, humid, and wind free inside the greenhouse.

Such an environment promotes the fast growth of most crops, but it is also ideal for the development of bacterial and fungal diseases (Jarvis 1992), of insects vectoring viruses, and of herbivorous insects. For bacteria and many fungi (causal agents of rusts, downy mildews, anthracnose, gray mold, etc.), infection is usually accomplished in a film or drop of water on the plant surface. Unless temperature, humidity, and ventilation are well regulated, this surface water can persist in the greenhouse until infection becomes assured.

Many of the energy-saving procedures adopted during the past three decades are favorable to disease development, since they favor increases in relative humidity (Jarvis 1992), but they may lead to pest suppression as temperatures are generally somewhat lower.

Most greenhouse crops are labor-intensive and for long periods require daily routine operations (such as tying, pruning, and harvesting). The risks of spreading pathogens through workers and machinery are increased by the risks deriving from accidental wounds and from the exposure of large areas of tissues by pruning.

Greenhouses are designed to protect crops from many adverse conditions, but most pathogens and several pests are impossible to exclude. Windblown spores and aerosols containing bacteria enter doorways and ventilators; soilborne pathogens enter in windblown dust and adhere to footwear and machinery. Aquatic fungi can be present in irrigation water; insects that enter the greenhouse can transmit viruses and can carry bacteria and fungi as well. Once inside a greenhouse, pathogens and pests are difficult to eradicate.

3.5 Managing Crop Pests and Diseases

3.5.1 Sanitation

After genetic resistance, prophylaxis is by far the most effective and cheapest way of escaping major disease epidemics and pest infestations. It reduces the need for multiple applications of pesticides (which stress the crop), the risks of pesticide resistance, and pesticide contamination of the produce, the operator, and the environment. Physical screening against immigrant pests, which is coupled with aggressive control of insects in the environs of the greenhouse and in adjacent weeds and field crops, is very effective prophylaxis against both direct damage and insect-transmitted diseases. Some growers rely on old crop prunings to perpetuate populations of biocontrol insects. This is not a good practice because they constitute a reservoir of pathogens and non-parasitized pests. New introductions of biocontrol insects are a better practice.

Reducing inoculum is also important in early crop management (Jarvis 1992), with such tactics as quarantine, seed disinfestations, the use of healthy mother plants for cuttings, micropropagation, removing and properly disposing of all previous crop debris, pasteurizing or solarizing soil and soilless media, and disinfesting the greenhouse structure, benches, trays, stakes, and other materials.

Disinfectants include formaldehyde (as formalin) and hypochlorites, but both materials are hazardous to humans and residues are phytotoxic. A persulfate oxidizing agent (Virkon; Antec International), however, destroys viruses and microorganisms without such side effects (Avikainen et al. 1993).

3.5.2 Crop Scheduling

Where there is risk of diseases more destructive in cool soils, for example, *Fusarium* crown and root rot and corky root rot of tomatoes, transplanting should be delayed until the root zone has warmed up, and insulating mulch materials put down later.

Where two or more monocrops are grown each year, overlapping of transplant production and marketable crop production means that pest and pathogen populations are perpetuated unless special care is taken to keep the young and cropping plants entirely separate. There is further risk if adjacent field crops constitute a reservoir of pathogens and pests.

3.5.3 Spacing

Close horizontal and vertical spacing of plants both on the bench and in the ground bed invites rapid plant-to-plant spread of walking insects and of pathogens as diverse as *Pythium* spp., tomato mosaic virus, *Clavibacter michiganensis* ssp. *michiganensis* (= *Corynebacterium michiganense* ssp. *michiganense*), the downy mildews, and *B. cinerea* (Burdon et al. 1989). The agents of virus spread are mainly water and soil splash, insects, and workers handling plants with contaminated tools and fingers (Thresh 1982). Since air movement is restricted in dense plantings, the movement of airborne propagules is restricted, giving patchy distribution of diseases (Burdon et al. 1989) and insects.

3.5.4 Growing Medium

Soilborne diseases are no less prevalent in soilless substrates than in soil (Jarvis 1992). All substrates must be substantially free of insects and pathogens at planting and must be kept so throughout the life of the crop, thus demanding a high standard of hygiene.

Soils are usually heavily amended with peat, farmyard manure, straw, or crop residues. Plowing or rotovating the soil should be done in order to comminute plant root debris and other organic matter and so expose pathogen propagules to natural biological control. Getting the soil into good tilth with optimum temperature, water content, and aeration promotes this microbial activity.

Soils also harbor several insects, such as pupae of leaf miners and thrips, as well as fungus gnat and shore fly larvae, both of which vector *Pythium* and *Fusarium* spp. Their populations, as well as populations of predatory microarthropods, are determined by soil organic matter, soil type, and pore size (Vreeken-Buis et al. 1998). Populations of omnivorous Collembola and non-cryptostigmatic mites, for example, are enhanced by the organic matter usually plentifully added to greenhouse soils. Fungal parasites of insects and nematodes are also encouraged in soils of good tilth. The root-knot nematode *Meloidogyne incognita*, however, survives at 1–2 m well below soil disturbance levels (Johnson and McKeen 1973).

Most substrates can be fumigated or heat-sterilized, but pasteurization to about 70 °C (Baker 1957) or solarization to about 40–55 °C (Katan 1981) is preferred over total steam sterilization to 100 °C because it preserves thermophilic biocontrol organisms. The whole greenhouse can be closed in sunny conditions for solarization of both substrate and superstructure (Shlevin et al. 1995). High temperature and vapor pressure deficit in closed greenhouses can kill the western flower thrips (*F. occidentalis*) but unfortunately also its predator *Neoseiulus* (= *Amblyseius*) *cucumeris* (Shipp and van Houten 1996).

As with the original ideas that soilless cultivation would eliminate soilborne pathogens, crops in rock wool or other inert substrates or in NFT are no less free of soilborne arthropods. Fungus gnats, leaf miners, and thrips are numerous in rock wool, and shore flies are always present in pools of water on plastic sheets. Even if soil is covered with plastic sheet, there are always gaps around stems, and tears and displacement of the cover readily permit insect access.

3.5.5 Nutrition

Deficiencies and excesses of macro- and micronutrients and imbalances in relative amounts of fertilizers can predispose plants to most diseases (Jarvis 1992). In addition, fertilizers that increase foliage density at the expense of flowers and fruit not only reduce yield but tend to lower the vapor pressure deficit (VPD) in the boundary layer by restricting transpiration and wind-assisted evaporation and consequently increase the risks of infection.

High nitrogen rates in fertilizers generally increase foliage density and softness, with increasing susceptibility to leaf and flower pathogens. For example, Hobbs and Waters (1964) found a quadratic increase in gray mold (*B. cinerea*) in chrysanthemum flowers (*Dendranthema grandiflora*) with nitrogen supplied with 1.5, 3.8, and 6 g/m^2. Nitrate nitrogen combined with liming gives excellent control of *Fusarium* wilt of several crops (Jones et al. 1989). Because of its role in the integrity of cell walls, calcium imparts resistance if balanced with potassium in a high ratio. A low Ca/K ratio permits susceptibility to *B. cinerea* in tomato (Stall et al. 1965). The K/N ratio is important in the susceptibility of tomato stems to the soft rot bacterium *Erwinia carotovora* ssp. *carotovora* (Dhanvantari and Papadopoulos 1995). The incidence of soft rot was low at a K/N ratio of 4:1, increasing at 2:1 and 1:1. Verhoeff (1968) noted similar trends in tomato stems infected by *B. cinerea.* Paradoxically, Verhoeff (1968) noted that high soil nitrogen can delay the development of latent lesions of *B. cinerea* in tomato, possibly because stem senescence is delayed.

Over-luxuriant foliage is conducive to greater damage by sap-sucking insects such as aphids (Scriber 1984).

3.5.6 Pruning and Training

Pruning and training tall staked and wire-supported crops like peppers, tomatoes, and cucumbers not only modify the microclimate by altering spacing and the disease susceptibility of various tissues.

Removal of leaves bearing prepupal and pupal stages of pests can reduce their populations, but premature removal of leaves bearing parasitized stages can result in loss of biocontrol.

3.5.7 Fruit Load

Closely related to the management of pruning is the distribution of photosynthates in heavily cropping plants (Gifford and Evans 1981) in relation to the susceptibility of tissues to fungal and bacterial pathogens (Grainger 1968). As Jarvis (1989) pointed out, modern technology has increased yields of greenhouse vegetables severalfold in the last two decades, with accompanying source–sink stresses on cultivars that have not changed very much. Thus, diseases such as *Fusarium* crown and root rot *(Fusarium oxysporum* f. sp. *radicis-lycopersici)* of tomatoes and *Penicillium* stem and fruit rot *(Penicillium oxalicum)* of cucumbers have become serious in that same period. Both have been shown to be stress related (Jarvis 1988), and there has been resurgence in the incidence of corky root rot *(Pyrenochaeta lycopersici)* of tomatoes that might be related to a diminished flow of photosynthates to roots. Grainger (1968) referred the "plunderable" carbohydrates available to certain pathogens – the so-called high-sugar pathogens (Horsfall and Dimond 1957) – which include *B. cinerea*, whereas other pathogens, notably *Fusarium* spp., are classed as low-sugar pathogens principally attacking tissues starved of photosynthates. It is therefore incumbent on the grower to manage the nutrition, light, and pruning of fruit and foliage so that a balanced partition of assimilates is attained without unduly compromising yield.

3.5.8 Managing Pesticides

Pesticides are a component of integrated pest management systems but are used too freely as insurance applications rather than judiciously as almost agents of last resort. Pesticides are significant agents of stress (Schoenweiss 1975) whose overuse leads to problems of resistance (van Lenteren and Woets 1988), to interference with microbial, insect biocontrol organisms and bee pollinators, and so to an increase in iatrogenic diseases, diseases normally held in check by indigenous biological controls (Griffiths 1981).

Unlike the pesticides on crops outdoors, pesticides in the greenhouse remain unweathered and persist longer, thus putting edible produce at risk of exceeding legally tolerated residues and exposing workers to higher concentrations for longer. There are no well-established economic threshold populations of insect pests and pathogens, and the grower must thus rely largely on his own experience and on the experience of his advisors. It is at present difficult, if not impossible, to predict the course of disease epidemics in the greenhouse because the complex sequence of events in the life cycles of pathogens is dependent on a succession of different microclimates occurring in the correct order. At best, therefore, fungicides can be used only in expensive and often unnecessary insurance programs or within a very few hours of the requisite microclimate for spore germination occurring. On foliage this can usually simply mean leaf wetness.

3.6 Managing the Crop Environment

3.6.1 Temperature

In very general terms, diseases as well as arthropods can be said to have optimum temperatures for their dispersal and development (Jarvis 1992), but these cardinal points are the integral of the optima of several growth phases of the pathogen as well as of different defense reactions of the host. Jarvis (1992) cited different temperature optima for different growth processes in the gray mold pathogen *B. cinerea*: mycelium growth, sporulation, conidium germination, germ tube growth, appressorium formation, sclerotial formation, and sclerotial germination. All have different temperature optima, most of which lie above the general optimum range for gray mold development, 15–20 °C. In most of its hundreds of hosts, resistance to *B. cinerea* is probably least within that range.

Temperatures of leaves, flowers, and fruit can be considerably decreased by shading from direct sun and by increasing evaporative cooling by adequate ventilation and forced airflow (von

Zabeltitz 1976). Eden et al. (1996) discussed the possibilities of raising flower truss temperatures in tomato crops to avoid gray mold. Higher temperatures resulted in increased numbers of flowers infected by *B. cinerea*; the fungus was less likely to grow proximally to the main stem where the damage would be far more severe than one infected flower. On the other hand, higher temperatures (20–25 °C) resulted in fewer infections of stem wounds than at 15 °C. Eden et al. (1996) interpreted these results in terms of changing balances between fungal aggression and host defense reactions.

Just as with diseases of shoots, temperatures can be to some extent selected to minimize diseases of roots; for example, corky root rot *(P. lycopersici)* of tomato can be largely avoided by transplanting into warm media at 20 °C (Last and Ebben 1966), as can *Fusarium* crown and root rot (*F. oxysporum* f. sp. *radicis-lycopersici)* (Jarvis 1988). By contrast, the optimum temperature for the expression of *Fusarium* wilt (*Fusarium oxysporum* f. sp. *lycopersici*) is 27 °C. Similarly, *Pythium aphanidermatum* is most pathogenic to spinach in hydroponic culture at 27 °C, whereas *P. dissotocum* is most pathogenic at 17–22 °C (Bates and Stanghellini 1984). It is therefore important to know exactly which of closely related pathogens is present.

Insects and mites, like diseases, have also an optimum temperature for their activity, dispersal, and development. Generally, greenhouse pests are thermophilic and perform best within 20–30 °C night–day ambient temperatures. The preferred temperature for aphids and the greenhouse whitefly is somewhat lower, 15–25 °C. The interaction between temperature and VPD on the survival of western flower thrips was determined by Shipp and Gillespie (1993).

Of course, temperature affects not only arthropod pests but also their natural enemies. Natural enemies may perform poorly if temperatures are too high or too low which may occur during summer and winter, respectively, in the Mediterranean area. Then, the more temperature-tolerant *Diglyphus isaea* or *Dacnusa sibirica* can be used according to thermal regimes expected in greenhouses. Excessive heat, combined with high VPD, is a serious constraint for *P. persimilis* in warmer Mediterranean areas. Shipp and van Houten (1996) determined optimum temperatures and VPD for the use of *N. cucumeris* in Canadian cucumber houses, and these types of studies serve as guides to more intelligent biological control.

Much less exploited are the effects of temperature and water pressure deficit on biocontrol agents, although the first models, resulting in advice for optimal climate control for insect natural enemies, are now becoming available (van Roermund and van Lenteren 1998). In the case of biological control of plant pathogens, most of the studies carried out are related to the effect of environmental conditions on *Trichoderma harzianum*, used as biocontrol agent of *B. cinerea* and of several hyperparasites of *Sphaerotheca fusca* (= *Sphaerotheca fuliginea*). In the case of *T. harzianum*, populations of the antagonist are promoted by low vapor pressure deficit; in commercial greenhouses significant control of gray mold of cucumber has been correlated with low water pressure deficit but not with conditions of air saturation and dew deposition (Elad and Kirshner 1993). In the case of *Ampelomyces quisqualis*, hyperparasite of *S. fusca*, a period of 24 h with low vapor pressure deficit is necessary. Low vapor pressure deficit also favors the activity of *Sporothrix flocculosa* (Hajlaoui and Bélanger 1991). More studies in this field are necessary, both in order to keep conditions close to the optimum for biocontrol agents within the greenhouse and for selecting biocontrol agents more adapted to the greenhouse environment (Elad et al. 1996).

3.6.2 Humidity

Predicting the onset of condensation and its evaporation is even more difficult using atmospheric variables such as relative humidity, temperature, air speed, and radiation. Most predictions have errors in excess of 0.8 h and as much as ±3 h (Wei et al. 1995). Clearly this is unacceptable in a cucumber house where infection of flowers by *Didymella bryoniae* can occur in 1–2 h (van Steekelenberg 1985).

While free water and low VPD are to be avoided if pathogens are present, those very conditions are needed to establish epidemics of fungal pathogens of insects, such as *Verticillium lecanii*, *Beauveria bassiana*, and *Paecilomyces fumosoroseus* (Quinlan 1988).

Similar contrary indications have been obtained for arthropod pests and their predators. While spider mites are most active at relatively high temperatures and low VPDs, their predator *P. persimilis* is inhibited in those same conditions. Optimum humidity conditions for the predatory activity of *N. cucumeris* have been established by Shipp and van Houten (1996).

3.6.3 Water Stress

Guttation results when the rate of water supply osmotically pumped by the roots exceeds the rate of water lost by transpiration and used in growth (Hughes and Brimblecombe 1994). Wilson (1963) described how reversal of transpiration flow permits conidia of *B. cinerea* to enter tomato stem xylem, there to remain a latent inoculum.

Water accumulating and evaporating alternately from hydathodes leaves toxic deposits of salts (Ivanoff 1963), with a ready entry point for necrotrophic pathogens (Yarwood 1959). Lesions of gummy stem blight (*D. bryoniae*) are frequently seen originating from such points on cucumber leaves.

3.6.4 Light

Low and high light intensities are important agents of stress in crops (Schoenweiss 1975) that induce physiologic strains predisposing the crops to disease. Particularly important are the effects of light combined with crop management procedures, such as plant spacing; row orientation; training and pruning systems; irrigation and nutrition, on the partition of assimilates; and the relative susceptibility of different tissues and organs to disease (Jarvis 1989, 1992).

Day length, however, is important in determining diapauses in both arthropod pests and their predators. Early diapauses may be a major constraint in their use. Nondiapausing strains can, to some extent, overcome this problem.

Light also has direct effects on fungal sporulation, germination, and sclerotial formation. In *B. cinerea*, most isolates are stimulated to form conidia by light in the near-UV band (320–380 nm), an effect temporarily reversed by blue light (Epton and Richmond 1980). Some isolates, however, form conidia in the dark (Stewart and Long 1987). All fungi grow mycelium in the dark, and *B. cinerea* forms its sclerotia in darkness, or in yellow or red light, or when irradiated for less than 30 min with near-UV light (Tan and Epton 1973).

The requirement of *B. cinerea* and some other fungi for near-UV light for sporulation has led to the development of greenhouse covering materials that screen out that band as a means of disease control. Tuller and Peterson (1988) found fiberglass to transmit much less light of 315–400 nm than did polyethylene, but in a comparative assessment of gray mold in Douglas fir seedlings (*Pseudotsuga menziesii*), it was concluded that the principal effect of low irradiance transmitted by fiberglass was in inducing needle senescence in dense canopies and thus susceptibility to gray mold, rather than on any direct effect on fungal sporulation. In both types of greenhouse, the mean intensity of light that inhibited sporulation (430–490 nm) exceeded that promoted sporulation (300–420 nm). In those greenhouses, too, predisposing conditions of temperature (15–20 °C) and humidity (>90 % RH) persisted 14.5 times longer in fiberglass than in polyethylene-covered houses.

Humidity effects also seem to have outweighed effects of light wavelength in a series of trials with colored clothes covering strawberries (Jordan and Hunter 1972). Gray mold was most severe under pink and blue plastic covers, where VPDs were lower (0.41 and 0.64 kPa, respectively), than under clear plastic (1.14 kPa) or under glass (1.74 kPa). The effects of light are evidently not simple. Nevertheless, attempts have been made to filter out the near-UV light that induces sporulation in some fungi. Reuveni et al. (1989) incorporated hydroxybenzophenone into polyethylene, which increased the ratio of inhibitory blue light (480 nm) to UV (310 nm) and reduced the sporulation of *B. cinerea* in polystyrene Petri dishes. Under the treated plastic, gray mold lesions were fewer in tomato and cucumber (17 and 15, respectively) than under untreated plastic (41 and 29, respectively) (Reuveni et al. 1988). Similarly, plastic coverings absorbing light at 340 nm inhibited the sporulation and reduced the incidence of gray mold lesions on cucumber and tomato (Honda et al. 1977) as well as white mold lesions caused by *Scerotinia sclerotiorum* (Honda and Yunoki 1977). Many isolates of *Alternaria solani* also depend on near-UV light for sporulation, and Vakalounakis (1991) used vinyl films filtering out light of <385 nm to reduce the incidence of early blight in tomato greenhouses to less than 50 % of that under unamended vinyl film.

Light has little direct effect on the rhizosphere microflora, except as an agent of stress on the host.

3.6.5 Carbon Dioxide and Oxygen

Carbon dioxide enrichment is a standard procedure in many commercial greenhouses (Porter and Grodzinski 1985). The concentrations of CO_2 that impair the growth of *B. cinerea* are 2–3 orders of magnitude greater than those found even in CO_2-enriched greenhouses (Svircev et al. 1984), and so reports, for example, of Winspear et al. (1970), of increased incidences of gray mold in CO_2 – enriched greenhouses can be interpreted in terms of enhanced levels of assimilates (Grainger 1968) or a denser canopy, with its increased risks of disease-susceptible wet plants (Grange and Hand 1987).

While CO_2 is a prominent component of the rhizosphere atmosphere as a product of root and microbial respiration, it has little direct effect on pathogens. Low oxygen tension has been advanced as an explanation for physiological root death (van der Vlugt 1989) as well as decreased host resistance to root pathogens.

3.6.6 Air Movement

The primary purposes of directing and regulating air movement in the greenhouse are (i) to reduce the steepness of gradients in temperature, vapor pressure deficits, and CO_2, (ii) to assist in the evaporation of infection droplets, and (iii) to induce thigmomorphogenesis in bench-grown crops. This last results in sturdier plants (Biro and Jaffe 1984) and resistance to *Fusarium* wilts (Shawish and Baker 1982).

Counter to the generally beneficial effects of air movement are its effects on pathogen spore dispersal. Most fungi sporulate best in still air at VPD of 1.2–0.6 kPa, but fungi of the Peronosporales, like *B. lactucae* and *Pseudoperonospora cubensis*, sporulate on wet surfaces (Crute and Dixon 1981). Airborne conidia are often liberated from conidiophores by hygroscopic mechanisms (Ingold 1971) and are dispersed by air currents. Both mechanisms rely on disturbance of the microenvironment such as is readily provided by worker activity (Hausbeck and Pennypacker 1991).

The same mechanisms that control the liberation and dispersal of pathogen spores also apply to spores of biocontrol fungi when control is by enhancement of indigenous populations (Jarvis 1992).

Air movement also affects the passive transport of spider mites on webs floating through the air and being trapped on neighboring plants (Avidov and Harpaz 1969). Forced airflows can transport larger insects into the greenhouse, even through some screens. Aggregation of insects is controlled by airborne semiochemicals, while the dispersal of pheromones on excessive air currents can interfere with mating disruption as a means of biological control or attraction into sticky traps.

3.6.7 Integration of Environmental Factors

Epidemics of diseases are the result of a complex sequence of biological events each with a different set of permissive environments that have to occur in sequence and coupled with hosts in a receptive state. Jarvis (1992) outlined the complexity of those events in the case of gray mold epidemics (Fig. 3.1). Beginning with sporulation, conidia are formed at temperatures around 15 °C and in moderate VPD; they are liberated by hygroscopic movements of the conidiophore in rapidly changing conditions of humidity and are dispersed on air currents or by watersplash; infection occurs on wet surfaces at 15–20 °C; and colonization of the host is fastest at 25–30 °C. Marois et al. (1988) found that epidemics of gray mold on rose depend as well on inoculum concentration, a relationship that was different in winter and summer and affected by temperature, relative humidity, and VPD, the latter the far more meaningful parameter for describing epidemiology of *B. cinerea* in roses.

It has been possible to construct working models of gray mold epidemics in cucumber (Steinberg and Elad 1997), tomatoes (Steinberg and Elad 1997), and gerbera and rose (Kerssies 1992). The value of epidemic models such as BOTMAN (Steinberg and Elad 1997), an integrated chemical and biological control program, in predicting the onset and course of epidemics, however, is severely compromised by the rapidity with which infection occurs (9–10 h for gray mold (Yunis et al. 1994) and only 1 h for gummy stem blight in cucumber flowers (Arny and Rowe 1991)) – and by the wide variability of the greenhouse climate typically served by only one psychrometer in several 100 m^3 of space (Jarvis 1992). Steinberg and Elad (1997) found that over 3 years, a rain forecasting system did not enable BOTMAN to perform significantly better than a weekly fungicide insurance program in unheated tomato and cucumber crops. However, a 4-day weather forecast proved more useful than immediate past records of weekly averages of surface wetness (calculated from dew point) of 7 h/d and 9.5 h/d at night temperatures between 9°C and 21 °C. By the time the requisite data have been collected and analyzed, infection has already begun and is an irreversible action even with the use of fungicides, which act mostly on germinating

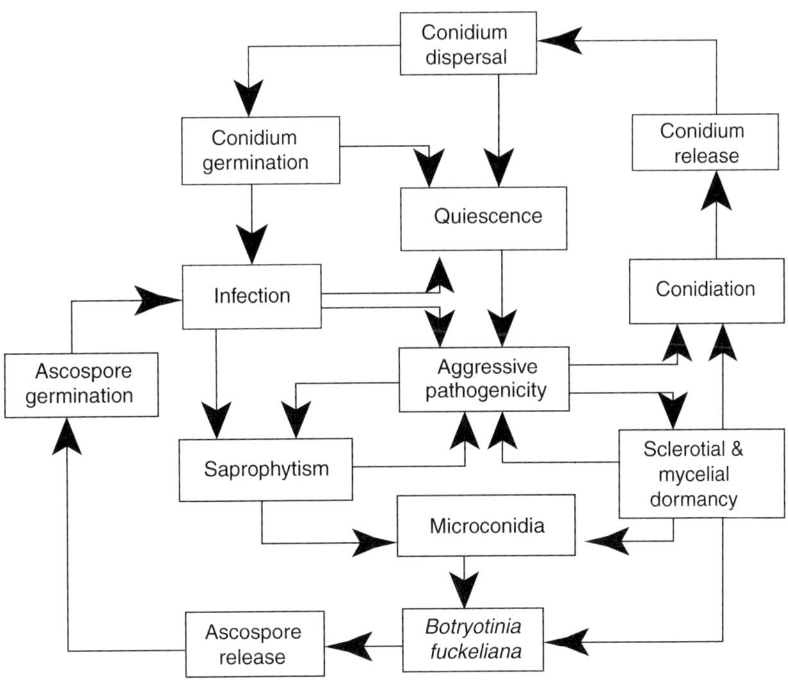

Fig. 3.1 The life cycle of *B. cinerea*. Each stage is differently affected by microclimate factors, and control of gray mold is achieved by interrupting as many pathways as possible with environmental and cultural manipulations (Jarvis 1992)

spores and thus too late to stop infection. Surface wetness is the key factor in all infections, and so its prediction from rates of change in surface and ambient air temperatures combined, by data processor, with simultaneous rates of change in VPD would be timelier in the immediate application of environmental control measures.

Powdery mildew epidemics have a somewhat less complicated sequence of events prior to infection than gray mold epidemics, but they, too, are ultimately dependent on the deposition of dew (Powell 1990).

Control of any fungus-incited disease is achieved by breaking any of the pathways in life cycles similar to those of Fig. 3.1 (Jarvis 1992), but the denial of water to germinable spores is the most important.

Computer models can be used to optimize greenhouse climate for both crop production and pest and disease control. For example, in the Netherlands, a climate management program was developed for optimal production of tomatoes and is linked to a model for biological control of greenhouse whitefly by *Encarsia formosa* (van Roermund et al. 1997). Further, the model can be extended with a humidity management module which prevents the development of fungal diseases.

Integration of pest and disease control primarily by manipulating the environment is a highly complex problem (Shipp et al. 1991). Clarke et al. (1994), in describing a computer-managed system, considered the holistic production system as a six-hierarchy complex of factors in which any change at one level affected the other five levels. Thus, any change in greenhouse climate, whether engineered or not, effects changes in pesticide efficacy, biological control agents, pests and disease vectors, diseases, and ultimately productivity and profit.

There are a number of electronic decision support systems for various facets of greenhouse pest and disease control and production strategy (Papadopoulos et al. 1997). Jones et al. (1988) described an expert system with grower selection of climate set points based on his experience; Jacobson et al. (1987) further developed an expert system with pre-set points for tomato production; and Dayan et al. (1993) developed TOMGRO that modeled physiological processes in tomato. Only Martin-Clouaire et al. (1993) considered disease escape in their model for tomato. Van Roermund et al. (1997), however, described the opposition of a whitefly control model to a production model, to which can be added a disease avoidance model. Clarke et al. (1994) and Jewett et al. (1996) described a holistic Harrow Greenhouse Crop Management System (HGCMS) for both greenhouse tomato and cucumber. In addition to providing blueprints for production in which the grower has his own input, HGCMS provides user-friendly diagnoses for diseases, pests, biological controls, and physiological disorders. It accepts climate monitoring. In addition, HGCMS allows the grower to enter economic data and will analyze it for him. Conflict resolution, as far as can be agreed among experts, is a feature of HGCMS, but ultimately the grower can accept or reject the advice of HGCMS.

The use and analysis of computer models and controls depends, of course, on a reasonable degree of computer literacy among growers, together with a basic understanding of plant growth and pest and disease biology. Otherwise, reliance on expert advisory services is obligatory.

3.6.8 Environments for Microbial Controls

In general, the microclimates for the successful deployment of fungal antagonists and parasites are close to those that promote plant infection by pathogens. Ideally, then, preemptive colonization of the phylloplane, as it is for rhizosphere, is the preferred strategy (Andrews 1992). Adaptation to that microenvironment is a prerequisite (Dickinson 1986). This colonization can also be achieved by enhancing indigenous populations of phylloplane antagonists (Jarvis 1992). Similarly, the use of green manures and composts can achieve control in the rhizosphere without the necessity of isolating, registering, and redeploying specific antagonists (Ebben 1987).

3.7 Cultural Techniques

In most greenhouses of northern Europe, continuous cropping is practiced, without a fallow crop-free interval. This has profound implications for diseases and pests. In the case of plant pathogens, it leads to the buildup of soilborne pathogens and an increased importance of foliar pathogens with a broad host spectrum (*B. cinerea*). The same can be said for insects that pupate in the soil such as leaf miners and thrips.

Greenhouse crops are grown in various soils and soilless media whose physical and chemical properties are adjusted to obtain maximum productivity. These properties, such as heat conservation, water holding capacity, fertilizer levels, and pH can also be manipulated to reduce the amount of inoculum of pests and pathogens and the probability of infection (Jarvis 1992). Systems for growing crops in the greenhouse vary widely in terms of complexity. The most common rooting media are soil and various soil mixtures, incorporating peat, vermiculite, perlite, and several other materials which are added to the soil in order to modify its structure.

In the 1960s, bench cultivation was adopted for high-value crops (carnations), permitting better results in soil disinfestations. In the 1980s and 1990s, soilless substrates gained more and more importance, particularly in the northern European countries, because they eliminate or reduce the need for soil disinfestations. Among soilless substrates, rock wool has been widely used in northern Europe, while in the tropics and subtropics, cheaper substrates have been exploited. The nutrient film technique, originally devised to improve precision in crop nutrition, reduces soilborne diseases and removes the cost of soil disinfestations. In fact, it confers relative freedom from diseases, although severe epidemics can still occur (Stanghellini and Rasmussen 1994).

During the past two decades, various composted organic wastes and sewage sludges have partially replaced peat in container media used for production of ornamentals. Recycling of these wastes has been adopted for economic and production reasons. The cost of these composts can be lower than peat. Production costs may also be decreased because some of the compost-amended media, particularly those amended with composted bark, suppress major soilborne plant pathogens, thus reducing plant losses (Hoitink and Fahy 1986). Disease suppressiveness is affected not only by chemical and physical but also by biotic factors. The low pH of sphagnum peat, pine bark, and composts could theoretically have beneficial side effects for some plants. For example, *Phytophthora* root rot of rhododendron (*Phytophthora cinnamomi*) is suppressed at pH < 4.0, because the low pH reduces sporangium formation, zoospore release, and motility. This may be important during propagation of rhododendron cuttings under mist. Moreover, chemical inhibitors of *Phytophthora* spp. have been identified in composted hardwood bark. These inhibitors do not affect *Rhizoctonia solani* (Hoitink and Fahy 1986).

Soilless cultivation can affect pests that need the soil/substrate to complete their development, as in the case of leaf miners or thrips.

The thermal and gas exchange properties of rooting media affect the growth of roots as well as the activities of pathogens. Peat, a common rooting medium, used either alone or in mixture, often suppresses pathogen activity, depending on its origin (Tahvonen 1982). However, pathogens, including species of pathogenic *Pythium* and *Fusarium* (including *Fusarium oxysporum* f. sp. *radicis-lycopersici*) have been isolated from commercial peat compost (Gullino and Garibaldi 1994).

Correct spacing prevents the establishment of a microclimate conducive to foliar diseases and the rapid spread of pathogens and pests from plant to plant in crops grown in ground beds. Altered greenhouse and bench design can improve air movement, thus reducing the risk of diseases. Bottom heating of benches, a traditional means of avoiding *Phytophthora*, *Pythium*, and *Rhizoctonia* root rots, is enhanced in cutting and seedling trays with upward air movement between the young plants. Through-the-bench air movement is perhaps the most neglected and simplest means of reducing seedling rots in tangled plant masses (Jarvis 1989).

Every crop species and cultivar requires a special fertilizer regime in order to obtain maximum productivity and to prevent stress on the plant. Fertilizer requirements change as the crop ages from seeding to harvest. In general, excessive nitrogen leads to excessive foliage that is intrinsically more succulent and susceptible to damage by necrotrophic pathogens, such as *B. cinerea*, and also stimulates development of pests such as aphids and leaf miners. Nitrogen generally has to be balanced with potassium; for many diseases, susceptibility decreases as the potassium–nitrogen ratio increases. Calcium generally enhances resistance, due to its role in the integrity of the cell wall.

No general practical recommendations can be made for controlling diseases by adjusting the fertilizer levels supplied to plants: each host–pathogen combination reacts differently. However, optimal (instead of maximal) fertilization results in lower pest and disease pressure. General recommendations can be given concerning irrigation. First of all, the factors that determine irrigation demand in greenhouse crops can all be closely regulated. From a general point of view, overhead irrigation must be carried out early in the day and should be limited late in the afternoon in order to avoid long periods of leaf wetness, which favor diseases such as downy mildews, rusts, gray mold, leaf spots, etc. When it is necessary to wet foliage for any reason (including pesticide spraying), it is always essential to maintain environmental conditions under which the foliage can dry out within a very short period of time. Also, it is important to avoid excess water in the soil: this creates conditions that are very favorable for the development of root rots. The effects of irrigation on pests are mainly through the relative humidity of the environment or through the water status of the plants. For instance, plants under stress are more easily colonized by thrips and spider mites.

3.8 Factors Stimulating Sustainable Crop Protection

Protected cultivation is an extremely high-input procedure to obtain food and other agricultural products per unit of land, although inputs are the lowest when related to the yield per area. Crop protection activities contribute to the total input in variable proportions mainly through the application of pesticides. Several features of protected cultivation are delaying the adoption of more sustainable ways to control pests and diseases. In areas where protected cultivation is most intensive, crop protection costs rarely exceed 5 % of the total production costs. In these circumstances, growers are not stimulated to make decisions based on economically founded criteria, and chemicals are frequently applied to prevent pest occurrence rather than to control real pest problems. This is particularly true in ornamental and flower crops, which can lose their value at extremely low pest densities. In addition, pesticides may be applied easily, and little expertise is needed to spray or to recommend pesticides so that no specialized advisory personnel is usually employed by growers who rely on this "simple" technology.

Consequently, innovative crop protection methods become difficult to implement in practice. From a general point of view, vegetable crops, due to their limited diversity, are most suitable for IPM. In the case of ornamentals, the enormous crop diversity and the many cultivars per species grown make the development of IPM strategies more complicated.

Several stimuli are pushing growers to use fewer pesticides and to adopt more sustainable ways to protect crops from noxious organisms as world marketing becomes more global. Among the factors stimulating sustainable forms of crop protection are the following:

3.8.1 Consumer Concern about Chemical Residues

This is a general stimulus for growers wishing to adopt IPM systems (Wearing 1988), but it is particularly relevant in fresh-consumed products like the majority of vegetables grown in greenhouses. Consumers not only demand high-quality products but are also concerned with how they are grown to judge them from the environmental aspects. Food marketers and European regional administrations are developing auditing procedures to sell vegetables under IPM or Integrated Production (IP) labels. In some cases, a surplus price is achieved by growers who produce vegetables under established IPM/IP technology.

3.8.2 Pesticide Resistance in Pests and Pathogens

As protected cultivation allows pest and pathogen populations to increase faster than in the open air, and as protected crops receive a great number of pesticide treatments, pesticide resistance develops rapidly. Dozens of greenhouse pests or pathogens are suspected to have developed resistance to the most common active ingredients, and this has been observed in many pests (aphids, whiteflies) and pathogens such as *B. cinerea*.

3.8.3 Side Effects of Chemical Application

Because society in general and governments in particular are aware of the impact of chemicals on soil, water, and air, several initiatives to restrict the use of chemicals in Europe and North America are being undertaken (van Lenteren 1997).

3.8.4 Efficacy

Some pests and diseases are difficult – sometimes impossible – to control if an integrated approach is not adopted. On the other hand, natural control can prevent several pests from building up high populations under the action of predators, parasitoids, and entomopathogens that naturally establish on greenhouse crops if chemicals are not intensively applied, and several cultural practices allow enhancement of their effectiveness. A first step toward sustainability in greenhouse crop protection is to analyze why and which phytophages and pathogens are able to increase their population densities until reaching damaging levels. Methods to improve the accuracy and speed of diagnosis are needed, particularly for diseases, and may be one of the most useful applications of biotechnology. Once the pest or disease is correctly diagnosed, environmental factors that allow or prevent such a pest or pathogen to reach economic injury levels should be identified.

Such knowledge may help us to design integrated methods to take advantage of the whole environment. If an action threshold is determined, accurate techniques for pest and disease sampling and monitoring should permit intervention at the best moment and prevent unnecessary treatments. The identification of key factors governing pest or pathogen population dynamics may allow modification of the greenhouse and crop environment – including greenhouse surroundings – to adversely affect a pest or pathogen or to favor the effectiveness of the natural enemies or antagonists.

Sometimes this can be achieved cheaply – in both economic and energetic terms – by means of correct crop and management practices. As mentioned before, the most damaging pests and many pathogens in greenhouses are polyphagous; although they are able to develop on many host plants, their negative effect on yield varies with host plant species and cultivar. The development of cultivars which are less susceptible to pests and diseases or that favor the activity of pest natural enemies is undoubtedly one of the most sustainable ways to control diseases in greenhouses, and its potential for pests has been shown in a few but significant cases.

Many of the arthropod pests and diseases that affect greenhouse crops are exotic and became established in greenhouse growing areas from accidental importation of infested crops, mainly ornamentals. In some cases, as for *Liriomyza trifolii* and *L. huidobrensis*, native natural enemies have been able to greatly contribute to the natural control of these pests, but in other cases exotic parasitoids or predators have to be released in the environment to control them, as is done for *Trialeurodes vaporariorum* by means of *Encarsia formosa*. Natural and biological control is nowadays the basis of most of the integrated pest management strategies adopted in northern Europe (van Lenteren 1995), and its practical achievements are particularly emphasized in this book. The history of biological control of diseases in greenhouses is more recent, but significant advances have also been achieved here in the last few years. Given the very high cosmetic demands and the low pest and disease thresholds applied by greenhouse growers, the progress in application of integrated pest and disease management is remarkable. Until recently, biological and integrated control was seen as a cost factor. Nowadays, however, it is considered as a beneficial marketing factor.

3.9 Conclusions

The greenhouse industry faces many new crop protection problems as a consequence of modification of production procedures and crops. The major changes will include more widely adopted mechanization and automation systems for improved crop management and the use of biotechnology in plant production. These modifications will affect the severity of pests and diseases.

The primary objective of the commercial greenhouse grower is to obtain maximum yield per unit area of space with the least financial input. However, in order to achieve this, certain minimum standards in environment management have to be maintained in such matters as crop spacing, pruning and training, irrigation, fertilization, CO_2 supplies, and temperature and humidity regimes. While much is known about disease epidemiology and insect behavior, scant attention, however, has been paid to the manipulation of greenhouse environments expressly to avoid disease epidemics and insect infestations, which together can easily account for 30 % crop losses (Pimentel 1991). This is a significant factor in a grower's balance sheet which is often overlooked and usually dealt with simplistically by indiscriminate pesticide applications (Regev 1984).

Careful analyses of epidemiological and epizootic data can indicate environments to be avoided or encouraged in greenhouse operations, but integrating the desired environments into those wanted by the grower solely to maximize yields by physiological means is extremely difficult. The solution of these problems requires the consensus of several specialized experts, experienced crop advisors, and, not least, good growers, whose experience and intuition are not to be ignored. The construction of predictive models can provide valuable insight into how environments affect diseases and insects, but experts can differ widely on which environment is best to escape, for example, lettuce downy mildew, or gray mold, or whiteflies or thrips. Resolution of these apparent conflicts can now be attained or at least reasonable compromises achieved, by the inference engine in a computer expert system. One developed by Clarke et al. (1994) and Jewett et al. (1996) is a decision support system for greenhouse tomatoes and cucumbers that collates expert opinions on all aspects of crop production, including disease and pest management, the grower's own input, and internal and external environmental parameters.

It can also provide the financial consequences of various actions, as well as of no action. Ultimately, the grower, whose brain no one can replace, has the final decision.

Strong cooperation among plant pathologists, entomologists, and horticulturists is necessary in order to assure that new management practices have a beneficial effect on plant health. Methods to improve the accuracy and speed of diagnosis are needed, and monitoring and diagnosis systems to determine the degree of infestation and economic thresholds of pathogens and pests will enable rational management decisions. A high priority should be given to the production of pathogen- and pest-free propagation material, obtained through sanitation. The use of pest- and pathogen-free material and growing media disinfested with steam or naturally suppressive to soilborne pathogens will help to reduce the impact of important pests and diseases considerably. When all such measures are integrated with the use of resistant germplasm, with modern techniques for applying pesticides and with biological control of several diseases and pests, a greatly reduced input of chemicals becomes realistic for protected cultivation.

References

Andrews JH (1992) Biological control in the phyllosphere. Annu Rev Phytopathol 30:603–635

Antignus Y, More O, Ben Yoseph R, Lapidot M, Cohen S (1996) UV-absorbing plastic sheets protect crops from insect pests and from virus diseases vectored by insects. Environ Entomol 25:919–924

Arny CJ, Rowe RC (1991) Effects of temperature and duration of surface wetness on spore production and infection of cucumbers by *Didymella bryoniae*. Phytopathology 81:206–209

Avidov Z, Harpaz I (1969) Plant pests of Israel. Israel Universities Press, Jerusalem

Avikainen H, Kopenen H, Tahvonen R (1993) The effect of disinfectants on fungal diseases of cucumber. Agric Sci Finl 2:179–188

Baker KF (ed) (1957) The U.C. system for producing healthy container-grown plants. Manual 23. California Agricultural Experiment Station, Berkeley

Baker JR, Shearin BA (1994) Insect screening: ornamentals and turf. Cooperative Extension Service, NC State University, Raleigh, Information Note

Bates ML, Stanghellini ME (1984) Root rot of hydroponically grown spinach caused by *Pythium aphanidermatum* and *P dissotocum*. Plant Dis 68:989–991

Berlinger MJ, Lebiush-Mordech S (1995) Physical methods for the control *of Bemisia*. In: Ceding D, Mayer RT (eds) Bemisia 1995: taxonomy, biology, damage control and management. Intercept Ltd, Andover, pp 617–634

Berlinger MJ, Dalian R, Mordechi S (1988) Integrated pest management of organically grown tomato in Israeli greenhouses. Appl Agric Res 3:233–238

Berlinger MJ, San M, Leeper A (1991) Application of screens to prevent whitefly penetration into greenhouses in the Mediterranean Basin. IOBC/WPRS Bull 14(5):105–110

Berlinger MJ, Lebiush-Mordechi S, Fridja D, Neta M (1993) The effect of types of greenhouse screens on the presence of western flower thrips: a preliminary study. IOBC/WPRS Bull 16:13–16

Bethke JA, Pain TD (1991) Screen hole size and barriers for exclusion of insect pests of glasshouse crops. J Entomol Sci 26:169–177

Biro RL, Jaffe MJ (1984) Thigmomorphogenesis: ethylene evolution and its role in the changes observed in mechanically perturbed bean plants. Physiol Plant 62:289–296

Burdon JJ, Jarosz AM, Kirby GC (1989) Pattern and patchiness in plant-pathogen interactions – causes and consequences. Annu Rev Ecol Syst 20:119–136

Cabello T, Cañero R (1994) Technical efficiency of plant protection in Spanish greenhouses. Crop Prot 13:153–159

Carrai C (1993) Marciume radicale su lattuga allevata in impianti NFT. Colture Protette 22(6):77–81

Clarke ND, Shipp JL, Jarvis WR, Papadopoulos AP, Jewett TJ (1994) Integrated management of greenhouse crops – a conceptual and potentially practical model. Hortic Sci 29:846–849

Cohen S, Berlinger MJ (1986) Transmission and cultural control of whitefly-borne viruses. Agr Ecosyst Environ 17:89–97

Crute JR, Dixon GR (1981) Downy mildew diseases caused by the genus *Bremia*. In: Spencer DM (ed) The downy mildews. Academic, London, pp 421–460

Dayan E, van Keulen H, Jones JW, Zipori I, Shmuel D, Challa H (1993) Development, calibration and validation of a greenhouse tomato growth model: i description of the model. Agr Syst 43:145–163

Dhanvantari BN, Papadopoulos AP (1995) Suppression of bacterial stem rot (*Erwinia carotovora* subsp. *carotovora*) by high potassium-to-nitrogen ratio in the nutrient solution of hydroponically grown tomato. Plant Dis 79:83

Dickinson CH (1986) Adaptations of micro-organisms to climatic conditions affecting aerial plant surfaces. In: Fokkema NJ, van den Heuvel J (eds) Microbiology of the phyllosphere. Cambridge University Press, Cambridge, pp 77–100

Ebben MH (1987) Observations on the role of biological control methods with integrated system, with reference to three contrasting diseases of protected crops. In: Cavalloro R (ed) Integrated and biological control in protected crops. AA Balkema, Rotterdam, pp 197–208

Eden MA, Hill RA, Beresford R, Stewart A (1996) The influence of inoculum concentration, relative humidity, and temperature on infection of greenhouse tomatoes by *Botrytis cinerea*. Plant Pathol 45:795–806

Elad Y (1997) Effect of filtration of solar light on the production of conidia by field isolates of *Botrytis cinerea* and on several diseases of greenhouse-grown vegetables. Crop Prot 16:635–642

Elad Y, Kirshner B (1993) Survival in the phylloplane of an introduced biocontrol agent (Improved and widely used *Trichoderma harzianum* may be one of the best applications of biotechnology) and populations of the plant pathogen *Botrytis cinerea* as modified by abiotic conditions. Phytoparasitica 21:303–313

Elad Y, Malathrakis NE, Dik AJ (1996) Biological control of botrytis-incited diseases and powdery mildews in greenhouse crops. Crop Prot 15:229–240

Epton HAS, Richmond DV (1980) Formulation, structure and germination of conidia. In: Coley-Smith JR, Verhoeff K, Jarvis WR (eds) The biology of botrytis. Academic, London, pp 41–48

Gifford RM, Evans LT (1981) Photosynthesis, carbon partitioning and yield. Annu Rev Plant Physiol 32:485–509

Grainger J (1968) C/R and the disease potential of plants. Hortic Res 8:1–40

Grange RK, Hand DW (1987) A review of the effects of atmospheric humidity on growth of horticultural crops. J Hortic Sci 62:125–130

Griffiths E (1981) Iatrogenic plant diseases. Annu Rev Phytopathol 19:69–82

Gullino ML (1992) Integrated control of diseases in closed systems in the sub-tropics. Pestic Sci 36:335–340

Gullino ML, Garibaldi A (1994) Influence of soilless cultivation on soil-borne diseases. Acta Horticult 361:341–354

Gullino ML, Aloi C, Garibaldi A (1991) Integrated control of grey mould of tomato. IOBC/WPRS Bull 14:211–215

Hajlaoui MR, Bélanger RR (1991) Comparative effects of temperature and humidity on the activity of three potential antagonists of rose powdery mildew. Neth J Plant Pathol 97:203–208

Hanan JJ, Holley WD, Goldsberry KL (1978) Greenhouse management. Springer, Berlin

Hausbeck MK, Pennypacker SP (1991) Influence of grower activity and disease incidence on concentrations of air-borne conidia of *Botrytis cinerea* among geranium stock plants. Plant Dis 75:798–803

Hobbs EL, Waters WE (1964) Influence of nitrogen and potassium on susceptibility of *Chrysanthemum morifolium* to *Botrytis cinerea*. Phytopathology 54:674–676

Hoitink HAJ, Fahy PC (1986) Basis for the control of soilborne plant pathogens with composts. Annu Rev Phytopathol 24:93–114

Honda Y, Yunoki T (1977) Control of Sclerotinia disease of greenhouse eggplant and cucumber by inhibition of development of apothecia. Plant Dis Rep 61:1036–1040

Honda Y, Toki T, Yunoki T (1977) Control of gray mold of greenhouse cucumber and tomato by inhibiting sporulation. Plant Dis Rep 61:1041–1048

Horsfall JG, Dirnond AE (1957) Interaction of tissue sugar, growth substances, and disease susceptibility. Zeitschrift für flanzenkrankheiten Pflanzenpathol Pflanzenschutz 64:415–421

Hughes RN, Brimblecombe P (1994) Dew and guttation: formation and environmental significance. Agr Forest Meteorol 67:173–190

Ingold CT (1971) Fungal spores: their liberation and dispersal. Clarendon, Oxford

Ivanoff SS (1963) Guttation injuries of plants. Bot Rev 29:202–209

Jacobson BK, Jones JW, Jones P (1987) Tomato greenhouse environment controller: Real-time expert system supervisor. Paper no. 87–5022, American Society of Agricultural Engineers, Univ Florida, Gainesville

Jarvis WR (1988) Fusarium crown and root rot of tomatoes. Phytoprotection 69:49–64

Jarvis WR (1989) Managing diseases in greenhouse crops. Plant Dis 73:190–194

Jarvis WR (1992) Managing diseases in greenhouse crops. American Phytopathological Society Press, St Paul

Jewett TJ, Short TH (1992) Computer control of a five stage greenhouse shading system. Trans ASAE 35:651–658

Jewett TJ, Clarice ND, Shipp JL, Papadopoulos AP, Jarvis WR (1996) Greenhouse climate control: past, present and future (Abstr). Can Agric Eng 38:239

Johnson PW, McKeen CD (1973) Vertical movement and distribution of *Meloidogyne incognita* (Nematodea) under tomato in a sandy loam greenhouse soil. Can J Plant Sci 53:837–841

Jones P, Jacobson BK, Jones JW (1988) Applying expert systems concepts to real-time greenhouse controls. Acta Horticult 230:201–208

Jones JP, Engelhard AW, Woltz SS (1989) Management of fusarium wilt of vegetables and ornamentals by macro- and micro-element nutrition. In: Engelhard AW (ed) Soil-borne plant pathogens: management of diseases with macro- and micro-elements. American Phytopathological Society, St Paul, pp 18–32

Jordan VWL, Hunter T (1972) The effects of glass cloche and coloured polyethylene tunnels on microclimate, growth, yield and disease severity of strawberry plants. J Horticult Sci 47:419–426

Katan J (1981) Solar heating (solarization) of soil for control of soil-borne pests. Annu Rev Phytopathol 19:211–236

Katan J (1984) The role of soil disinfestation in achieving high production in horticulture crops. In: Proceedings of brighton crop protection conference, BCPC, Farnhan 3, pp.1189–1196

Kennedy R, Pegg GF (1990) *Phytophthora cryptogea* root rot of tomato in rock wool nutrient culture. III. Effect of root zone temperature on infection, sporulation and symptom development. Ann Appl Biol 117:537–551

Kerssies A (1992) Epidemiology of *Botrytis cinerea* in gerbera and rose grown in glasshouses. In: Verhoeff K, Malathrakis NE, Williamson B (eds) Recent advances in *Botrytis* research. Pudoc Scientific Publishing, Wageningen, pp 159–166

Last FT, Ebben MH (1966) The epidemiology of tomato brown root rot. Ann Appl Biol 57:95–112

Marois JJ, Redmond JC, MacDonald JD (1988) Quantification of the impact of the environment on the susceptibility of *Rosa hybrida* flowers to *Botrytis cinerea*. J Am Soc Hortic Sci 113:842–845

Martin-Clouaire R, Kovats K, Cros M (1993) Determinations of greenhouse set points by SERRISTE: the approach and its object-oriented implementation. Al Appl 7:1–15

McIntyre JA, Hopper DA, Cranshaw WS (1996) Aluminized fabrics deter thrips from entering greenhouses. Southwestern Entomol 21:135–140

Morgan WM (1984) Integration of environmental and fungicidal control of *Bremia lactucae* in a glasshouse lettuce crop. Crop Prot 3:349–361

Papadopoulos AP, Pararajasingham S, Shipp JL, Jarvis WR, Jewett TJ, Clarke ND (1997) Integrated management of vegetable crops. Hortic Rev 21:1–39

Pimentel D (ed) (1991) Handbook of pest management in agriculture. CRC Press, Boca Raton

Porter MA, Grodzinski B (1985) Enrichment of protected crops. Hortic Rev 7:345–398

Powell CC (1990) Studies on the chemical and environmental control of powdery mildew on greenhouse roses. Roses Inc Bull Sept: 57–65

Quinlan RJ (1988) Use of fungi to control insects in glasshouses. In: Burge MN (ed) Fungi in biological control systems. Manchester University Press, Manchester, pp 19–36

Regev U (1984) An economic analysis of man's addiction to pesticides. In: Conway GR (ed) Pest and pathogen control: strategic, tactical and policy models. Wiley, Chichester, pp 441–453

Reuveni R, Raviv M (1992) The effect of spectrally-modified polyethylene films on the development of *Botrytis cinerea* in greenhouse grown tomato plants. Biol Agric Hortic 9:77–86

Reuveni R, Raviv M (1997) Control of downy mildew in greenhouse-grown cucumbers using blue photoselective polyethylene sheets. Plant Dis 81:999–1004

Reuveni R, Raviv M, Allingham Y, Bar R (1988) Development of a solar radiation-selective greenhouse cover for foliar plant pathogen control (Abstr). Phytoparasitica 16:74

Reuveni R, Raviv M, Bar R (1989) Sporulation of *Botrytis cinerea* as affected by photoselective sheets and filters. Ann Appl Biol 115:417–424

Sasaki T, Honda Y, Umekawa M, Nemoto M (1985) Control of certain diseases of greenhouse vegetables with ultraviolet-absorbing vinyl film. Plant Dis 69:530–533

Schoenweiss DF (1975) Predisposition, stress, and plant disease. Ann Rev Plant Pathol 13:193–211

Scriber JM (1984) Nitrogen nutrition of plants and insect invasion. In: Hauck RD (ed) Nitrogen in crop production. American Society of Agronomy, Madison, pp 441–460

Shawish O, Baker R (1982) Thigmomorphogenesis and predisposition of hosts to Fusarium wilt. Phytopathology 72:63–68

Shipp JL, van Houten YM (1996) Effects of temperature and vapor pressure deficit on the rate of predation by the predatory mite. Amblyseius cucumeris, Entomol Exp Appl 78:31–38

Shipp JL, Boland GJ, Shaw LA (1991) Integrated pest management of disease and arthropod pests of greenhouse vegetable crops in Ontario: current status and future possibilities. Can J Plant Sci 71:887–914

Shipp JL, Gillepsie TJ (1993) Influence of temperature and water vapor pressure deficit on survival of *Frankliniella occidentalis* (Thysanoptera:Thripidae). Environ Entomol 22(4):726–732

Shlevin E, Katan J, Mahrer Y (1995) Space solarization for sanitation of inocula of plant pathogens in the greenhouse structure (Abstr). Phytopathology 85:1209

Stall RE, Hortenstine CC, Iley JR (1965) Incidence of Botrytis gray mold of tomato in relation to a calcium-phosphorus balance. Phytopathology 55:447–449

Stanghellini ME, Rasmussen SL (1994) Hydroponics a solution for zoosporic pathogens. Plant Dis 78:1130–1137

Stenseth C (1979) The effect of temperature and humidity on the development of *Phytoseioulus persimilis* and its ability to regulate populations of *Tetranychus urticae* (Acarina: Phytoseiidae, Tetranychidae). Entomophaga 24:311–317

Stewart TM, Long PG (1987) Sporulation of *Botrytis cinerea* in the dark. N Z J Exp Agric 5:389–392

Svircev AM, McKeen WE, Berry JW (1984) Sensitivity of *Peronospora hyoscyami* f sp *tabacina* to carbon dioxide, compared to that of *Botrytis cinerea* and *Aspergillus niger*. Phytopathology 74:445–447

Tahvonen R (1982) The suppressiveness of finnish light colored sphagnum peat. J Sci Agric Soc Finland 54:345–356

Tan KK, Epton HAS (1973) Effect of light on the growth and sporulation of *Botrytis cinerea*. Trans Br Mycol Soc 61:145–147

Thresh JM (1982) Cropping practices and virus spread. Ann Rev Plant Pathol 20:195–218

Tuller SE, Peterson MJ (1988) The solar radiation environment of greenhouse-grown Douglas-fir seedlings. Agric Forest Meteorol 44:49–65

Vakalounakis DJ (1991) Control of early blight of greenhouse tomato, caused by *Alternaria solani*, by inhibiting sporulation with ultraviolet absorbing vinyl films. Plant Dis 75:795–797

van der Vlugt JLF (1989) A literature review concerning root death in cucumber and other crops. Nor J Agric Sci 3:265–274

van Lenteren JC (1995) Integrated pest management in protected crops. In: Dent D (ed) Integrated pest management. Chapman & Hall, London, pp 311–343

van Lenteren JC (1997) Benefits and risks of introducing exotic macrobiological control agents into Europe. OEPP/EPPO Bull 27:15–27

van Lenteren JC, Woets J (1988) Biological and integrated pest control in greenhouses. Annu Rev Entomol 33:239–269

van Roermund HJW, van Lenteren JC (1998) Simulation of the whitefly *Encarsia formosa* interaction, based on foraging behaviour of individual parasitoids. In: J Baumgaertner J, Brandmayr P, Manly BFJ (eds) Population and community ecology for insect management and conservation. Proceedings of international congress of entomology, Florence/Balkema, pp 175–182

van Roermund HJW, van Lenteren JC, Rabbinge R (1997) Biological control of greenhouse whitefly with the parasitoid *Encarsia formosa* on tomato: an individual-based simulation approach. Biol Control 9:25–47 van Steekelenberg NAM (1985) Influence of humidity on incidence of *Didymella bryoniae* on cucumber leaves and growing tips under controlled conditions. Neth J Plant Pathol 91:277–283

Verhoeff K (1968) Effect of soil nitrogen level and methods of deleafing upon the occurrence of *Botrytis cinerea* under commercial conditions. Neth J Plant Pathol 74:184–194

von Zabeltitz C (1976) Temperaturverteilung bei einer Luft-Vegetationsheizung aus PE-Schläuchen. Deutscher Gartenbau 30:7–8

Vreeken-Buis MJ, Hassink J, Brussaard L (1998) Relationships and soil microarthropod biomass with organic matter and pore size distribution in soils under different land use. Soil Biol Biochem 30:97–106

Wearing CH (1988) Evaluating the IPM implementation process. Annu Rev Entomol 33:17–38

Wei YQ, Bailey BJ, Thompson AK, Stenning BC (1995) Prediction of condensation on tomatoes. In Proceeding of the IFAC/CIGR/EURAGENG/ISHS workshop: control applications in post-harvest and processing technology. Ostend, pp 51–56

Wilson AR (1963) Some observations on the infection of tomato stems by *Botrytis cinerea* (Abstr). Ann Appl Biol 51:171

Winspear KW, Postlethwaite JD, Cotton RE (1970) The restriction of *Cladosporium fulvum* and *Botrytis cinerea,* attacking glasshouse tomatoes, by automatic humidity control. Ann Appl Biol 65:75–83

Wittwer SH, Castilla N (1995) Protected cultivation of horticultural crops worldwide. HortTechnology 5:6–23

Yarwood CE (1959) Microclimate and infection. In: Holton CS, Fischer GW, Fulton RH, Hart H, McCallan SEA (eds) Plant pathology: problems and progress 1908–1958. University of Wisconsin Press, Madison, pp 548–556

Yunis H, Shtienberg D, Elad Y, Mahrer Y (1994) Qualitative approach for modeling outbreaks of grey mould epidemics in non-heated cucumber greenhouses. Crop Prot 13:99–104

4 Disinfestation of Soil and Growth Media for Management of Soilborne Diseases

Abstract

Soilborne plant pathogens constitute a major problem of plant protection in greenhouses. The management of soilborne pathogens in glass or plastic greenhouses can be achieved through a wide range of chemical and physical treatments as well as soil solarization. Soil solarization applied to soil or growth media alone or in combination with reduced doses of soil fumigants or other amendments can also control most soilborne plant pathogens effectively. Methods of soil disinfestation, combining disinfestation methods, limitations, and difficulties in soil disinfection, are discussed.

Keywords

Soilborne plant pathogens • Soil solarization • Soil fumigation • Bioagents

4.1 Introduction

Several methods have been developed for the management of soilborne diseases incited by various plant pathogens, which include fungicidal application, breeding for disease resistance, sanitation, crop rotation, biological control, and soil disinfestations. The need for different methods of plant disease management stems from the fact that usually none of them is perfect nor can anyone be used under all circumstances. Moreover, the life cycles of pathogens may vary in different crop systems, thus requiring different management strategies. Therefore, any new method of disease management is of value since it adds to our rather limited arsenal of control methods. This is particularly true with innovative non-chemical approaches which are needed to replace hazardous chemicals.

The concept of managing soilborne pathogens has now changed. In the past, control of these pathogens concentrated on eradication. Later, it has been realized that effective control could be achieved by interrupting the disease cycle, plant resistance, or the microbial balance leading to disease reduction below the economic injury

P.P. Reddy, *Sustainable Crop Protection under Protected Cultivation*,
DOI 10.1007/978-981-287-952-3_4

level, rather than absolute control. The integrated pest management concept encompasses many elements.

Soilborne plant pathogens constitute a major problem of plant protection in greenhouses. This is basically due to the pathogens' ability to survive for several years in the soil (or in used container media) as dormant resting structures (sclerotia or microsclerotia, chlamydospores, and resting mycelia) until a susceptible crop is introduced again into the same field. These structures are able to withstand adverse environmental conditions and chemical applications, thus creating major control problems in the world agriculture. The same holds true for other soilborne pests such as arthropods, nematodes, parasitic plants, and weeds, although different mechanisms of persistence are involved. To date, fumigation (or steaming) is the most effective approach to control soilborne pests. Soil solarization (SSOL), applied to soil or growth media alone or in combination with reduced doses of soil fumigants or other amendments, can also control most soilborne plant pathogens effectively.

The management of soilborne pathogens in glass or plastic greenhouses can be achieved through a wide range of chemical and physical treatments as well as SSOL, taking into consideration the ban on the use of methyl bromide (MBr) and the current lack of alternatives for some of its current uses {although some combinations of those chemicals [e.g., ethyl dipropil thiolcarbamate (EPTC)] with SSOL have been found to be highly effective}.

4.2 Methods of Soil Disinfestation

4.2.1 Steaming

Steaming, aerated steam (Dawson and Johnson 1965), overheated, and hot water treatments are used in greenhouses, especially when container (growth) media are used. Steam has been applied for soil disinfestation for almost a century. Plant pathogens (as well as other pests) are eliminated by steaming due to heating to lethal levels or to physical damages incurred to their resting structures, even in cases of heavy soil contamination. Moreover, steaming usually shows a growth stimulation effect on the following crop.

The "classic" steaming by pipes, dug into the soil, is no longer used. This holds true also for heating the soil to 80–100 °C as this treatment results, in many cases, in a biological vacuum in the treated soil. Heating the soil or growth substrate to 70 °C – mainly by aerated steam – is now favored; this treatment leaves part of the saprophytic population uncontrolled (Bollen 1985).

Careful soil preparation is essential for good steam penetration. The soil should be tilled as deep as possible, preferably by a shovel plow, and then left for complete drying before steaming. It is important to reduce amount of plant debris, especially when steaming growth medium. Good preparation permits good steam penetration and enables pest control in heavy soils, but might still result in only partial control in very light sandy soils. Steaming of aerated growth substrates, such as tuff stones, vermiculite, etc. is usually good, but peat soils pose difficulties due to their high water content.

Soil steaming is done either by "passive" or "active" techniques. In passive steaming, steam is blown to the surface, under a covering sheet, and left to heat the upper layer. Lower layers are then heated by heat transmission. This process continues until 100 °C is reached at a depth of 10 cm (Runia 1983). Disinfestation of deep layers, especially in sandy soil, might be only partial.

Active steaming can be done by either positive or negative pressure. Both techniques employ drainage systems, based on pipes laid at a 50–70 cm depth and approximately 80 cm apart. With the "positive pressure" technique, the steam is blown through holes located along the pipes. The "negative pressure" involves an improved technique, utilizing the advantages of the two abovementioned application methods. The steam is released over the treated area under plastic sheeting, as for passive steaming, assuring rapid and even distribution throughout the plot surface, followed by active suction to the deeper layers of the soil, achieved by negative pressure applied through the drainage system. This technique, widely used in the Netherlands, is much cheaper than the two others, due to energy saving caused by the faster heat transfer (Runia 1983). Despite this, steaming treatments are expensive and are feasible mainly in places where there are heating systems (used mainly for heating the greenhouse during the cold season) or if applied by contractors (Anonymous 1994). Steaming, however, can be useful and economical for disinfestation of shallow layers of growth media placed on tables, as is usually done in nurseries.

4.2.2 Soil Fumigation

Soil fumigation is done by applying toxic pesticides to the soil by various means, and these fumigants move down and across the soil profile and reach the target organisms directly or by a very efficient secondary distribution due to their relatively high vapor pressure. Methyl bromide (MBr) is by far the most effective fumigant (Klein 1996). However, current concerns regarding the possible role of MBr in ozone depletion and its forthcoming phaseout have triggered research efforts to develop optional methods for soil disinfestation. Other soil fumigants used for greenhouses include methyl isothiocyanate (MIT), CS_2-releasing compounds, formaldehyde, dichloropropene, etc. (Anonymous 1994; Ristaino and Thomas 1997).

4.2.2.1 Fumigation with Methyl Bromide (MBr)

MBr is the most powerful soil fumigant with a very broad spectrum of activities. Many soilborne fungi (e.g., *Rhizoctonia* spp., *Pythium* spp., *Phytophthora* spp., *Sclerotinia sclerotiorum*, *S. minor*, *Sclerotium rolfsii*, *Verticillium* spp., and many *Fusarium* spp.) are sensitive to MBr. In contrast, some soilborne bacteria, such as *Clavibacter michiganensis* ssp. *michiganensis* (=*Corynebacterium michiganense* ssp. *michiganense*), are not satisfactorily controlled at regular (commercial) rates of application (Antoniou et al. 1995a). The effectiveness of MBr fumigation also depends on proper soil preparation, irrigation reaching approximately 60 % of "field capacity," and a tight covering of the fumigated soil with plastic (mostly polyethylene) sheeting. MBr is applied to the soil at a rate of 50–150 g/m^2 either by injection as a cold liquid just before covering or by distribution as a cold or hot gas under the mulch released through a manifold of perforated pipes or from 0.3 to 1 L disposable containers which are opened under the mulch. The duration of the application depends on soil temperature (1–2 days at 18 °C, 3 days at 10–15 °C at the 0–20-cm-deep soil layer, but more than 4 days at 8–10 °C at the same depth) (Klein 1996). Possible problems due to the toxicological hazards of MBr are related mainly to the health danger for applicators and to the increase in inorganic bromine residues in edible plant products. MBr was found in a few cases in water near greenhouses in the Netherlands, where PVC water pipes were improperly placed only 10 cm deep in the ground.

In 1992, MBr was listed by the Montreal Protocol as an ozone-depleting material, and a procedure for banning its use was initiated (Gamliel et al. 1997b). According to this decision, MBr will not be available in developed countries after 2005, and its consumption will be gradually reduced during the period remaining until the ban goes into effect (Anonymous 1997).

There are some MBr uses without any known substitute yet (Anonymous 1994). Continuous efforts are now under way, to reduce MBr dosages and minimize its emission and negative side effects on the environment. Most solutions are based on using improved, virtually impermeable mulching films. Common low- and high-density polyethylene films are poor barriers and allow the escape of MBr at very high rates, especially where the film temperature is higher than 40 °C (as is the case in most greenhouses when the film is exposed to solar irradiation). The permeability of MBr through impermeable film (normally co-extruded with a barrier layer protected by polyethylene coating from both sides) is only 0.001–0.0001 g/m^2/h depending on the barrier formula, compared with emission of 5 g/m^2/h for regular low-density polyethylene. Control of a pest is a factor of pesticide concentration (C) and exposure time (T). Thus, extending MBr retention in soil under impermeable films for a longer period allows the use of reduced MBr dosages with the same CT values, without reducing control efficacy. Fungal pathogens such as *Fusarium oxysporum* f. sp. *dianthi*, *F. oxysporum* f. sp. *radicis-lycopersici*, *F. oxysporum* f. sp. *cucumerinum*, etc. were controlled by reduced dosage of MBr at 25–50 % of the recommended dose under impermeable films (Antoniou et al. 1997; Gamliel et al. 1997b, c). Further reduction is possible by deeper burying of the film edges into the soil and by continuous mulching or by combination with SSOL (Grinstein et al. 1995; Antoniou et al. 1997; Gamliel et al. 1997b).

4.2.2.2 Fumigants with Methyl Isothiocyanate (MIT)

4.2.2.2.1 Dazomet [3,5,-Dimethyl-tetrahydro-l,3,5,-(2H)-thiodiazino-thione]

Dazomet is a product formulated either as a powder (85 % a.i.) or as granules (98 % a.i.). The chemical is gradually hydrolyzed to at least four subproducts, MIT being the main one. Dazomet is effective against *Verticillium dahliae*, *V. alboatrum*, *Rhizoctonia solani*, *Sclerotinia sclerotiorum*, *Phytophthora* spp., and *Pythium* spp. at 400–600 kg a.i./ha. The fumigant can be used for the control of several diseases in seedbeds and greenhouses. It is applied to the soil by spreading or irrigating followed by mechanical mixing (such as rotovator cultivation or shovel plow) into the soil. The chemical, which is not applicable at temperatures lower than 8 °C, is also partially effective against insects, various nematodes, and weed seeds. One of the disadvantages of dazomet is the long period (3 weeks) needed after application of the chemical before planting or sowing is permissible (Anonymous 1994; Middleton and Lawrence 1995).

4.2.2.2.2 Metam Sodium (Sodium Methyldithiocarbamate) (MES)

MES is effective against several soilborne pathogens in both covered and open outdoor cultivation. In water solutions MES rapidly changes to methyl isothiocyanate (MIT). The broad spectrum of controlled pathogens includes pythiaceous fungi; races of *Fusarium oxysporum*, *S. sclerotiorum*, *S. rolfsii*, and *V. dahliae*; and species of *Phialophora*, *Phoma*, *Botrytis*, etc. Since resting structures are present mainly in the upper 40 cm of the soil profile, MES (100 % water soluble) is most effective when applied via the sprinkler irrigation system. The chemical is used at various doses according to the target pathogen and/or the soil type to be disinfected. The recommended dosages for sandy, heavy, and very heavy soils are 490–650, 800, and 1000 l/ha, respectively. Soil temperature is also a critical factor in the effective application of the chemical: fluctuating between 10 °C and 30 °C at a soil depth of 10 cm is best. Use of MES for chemigation is an effective procedure against soilborne pathogens. However, side effects may arise under certain conditions, such as when the irrigation water is contaminated with urban sewage. *Fusarium oxysporum* f. sp. *cepae* on onion has been controlled by MES application, but fumigation resulted in the eradication of endomycorrhizal fungi, reduced onion growth, and increased the population of another bacterial pathogen of onions, *Pseudomonas gladioli* pv. *allicola*, which replaced fusaria and caused very serious damage (Kritzman and Ben-Yephet 1990).

4.2.2.3 Alternatives to Methyl Bromide (MBr)

Fumigants other than MBr, having a much narrower range, are registered and used in various cropping systems. These include nematicides (dichloropropene), fungicides (CS_2-releasing pesticides), and others. These are used on relatively small scale (Anonymous 1994). It is clear that with the currently available fumigants, there is no satisfactory replacement to MBr. The use of other fumigants involves the identification of causal agent and in many cases the use of a mixture of two or more chemicals to control a wider range of disease agents, pests, and weeds in the treated plot (Anonymous 1994). Di-Trapex (methyl isothiocyanate 20 + dichloropropane–dichloropropene 80) may serve as an example to this tendency, as this pesticide was formulated to control both pest controlled by MES and the root-knot nematode. Furthermore, data regarding residual effect of the abovementioned fumigants before planting is needed while their environmental impact is not yet fully clear.

4.2.3 Soil Solarization (SSOL)

Soil solarization is a nonchemical soil disinfestation method applied worldwide for the control of soilborne plant pathogens, weeds, and nematodes. In Israel, extension workers and growers suggested that the intensive heating that occurs in mulched soil might be used for disease control. By mulching the soil with transparent polyethylene sheets in the hot season prior to planting, a team of Israeli workers developed a solar heating approach for soil disinfestation. Soil solarization is a method of controlling soilborne pests and pathogens by raising the temperature of the soil through application of transparent polyethylene sheet to a moist soil surface. With solarization, vast possibilities for disease control are possible. Soil solarization as a disinfestation method has potential advantages. It is a nonchemical method which is not hazardous to the user and does not involve substances toxic to the consumer, to the host plant, or to other organisms. In the right perspective, it is less expensive than other methods. This technology can easily be transmitted to the farmers and can be applied in large areas manually and mechanically. It may have a long-term effect, since effective disease control lasts for more than one season. This method has the characteristics of an integrated control, since physical, chemical, and biological mechanisms are involved and because the control of a wide variety of pests is achieved.

Use of this method has been reported to reduce the population of many soilborne pathogens including fungi, bacteria, and nematodes as well as weeds (Pullman et al.1981; Barbercheck and von Broembsen 1986; Verma et al. 2005). Soil solarization applied singly or in combination with biocontrol agents or reduced doses of soil fumigants/fungicides has shown a remarkable destructive effect on most soilborne plant pathogens. Various terms like solar heating, plastic or polyethylene tarping, and polyethylene or plastic mulching of soil have been used to describe this method. Since this method involves repeated daily heating at relatively mild temperatures, the term solar pasteurization has also been suggested.

SSOL represents one of the very few cases where a new nonchemical control procedure has been adopted by greenhouse growers in several parts of the world, within a relatively short period of time (Katan et al. 1987). SSOL is based on trapping solar irradiation by tightly covering the wet soil, usually with transparent polyethylene or other plastic sheets (Grinstein and Hetzroni 1991). This results in a significant elevation (10–15 °C above normal, depending on the soil depth) of soil temperatures up to the point where most pathogens are vulnerable to heat when applied for 4–6 weeks and controlled either directly by the heat or by chemical and biological processes generated in the heated soil (DeVay and Katan 1991).

4.2.3.1 Principles

Heat is used as a lethal agent for the control of plant pathogenic organisms through the use of transparent polyethylene soil mulches (tarps) for capturing solar energy. Polyethylene covering of soil induces greenhouse effect and raises soil temperature. The following recommendations are made to bring about effective solar heating of soil:

- Transparent (clear), not black, polyethylene should be used since it transmits most of the solar radiation that heats the soil. Black polyethylene, though it is greatly heated by itself, is less efficient in heating the soil than transparent sheet.
- Soil mulching should be carried out during the period of high temperatures and intense solar irradiation.
- Soil should be kept wet during mulching to increase thermal sensitivity of resting structures such as sclerotia, chlamydospores, etc. and to improve heat conduction.
- The thinnest possible polyethylene tarp (25–30 μm) is recommended, since it is both cheaper and more effective in heating, due to

better radiation transmittance, than the thicker one. Polyethylene reduces heat convection and water evaporation from the soil to the atmosphere. As a result of the formation of water droplets on the inner surface of the polythene film, its transmissivity to long-wave radiation is highly reduced, resulting in better heating due to an increase in its greenhouse effect. Ideal plastic mulch should be 100 % transparent to solar radiation and completely opaque to long-wave radiation. This ideal mulch can increase soil temperature by 6–8 °C over ordinary polyethylene.

- Since temperatures at the deeper soil layers are lower than at the upper ones, the mulching period should be sufficiently extended, usually 4 weeks or longer, in order to achieve pathogen control at all desired depths.

The solar heating method for disease control is similar, in principle, to that of artificial soil heating by steam or other means. There are, however, important biological and technological differences:

- With soil solarization there is no need to transport the heat from its source to the field.
- Solar heating is carried out at relatively low temperatures as compared to artificial heating; thus, its effects on living and nonliving components are likely to be less drastic.
- Negative side effects observed with soil steaming such as phytotoxicity due to release of manganese or other toxic products and a rapid soil reinfestation due to the creation of a biological vacuum have not been reported so far with solar heating.

Absorption of solar radiation in different soils varies according to the color, moisture, and texture of the soil. In general, the soil has high thermal capacity and is a poor heat conductor thus resulting in a very slow heat penetration in soil. The energy is lost from the soil in the form of long-wave radiation through conduction, convection, and water evaporation. The principles of solar heating in polyethylene-mulched soil were demonstrated by Waggoner et al. (1960). If thermal processes occurring in mulched soil are considered, then soil temperatures at the desired depth can be predicted. Mahrer (1979) developed a one-dimensional numerical model for such predictions. As per this model in wet, polyethylene-mulched soil, increased temperatures are due primarily to the elimination of heat loss by evaporation and heat convection during the day time and partially to the greenhouse effect (preventing part of the long-wave radiation from leaving the ground). By predicting the temperatures at any depth of the mulched soil, the model enables us to select the suitable climatic regions and the time of year most adequate for solarization of soil, provided data on the heat sensitivity of the pathogens and their population density at various depths are available. Relative importance of type of mulching material, soil type, moisture, and climatic factors can also be evaluated. Analysis of the spatial soil temperature regimes in mulched soil showed that heating at the edges of the mulch is lower than at the center and that a narrow mulch strip is less efficient in heating than a wider one (Mahrer and Katan 1981).

4.2.3.2 Mechanisms

Reduction in disease incidence occurring in solarized soils results from the effects exerted on each of the three living components involved in disease (host, pathogen, and soil microbiota) as well as the physical and chemical environment which in turn affects the activity and interrelationships of the organisms. Although these processes occur primarily during solarization, they may continue to various extents and in different ways, after the removal of the polyethylene sheets and planting. The most pronounced effect of soil mulching with polyethylene is a physical one, i.e., an increase in soil temperatures, for several hours of the day. However, other accompanying processes such as shifts in microbial populations, changes in chemical composition and physical structure of the soil, high moisture levels maintained by the mulch, and changes in gas composition of the soil should also be considered while analyzing mechanisms of disease control.

The following equation proposed by Baker (1968), for relating the various factors involved in biological control, should be adopted for this analysis:

- Disease severity = inoculum potential × disease potential
- where inoculum potential is the energy available for colonization of a substrate (infection court) at the surface and disease potential is the ability of the host to contract disease.
- More specifically, the equation becomes:
- Disease severity = (inoculum density × capacity) × (proneness × susceptibility)
- where capacity is the effect of the environment on energy for colonization and proneness is the effect of the environment on the host. Of these four components, inoculum density (ID) is the one most affected by solarization either through the direct physical effect of the heat or by microbial processes induced in the soil. The other components, however (except for susceptibility which is genetically determined), might also be affected.

Whenever microorganisms are subjected to moist heat, at temperatures exceeding the maximum for growth, their viability is reduced. The thermal death rate of a population of an organism depends on both the temperature level and exposure time, which are inversely related. At a given temperature and time of exposure, mortality rate is related to the inherent heat sensitivity of the organisms and to the prevailing environmental conditions. In general, populations of soilborne fungal pathogens are drastically reduced at temperatures of 40–50 °C, exposure time ranging from minutes to hours for the higher temperatures and up to days for the lower temperatures. The response of the population to elevated temperatures depends on propagule type, age, and environmental factors like pH, presence of ions, etc. Presence of moisture is a crucial factor since microorganisms are much more resistant to heat under dry conditions. The effect of water can be explained by the dependence of the heat stability of proteins on hydration. In the presence of water, less energy is required to unfold the peptide chain of proteins, resulting in a decreased heat resistance. Heating dry soils is therefore not effective in pathogen control (Katan et al. 1976).

Microbial processes, induced in the soil by solarization, may contribute to disease control, since the impact of any lethal agent in the soil extends beyond the target organisms. If induced by solarization, biological control may affect the pathogen by increasing its vulnerability to soil microorganisms or increasing the activity of soil microorganisms toward pathogen or plant, which will finally lead to a reduction in disease incidence, pathogen survivability, or both. Thus, both short- and long-term effects might be expected. Biological control may operate at any stage of pathogen survival or disease development during or after solarization, through antibiosis, lysis, parasitism, or competition.

4.2.3.3 Disease Management

Protected structures on the one hand provide ambient growing conditions to the plant, and on the other hand this condition is favorable to the plant pathogens also. Though protected farming has advantage that pathogens do not enter easily from outside, once a pathogen has been introduced, it is very difficult to manage. Control of particularly soilborne plant pathogens is a challenge, as traditional practices like crop rotation, fallow, mixed cropping, etc. usually cannot be applied under polyhouse conditions. Solarization appears to be of major use in greenhouse culture. Many greenhouse and nursery crops worldwide now utilize solarization. The ability of greenhouse operators to close up greenhouses during the hot summer months allows higher solarization temperatures than achievable in treatment of open fields. Another application for which solarization may come into common use, particularly in developing countries, is for disinfestation of seedbeds, containerized planting media, and cold frames. As with use in greenhouses, these are ideal niches for solarization, since individual areas to be treated are small, soil temperature can be greatly increased, the cost of application is low, the value of the plants produced is high, and the production of disease-free planting stock is critical for producing healthy crops.

4.2.3.3.1 Effect of SSOL on Fungal Diseases

Ecological observations and quantitative measurements carried out after the application of the technique have differentiated the pathogens into two main categories. It should be pointed out that a pathogen could be effectively controlled by solarization in one region but less effectively in another depending on environmental and cultural parameters. A partial list of soilborne pathogens and pests which are controlled by solarization as reported for greenhouses and open fields is listed in Table 4.1. It is important to mention that application of SSOL in a close greenhouse or by employing two-layer mulch further improves its effects (Kodama and Fukui 1982; Garibaldi and Tamietti 1984; Garibaldi and Gullino 1991).

Soil solarization has been demonstrated to control diseases caused by many fungal pathogens such as *Rhizoctonia solani*, *Fusarium* spp., *Pythium* spp., *Phytophthora* spp., *Verticillium* spp., *Sclerotium rolfsii*, etc.

Table 4.1 Soilborne fungal pathogens controlled by soil solarization

Pest	Crop	Selected references
Fusarium spp.	Cucumber, melon, tomato, strawberry, watermelon	Kodama et al. 1980; Kodama and Fukui 1982; Tjamos and Malcrynakis 1990; Grinstein and Ausher 1991; Oliveira 1992; Antoniou et al. 1997
Phoma lycopersici	Tomato	Cartia 1989
Phytophthora spp.	Tomato	Cartia 1989; Grinstein and Ausher 1991
Pyrenochaeta spp.	Onion, tomato	Malathrakis et al. 1983; Garibaldi and Tamietti 1984; Tjamos 1984; Grinstein and Ausher 1991
Pythium spp.	Various	Hilderbrand 1985; Stapleton and Garza-Lopez 1988
Rhizoctonia solani	Various	Grinstein and Ausher 1991
Sclerotinia spp.	Lettuce	Porter and Merriman 1985; Materrazzi et al. 1987; Vannacci et al. 1988; Phillips 1990
Sclerotinia cepivorum	Onion	Abdel Rahim et al. 1983; Porter and Merriman 1985
Verticillium spp.	Eggplant, tomato, strawberry	Katan et al. 1976; Cartia 1989; Tjamos et al. 1989; Grinstein and Ausher 1991

4.2.3.3.2 Partial Control of Fungal Diseases by SSOL

The heat-tolerant *Monosporascus* sp. and *Macrophomina phaseolina*, root-knot nematode *Meloidogyne* spp., and some weeds, e.g., *Cyperus rotundus* and the annual weed *Melilotus sulcatus*, are only partially controlled by SSOL. *Fusarium oxysporum* f. sp. *dianthi* is also considered as one of the wilt pathogens not easily controlled by SSOL (Rubin and Benjamin 1983; Gamliel and Stapleton 1997).

4.2.3.3.3 Bacterial Diseases Controlled by SSOL

Relatively, only few reports about SSOL and bacterial diseases were published. Application of SSOL (1–2 months soil mulching with transparent polyethylene films) in tomato plastic houses drastically reduced symptoms caused by *Corynebacterium michiganense* ssp. *michiganensis* (Antoniou et al. 1995b), while MBr (68 g/m^2) was ineffective in controlling the disease. Populations of Gram-positive bacteria were reduced by 64–99 % by SSOL (Stapleton and Garza-Lopez 1988). Bacterial populations of cultures of *C. michiganensis* ssp. *michiganensis* infiltrated into tomato stem segments were embedded at various soil depths prior to the application of SSOL. A sharp decrease or elimination of the pathogen in solarized compared to MBr-treated plots was observed. *Streptomyces* spp., causing deep-pitted scab of potatoes, was successfully controlled (Grinstein et al. 1995). Negative effects, due to control of beneficial Rhizobia, were also reported (Abdel-Rahim et al. 1987).

Soil solarization has been demonstrated to control diseases caused by many bacterial pathogens such as *Agrobacteria* and *Pseudomonas*.

4.2.3.3.4 Nematode Diseases Controlled by SSOL

Soil solarization has been demonstrated to control diseases caused by many species of nematodes. Diseases caused by *Meloidogyne* spp., *Heterodera* spp., etc. have been successfully controlled by soil solarization (Rao and Krishnappa 1995; Grinstein et al. 1995).

4.2.3.3.5 Biological Control Aspects of SSOL

Disturbances in the biological equilibrium of the soil microflora, following soil fumigation or steaming, are known to be drastic and undesirable. Application of SSOL, however, favors the survival and increase of several heat-tolerant microorganisms able to act as antagonists against soilborne pathogens, such as *Talaromyces flavus*, *Aspergillus terreus*, fluorescent pseudomonads, and others (Tjamos et al. 1991). Solarization favors establishment of added antagonists such as *Trichoderma* spp. and *A. terreus*, saprophytic fusaria, and others (Martyn and Hartz 1986; Triolo et al. 1988).

The survival of thermophilic genera of *Bacillus* and *Actinomyces*, as well as the buildup of fluorescent pseudomonads and other populations of rhizosphere bacteria, was reported (Gamliel and Katan 1991; Antoniou et al. 1995a). The effect of SSOL can be improved also by combination with no pesticide organic amendments incorporated into the soil before mulching. This can be related both to the release of toxic materials by combination of heating and biological activity and to positive changes in soil microflora. Gamliel and Stapleton (1997) reported control of root-knot nematodes by mixing chicken manure or dry cabbage leaves in the plot (biofumigation) before mulching.

4.2.3.4 Increased Growth Response

Plant growth in solarized infested soil is enhanced as compared to untreated, infested soil as a result of pathogen control, but solarization of soil which is apparently free of known pathogens often results in improved plant growth. This could be attributed to increased micro- and macronutrients in soil solution, elimination of minor or unknown pathogens, destruction of phytotoxic substances in the soil, release of growth regulator-like substances, and stimulation of mycorrhiza, PGPR, and other beneficial microorganisms. The effect of soil solarization on earthworm's population has not received much attention, but it is thought that they retreat to lower depths to escape the effect of soil heating. The increased growth response of plants in solarized soil is a well-documented phenomenon and has been verified both in greenhouse experiments and under field conditions (Katan 1987; Chen et al. 1991).

4.3 Combining Disinfestation Methods

One of the major limitations of SSOL is its climate dependence. Another problem diverts from the need to keep the treated area for 35–60 days without any crop. Solarization use in many places is limited by partial control of some pests, as well as reduced efficacy in marginal seasons. These constraints can be reduced, or solved, by combining solarization with other control measures at reduced dosages. The control efficacy may be increased due to additive effect. More likely, it is due to synergistic effect caused by the hotter environment which increases vapor pressure and chemical activity of the added pesticide. Another reason for the improved activity of the pesticide is the weakening of the resting structure by the heat (Freeman and Katan 1988).

Despite the successes achieved with solarization when used singly, this method may be usefully aided by combination with other methods of disinfestation. As soil solarization is dependent upon local climatic conditions, sometimes even during conducive periods of the year, local weather conditions will not permit an effective solarization treatment. Therefore, we must come up with integrated uses of solarization in order to increase the predictability of the treatment and thus make it more acceptable to growers. Combining solarization with pesticides, organic amendments, or biocontrol agents improves disease control. Whenever a pathogen is weakened by heating, even reduced dosages might suffice for improved control combining with biocontrol agents, organic amendments, etc.

4.3.1 SSOL and Reduced Rate of Pesticides

Low application rates of fungicides, fumigants, or herbicides have been successfully combined with soil solarization to achieve better pest control (Hartz et al. 1993). Simultaneous application of chemicals and tarping the soil for solarization have been shown to increase the effectiveness of both the methods because of synergism (Ben-Yephet et al. 1988; Tjamos 1984).

Reduced doses of metam sodium (12.5 or 25 ml/m^2) applied singly or in combination with soil solarization synergistically reduced to 1 week the time needed to kill sclerotia of *Sclerotinia sclerotiorum* in the top 10 cm of soil in a lettuce field and reduced apothecia production. Carbendazim has shown slower degradation rates after solarization, possibly because of changes in the populations of soil microorganisms after solarization.

Reduced doses of MBr, impermeable plastics, and solarization were applied against a variety of diseases, e.g., *F. oxysporum* f. sp. *cucumerinum* of cucumbers, *C. michiganensis* ssp. *michiganensis* of tomatoes (Antoniou et al. 1997), the melon sudden wilt (Gamliel et al. 1997b), *Verticillium* of potatoes (Grinstein et al. 1979), deep-pitted scab of potatoes, *Fusarium* crown rot in tomatoes, and soil sickness of *Gypsophila* sp. Reduced rates of MBr (34 g/m^2) combined with simultaneous solarization effectively controlled corky root rot disease of tomatoes (Tjamos 1984) and *Verticillium* wilt of globe artichoke (Tjamos and Paplomatas 1987).

Reduced doses of chemicals are recommended as an alternative approach to the acute toxicity of full fumigation. However, their effectiveness is dependent on combinations with other pesticides or with nonchemical procedures. Sublethal fumigation is considered here in combination with SSOL (Gamliel et al. 1997b). Combining sublethal fumigation with solarization could be focused on the following:

- MBr fumigation followed immediately by solarization
- Simultaneous application of solarization with reduced doses of various fumigants
- Solarization followed by fumigant for pathogens that are heat tolerant

Recent studies show that the control efficacy of reduced dose of MBr combined with solarization was highly increased when applied after a short heating period, 2–3 days after the mulching (Gamliel et al. 1997a). Application of MBr after the termination of the SSOL, however, can control some of the beneficial microorganism populations which remain in the solarized plot and has to be considered carefully.

Current reports mainly deal with combinations of chemicals with SSOL. They include MES for the control of *V. dahliae* and *Fusarium oxysporum* f. sp. *vasinfectum* (Ben-Yephet 1988), dazomet either alone or in combination with SSOL to control *Phoma terrestris* on onions (Porter and Merriman 1985), and MBr and SSOL for the control of *Pyrenochaeta lycopersici* on tomatoes (Tjamos 1984). Dazomet (750 kg/ha) either alone or in combination with solarization reduced disease incidence and severity of pink root rot (caused by *P. terrestris*) and white rot (caused by *Sclerotium cepivorum*) of onions and increased yield by at least 100 %. Reduced rates of MBr (34 g/m^2) combined with simultaneous solarization effectively controlled corky root rot disease of tomatoes (Tjamos 1984) and *Verticillium* wilt of globe artichokes (Tjamos and Paplomatas 1987).

4.3.2 SSOL and Organic Amendments

Solarization may also be combined with application of crop residues and green and farmyard manures. There is increasing evidence that these materials release volatile compounds in the soil that kill pests and help stimulate the growth of beneficial soil organisms (Deadman et al. 2006; Gamliel and Stapleton 1993).

4.3.3 SSOL and Bioagents

Soil solarization has also been successfully combined with biological control. The use of *Trichoderma harzianum* with solarization in fields infested with *Rhizoctonia solani* has been shown to improve disease control while delaying the buildup of inoculum (Chet et al. 1982). Greenberger et al. (1987) concluded that solarized soils are frequently more suppressive and less conducive to certain soilborne pathogens than non-solarized soils. An increase in population of green fluorescent pseudomonads along with an increase of *Penicillium* and *Aspergillus* spp. following solarization has been demonstrated (Stapleton and DeVay 1982).

Synergism in reducing disease incidence can be observed between fumigants and fungal antagonists of soilborne pathogens. Solarization in combination with *Gliocladium virens* proved to be a potential control strategy against southern blight of tomatoes (Ristaino et al. 1991).

4.4 Limitations

Solarization involves the following limitations, difficulties, and possible negative effects:

- It is weather dependent and can only be used in regions where the climate is suitable (hot) and the soil is free of crops for about 1 month or more at a time of tarping with PE sheets. The soil heating effect may be limited on cloudy days. Wind or air movement across the plastic sheet rapidly dissipates the trapped heat. Strong winds may also lift or tear the sheets.
- It is too expensive for some crops and ineffective in the control of certain diseases.
- Heat-tolerant pathogens might develop after repeated application, though selection for tolerance to lethal agents is not likely to develop with disinfestation methods which are not target specific.
- Another possibility would be an increase in pathogen population due to a harmful effect on its antagonists.

4.5 Difficulties in Soil Disinfection

Soil fumigation with chemicals may have negative effects on the environment, could be extremely dangerous to humans, and may leave toxic residues in plant products. Thus, innovative approaches are desperately needed by the farmers and are under great demand by the consumers. Research toward exploiting SSOL by combining reduced doses of allowed fumigants, or various antagonists, could be one of the most promising approaches. This could also result in reducing duration of solarization thus making the method more acceptable by the farmers. Furthermore, sublethal fumigation in combination with solarization could solve many problems, since the combination is suitable for areas marginal for the application of solarization and is able to reduce the duration of solarization to one half. SSOL in combination with biocontrol agents could exploit the weakening effect imposed by solar heating and could prolong its effectiveness.

4.6 Future Thrusts

4.6.1 Economics

The economic profitability of disease control depends on the additional income obtained and the cost of application. The additional income obtained through solarization far exceeds with high-value crops, but with other crops, the situation may not be the same. There are several possibilities for reducing the cost of mulching:

- Used polyethylene may be as effective as the new, thus reducing the cost to nearly zero.

- Reusing the polyethylene is possible, provided it is durable.
- If required during the growing season, durable sheets may be used for both solarization and mulch.
- The production of thinner polythene sheets (of an adequate strength) will reduce the amount needed per hectare.

4.6.2 Development in Plastic Technology

Developments in this field may provide improved and economical mulching materials with greater heating efficiency and increased durability. This may include:

- Biodegradable plastic that decomposes in the natural environment.
- Further development of polyethylene recycling processes.
- Developing economic, novel plastic or other materials more efficient than polythene in trapping solar energy, thus reducing our dependence on climate and making this available to cooler regions.
- Possibility of plastic material that can be sprayed on the soil, instead of polyethylene mulching, should be explored.

At present, biodegradable plastic products available in the market are more expensive than traditional plastics. Their cost needs to be reduced to make them economical.

References

Abdel-Rahim MF, Satour MM, Mickail KY, El-Eraki SA, Grinstein A, Chen Y, Katan J (1987) Effectiveness of soil solarization in furrow-irrigated Egyptian soils. Plant Dis 72:143–146

Anonymous (1994) Montreal protocol on substances which deplete the ozone layer. Report of the methyl bromide technical options committee, Nairobi

Anonymous (1997) Ninth meeting of the ponies to the montreal protocol on substances that deplete the ozone layer. UN environmental program, UNEP/OzL. Pro 9/12

Antoniou PP, Tjamos EC, Andreou MT, Panagopoulos CG (1995a) Effectiveness, modes of action and commercial application of soil solarization for control of *Clavibacter michiganensis* s sp *michiganensis* of tomatoes. Acta Hortic 382:119–124

Antoniou PP, Tjamos EC, Panagopoulos CG (1995b) Use of soil solarization for controlling bacterial canker of tomato in plastic houses in Greece. Plant Pathol 44:438–447

Antoniou PP, Tjamos EC, Panagopoulos CG (1997) Reduced doses of methyl bromide, impermeable plastics and solarization against *Fusarium oxysporum* f sp *cucumerinum* of cucumbers and *Clavibacter michiganensis* s sp *michiganensis* of tomatoes. In: Proceedings of the 10th congress of the mediterranean phytopathological union, Montpellier, pp 83–86

Baker R (1968) Mechanisms of biological control of soil-borne pathogens. Annu Rev Phytopathol 6:263–294

Barbercheck ME, von Broembsen SL (1986) Effect of soil solarization on plant parasitic nematodes and *Phytophthora cinnamomi* in South Africa. Plant Dis 70:945–950

Ben-Yephet Y (1988) Control of sclerotia and apothecia of *Sclerotinia sclerotiorum* by metham-sodium, methyl bromide and soil solarization. Crop Prot 7:25–27

Ben-Yephet Y, Melero-Varo JM, DeVay JE (1988) Interaction of soil solarization and metham-sodium in the destruction of Vertcillium dahliae and Fusarium oxysporum f. sp. vasinfectum. Crop Protect 7:327–331

Bollen GJ (1985) Lethal temperatures of soil fungi. In: Rovira AD, Moore KJ, Wong PTW, Parker CA (eds) Ecology and management of soilborne plant pathogens. APS Press, St Paul, pp 191–193

Cartia G (1989) La solarizzazione del terreno: Esperienze maturate in Sicilia [Soil solarization: experiments in Sicily]. Informatore Fitopatologico 39:49–52

Chen Y, Gamliel A, Stapleton JJ, Aviad T (1991) Chemical, physical and microbial changes related to plant growth in disinfested soils. In: Katan J, De Vay JE (eds) Soil solarization. CRC Press, Boca Raton, pp 103–129

Chet I, Elad Y, Kalfon A, Hadar Y, Katan J (1982) Integrated control of soilborne and bulbborne pathogens in iris. Phytoparasitica 10:229

Dawson JR, Johnson RAH (1965) Influence of steam air-mixtures, when used for heating soil, on biological and chemical properties that affect seedling growth. Ann Appl Biol 56:243–247

DeVay JE, Katan J (1991) Mechanisms of pathogen control in solarized soils. In: Katan J, DeVay JE (eds) Soil solarization. CRC Publications, Boca Raton, pp 87–102

Deadman M, Al Nasani H, Al Sa'di A (2006) Solarization and biofumigation reduce *Pythium aphanidermatum* induced damping-off and enhance vegetative growth of greenhouse cucumber in Oman. J Plant Pathol 88:335–337

Freeman S, Katan J (1988) Weakening effect on propagules of Fusarium by sublethal heating. Phytopathology 78:1656–1661

Gamliel A, Katan J (1991) Involvement of fluorescent Pseudomonas and other micro-organisms in increased growth response of plants in solarized soils. Phytopathology 81:494–502

Gamliel A, Stapleton J (1993) Effect of chicken compost or ammonium phosphate and solarization on pathogen control, rhizosphere organisms, and lettuce growth. Plant Dis 77:886–891

Gamliel A, Stapleton JJ (1997) Improved soil disinfestation by biotoxic volatile compounds generated from solarized, organic amended soil. In: A Grinstein, KRS Ascher, G Mathews, J Katan, A Gamliel (eds) Improved application technology for reduction of pesticide dosage and environmental pollution. Phytoparasitica 25S:31–38

Gamliel A, Grinstein A, Eshel D, Di Primo P, Katan J (1997a) Combining solarization and fumigants at reduced dosages for effective control of soil-borne pathogens: controlled environment study I. Phytoparasitica 25:252–253

Gamliel A, Grinstein A, Katan J (1997b) Improved technologies to reduce emissions of methyl bromide from soil fumigation. In: A Grinstein, KRS Ascher, G Mathews, J Katan, A Gamliel (eds) Improved application technology for reduction of pesticide dosage and environmental pollution. Phytoparasitica 25S:21–30

Gamliel A, Grinstein A, Peretz Y, Klein L, Nachmias A, Tzror L, Katan J (1997c) Reduced dosage of methyl bromide for controlling Verticillium wilt of potato in experimental and commercial plots. Plant Dis 81:469–474

Garibaldi A, Gullino ML (1991) Soil solarization in southern European countries, with emphasis on soil-borne disease control of protected crops. In: Katan J, DeVay JE (eds) Soil solarization. CRC Publications, Boca Raton, pp 228–235

Garibaldi A, Tamietti G (1984) Attempts to use soil solarization in closed glasshouses in northern Italy for controlling corky root of tomato. Acta Horticult 152:237–243

Greenberger A, Yogev A, Katan J (1987) Induced suppressiveness in solarized soils. Phytopathology 77:1663–1667

Grinstein A, Ausher R (1991) The utilization of soil solarization in Israel. In: Katan J, DeVay JE (eds) Soil solarization. CRC Publications, Boca Raton, pp 193–204

Grinstein A, Hetzroni A (1991) The technology of soil solarization. In: Katan J, DeVay JE (eds) Soil solarization. CRC Publications, Boca Raton, pp 159–170

Grinstein A, Katan J, Abdul-Razik A, Zeidan O, Elad Y (1979) Control of *Sclerotium rolfsii* and weeds in peanuts by solar healing of the soil. Plant Dis Rep 63:991–994

Grinstein A, Kritzman G, Hetzroni A, Gamliel A, Mor M, Katan J (1995) The border effect of soil solarization. Crop Prot 14:315–320

Hartz T, DeVay J, Elmore C (1993) Solarization is an effective soil disinfestation technique for strawberry production. Hort Sci 28(2):104–106

Hilderbrand DM (1985) Soil solar heating for control of damping-off fungi and weeds at the Colorado State Forest Service Nursery. Tree Planters' Notes 36:28–34

Katan J (1987) Soil solarization. In: Chet I (ed) Innovative approaches to plant diseases control. Wiley, New York, pp 77–105

Katan J, Greenberger A, Alon H, Grinstein A (1976) Solar heating by polyethylene mulching for control of diseases caused by soil-borne pathogens. Phytopathology 66:683–688

Katan J, Grinstein A, Greenberger A, Yarden O, DeVay JE (1987) First decade (1976–1986) of soil solarization (solar heating) – a chronological bibliography. Phytoparasitica 15:229–255

Klein L (1996) Methyl bromide as a soil fumigant. In: Bell CH, Price N, Chakrabarti B (eds) The methyl bromide issue. Wiley, Chichester, pp 191–235

Kodama T, Fukui T (1982) Solar heating in closed plastic house for control of soil-borne diseases. V. Application for control of Fusarium wilt of strawberry. Ann Phytopathol Soc Jpn 48:570–577

Kodama T, Fukui T, Matsumoto Y (1980) Soil sterilization by solar heating against soil-borne diseases in a closed vinyl house. III. Influence of the treatment on the population level of soil microflora and on the behavior of strawberry yellows pathogen, *Fusarium oxysporum* f. sp. *fragariae*. Bull Nara Agric Exp Station 11:41–52

Kritzman G, Ben-Yephet Y (1990) Disease exchange in onions as influenced by metham-sodium. Hassadeh 1:26–29

Mahrer Y (1979) Prediction of soil temperature of a soil mulched with transparent polyethylene. J Appl Meteorol 18:263–267

Mahrer Y, Katan J (1981) Spatial soil temperatures regime under transparent polyethylene mulch-numerical and experimental studies. Soil Sci 131:82–87

Malathrakis NE, Kapetanakis GE, Linardakis DC (1983) Brown root rot of tomato and its control in Crete. Ann Appl Biol 102:251–256

Martyn RD, Hartz TK (1986) Use of soil solarization to control Fusarium wilt of watermelon. Plant Dis 70:762–766

Materrazzi A, Triolo E, Vannacci G, Scaramuzzi G (1987) The use of soil solar heating for controlling neck rot of greenhouse lettuce. Colture Protette 16:51–54

Middleton LA, Lawrence NJ (1995) The use of dazomet via the "planting through" technique in horticultural crops. Acta Hortic 382:86–103

Oliveira H (1992) Evaluation of soil solarization for the control of Fusarium wilt of tomato. In: Tjamos EC, Papavizas GC, Cook RJ (eds) Biological control of plant diseases: progress and challenges for the future. NATO ASI series A, Life Sciences, 230. Plenum Press, New York, pp 69–73

Phillips AJL (1990) The effects of soil solarization on sclerotial populations of *Sclerotinia sclerotiorum*. Plant Pathol 39:38–43

Porter IJ, Merriman PR (1985) Evaluation of soil solarization for control of root diseases of row crop in Victoria. Plant Pathol 34:108–118

Pullman GS, Devay JE, Garber RH, Weinhold AR (1981) Soil solarization on Verticillium wilt of cotton and soil-borne population of *Verticillium dahliae*, *Pythium* spp., *Rhizoctonia solani* and *Thielaviopsis basicola*. Phytopathology 71:954–959

Rao VK, Krishnappa K (1995) Soil solarization for the control of soil-borne pathogen complexes with special reference to *M. incognita* and *F. oxysporum* f sp *ciceri*. Indian Phytopath 48:300–303

Ristaino JB, Thomas W (1997) Agricultural MBr, and the ozone hole: can we fill the gaps? Plant Dis 81:964–977

Ristaino JB, Perry KB, Lumsden RD (1991) Effect of solarization and *Gliocladium virens* on sclerotia of *Sclerotium rolfsii,* soil microbiota and incidence of southern blight of tomato. Phytopathology 81:1117–1124

Rubin B, Benjamin A (1983) Solar heating of the soil: effect on soil incorporated herbicides and on weed control. Weed Sci 31:819–825

Runia WT (1983) A recent developments in steam sterilization. Acta Hortic 152:195–200

Stapleton JJ, DeVay JE (1982) Effect of soil solarization on populations of selected soil-borne microorganisms and growth of deciduous fruit tree seedlings. Phytopathology 72:323–326

Stapleton JJ, Garza-Lopez JG (1988) Mulching of soil with transparent (solarization) and black polyethylene films to increase growth of annual and perennial crops in Southern Mexico. Trop Agric 65:29–35

Tjamos EG (1984) Control of *Pyrenochaeta lycopersici* by combined soil solarization and low dose of methyl bromide in Greece. Acta Hortic 152:253–258

Tjamos EC, Malcrynakis N (1990) Control of fungal wilt diseases of melon by application of soil solarization in the field. In: Proceedings of the 8th congress of the mediterranean phytopathological union, agadir, Morocco, pp 423–425

Tjamos EC, Paplomatas EJ (1987) Effect of soil solarization on the survival of fungal antagonists of *Verticillium dahliae*. Eur Medit Plant Protect Organ Bull 17:645–653

Tjamos EC, Karapappas V, Bardas D (1989) Low cost application of soil solarization in covered plastic houses for the control of Verticillium wilt of tomatoes in Greece. Acta Hortic 255:139–149

Tjamos EC, Bins DA, Paplomatas EJ (1991) Recovery of Verticillium wilted olive trees after individual application of soil solarization in established olive orchards. Plant Dis 75:557–562

Triolo E, Vannacci G, Materazzi A (1988) Solar heating of the soil in vegetable production. Part 2. Studies of possible mechanisms of the effect. Colture Prottete 17:59–62

Vannacci G, Triolo E, Materazzi A (1988) Survival of *Sclerotinia minor* sclerotia in solarized soil. Plant Soil 109:49–55

Verma RK, Singh Y, Soni KK, Jamalluddin (2005) Solarization of forest nursery soil for elimination of root pathogens and weeds. Indian J Trop Biodiv 13:81–86

Wagonner PE, Miller PM, DeRoo HC (1960) Plastic mulching principles and benefits. Conn Agric Exp Stn Bull 623, 44 pp

5 Biological Control of Plant Pathogens

Abstract

The protected nature, expensive crops, and microclimatic conditions in the greenhouse make it rather imperative to suitably alter the disease management strategies. There is lack of specific chemicals suited for greenhouse conditions. Hence, agriculture in greenhouses and protected structures offers a unique niche for the development and use of biological control agents. Biological control of greenhouse diseases and commercial products registered for biological control are discussed.

Keywords

Biological control • Greenhouse diseases • Commercial bioagents

5.1 Introduction

The ideal and stable environment with warm, humid, and abundant food under greenhouse provides an excellent platform for the development of diseases often more than field conditions. Most pathogens cannot be excluded from the greenhouse environment, for example, airborne spores enter through doors and screens; soilborne pathogens enter through dust or contaminated soil on shoes, tools, or equipment; and many pathogens are introduced on seeds or contaminated propagating materials. Zoosporic pathogens get entry through irrigation water, and insects transmit both fungal inoculum and viruses. The optimum conditions like temperature, light, and fertilizer maximize not only plant growth but also favorable conditions for pathogens. Moreover, warmth and humidity, due to the water vapor transpired by the plants and the lack of air exchange with the outside, provide ideal conditions for foliar pathogens such as *Botrytis* and powdery mildews. Because of high-energy costs, ventilation is often reduced to prevent loss of heat. Disinfested soil or soilless substrates such as peat or rock wool lack the microbial diversity and biological buffering present in a natural soil. In this biological vacuum, soilborne pathogens such as *Pythium* and *Rhizoctonia* can quickly grow and spread. High-density planting of greenhouse crops increases the relative humidity and the chances of disease spread, and

P.P. Reddy, *Sustainable Crop Protection under Protected Cultivation*,
DOI 10.1007/978-981-287-952-3_5

management practices such as pruning and harvesting increase the spread and infection through wounds. Hydroponic systems (rock wool, nutrient film, or ebb and flow) present another set of disease problems. In this closed recirculating system, zoosporic pathogens can easily spread in the water system.

5.2 Biocontrol in Greenhouses

The protected nature, expensive crops, and microclimatic conditions in the greenhouse make it rather imperative to suitably alter the disease management strategies as per the requirements. There is lack of specific chemicals suited for greenhouse conditions. High registration and development costs and the lack of return on investment act as deterrents to chemical companies in registering products for the relatively small greenhouse market. Containment of toxic fumes from pesticides is hazardous to greenhouse workers. Workers are at greater risk of fungicide exposure in the greenhouse because of the intensive nature of crop management.

Most fungicides require a reentry period before the workers can return to a treated crop and there is a waiting period, a period of time between the last application and harvest. Moreover, many greenhouse crops are continuously harvested and therefore cannot use most fungicides. Due to frequent harvesting and direct table value of the produce, pesticides with lower waiting periods only can be used. Breakdown, weathering, and wash-off of chemicals on the leaves or in substrates are all lower in greenhouses than in the field, so fungicides may have a longer residual activity. Finally, the development of fungicide resistance in the pathogen may be exacerbated by the intensive use and limited choice of fungicides in the greenhouse. There is increasing societal concerns about the environmental and health effects of fungicides. Large sprayers cannot be used in greenhouses due to the limitation of space. Therefore, the grower must be extra careful in the long-range planning, selection, use, and application of pesticides in greenhouse crop cultivation systems. That's why agriculture in greenhouses and protected structures offers a unique niche for the development and use of biological control agents.

5.2.1 Suitability of Biological Control for Greenhouses

Of the commercial biocontrol products, over half have applications in nurseries or greenhouses, and many were specifically developed against the soilborne pathogens *Pythium* and *Rhizoctonia*, which are major greenhouse pathogens. The use of biocontrol is more prevalent in greenhouse and protected structures than in field crops, even though greenhouses account for only 0.02 % of the area used in agriculture. Some of the very conditions that favor disease also favor the management of diseases with biological control agents. Environmental conditions such as temperature and relative humidity can be tightly controlled. Like the pathogen, biocontrol agents are also sensitive to environmental conditions, and an unfavorable environment in the field has been cited as a reason for failure or inconsistent performance. Conditions in the greenhouse can be optimized for the biocontrol agent. For instance, biocontrol agents of powdery mildews are much more efficient when relative humidity can be maintained above 80 %, a condition that is easily monitored under glasshouse conditions. The biological vacuum in soil substrates can also favor the establishment of biocontrol agents, provided they are applied before pathogen introduction.

5.3 Products Registered for Biological Control

Nematodes and fungi are the most serious disease-causing organisms of the various vegetables and fruits grown under protected cultivation. Nematodes can be very abundant in soils; there are thousands of known species that cause diseases. Plant parasitic nematodes also often limit crop productivity by predisposing plants to attack by fungi or bacteria and by serving as vectors of a number of plant viruses. Yield losses worldwide from plant parasitic nematodes have been estimated to range from 5 % to 12 % of annual production (Sasser and Freckman 1987).

Several formulations of either of the fungi *Gliocladium* and *Trichoderma* or the bacteria *Pseudomonas* and *Bacillus* have been widely used for biocontrol of soilborne pathogens. These products are not only registered as biofungicides but also used as plant strengtheners. In European countries, plant strengtheners include inorganic compounds such as SiO_2, $NaHCO_3$, organic constituents such as compost, homeopathic compounds, and some containing microorganisms such as *Trichoderma harzianum*, *Bacillus subtilis*, *Pseudomonas*, and *Pythium oligandrum*.

5.3.1 *Coniothyrium minitans*

Coniothyrium minitans is a mycoparasite which destroys sclerotia of *Sclerotinia sclerotiorum* and *S. minor.* It is used for the control of *Sclerotinia* wilt of lettuce in greenhouse. *C. minitans* reduced the sclerotial populations at the soil surface, survived at least 39 weeks at a density of 10^4–10^5 CFU/g, and spread to infect sclerotia in control plots.

5.3.2 *Gliocladium virens (Trichoderma virens)*

It is a soilborne fungus, developed for control of *Pythium ultimum* and *Rhizoctonia solani* in soilless mixes. The fungus produces two fungitoxic compounds, glioviren and gliotoxin compounds.

5.3.3 *Trichoderma harzianum*

Trichoderma harzianum reduces *Fusarium* crown and root rot of tomatoes grown in potting mix containing *T. harzianum* and transplanted into the field. In greenhouse trials, *T. harzianum* controlled *R. solani* in poinsettia, geraniums, and *Catharanthus* and *Pythium* on geraniums, impatiens, and petunias. It inhibits the pathogens by mycoparasitism via production of chitinases, β-1,3-glucanases and β-1,4-glucanases, antibiotics, competition, solubilization of inorganic plant nutrients, induced resistance, and inactivation of the pathogen's enzymes involved in the infection process.

5.3.4 *Streptomyces griseoviridis*

This culture is marketed in Europe and the USA, under the name of Mycostop. It was originally isolated from sphagnum peat and was tested as a biocontrol agent against *Fusarium* wilt of carnations in commercial greenhouses.

5.3.5 *Gliocladium catenulatum*

It is a mycoparasite effective against damping-off, seed rot, root rot, and wilt pathogens. It is sold as a wettable powder that can be applied to the soil, roots, or foliage. In glasshouse trials with ornamental bedding plants, application by incorporation into the growing mix or drench reduced damping-off caused by *Pythium* and *Rhizoctonia.* In some cases, *G. catenulatum* was as effective as the fungicides propamocarb or tolclofos.

5.3.6 Nonpathogenic *Fusarium oxysporum*

It is effective against *Fusarium* wilt diseases on carnation, tomato, cyclamen, and *Fusarium* crown and root rot on tomato. Mechanisms of action include competition for carbon, direct competition with pathogenic strains, and induction of host defenses.

5.3.7 *Bacillus subtilis* var. *amyloliquefaciens*

Several strains of *B. subtilis* were tested on cucumber and tomato against *Pythium aphanidermatum* and *Phytophthora nicotianae* in a series of greenhouse trials. It shows growth promotion effect on corn and radish. *FZB C* and *FZB G* strains of *B. subtilis* produced peptide antibiotics active against *F. oxysporum* f. sp. *radicis-lycopersici*.

5.3.8 Opticure

Russell IPM has developed a biological pesticide, Opticure. It is a natural extract, including dairy derivatives, which is treated with a microbial flora exclusively developed by Russell IPM. Opticure includes five types of microorganisms, each with specific activity against nematode or fungal pests:

- **Natural *Lactobacillus***: Improves the microbial activity of the soil when applied to the soil.
- **Nematode larvae–parasitic bacteria**: Inhibit the development of young larvae (stages 1, 2, and 3) and prevent their development to adulthood in crops in temperate and warm areas.
- **Two strains of parasitic fungi of the genus *Trichoderma***: This type of fungus easily colonizes plant roots. It can also act as biocontrol agent by attacking and parasitizing other fungi, especially in soil. The mechanism involved includes micro-parasitism; antibiosis; secreting-specific enzymes that attack fungi, competition for nutrients and space; stress tolerance in the plant, aiding the development of the root system; absorption of inorganic nutrients; induced resistance; and deactivation of pathogen enzymes.

The active ingredients of Opticure together with amino acids and its special formulation improve soil structure, promote development of the soil microbial flora, and reestablish the natural balance of the soil. Opticure acts as root stimulant and increases plant root resistance to nematodes and has been proven effective against soil fungi such as *Armillaria*, *Pythium*, *Rhizoctonia*, *Alternaria*, *Phytophthora*, and *Fusarium*.

The efficacy of Opticure against *Meloidogyne* sp. in tomato was compared with a chemical nematicide (Vydate) and an untreated control. Of the three treatments, Vydate showed highest efficacy. Opticure was also found to be effective against *Meloidogyne* sp. and significantly reduced number of infected plants in comparison to the control treatment, although it was less effective than the synthetic nematicide (Fig. 5.1).

Opticure was also tested against *Fusarium* sp. in tomato (Fig. 5.2). Two consecutive applications of Opticure reduced the number of dead plants compared with an untreated control. The number of dead plants was reduced by a single application, but a second application, 15 days after the first, significantly reduced the number of dead plants. Because Opticure is nontoxic, crops can be harvested the day after the product has been applied. There is a zero waiting time to harvest once applied.

Biological control agents formulated for specific diseases are listed in Table 5.1.

5.4 Biological Control of Greenhouse Diseases

In India, integrated pest management research has not matched the progress and problems of plant protection at large and greenhouse cultivation in particular. *Trichoderma harzianum* and *T. viride* are the most common fungal bioagents used for suppression of soilborne plant pathogens. *Pseudomonas fluorescens* is used as addendum to FYM, effective against *Fusarium oxysporum*, *Pythium* spp., *Rhizoctonia solani*, *Phytophthora* spp., *Sclerotium rolfsii*, and nematode parasites. The following table shows important diseases of greenhouse crops in India and their biocontrol (Table 5.2).

5.4.1 Ecology of Biocontrol Agents

The use of living organisms to combat other living organisms presupposes a thorough knowledge of their ecology. Efficient root colonization and establishment of biocontrol bacteria is of key importance for effective suppression of deleterious organisms. Once biocontrol bacteria are

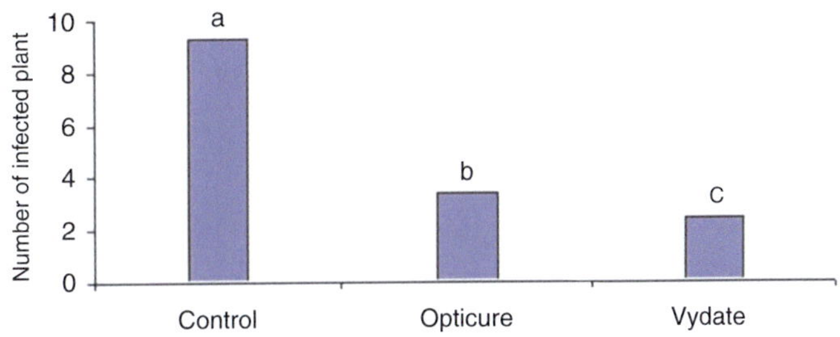

Fig. 5.1 The efficacy of Opticure against *Meloidogyne* sp. in tomato. Common letters indicate treatments did not differ from each other at $P \leq 5$ % (ANOVA, LSD test)

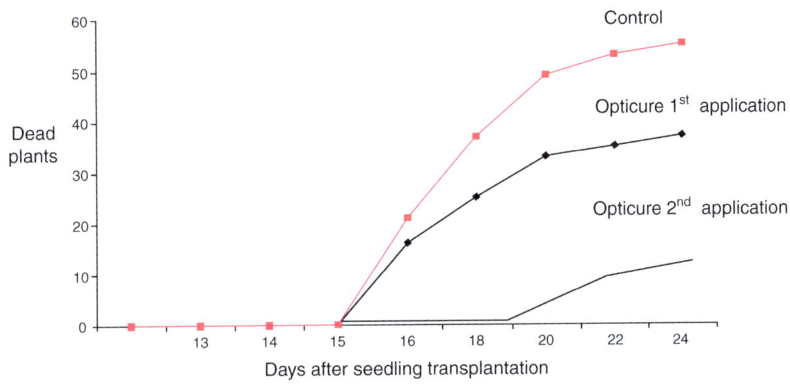

Fig. 5.2 The efficacy of Opticure on *Fusarium* sp. on tomato

established in the rhizosphere, a wide variety of mechanisms can result in suppression of plant pathogens. Accordingly, except in the case of induced resistance, a biocontrol agent must occupy an ecological niche similar to that of the plant pathogen, and its mode of action (competition, parasitism, antibiosis, induction of SAR) must interfere both spatially and temporally with precise steps in the development of the pathogen.

5.4.2 Interactions between Plants and Rhizobacteria

The rhizosphere is a narrow zone of soil that is influenced by root secretions. The structure of rhizobacterial communities is determined by the plant species, and differences in the composition and amounts of root exudates probably account for the differences in microbial populations. Root exudates offer a carbon-rich diet to the rhizosphere microorganisms which includes organic acids such as citrate, malate, succinate, pyruvate, fumarate, oxalate, and acetate and sugars such as glucose, xylose, fructose, maltose, sucrose, galactose, and ribose and constitute the "main course," whereas variable amounts of amino acids, nucleobases, and vitamins (such as thiamin and biotin) provide the "entrée" or "dessert." The ability of rhizobacteria to use organic acids as carbon sources correlates with rhizosphere competence. The rhizosphere, which is the narrow zone of soil that is influenced by root secretions, can contain up to 10^{11} microbial cells per gram root and more than 30,000 prokaryotic species. The collective genome of this microbial community is much larger than that of the plant and is also referred to as the plant's second genome. Chemotaxis, flagellar mobility, lipopolysaccharide (LPS) structure, the outer membrane protein OprF, and, to a lesser extent, pili are all important for competitive root colonization. A root glycoprotein complex – agglutinin – is involved in the short-term adherence of pseudomonads to the plant roots. Once biocontrol pseudomonads have moved and attached to a root zone, microcolonies

Table 5.1 Commercial products for biological control of pathogens under protected conditions

Biorational pesticides	Supplier	Pests controlled	REI[a]	Application/comments
AQ10 (*Ampelomyces quisqualis*)	Plant Health Care	Powdery mildew	4 h	Begin application as soon as host tissue emerges. Apply at least two sequential sprays 7–14 days apart. Works best under conditions of high humidity
SoilGard (*Gliocladium virens*)	Thermo Trilogy	*Rhizoctonia*, *Fusarium*, *Pythium*, *Phytophthora*, *Thielaviopsis*	4 h	Incorporate into soil before planting
DiTera (*Myrothecium verrucaria*)	Abbott	Plant-pathogenic nematodes	4 h	Can be used before or after planting
Sporodex (*Pseudozyma flocculosa*)		Powdery mildew		Cucumber, tomato, rose
Bio-Trek HB (*Trichoderma harzianum*)	Wilbur-Ellis	*Fusarium*, *Pythium*, *Rhizoctonia*	12 h	Apply to seed
Bio-Trek Nursery Drench (*Trichoderma harzianum*)	Wilbur-Ellis	*Fusarium*, *Pythium*, *Rhizoctonia*,	12 h	–
RootShield (*Trichoderma harzianum*)	BioWorks	*Fusarium*, *Pythium*, *Rhizoctonia*, *Sclerotinia*	–	–
Trichodex (*Trichoderma harzianum*)	Abbott	*Botrytis*	–	–
TopShield (*Trichoderma harzianum*)	BioWorks	*Botrytis*, powdery mildew	–	–
PlantShield (*Trichoderma harzianum*)		Gray mold		Vegetables, ornamentals
Ketomium (*Colletotrichum globosum*, *C. cuperum*)		*Colletotrichum* leaf spot		Vegetables
Prosper Nema (pathogenic fungi)	Arbico	Nematodes	0 h	Apply as needed to maintain control
Deny [*Pseudomonas* (*Burkholderia*) *cepacia*]	Stine Seed	*Rhizoctonia*, *Fusarium*, *Pythium*	–	–
Intercept [*Pseudomonas* (*Burkholderia*) *cepacia*]	Soil Technologies	*Rhizoctonia*, *Fusarium*, *Pythium*, nematodes	–	–
Cease (*Bacillus subtilis*)		Anthracnose, gray mold, powdery mildew, leaf spots (*Xanthomonas*, *Pseudomonas*) (vegetables, ornamentals)		Spore powder
Serenade (*Bacillus subtilis*)		Powdery mildew, downy mildew, *Cercospora* leaf spot, early and late blight (cucurbits, grapes, vegetables, ornamentals)		Spore powder
Bio-save (*Pseudomonas syringae*)		Gray mold (ornamentals, vegetables)		
BlightBan (*Pseudomonas fluorescens*)		*Erwinia amylovora*, russet-inducing bacteria (strawberry, tomato)		
Mycostop (*Streptomyces griseoviridis*)	Green Spot	*Fusarium*, *Alternaria*, *Phomopsis*	4 h	Can be incorporated into medium or applied to seed
Actinovate (*Streptomyces lydicus*)	Green Spot	*Pythium*, *Fusarium*, *Phytophthora*, *Sclerotinia*	–	Can be incorporated into medium or applied to seed

[a]REI – reentry intervals

Table 5.2 Biological control diseases of greenhouse crops

Diseases	Crops	Management
Powdery mildew	Capsicum, cucumber	*Bacillus subtilis* (10 g/m^2)
Downy mildew	Cucumber	*Bacillus subtilis* (10 g/m^2)
Fusarium wilt	Tomato, cucumber	*Trichoderma* spp. (10 g/m^2) *Pseudomonas fluorescens* (10 ml/m^2)
Damping-off	Capsicum, cucumber	*Trichoderma* spp. (10 g/m^2) *Pseudomonas fluorescens* (10 ml/m^2)
Fruit rot	Tomato	*Bacillus subtilis* (10 g/m^2)
Root-knot nematode	Capsicum, cucumber, tomato	*Trichoderma* spp. (10 g/m^2) *Pseudomonas fluorescens* (10 ml/m^2)

form in a few days. Other bacteria can reach the same site at a later time and intermingle with pre-existing microcolonies. The root collar, where the root joins the main stem, is a site of intense exudation and is more strongly colonized by bacteria than is the root tip.

5.4.3 The Rhizosphere Microbiome

The microflora of most soils is carbon starved. Because plants secrete up to 40 % of their photosynthates into the rhizosphere, the microbial population densities in the rhizosphere are much higher than in the surrounding bulk soil. This phenomenon is known as the "rhizosphere effect." In general, rhizosphere microbial communities are less diverse than those of the bulk soil. It appears that, from the reservoir of microbial diversity that the bulk soil comprises, plant roots select for specific microorganisms to prosper in the rhizosphere. Together with the plant genotype, soil type is an important driver of the microbial community composition in the rhizosphere. Plants can determine the composition of the root microbiome by active secretion of compounds that specifically stimulate or repress members of the microbial community. Recent evidences suggest that plant genotype also determines microbiome composition. Differences in a single gene between plant genotypes can have a significant impact on the rhizosphere microbiome. The production of a single exogenous glucosinolate significantly altered the microbial community on the roots of transgenic *Arabidopsis*.

5.4.4 The Root Microbiome to the Rescue

Microbiome changes upon defense activation, i.e., the interactions between a plant and its root microbiome might change when the plant is attacked. Recently, it was demonstrated that infection of citrus by *Candidatus Liberibacter asiaticus*, associated with huanglongbing, drastically altered the composition of citrus rhizosphere communities. Changes in rhizosphere composition upon infection might be the result of the induced excretion of antimicrobial compounds by infected roots. However, infection does not only lead to the secretion of pathogen deleterious compounds. It is also found that infected roots also induced excretion of pathogen spore germination stimulators. For example, infection of watermelon plants by *F. oxysporum* enhanced the stimulation of *Fusarium* spore germination by root exudates.

In addition to this, recent researches show that plants recruit plant-beneficial microbes to their roots in response to the attack. For example, colonization of the roots of *Arabidopsis* by the plant-beneficial soil bacterium *Bacillus subtilis* FB17 was greatly improved when aboveground plant tissues were infected by *Pseudomonas syringae* pv. *tomato*. A mutually beneficial relationship exists between *Arabidopsis* and FB17: *FB17* is recruited to aid in plant defense, and the plant provides the bacterium with maleic acid. Activation of beneficials and inoculation of strawberry plants with *Verticillium dahliae* stimulated the expression of cyanide biosynthetic genes in the biocontrol bacterium *Pseudomonas* sp. *LBUM300*. These changes in gene expression could be a result of nutrients leaking from damaged roots.

5.4.5 Limitations with the Use of Biocontrol

Because of the high value of the crop and emphasis on quality in floriculture, vegetable crops, and ornamentals, there is less acceptance of damage and thresholds for disease are very low. If biocontrol agents cannot perform with the consistency and efficacy of fungicides in these crops, they may not be adopted. The success of bioagents depends on the quality, timely availability, and appropriate release timing and methodology of bioagents. Given that a biological control strategy is scientifically feasible, the successful grower is most likely to adopt the strategy if it is an economically feasible decision for the greenhouse. The adoption decision is economically feasible for the grower if expected profits are greater than zero (i.e., expected revenues are greater than expected costs). If growers habitually make decisions where expected profits are less than zero, the greenhouse will not be a viable business in the longer run.

5.4.6 Biocontrol of Foliar Plant Pathogens

Insects and diseases are the major challenges under protected cultivation as it provides favorable environmental conditions. The greenhouse environment presents a unique situation that makes conditions more favorable for foliar diseases especially powdery mildew, downy mildew, and gray mold caused by fungi and bacterial diseases (*Xanthomonas* and *Pseudomonas*). Some of the conditions that favor diseases also favor the biocontrol agents for the management of foliar diseases. Biocontrol in greenhouses may have a greater potential because:

- Of the development of fungicide resistance in the pathogen.
- Workers are at greater risk of fungicide exposure in the greenhouse as most fungicides require a reentry period.
- Fungicides have a longer residual activity.
- Controlled environment of greenhouses favors the survival and growth of biocontrol agent.
- Of the availability of limited number of registered fungicides.
- Most of the fungicides cannot be used due to preharvest waiting period of most of the vegetables.
- Of environmental awareness and consumer concern due to the ill effect of the fungicides and for the production of organic and export-oriented products.
- Of reduced area and high density of planting.
- It is safe and nontoxic to humans and animals.
- It is rapidly biodegradable.

Biopesticides currently represent about 1.5–2.0 % of the world pesticide market. However, the growth rate of biopesticide over the next 10 years has been forecast at 10–15 % annually in comparison to 2 % of chemical pesticides. During the last 10 years, over 80 biocontrol products have been marketed worldwide. A large of these has been developed for greenhouse crops. However, not all of these products are registered as biocontrol agents, but some are also marketed as plant growth promoters, plant strengtheners, or soil conditioners. Most of these products are formulations of either of the fungi *Trichoderma* and *Ampelomyces* or the bacteria *Pseudomonas* and *Bacillus*. Because of the high value and emphasis on quality of greenhouse crops, a biocontrol agent should be recommended, if it can perform with the consistency and efficacy of fungicides.

The increased use of biological controls has led to a reduction in pesticide applications and leads to environmentally responsible, intensive crop production. The primary strategy of biological control for greenhouse plant diseases is to introduce bacterial or fungal parasites to control disease-causing bacteria or fungi in the greenhouse environment so that pathogens are unable or have a reduced ability to infect the plants. Research in developing biological controls for greenhouse crop diseases is ongoing, and it is likely that more potential products of the biological control will be available in the near future for greenhouse diseases. Biocontrol agents should be used as preventative measures and/or in combination or alternating with compatible fungicides for the effective management of foliar diseases.

There is a growing demand for sound, biologically based pest management practices. Recent surveys of both conventional and organic growers indicate an interest in using biocontrol products, suggesting that the market potential of biocontrol products will increase in the coming years. Applications of diverse biological control strategies have been successful in the greenhouse industry and continue to increase. The Biopesticide Industry Alliance has been formed, and it is now actively promoting the value and efficacy of biopesticides. Clearly, the future success of the biological control industry will depend on innovative business management, product marketing, extension education, and research.

5.5 Future Prospects

In Britain, although protected crops represent a small fraction of the total area, they account for two-thirds of all biologicals. At the same time, the use of pesticides in greenhouses has declined from 4866 treated ha in 1981–2292 ha in 1995. A combination of economic, political, and environmental factors has probably contributed to the transition to biological due to the loss of insecticide registrations, insect resistance, and concern for worker safety.

More scientific efficacy trials with proper replication and statistical analysis are needed under commercial or near-commercial conditions. Biocontrol registrations now request data on the interaction of pesticides with beneficial insects. Growers need to know whether new products are compatible with their current pest management strategies. Knowledge of epidemiology and ecology of pathogens in the greenhouse, which may be different from the field, is required. Low-cost methods for rapid detection of pathogens in the greenhouse are important research priorities. Finally, the challenge of production and formulation of biocontrol agents remains, with each organism bringing its own set of problems. Effective production and formulation protocols are usually proprietary, involving substantial investment to develop economic production and a formulation with adequate shelf life, stability, and titer.

Greenhouses offer a privileged environment for disease biocontrol, but implementation is still very limited. However, if we have anything to learn from our crop protection scientists, it is that this will change. Indeed, from a modest and uneven start in the early 1970s, biocontrol has grown to a standardized approach throughout the greenhouse market. Plant pathologists and companies investing in biocontrol products should likewise view the future of biological control of plant diseases in greenhouse systems with optimism. A few products have already been registered and several more should be commercialized within the next few years. Success stories against a number of diseases will be important both to validate biocontrol of plant diseases and, most importantly, to gain acceptance by growers.

Reference

Sasser JN, Freckman DW (1987) A world perspective on nematology: the role of the society. In: Veech JA, Dickson DW (eds) Vistas on nematology. Society of Nematologists, Hyattsville, pp 7–14

Compost in Disease Management 6

Abstract

The main advantage of exploring the role of microbiologically rich compost in reducing biotic and abiotic stresses is that it is a novel way to imitate the natural system occurring in ecosystems. Utilization of compost with disease-suppressive properties is a relatively new biological way of decreasing stress in plant production. The compost-enriched soil can also help suppress diseases and ward off pests. These beneficial uses of compost can help growers save money, reduce their use of pesticides, and conserve natural resources. Disease and nematode management, mortality composting, how compost works, enrichment of compost with biopesticides, and commercial products are discussed.

Keywords

Compost • Diseases • Nematodes • Enriched compost • Mortality composting

6.1 Introduction

Plant abiotic and biotic stresses represent the principal cause of crop failure, decreasing average yields of major crops by more than 50 %. Compost can be considered as a soil conditioner that contributes to soil fertility, structure, porosity, organic matter, water holding capacity, and disease suppression (Fig. 6.1). Composts suppress soilborne diseases, and this suppression has been widely reported for *Pythium* spp., *Phytophthora* spp., *Rhizoctonia* spp., and *Fusarium* spp. Compost amendments also contribute to controlling foliar diseases, such as *Puccinia* spp., *Alternaria solani*, and *Pseudomonas syringae* pv. *syringae*. Disease suppression by composts has been explained mainly by biotic mechanisms. The severity of soilborne plant diseases is often reduced when microbiologically improved compost is used as growth media. A wide variety of rhizosphere microorganisms have been isolated and used as microbiological inoculants for improving plant growth and health. These include arbuscular mycorrhizal fungi (AMF), plant growth-promoting rhizobacteria (PGPR), nitrogen fixers,

P.P. Reddy, *Sustainable Crop Protection under Protected Cultivation*,
DOI 10.1007/978-981-287-952-3_6

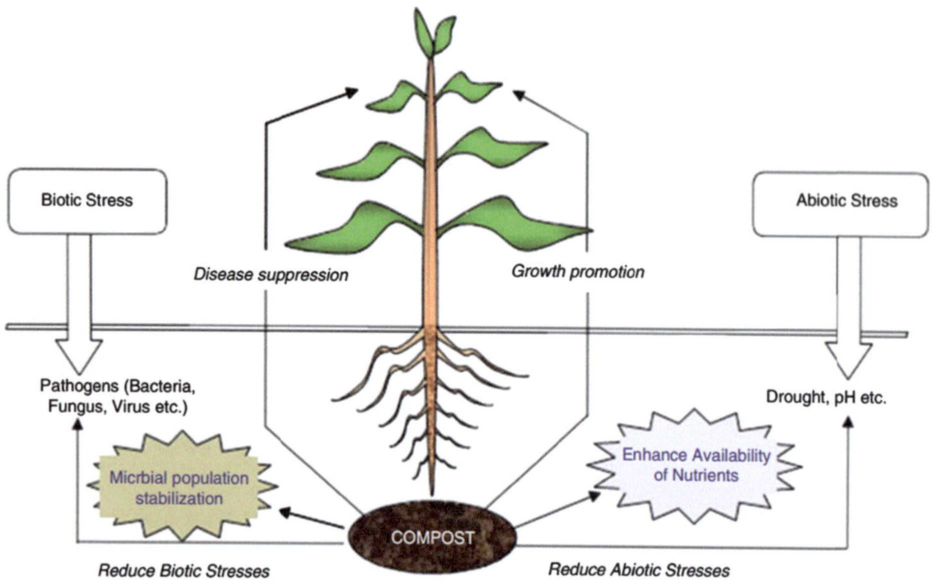

Fig. 6.1 Compost and its possible responses toward reducing plant stresses and in improvement of soil quality by enhancing nutrient availability and stabilizing microbial population

and phosphorus-solubilizing microorganisms (PSM). Utilization of compost with disease-suppressive properties is a relatively new biological method of decreasing biotic stress in plant production. The main advantage of exploring the role of microbiologically rich compost in reducing biotic and abiotic stresses is that it is a novel way to imitate the natural system occurring in ecosystems. Under saline conditions, compost was not completely efficient in correcting the detrimental effects of salt but was able to mitigate them. Such an amendment may be used to enhance crop yield irrigated with saline waters or grown on saline soils. Compost amendments in soil are also efficient in combating the pH stress. The amendment of alkaline soils with composts has a positive effect on plant vigor. Molecular studies also showed a different profile of microorganisms in disease-suppressive composts. This study shows a good approach to add up value to compost and make it more efficient in reducing biotic and abiotic stresses (Mehta et al. 2012).

Compost is organic matter that has been decomposed and recycled as a fertilizer and a key ingredient in organic farming. Composting is a biological process in which various organic biodegradable wastes are converted into hygienic, humus-rich product (compost) which could be used as a soil conditioner and an organic fertilizer (Popkin 1995). Composts of various origins are also used to provide biological control against various plant pathogens (Hoitink and Grebus 1994). The addition of composts to agricultural soils has beneficial effects on crop development and yields by improving soil physical and biological properties (Zheljazkov and Warman 2004). Poultry manure application in *Java citronella* plants significantly increased the herbage, essential oil content, and dry matter yield (Adholeya and Prakash 2004). Compost is also an inhabiting place for various microbes due to which biological activities are markedly enhanced in the rhizosphere of plants (Tilak and Reddy 2006). Such syntrophic associations are of ecological importance with implied agricultural significance. Aqueous extracts of compost have also been efficient in replacing synthetic fungicides (Zhang et al. 1998). Utilization of compost with disease-suppressive properties is a relatively new biological way of decreasing stress in plant production.

Compost technology is a valuable tool already being used to increase yields by farmers interested in sustainable agriculture. Now, professional growers are discovering that compost-enriched soil can also help suppress diseases and ward off pests. These beneficial uses of compost can help growers save money, reduce their use of pesticides, and conserve natural resources. In the poultry industry, composting has also become a cost-effective method of mortality management. It destroys disease organisms and creates a nutrient-rich product that can be used or sold.

Compost has been used effectively in the nursery industry, in high-value crops, and in potting soil mixtures for control of root rot diseases.

6.2 Disease Management

Composts have long been known to improve soil fertility and plant disease management. Previously it has been shown that components of composts improve the ability of plants to resist disease caused by root pathogens, like *Fusarium*, *Phytophthora*, *Pythium*, *Rhizoctonia*, etc., and foliar pathogens, like *Pseudomonas*, *Colletotrichum*, *Xanthomonas*, etc.

Among soilborne root pathogens, suppression of *Fusarium* using composts has been reported by several researchers (Punja et al. 2002). The severity of various diseases caused by *Fusarium* has been reduced between 20 % and 90 % using compost amendments. Microbial activity has been considered a key factor in suppression of *Fusarium* wilt. Composts increased microbial populations and microbial activity in composts and composted peat mixes increased by 50 % (Cotxarrera et al. 2002). Several microorganisms or biocontrol agents have been isolated from composts or shown to contribute to suppression of *Fusarium* spp. including populations of non-pathogenic strains of *Fusarium oxysporum* and fluorescent *Pseudomonas* spp. (Kannangara et al. 2000) and *Trichoderma* and *Flavobacterium* (Hoitink and Fahy 1986).

Phytophthora is another soilborne pathogenic fungus that causes a variety of problems including root rot, a form of "dieback," crown rot, or *Phytophthora* blight of plants. Composts have been used successfully for suppression of *Phytophthora* crown and root rots of nursery and fruit crops produced in container media (Aryantha et al. 2000) and field soils (Downer et al. 2001). Addition of compost serves two possible purposes significant for the biological control of *Phytophthora cinnamomi*: it provides a substrate for the growth of fungal antagonists and creates an environment that promotes enzyme activity (Downer et al. 2001). Several potential biocontrol agents of *Phytophthora* root and crown rots have been identified from compost-amended growing media, including *Pseudomonas* spp. (Aryantha et al. 2000), *Pantoea* spp. (formerly *Enterobacter* spp.) (Krause et al. 2003), *Penicillium* and *Aspergillus* spp. (Downer et al. 2001), actinomycetes (Aryantha et al. 2000), and *Trichoderma* spp. (Downer et al. 2001).

Pythium is also a destructive, soilborne parasitic root fungus, which causes damping-off disease in seedlings and root and crown rot of plants. The suppression of diseases caused by *Pythium* spp. has been well documented (Stone et al. 2003). The severity of diseases caused by the fungus *Pythium* was reduced by 30–70 % when growing media were amended with various compost products. Adequately mature composts have large microbial populations and high microbial activity, which have been directly linked to the suppression of *Pythium* (Ringer et al. 1997), but this may not hold true for all composts (Craft and Nelson 1996).

Similarly, *Rhizoctonia* is a soilborne pathogenic fungus which causes a range of soilborne diseases as well as diseases of aerial parts of plants. Composts have been used with varying success to suppress *Rhizoctonia* diseases of several crops (Tuitert et al. 1998). Amendment of soil or container media with composts reduced diseases caused by the soilborne pathogenic fungus *Rhizoctonia* by up to 70 %. Higher levels of microbial biomass and microbial activity have been reported to suppress *Rhizoctonia* by increasing competition between compost-inhabiting microorganisms and *R. solani* for cellulose or other available nutrients (Diab et al. 2003). The increase in microbial count (MBC) and microbial number (MBN) indicated growth of saprophytic microorganisms. The high population density of fluorescent pseudomonads, actinomycetes, and heterotrophic fungi in growing media amended with adequately matured compost has been found to better suppress *Rhizoctonia* spp. than the use of less matured (immature) composts (Diab et al. 2003).

Composts are considered able to induce systemic resistance, which can reduce the severity of some plant foliage diseases (Stone et al. 2003). However, Krause et al. (2003) report that only a small proportion of composts have the ability to suppress foliar diseases. Some pathogenic *Pseudomonas* species cause several plant foliar diseases including bacterial speck and bacterial canker. Compost application has resulted in reduced bacterial speck of *Arabidopsis* (mustard family) and tomatoes. In tomatoes, different types of paper mill residue-based composts resulted in the reduction of bacterial speck between 47 % and 62 % (Vallad et al. 2003).

Similarly pathogenic *Xanthomonas* spp. causes many plant diseases, including such foliar diseases as bacterial leaf spot or speck, bacterial blight or angular leaf spot, and stem rot. Two studies report a reduction in bacterial leaf and fruit spot of vegetables (radish and tomato) caused by *Xanthomonas* spp. (Krause et al. 2003). Abbasi et al. (2002) found that, in tomato production, application of composted garden organics resulted in reduced bacterial spot incidence on fruit by 28–33 %, although the severity of disease on foliage was increased. *Bacillus* spp. has been identified as most effective in suppressing bacterial leaf spot (Krause et al. 2003).

Some fungi like *Colletotrichum* also cause foliar diseases such as anthracnose fruit rot or lesions. Research has shown that amendment of soil or container media with composts can reduce the severity of anthracnose (Abbasi et al. 2002). Composted cannery waste applied at a high rate (24–30 t ha^{-1}) reduced the incidence of anthracnose in organic tomatoes by 40 % (Abbasi et al. 2002).

These studies indicate that compost amendments play a valuable role in reducing disease. However, the variability among composts makes it difficult for frequent use of composts in agriculture field. In the future, a lot of studies are needed to understand the role and mechanism of compost against plant diseases so that a strategy can be made to minimize the compost variability. In the future, attention also needs to be paid to non-culturable members of the root-associated and soil communities because these microorganisms may be numerically dominant and have not been studied. Molecular methods developed for the study of microorganisms in their environments are key tools for the study of the influences of the microbial community on biocontrol, and also in the future, there is a need to understand the mechanism behind the compost-based disease suppression.

Each year, more than 10 % of the vegetables planted in the USA are lost to root rot alone, according to researchers at the University of Florida's Tropical Research and Education Center. Additional crop losses are caused by other soilborne plant pathogens, such as the microorganisms that cause ashy stem blight and chili pepper wilt. Compost can help control plant disease and reduce crop losses. Disease control with compost has been attributed to four possible mechanisms:

- Successful competition for nutrients by beneficial microorganisms
- Antibiotic production by beneficial microorganisms
- Successful predation against pathogens by beneficial microorganisms
- Activation of disease-resistant genes in plants by composts

Scientists have enhanced the natural ability of compost to suppress diseases by enriching it with specific disease-fighting microorganisms or other amendments. This amended or "tailored" compost can then be applied to crops infected by known diseases. Research has shown that tailored compost significantly reduced or replaced the application of pesticides, fungicides, and nematicides – which could adversely affect water resources, food safety, and worker safety.

The use of tailored compost can also be more cost-effective than chemical soil treatments, such as methyl bromide. Soil treated with compost retains irrigation water better, which lowers water costs. Chemicals also must be applied more often than compost. In addition, some chemicals have reentry requirements that prohibit workers from entering the greenhouse immediately after chemicals have been applied, reducing worker productivity.

6.2.1 *Pythium* Root Rot

Hoitink and Fahy (1986) of the Ohio State University has conducted compelling research on compost's effects on plants afflicted with *Pythium* root rot. As the photo below illustrates, the application of tailored compost had a dramatically positive effect on plant growth and impeded the spread of the disease (Fig. 6.2).

Fig. 6.2 The plant on the right was treated with compost. The plant on the left was not and suffered the effects of *Pythium* root rot. Photo courtesy of Dr. Harry Hoitink, University of Ohio

6.2.2 Chili Wilt

Researchers from New Mexico State University applied compost made from municipal wastewater sludge and yard trimmings to chili crops in a field known to be infested with *Phytophthora* root rot or chili wilt. The study showed that the 10-t and 20-t compost applications provided the greatest suppression of chili wilt and the highest yields.

6.2.3 Ashy Stem Blight and Root Rot

University of Florida researchers tested the effects of Agrisoil (compost made of mixed municipal solid waste) and Daorganite (a heat-treated biosolids mix), on test plots in a field in Homestead, Florida. The Agrisoil compost was applied at rates of 36 and 72 t/ac, and the Daorganite sludge was applied at rates of 0.67 and 1.33 t/ac. Sections of the field also were left untreated as a control. Six weeks after the materials were incorporated into the soil, researchers planted bush beans throughout the field. A second crop, black-eyed peas, was planted following the bean harvest, and Agrisoil compost and Daorganite were applied at the same rates as in the bush bean project. The field was also fertilized according to accepted local agricultural practices.

The health and yields of the bush bean crops were significantly improved by compost. Beans grown in the Agrisoil compost were larger and healthier. Yields from the 36 and 72 t/ac application areas were both 25 % higher than control area yields. Beans grown in the Daorganite mix showed low yields similar to those grown in the control areas. In addition, ashy stem blight severely affected beans grown in both the control and Daorganite-treated areas but not the plots with Agrisoil compost.

The health and yields of the black-eyed pea crops grown in compost were also significantly improved. These crops had greener foliage and were larger than those grown in the control or Daorganite-treated plots. Yields from the compost-enriched areas were more than double the control yields in the 72 t/ac application sections and also significantly higher in the 36 t/ac sections. By comparison, yields in the Daorganite-treated areas were only slightly higher or comparable to those in the control sections. *Rhizoctonia* root rot severely affected plants in the Daorganite-treated and control areas, but the disease was considerably less prevalent in the compost-enriched areas.

In this particular study, yields and disease infection proved to be directly related in both the bean and pea crops. Mature Agrisoil compost was more effective at disease suppression than the Daorganite heat-treated biosolids mix. Thus, yields were uniformly higher in the Agrisoil-treated areas than in the Daorganite-treated and control areas.

Successful disease suppression by compost has been less frequent in soils than in potting mixes. This is probably why there has been much more research (and commercialization) concerning compost-amended potting mixes and growing media for greenhouse plant production than research on compost-amended soils for field crop production. Below is a table that outlines some of the (mostly) field research done on compost-amended soils and the effects on plant disease (Table 6.1).

In some further research, University of Florida field trials (Ozores-Hampton et al. 1994) showed disease-suppressive effects of compost and heat-treated sewage sludge on snap beans and southern peas (black-eyed peas). The compost was applied at 36 or 72 t/ac and the sludge at 0.67 and 1.33 t/ac. Bush beans were planted 6 weeks after the organic treatments were applied and tilled in. After the bush beans were harvested, a second crop of southern peas was planted. A standard fertilizer program was used. Plant damage from ashy stem blight was given a rating of slight, moderate, or severe. *Rhizoctonia* root rot disease ratings were made using a scale from 0 to 10, where 10 represented the most severe symptoms.

Bean size from the compost treatment, at both application rates (36 and 72 t/ac), was larger and yields 25 % higher than those from areas receiving no organic amendment. Ashy stem blight was severe in areas with no compost applied. The disease was reduced under the sludge treatment but almost eliminated where compost had been applied. Leaf wilting and leaf death were pronounced in that portion of the field where compost was not applied.

Southern peas as a second crop had greener foliage and larger plants under both rates of compost. Pea yields were significantly higher with 36 t of compost. Where 72 t of compost when used,

Table 6.1 Compost treatment and disease management

Vegetable	Pathogen/disease	Treatment	Comments
Beans (CA black-eyed pea No. 5)	*Rhizoctonia* sp.	Compost added to soil at varying rates (36–72 t/ac)	Disease reduced to 80 % in areas with highest compost rates and 40 % where intermediate rates applied. Control plots yielded 75 bushels/acre, and compost plots yielded 200 bushels/acre (Hudson 1994)
Cucumber	Powdery mildew, *Sphaerotheca* sp.	Young cucumber plants grown in soil/compost mix at variable rates	1:1 soil/compost mix decreased powdery mildew by 20 % over control; 1:3 mix decreased infection by 40 % (Trankner 1992)
Pea (*Pisum sativum*)	Damping-off, *Pythium* sp.	Seed treatment; seeds soaked in dilute compost extract, dried before sowing	Pea seed treatment with compost extract germinated significantly better than untreated seed in soil artificially inoculated with *Pythium ultimum* (Trankner 1992)
Bell pepper	*Phytophthora* sp.	40 tons of compost/acre	Compost in combination of hilling plant rows is best practice to reduce *Phytophthora* (Hudson 1994)

yields were more than double the non-amended plots. With the sludge treatment, yields were comparable or slightly higher than where no amendment was added. *Rhizoctonia* root rot caused severe infections, plant stunting, and premature death where no compost was applied. Plants growing under the sludge treatment suffered severe root infection. Disease was reduced considerably as compost rates increased from 36 t/ac to 72 t/ac (Ozores-Hampton et al. 1994).

6.3 Nematode Management

Compost also can eradicate some types of pests, such as parasitic nematode infections, in addition to its use in controlling diseases. Specially formulated (tailored) compost can include chemicals that actually kill nematodes or prevent their eggs from hatching. Most types of compost help control parasitic nematodes by providing nutrients to the soil, which encourage the growth of fungi and other organisms, which, in turn, compete with or destroy nematodes. Compost also contributes to plants' basic health, making them less susceptible to pests.

Compost's ability to halt soil nematode invasion was identified by the staff of Dr. Herbert Bryan of the University of Florida. While studying plant response to different compost applications and irrigation rates, the staff, who had a background in nematology, noted the unexpected results while conducting routine observations. "Where compost was used, even without a fumigant, there was a significant reduction in root-knot nematodes," said Dr. Bryan.

Later research by Dr. Tom Obreza, a soil and water scientist at the Southwest Florida Research and Education Center, turned up with similar results. Dr. Obreza's experiment consisted of growing tomatoes in composts from several different sources and comparing them to control plots treated with the usual fertilizers. Dr. Obreza found no disease problems in any of the plots except one. Dr. Obreza noted, "We had a little invasion of root-knot nematode in one corner of the field. The infection was evident in the plants right up to the compost treated plots and stopped right there. The difference was as plain as night and day."

6.4 Mortality Composting

More than 7.3 billion chickens, ducks, and turkeys are raised for commercial sale in the USA each year, according to the US Department of Agriculture's National Agricultural Statistics Service. About 37 million birds (18–25 %) die from disease or other natural causes before they are marketable. As more poultry is consumed, these numbers are expected to climb.

Composting is a viable and cost-effective option for disposing of poultry mortalities as compared to incineration or burial. Pathogens in poultry carcasses are destroyed during composting by the high temperatures (130–155 °F) inherent in the process.

During composting, various odor control techniques can be used. As a result, this type of compost is not only safe for crop application, but it also can be safely sold by farmers. In fact, selling excess compost could even be a source of additional income for farmers. Markets for high-quality compost include professional growers (such as horticultural greenhouses and nurseries), homeowners, turf growers, and crop farmers (such as corn and wheat farmers). Professional growers alone purchase $250 million per year in compost products.

6.5 How Compost Works

Compost is effective because it fosters a more diverse soil environment in which a myriad of soil organisms exist. Compost acts as a food source and shelter for the antagonists that compete with plant pathogens, for those organisms that prey on and parasitize pathogens, and for those beneficials that produce antibiotics. Root rots caused by *Pythium* and *Phytophthora* are generally suppressed by the high numbers and diversity of beneficial microbes found in the compost. Such beneficials prevent the germination of spores and infection of plants growing on the amended soil (Harrison and Frank 1999). To get more reliable results from compost, the compost itself needs to be stable and of consistent quality.

Systemic resistance is also induced in plants in response to compost treatments. Hoitink has now established that composts and compost teas indeed activate disease resistance genes in plants (Goldstein 1998). These disease resistance genes are typically "turned on" by the plant in response to the presence of a pathogen. These genes mobilize chemical defenses against the pathogen invasion, although often too late to avoid the disease. Plants growing in compost, however, have these disease-prevention systems already running (Goldstein 1998). Induced resistance is somewhat pathogen specific, but it does allow an additional way to manage certain diseases through common farming practices.

It has become evident that a "one size fits all" approach to composting used in disease management will not work. Depending on feed stock, inoculum, and composting process, composts have different characteristics affecting disease management potential. For example, high carbon-to-nitrogen ratio (C/N) tree bark compost generally works well to suppress *Fusarium* wilts. With lower C/N ratio composts, *Fusarium* wilts may become more severe as a result of the excess nitrogen, which favors *Fusarium* (Hoitink et al. 1997). Compost from sewage sludge typically has a low C/N ratio. Some of the beneficial microorganisms that reinhabit compost from the outside edges after heating has subsided include several bacteria (*Bacillus* species, *Flavobacterium balustinum*, and various *Pseudomonas* species) and several fungi (*Streptomyces*, *Penicillium*, *Trichoderma*, and *Gliocladium virens*). The moisture content following peak heating of compost is critical to the range of organisms inhabiting the finished compost. Dry composts with less than 34 % moisture are likely to be colonized by fungi and, therefore, are conducive to *Pythium* diseases (Hoitink et al. 1997). Compost with at least 40–50 % moisture will be colonized by both bacteria and fungi and will be disease suppressive (Hoitink et al. 1997). Water is typically added during the composting process to avoid a dry condition. Compost pH below 5.0 inhibits bacterial biocontrol agents (Hoitink et al. 1991). Compost made in the open air near trees has a higher diversity of microbes than compost made under a roof or in vessel (Granatstein 1998).

Three approaches can be used to increase the suppressiveness of compost: first, curing the compost for 4 months or more; second, incorporating the compost in the field soil several months before planting; and third, inoculating the compost with specific biocontrol agents (Hoitink et al. 1997). Two of the more common beneficials used to inoculate compost are strains of *Trichoderma* and *Flavobacterium*, added to suppress *Rhizoctonia solani*. *Trichoderma harzianum* acts against a broad range of soilborne fungal crop pathogens, including *R. solani*, by production of antifungal exudates.

The key to disease suppression in compost is the level of decomposition. As the compost matures, it becomes more suppressive. Readily available carbon compounds found in low-quality, immature compost can support *Pythium* and *Rhizoctonia*. As these compounds are reduced during the complete composting process, saprophytic growth of these pathogens is dramatically slowed (Nelson et al. 1994). Beneficials, such as *Trichoderma hamatum* and *T. harzianum*, were unable to suppress *Rhizoctonia* in immature composts, which are extremely effective when introduced into mature composts.

For *Pythium* suppression, there is a direct correlation between general microbial activity, the amount of microbial biomass, and the degree of suppression. *Pythium* is a nutrient-dependent pathogen with the ability to colonize fresh plant residue, especially in soil that has been fumigated to kill all soil life. The severity of diseases caused by *Pythium* and *R. solani* relates less to the inoculum density than to the amount of saprophytic growth the pathogen achieves before infection (Cook and Baker 1983). Consequently, soils that are antagonistic to saprophytic growth of *Pythium* – such as soils amended with fully decomposed compost – will lower disease levels.

Rhizoctonia is a highly competitive fungus that colonizes fresh organic matter (Chung et al. 1988). Its ability to colonize decomposed organic matter is decreased or nonexistent. There is a direct relationship between the compost's level of decomposition and its suppression of *Rhizoctonia* – again pointing to the need for high-quality mature compost. Like immature compost, raw manure is conducive to diseases at first and becomes suppressive after decomposition. In other words, organic amendments supporting high biological activity (i.e., decomposition) are suppressive of plant root diseases, while raw organic matter will often favor colonization by pathogens.

6.6 Enrichment of Compost with Biopesticides (Tailored Compost)

Biopesticides are becoming effective alternatives to chemical pesticides. Biopesticides are made by adding controlled amounts of pest-fighting microorganisms to compost, which results in "tailored" compost with a specific pesticidal capability. Biopesticides must be registered with the US Environmental Protection Agency (EPA) and undergo the same level of testing as chemical pesticides to determine their effectiveness and their safety for public health and the environment. Although one type of tailored compost that is inoculated by a patented process is already registered as a biopesticide with EPA, many more are expected to follow suit.

6.7 Determining and Monitoring Compost Quality

It is clear that compost's maturity is a key factor in its ability to suppress disease. The challenge involved in achieving and measuring that maturity is the primary reason that compost is not more widely used. Certainly, immature compost can be used in field situations, as long as it is applied well ahead of planting, allowing for eventual stabilization. However, good disease suppression may not develop due to other factors. For example, highly saline compost actually enhances *Pythium* and *Phytophthora* diseases unless applied months ahead of planting to allow for leaching (Hoitink et al. 1997).

Dr. Harry Hoitink at Ohio State University has pioneered much of the work associated with disease-suppressive composts. He notes that success or failure of any compost treatment for disease control depends on the nature of the raw product from which the compost was prepared, the maturity of the compost, and the composting process used. Failure to assess compost quality may be responsible for some of the failures in using compost for disease suppression (Hoitink and Fahy 1986). High-quality compost should contain disease-suppressive organisms and mycorrhizal inoculum (Sances and Elaine 1997). Furthermore, high-quality compost should contain very few if any weed seeds.

6.8 Commercial Products

There are a number of commercial products containing beneficial, disease-suppressive organisms. These products are applied in various ways – including seed treatments, compost inoculants, soil inoculants, and soil drenches. Among the beneficial organisms available are *Trichoderma*, *Flavobacterium*, streptomycetes, *Gliocladium* spp., *Bacillus* spp., *Pseudomonas* spp., and others.

Trichoderma and *Gliocladium* are effective at parasitizing other fungi, but they stay alive only as long as they have other fungi to parasitize. So, these fungi do a good job on the pathogenic fungi that are present when you inoculate them, but then they run out of food and go to sleep. In soils with low fungal biomass (soils with low organic matter and plenty of tillage), these two beneficials have nothing to feed on. Compost is a great source of both the organisms and the food they need to do their jobs. A great diversity of bacteria, fungi, protozoa, and beneficial nematodes exists in good compost (Ingham and Molina 1991).

6.9 Conclusions

Soilborne diseases result from a reduction in the biodiversity of soil organisms. Restoring important beneficial organisms that attack, repel, or otherwise antagonize disease-causing soil organisms will reduce their populations to a manageable level. Beneficial organisms can be added directly, or the soil environment can be made more favorable for them with compost and other organic amendments. Compost quality determines its effectiveness at suppressing soilborne plant diseases.

References

Abbasi PA, Al-Dahmani J, Sahin F, Hoitink HAJ, Miller SA (2002) Effect of compost amendments on disease severity and yield of tomato in conventional and organic production system. Plant Dis 86:156–161

Adholeya A, Prakash A (2004) Effect of different organic compost/manures on yield and yield component of bean (*Phaseolus Vulgaris* L). Bioresour Technol Tanu 92:311–319

Aryantha IP, Cross R, Guest DI (2000) Suppression of *Phytophthora cinnamomi* in potting mixes amended with uncomposted and composted animal manures. Phytophthology 90:775–782

Chung YR, Hoitink HAJ, Lipps PE (1988) Interactions between organic-matter decomposition level and soil-borne disease severity. Agric Ecosyst Environ 24:183–193

Cook RJ, Baker KF (1983) The nature and practice of biological control of plant pathogens. APS Press, St Paul

Cotxarrera L, Trillas-Gay MI, Steinberg C, Alabouvette C (2002) Use of sewage sludge compost and *Trichoderma asperellum* isolates to suppress Fusarium wilt of tomato. Soil Biol Biochem 34:467–476

Craft M, Nelson EB (1996) Microbial properties of composts that suppress damping-off and root rot of creeping bentgrass caused by *Pythium graminicola*. Appl Environ Microbiol 62:1550–1557

Diab HG, Hu S, Benson DM (2003) Suppression of *Rhizoctonia solani* on impatiens by enhanced microbial activity in composted swine waste-amended potting mixes. Phytopathology 93:1115–1123

Downer A, Menge JA, Pond E (2001) Association of cellulytic enzyme activities in eucalyptus mulches with biological control of *Phytophthora cinnamomi*. Phytopathology 91:847–855

Goldstein J (1998) Compost suppresses disease in the lab and on the fields. BioCycle, Nov. pp 62–64

Granatstein D (1998) Suppressing plant diseases with compost. Good Fruit Grower, 1 May, pp 9–11

Harrison UJ, Frank JL (1999) Disease management through suppressive soils. Department of Plant Pathology, North Carolina State University (draft document), 23 Sept, pp 14

Hoitink HAJ, Fahy PC (1986) Basis for the control of soil-borne plant pathogens with composts. Annu Rev Phytopathol 24:93–114

Hoitink HAJ, Grebus ME (1994) Status of biological control of plant diseases with composts. Compost Sci Util 2:5–12

Hoitink HAJ, Inbar Y, Boehm MJ (1991) Status of composted-amended potting mixes naturally suppressive to soil-borne diseases of floricultural crops. Plant Dis 75:869–873

Hoitink HAJ, Stone AG, Han DY (1997) Suppression of plant diseases by composts. HortSci 32(2):184–187

Hudson BD (1994) Soil organic matter and available water capacity. J Soil Water Conserv, Mar–Apr, pp 189–194

Ingham ER, Molina R (1991) Interactions among mycorrhizal fungi, rhizosphere organisms, and plants. In: Barbosa P, Krischik VA, Jones CG (eds) Microbial mediation of plant-herbivore interactions. Wiley, New York, pp 169–197

Kannangara T, Utkhede RS, Paul JW, Punja ZK (2000) Effects of mesophilic and thermophilic composts on suppression of Fusarium root and stem rot of greenhouse cucumber. Can J Microbiol 46:1021–1028

Krause MS, De Ceuster TJJ, Tiquia SM, Michel FC Jr, Madden LV, Hoitink HAJ (2003) Isolation and characterization of rhizobacteria from composts that suppress the severity of bacterial leaf spot of radish. Phytopathology 93:1292–1300

Mehta CM, Gupta V, Singh S, Srivastava R, Sen E, Romantschuk M, Sharma AK (2012) Role of microbiologically rich compost in reducing biotic and abiotic stresses. In: Satyanarayana T, Johri BN, Prakash A (eds) Microorganisms in environmental management. Springer Dordrecht Heidelberg, London/New York, pp 113–134

Nelson EB, Burpee LL, Lawton MB (1994) Biological control of turf grass diseases. In: Leslie AR (ed) Handbook of integrated pest management for turf and ornamentals. Lewis Publishers, Ann Arbor, pp 409–427

Ozores-Hampton, Monica HB, McMillian R Jr (1994) Suppressing disease in field crops. BioCycle, July, pp 60–61

Popkin R (1995) Good news for waste watchers: recycling, composting show results for the future. Environ Prot Agency J 21:188–190

Punja ZK, Rose S, Yip R (2002) International Organisation for Biological and Integrated control of Noxious Animals and Plants (OIBC/OILB), West Palaearctic Regional Section (WPRS/SROP), Dijon, France 25: 93–96

Ringer CE, Millner PD, Teerlinck LM, Lyman BW (1997) Suppression of seedling damping-off disease in potting mix containing animal manure compost. Compost Sci Util 5:6–14

Sances FV, Elaine RI (1997) Conventional organic alternatives to methyl bromide on California strawberries. Compost Sci Util 5:23–37

Stone AG, Vallad GE, Cooperband LR, Rotenberg D, Darby HM, James RV, Stevenson WR, Goodman RM (2003) Effect of organic amendments on soil-borne and foliar diseases in field-grown snap bean and cucumber. Plant Dis 87:1037–1042

Tilak KVBR, Reddy BS (2006) *Bacillus cereus* and *B. circulans* novel inoculants for crops. Curr Sci 5:642–644

Trankner A (1992) Use of agricultural and municipal organic wastes to develop suppressiveness to plant pathogens. In: Tjamos EC, Papavizas GC, Cook RJ (eds) Biological control of plant diseases: progress and challenges for the future, vol 230, NATO ASI Series No. Plenum Press, New York, pp 35–42

Tuitert G, Szczech M, Bollen GJ (1998) Suppression of *Rhizoctonia solani* in potting mixtures amended with compost made from organic household waste. Phytopathology 88:764–773

Vallad EG, Cooperband L, Goodman RM (2003) Plant foliar disease suppression mediated by composted forms of paper mill residuals exhibits molecular features of induced resistance. Physiol Mol Plant Pathol 63:65–77

Zhang W, Han DY, Dick WA, Davis KR, Hoitink HAJ (1998) Compost and compost water extract-induced systemic acquired resistance in cucumber and Arabidopsis. Phytopathology 88:450–455

Zheljazkov VD, Warman PR (2004) Source-separated municipal solid waste compost application to Swiss chard and basil. J Environ Qual 33:542–552

7 Grafted Vegetables for Management of Soilborne Pathogens

Abstract

Grafting as a technique is gaining wide attention throughout the world, especially for greenhouse cultivation of vegetable crops, mainly the solanaceous (against bacterial wilt) and cucurbitaceous (against *Fusarium* wilt) ones, from the viewpoint of resistance against the soilborne pathogens in addition to obtaining better yield and quality. Cucurbits are commonly grafted to gourds or to interspecific hybrids. Grafting methods, grafting for management of soilborne pathogens, problems commonly associated with grafted plants, and grafting-conferred defense mechanisms are discussed.

Keywords

Grafting • Rootstocks • Cucurbits • Solanaceous vegetables • *Fusarium* wilt • Bacterial wilt

7.1 Introduction

Growing grafted vegetables was first launched in Japan and Korea in the late 1920s by grafting watermelons to gourd rootstocks. After the first trial, the cultivated area of grafted vegetables, as well as the kinds of vegetables being grafted, has been consistently increased. At present, most of the watermelons (*Citrullus lanatus*), melons (*Cucumis melo*), Oriental melons (*Cucumis melo* var. *makuwa*), greenhouse cucumbers (*Cucumis sativus*), and several solanaceous crops in Korea and Japan are grafted before being transplanted in the field or in greenhouses (Ito 1992; Kurata 1992). The purpose of grafting also has been greatly expanded, from reducing infection by soilborne diseases caused by pathogens such as *Fusarium oxysporum* (Itagi 1992; Yamakawa 1983) to increasing low-temperature (Tachibana 1989) and salt and wet-soil tolerance (Tachibana 1989; Park 1987), enhancing water and nutrient uptake (Heo 1991; Jang et al. 1992), and increasing plant vigor and extending the duration of economic harvest time (Itagi 1992; Ito 1992), among other purposes (Ali et al. 1991; Dole and Wilkins 1992; Matsuzoe et al. 1993). Growing grafted

P.P. Reddy, *Sustainable Crop Protection under Protected Cultivation*, DOI 10.1007/978-981-287-952-3_7

vegetables, compared to growing grafted trees, was seldom practiced in the USA or in other Western countries where land use is not intensive, i.e., proper crop rotation is being practiced. However, it is highly popular in some Asian and European countries where land use is very intensive and the farming area is small (Hartmann and Kester 1975).

7.2 Objectives of Grafting Vegetables

The main objective of vegetable grafting is to avoid soilborne diseases such as *Fusarium* wilt in Cucurbitaceae (cucumber, melons, etc.) and bacterial wilt in Solanaceae (tomato, pepper, etc.).

Grafting's early purpose was to avoid or reduce the soilborne disease caused by *F. oxysporum*, but the reasons for grafting, as well as the kinds of vegetables grafted, have increased dramatically. Watermelons, other melons (*Cucumis* spp.), cucumbers, tomatoes (*Lycopersicon esculentum*), and eggplants (*Solanum melongena*) are commonly grafted to various rootstocks. Numerous rootstocks also have been developed. Watermelons are commonly grafted to gourds (*Lagenaria siceraria*) or to interspecific hybrids (*C. maxima x C. moschata*). Cucumbers are frequently grafted to figleaf gourd (*C. ficifolia*) or interspecific hybrids (*C. maxima x C. moschata*).

7.3 Cultivation Statistics for Selected Vegetables and Grafted Seedlings

Table 7.1 shows the statistics of cultivated area under greenhouse conditions using grafted seedlings and the estimated number of grafted vegetable seedlings in Korea and Japan. About 192.6 million grafted seedlings are planted annually in Korea and 341.4 million in Japan in greenhouses. More than 95 % of the watermelons in both countries are grafted. The majority of greenhouse cucumbers are grafted. Most of the Oriental melons are grafted to squash (*Cucurbita* spp.) rootstock (Ito 1992; Jang et al. 1992). It is likely that these quantities will be increased soon with the introduction of new uses of grafting for other vegetables and the development of rootstock having desirable characteristics.

7.4 Grafting Methods

Grafting involves the union of suitable varieties of greenhouse crops used as scions over the resistant rootstocks usually from the same family. Trials have been conducted on grafting of tomato seedlings over the available rootstocks from different families. Although the success has been achieved in grafting of greenhouse tomato over interfamilial plants, however, tomato and cucur-

Table 7.1 Estimated cultivated area under greenhouse of some vegetables being grafted and the number of grafted seedlings needed annually in Japan and Korea (FAO 1992)

	Japan		Korea	
Vegetables	Cultivated area of grafted seedlings ('000 ha)	No. of grafted seedlings required[a] (millions)	Cultivated area of grafted seedlings ('000 ha)	No. of grafted seedlings required[a] (millions)
Watermelon	3.2	9.6	7.4	22.1
Cucumber	6.0	216.1	3.3	78.4
Oriental melon	2.0	35.3	5.0	89.4
Melons and muskmelons	0.5	23.8	0.02	1.0
Tomato	0.8	36.9	0.04	1.7
Eggplant	1.6	19.7	–	–
Total	14.1	341.4	15.7	192.6

[a]The quantity actually needed could be 15 % higher than the numbers listed

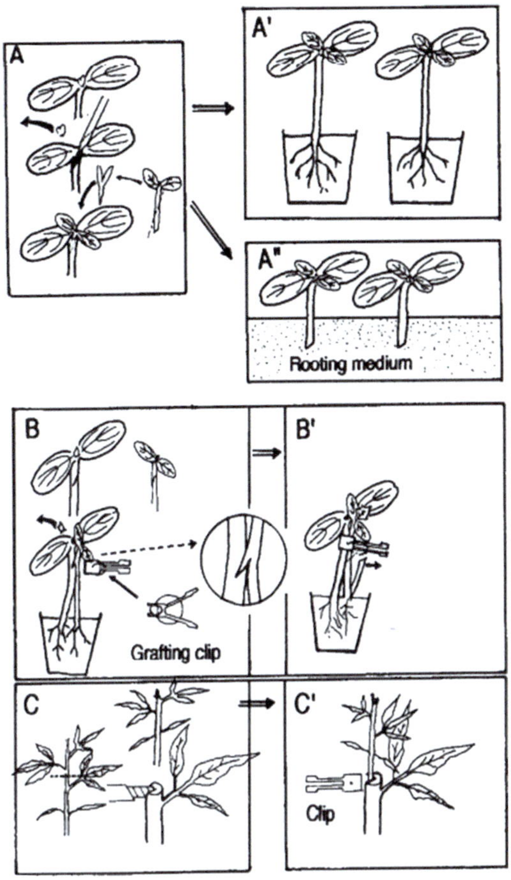

Fig. 7.1 Common grafting methods. (**a**) Hole insertion grafting. (**a**') Grafted seedlings are transplanted in for further growth. (**a**") Rootstocks are cut just above the root zone and then grafted as in (**a**). Grafted seedlings are then rooted in rooting medium while healing takes place. (**b**) Tongue approach grafting. (**b**') Grafting clip is used to hold the graft union. The scion's hypocotyl is cut below the graft union about a week after grafting. (**c**) Cleft grafting. (**c**') Specifically devised grafting clips are used to hold the scion and stock together tightly

bits only gave good results. Technology is underway and is proving to be promising for greenhouse cultivation in India.

Various grafting methods have been developed and growers must choose their favorite methods based on experiences and preferences. Grafting cucurbitaceous crops is commonly done when scion and rootstock seedlings are young, i.e., before the outgrowth of the first true leaf between the cotyledons. Some of the most widely practiced grafting methods are shown in Fig. 7.1.

Fig. 7.2 Vegetable grafting

Grafting methods vary considerably with the type of crops being grafted, and the sowing time for scion and stock seeds varies with grafting method and crop. For example, "hole insertion grafting" would be convenient for watermelons because of their small seedling size compared to the size of stock seedlings, such as gourd and squash. In cucumbers, however, "tongue approach grafting" has been used widely, mainly because of their large seedling size, including hypocotyl length and diameter, and grafting ease.

In the past, grafting was routinely carried out by the growers themselves, but it is now rapidly shifting to cooperative operations owing to the efficiency, ease of post-graft care, and recent expansion of the commercial seedling industry (such as sales of plug-grown seedlings) (Figs. 7.2 and 7.3). Even though the single-edged razor blade is still the most widely used grafting knife among farmers today, many other devices, such as specifically designed knives, clips, tubes, or glue, have now been developed for easier grafting and post-graft care (Kurata 1994). Semiautomatic grafting machines and/or fully automatic grafting machines using robots (Itagi 1992; Ito 1992) are being developed. By using the most sophisticated machine, grafting efficiency could be significantly increased from the present 150 seedlings/hour per expert.

The grafted seedlings are conditioned (hardened) in the dark and cool shade nets prepared especially for the purpose before planting in seedling trays and finally their transplantation in the greenhouses (Fig. 7.4). Grafting robots are

Fig. 7.3 Nursery production of grafted vegetables in greenhouse

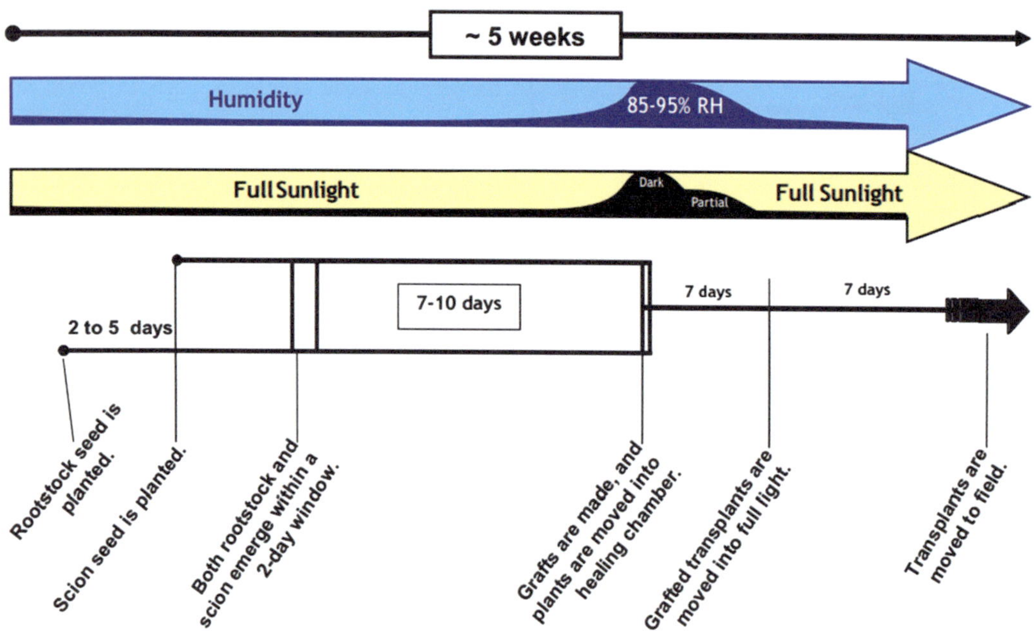

Fig. 7.4 A typical timeline for grafting (Rivard and Louws 2006)

being increasingly used for the commercial production of a healthy nursery.

The labor required for intensive post-graft care, mostly 7–10 days of careful management, could be markedly reduced by using specifically designed conditioning chambers (Ito 1992). Kurata (1994) describes four methods of robot grafting in detail.

7.5 Grafting for Management of Soilborne Pathogens

Under continuous cropping, soilborne diseases and nematode problems are likely to increase. Since soil sterilization can never be complete, grafting has become an essential technique for the production of repeated crops of fruit-bearing vegetables.

Fig. 7.5 Grafting of watermelon scion on bottle gourd rootstock to control *Verticillium* wilt (*L*, grafted plant; *R*, control)

Table 7.2 Rootstocks for tomato and their specific resistance to some selected soilborne diseases (Yamakawa 1983)

Rootstocks for tomato	Major tomato diseases					
	Ralstonia solanacearum	*Fusarium oxysporum*	*Verticillium dahliae*	*Pyrenochaeta lycopersici*	Root-knot nematode	Tomato mosaic virus
BF	R	R	S	S	S	S
LS89	R	R	S	S	S	S
PFN	R	R	S	S	R	R
PFNT	R	R	S	S	R	R
KNVF	S	R	R	R	R	R
KNVF Tm Signaal	S	R	R	R	R	R
KCFT-N	S	R	S	R	R	R

R resistant, *S* susceptible

Grafting as a technique is gaining wide attention throughout the world, especially for greenhouse cultivation of vegetable crops, mainly the solanaceous and cucurbitaceous ones, from the viewpoint of resistance against the soilborne pathogens in addition to obtaining better yield and quality. It is highly successful in cucurbits like melons for resistance against diseases (Fig. 7.5).

Many rootstocks having distinctive characteristics are available (Yamakawa 1983), and growers select the rootstocks they think are the most suitable for their growing season, cultivation methods (field or greenhouses), soil type, and type of crops and cultivars (Lee 1989). For example, cucumber grown in greenhouses during the cool season should be grafted onto figleaf gourd; however, those grown during the hot summer season are usually grafted onto Sintozwa rootstocks (interspecific hybrid) or others (Lee 1989).

The vigorous roots of the rootstock exhibit excellent tolerance to serious soilborne diseases, such as those caused by *Fusarium*, *Verticillium*, and *Ralstonia* (Table 7.2), even though the degree of tolerance varies considerably with the rootstock. The mechanism of disease resistance, however, has not been intensively investigated. The disease tolerance in grafted seedlings may be entirely due to the tolerance of stock plant roots to such diseases.

However, in actual plantings, adventitious rooting from the scion is very common (Lee 1989). Plants having the root systems of the scion and rootstock are expected to be easily infected by soilborne diseases. However, seedlings having dual root systems often exhibit excellent disease resistance, almost comparable to those having only rootstock roots. This observation partially supports the previous report that substances associated with *Fusarium* tolerance are synthesized in the root and translocated to the scion through the xylem (Biles et al. 1989). The activity of the substances related to disease resistance may vary during the developmental stages of the grafted plants (Heo 1991; Padgett and Morrison 1990). More research is needed in this regard. It is generally accepted that the disease-susceptible characteristics of the scion are not translocated to the rootstock.

Disease resistance researchers around the world have demonstrated that grafting can be effective against a variety of soilborne fungal, bacterial, viral, and nematode diseases. It has been used to eliminate *Verticillium* and *Fusarium* wilt in tomato (Table 7.3) and cucurbit production systems in Japan, Korea, and Greece. In New Zealand, it has been used to reduce levels of corky root rot. In Morocco and Greece, grafting is used to control root-knot nematodes (*Meloidogyne* species) in both tomatoes and cucurbits. Researchers have proposed using grafted plants instead of methyl bromide to manage soilborne diseases in these regions of the world.

Improved resistance to many soilborne fungal, bacterial, and nematode pathogens have been reported in grafted solanaceous and cucurbitaceous crops. Moreover, certain foliar fungal and viral diseases were suppressed when susceptible scions were grafted onto specific rootstocks (Louws et al. 2010). Diseases controlled by grafting in different vegetable crops due to fungi, bacteria, viruses, and nematodes are listed in Tables 7.4 and 7.5.

7.5.1 Soilborne Fungal Diseases

The earliest reported use of vegetable grafting for disease control was for management of *Fusarium* wilt in cucurbits (Sakata et al. 2005). Commonly used cucurbitaceous rootstocks are nonhosts to most *formae speciales* of *F. oxysporum*, and thus grafting has been successfully used to control *Fusarium* wilt in cucurbit production (Louws et al. 2010). *Verticillium* wilt, primarily caused by *Verticillium dahliae*, is another vascular wilt disease that often affects Solanaceae and Cucurbitaceae. Studies with plants grafted onto commercial rootstocks and subjected to infection with *V. dahliae* indicated that both scions and rootstocks contributed to disease resistance of the grafted combinations in watermelons, melons (*Cucumis melo*), cucumbers (*Cucumis sativus*), and tomatoes (*Solanum lycopersicum*) (Paplomatas et al. 2002). *Monosporascus* sudden wilt, caused by *Monosporascus cannonballus*, is an important soilborne disease of melon and watermelon in hot and semiarid areas. Grafting scions of susceptible melon cultivars onto *C. maxima* and *C. maxima* × *C. moschata* rootstocks improved the resistance of melon (Edelstein et al. 1999) although *Cucurbita* is normally regarded as a host for *M. cannonballus* (Mertely et al. 1993). However, the improved resistance and better yield with grafted plants was inconsistent. The variable results might be attributed to differences in rootstock–scion combinations and growing conditions. *Phytophthora* blight, caused by *Phytophthora capsici*, is regarded as one of the most destructive diseases in production of cucurbits. In *P. capsici*-infested fields, yields of cucumbers grafted on bottle gourd (*Lagenaria siceraria*), *C. moschata*, and wax gourd (*Benincasa hispida*) rootstocks were significantly increased, and vegetative growth was more vigorous (Wang et al. 2004). Watermelons grafted onto selected bottle gourd rootstocks also exhibited resistance to *P. capsici* (Kousik and Thies 2010). Corky root disease caused by *Pyrenochaeta lycopersici* is a severe problem for Solanaceae. Tomatoes grafted onto "Beaufort" rootstocks (*S. lycopersicum* × *S. habrochaites*) had lower disease incidence, higher yield, and larger fruit (Hasna et al. 2009). Similar results were also found in grafted eggplants (*S. melongena*) (Iouannou 2001).

7.5.2 Other Fungal Diseases

Other fungal diseases that have been controlled by grafting include target leaf spot (*Corynespora cassiicola*) on cucumbers, black root rot (*Phomopsis sclerotioides*) on cucumbers and melons, gummy stem blight (*Didymella bryoniae*) on melons, southern blight (*Sclerotium rolfsii*) on tomatoes, brown root rot (*Colletotrichum coccodes*) on tomatoes and egg-

Table 7.3 Disease resistance and vigor of commercial tomato rootstock cultivars (Ratings are based on seed supplier information. Vigor is measured on a scale of one to five, where one represents poor and five represents excellent)

Seed supplier/rootstock	Disease resistance							Vigor
	Tomato mosaic virus	Corky root	*Fusarium* wilt Race 1	*Fusarium* wilt Race 2	*Verticillium* wilt Race 1	Bacterial wilt	Nematodes	
De Ruiter seeds								
Maxifort	High	High	Moderate	High	High	Susceptible	High	5
Beaufort	High	High	Moderate	High	High	Susceptible	High	3
Takii seeds								
Anchor-T	High	Susceptible	High	High	High	Moderate	High	5
Survivor	High	Susceptible	High	High	High	Moderate	High	5
Aegis	High	Moderate	High	High	High	Moderate	High	4
Bruinsma seeds								
Body	High	High	Susceptible	High	High	Susceptible	High	5
Robusta	High	High	Susceptible	High	High	Susceptible	Susceptible	3

Table 7.4 Rootstock, major grafting methods, and purpose of grafting for some vegetables

Vegetables	Popular rootstock species	Grafting methods[a]	Purpose[b]
Watermelon	Gourd (*Lagenaria siceraria* var. *hispida*)	1	1
	Interspecific hybrids	1, 2	1
	Wax gourd (*Benincasa hispida*)	1, 3	1
	Pumpkin (*Cucurbita pepo*)	2, 3	1
	Squash (*Cucurbita moschata*)	1, 2	1
	Bur cucumber (*Sicyos angulatus*)	2	2
Cucumber	Figleaf gourd (*Cucurbita ficifolia*)	2	1
	Interspecific hybrids	1, 2	1
	F1 (*Cucurbita maxima* x *C. moschata*)	2	1
	Cucumber (*Cucumis sativus*)	2	1
	Bur cucumber (*Sicyos angulatus*)	2	2
Oriental melon	Interspecific hybrids	2	1
	Squash (*Cucurbita moschata*)	2	1
Melons	*Cucumis melo*	2, 3	1
Tomato	*Lycopersicon pimpinellifolium*	3	2
	Lycopersicon hirsutum	3	2
	Lycopersicon esculentum	3	2
Eggplant	*Solanum integrifolium*	2, 3	3
	Solanum torvum	2, 3	4

[a]Grafting methods: 1 = hole insertion; 2 = tongue approach; 3 = cleft grafting (see Fig. 7.1 for details)
[b]Purpose of grafting: 1 = *Fusarium* wilt control; 2 = nematode resistance; 3 = bacterial wilt control; 4 = virus infection reduction

plants, and *Rhizoctonia* damping-off (*Rhizoctonia solani*) on tomatoes (King et al. 2008; Louws et al. 2010) (Table 7.1). Grafting has also been reported to improve crop resistance to the foliar fungal diseases such as powdery mildew (*Podosphaera xanthii*) and downy mildew (*Pseudoperonospora cubensis*) on cucumbers, when certain rootstocks were used (Louws et al. 2010; Sakata et al. 2005).

7.5.3 Soilborne Bacterial Diseases

Tomato bacterial wilt, caused by *Ralstonia solanacearum*, is one of the most destructive diseases of tomato. Resistance to bacterial wilt in tomatoes is a quantitative trait and is closely associated with small fruit size (Louws et al. 2010). Thus, few resistant tomato cultivars are commercially available (King et al. 2008). Grafting scions of susceptible tomato cultivars onto resistant rootstocks has been successful for managing tomato bacterial wilt (Matsuzoe et al. 1993). The improved resistance in grafted plants may result from limited colonization in the lower stem rather than prevention of bacteria from invading xylem tissues (Fig. 7.6) (Grimault and Prior 1994).

Eggplants and relatives can be used as rootstocks for tomato. The Asian Vegetable Research and Development Center (AVRDC) has developed *S. melongena* rootstock resistant to waterlogging as well as bacterial wilt and other soilborne diseases for grafting tomatoes. Grafting has been essential in Asian horticultural production for eliminating bacterial wilt incidence in solanaceous crops, such as tomatoes and peppers. By the end of the season, none of the control plants had survived, while 100 % of the grafted plants continued to produce. Furthermore, the yield of the tomatoes with resistant rootstocks was four times that of the non-grafted susceptible plants. Yields increased by as much as 106 % with the use of certain rootstocks for watermelon production in Australia. Grafting can be a valuable

Table 7.5 Fungal, bacterial, viral, and nematode diseases reported to be controlled by grafting in different vegetable crops

Disease	Pathogen	Crops
Fungal diseases		
Fusarium wilt	*Fusarium oxysporum*	Tomato, pepper, watermelon, melon, cucumber
Fusarium crown and root rot	*Fusarium oxysporum*, *F. solani*	Tomato, pepper, watermelon
Verticillium wilt	*Verticillium dahliae*	Tomato, eggplant, watermelon, melon, cucumber
Monosporascus sudden wilt	*Monosporascus cannonballus*	Watermelon, melon
Phytophthora blight	*Phytophthora capsici*	Tomato, pepper, watermelon, cucumber
Corky root	*Pyrenochaeta lycopersici*	Tomato, eggplant
Target leaf spot	*Corynespora cassiicola*	Cucumber
Black root rot	*Phomopsis sclerotioides*	Cucumber, melon
Gummy stem blight	*Didymella bryoniae*	Melon
Southern blight	*Sclerotium rolfsii*	Tomato
Brown root rot	*Colletotrichum coccodes*	Tomato, eggplant
Rhizoctonia damping-off	*Rhizoctonia solani*	Tomato
Powdery mildew	*Podosphaera xanthii*	Cucumber
Downy mildew	*Pseudoperonospora cubensis*	Cucumber
Bacterial diseases		
Bacterial wilt	*Ralstonia solanacearum*	Tomato, pepper, eggplant
Viral diseases		
Melon necrotic spot virus	*Melon necrotic spot virus* (MNSV)	Watermelon
Tomato yellow leaf curl	*Tomato yellow leaf curl virus* (TYLSV)	Tomato
Tomato spotted wilt	*Tomato spotted wilt virus* (TSWV)	Tomato
Pepino mosaic virus	*Pepino mosaic virus* (PepMV)	Tomato
Nematode diseases		
Root knot	*Meloidogyne* spp.	Cucumber, melon, watermelon, tomato, eggplant, pepper

Fig. 7.6 Grafted plants on the *left* and non-grafted on the *right* in a field severely infested with bacterial wilt (Photo by Josh Freeman)

tool for eliminating bacterial wilt in tomato, pepper, and eggplant production.

7.5.4 Viral Diseases

Vegetable grafting research on resistance to viral diseases yielded mixed results because of the lack of systematic studies in this area. Wang et al. (2002) reported improved antivirus performance in grafted seedless watermelon plants. In Israel, use of resistant rootstocks for controlling the soilborne *melon necrotic spot virus* in cucurbits was a significant advantage over soil fumigation with methyl bromide, which does not control this viral disease (Cohen et al. 2007). Meanwhile, *tomato yellow leaf curl virus*, *tomato spotted wilt virus*, and *pepino mosaic virus* were also reported to be controlled by grafting (Louws et al. 2010). However, some reports indicated that grafted plants were more vulnerable to viral diseases, possibly as a result of graft incompatibility that weakened the scion plants (Davis et al. 2008).

7.5.5 Root-Knot Nematodes

Root galling of susceptible plants is a typical response to root-knot nematode (RKN) (*Meloidogyne* spp.) infection, resulting in poor absorption of water and nutrients. In cucurbits, resistance to *M. incognita* was identified in *Cucumis metuliferus*, *C. ficifolia*, and bur cucumber (*Sicyos angulatus*) (Gu et al. 2006). Using *C. metuliferus* as a rootstock to graft on RKN-susceptible melons led to lower levels of root galling and nematode numbers at harvest (Sigüenza et al. 2005). Moreover, *C. metuliferus* showed high graft compatibility with several melon cultivars (Trionfetti Nisini et al. 2002). Cucumbers grafted on the bur cucumber rootstock exhibited increased RKN resistance (Zhang et al. 2006). Promising progress has also been made in developing *M. incognita*-resistant germplasm lines of wild watermelon (*Citrullus lanatus* var. *citroides*) for use as rootstocks (Thies et al. 2010). However, at present, cucurbit rootstocks with resistance to RKN are not commercially available (Thies et al. 2010). The *Mi* gene, which provides effective control against RKN in tomato, has been introgressed into cultivated tomatoes and rootstock cultivars (Louws et al. 2010). Grafting of susceptible tomato cultivars on RKN-resistant rootstocks was effective in controlling RKN in fields naturally infested with RKN (Rivard and Louws 2006). However, as a result of temperature sensitivity of the *Mi* gene, such resistance may not be uniformly stable (Cortada et al. 2009). Pepper (*Capsicum annuum*) cultivars possessing the *N* gene, which controls resistance to RKNs (*M. incognita*, *M. arenaria*, and *M. javanica*), have been effective as rootstocks to control RKNs in pepper (Oka et al. 2004).

Several new rootstocks are being developed. For example, bur cucumber (*Sicyos angulatus*) collected near Andong, Korea, showed good compatibility with cucumbers and watermelons for early summer growth and good resistance to nematodes (Lee 1992; Lee et al. 1992).

7.6 Problems Commonly Associated with Grafted Plants

Various problems are commonly associated with grafting and cultivating grafted seedlings. Major problems are the labor and techniques required for the grafting operation and post-graft handling of grafted seedlings for rapid healing for 7–10 days. An expert can graft 1200 seedlings per day (150 seedlings per hour), but the numbers vary with the grafting method. Similarly, the post-graft handling method depends mostly on the grafting methods. The problems could be minimized or easily overcome by careful cultural management and wise selection of scion and rootstock cultivars.

The introduction of mechanization and automation technology will help to address large-scale production issues. A fully automated grafting robot for cucurbits was recently developed based on the 1993 version of Iseki's GR800 semiautomated robot. The robot (Fig. 7.7) has scion and rootstock feeders, which pick, orient,

Fig. 7.7 Grafting robot

and feed the scion and the rootstock shoots to the grafting processor, performing 750 grafts per hour with a 90 % success rate.

Since there is prohibition of methyl bromide for disinfecting soils, there is a renewed interest for grafting eggplant (and tomato) on rootstocks resistant to soilborne pathogens.

Even though there are many problems associated with cultivating grafted vegetable seedlings, the need for successful grafted seedlings is growing rapidly. Breeding multipurpose rootstocks and developing efficient grafting machines and techniques will undoubtedly encourage increased use of grafted seedlings in many countries. Large-scale commercial production of vegetable seedlings is expanding rapidly in many developed countries, and this will lead to an increased commercial supply and use of grafted vegetable seedlings throughout the world.

7.7 Grafting-Conferred Defense Mechanisms

7.7.1 Using Nonhost Resistance

Nonhost disease resistance refers to the resistance provided by all members of a plant species against all races of a certain pathogen, and it is often considered the most common and durable plant disease resistance (Mysore and Ryu 2004). Solanaceous vegetables and cucurbits are often grafted onto rootstocks that, although related, are different species or hybrids of different species. Taking advantage of the nonhost resistance in these rootstocks is a convenient approach to battling specific pathogens that infect the cultivated scion species but not the rootstock species. For example, *F. oxysporum* f. sp. *niveum* causes disease specifically on watermelon, and *F. oxysporum* f. sp. *lagenaria* typically infects bottle gourd (Namiki et al. 1994). Because all of the bottle gourd genotypes offer resistance to all races of *F. oxysporum* f. sp. *niveum* (races 0, 1, 2), nonhost resistance has been proven to be a viable mechanism to control *Fusarium* wilt in watermelons when they are grafted onto bottle gourd rootstocks (Yetisir et al. 2007).

7.7.2 Developing Transgenic Rootstocks

In addition to use of nonhost-resistant rootstocks as well as selection and breeding for host resistance from a wide range of germplasms, transgenic rootstocks with specific disease resistances have been developed. For example, a *cucumber green mottle mosaic virus* coat protein gene and a *cucumber fruit mottle mosaic tobamovirus* replicase gene were introduced into watermelon and cucumber rootstocks, respectively. Susceptible scions grafted onto transgenic rootstocks exhibited high resistance against viral pathogens (Yi et al. 2009).

7.7.3 Shift in Rhizosphere Microbial Diversity as a Result of Grafting

Rhizosphere microorganisms can play critical roles in suppressing soilborne diseases through a variety of mechanisms such as nutrient competition, antagonism, and parasitism. Exploring rhizosphere microbial diversity related to plant species and genotypes, therefore, is another approach to understanding soilborne disease incidence and severity (Yao and Wu 2010).

Studies on cucumbers grafted onto *C. moschata* indicated that grafting increased the population of bacteria and actinomycetes while reducing the total number of fungi in the rhizosphere (Dong et al. 2010). Research on grafted peppers also showed that actinomycete populations in the rhizosphere were higher in the resistant rootstock and grafted plants compared with the self-rooted scion control when plants were inoculated with *F. solani* (Jiang et al. 2010a). Incidence of *Verticillium* wilt was reduced when a susceptible eggplant scion was grafted onto *Solanum torvum* rootstock accompanied by enhanced ratios of bacteria and actinomycetes to fungi in the rhizosphere of grafted plants (Yin et al. 2008). Actinomycetes represent a group of soil microorganisms with great potential to protect plants against pathogens (*Doumbou* et al. 2001). Although it is still uncertain about the extent to which actinomycetes contribute to plant disease control, higher populations of actinomycetes detected in the rhizosphere of grafted plants could possibly play a role in disease suppression when susceptible scions are grafted onto certain resistant rootstocks.

7.7.4 Vigorous Root Systems of Grafted Vegetables to Plant Defense

Many rootstocks developed for vegetable grafting were selected or bred from wild genotypes. In addition to specific disease resistance, they are characterized by large and vigorous root systems (Lee 1994). Soilborne pathogens often infect and damage plant roots, and, as a result, plant nutrient and water uptake can be affected. Therefore, root system size and vigor may be associated with resistance to soilborne diseases. Moreover, vigorous roots help improve nutritional status and thus the overall health of plants, which may augment resistance against foliar diseases.

7.7.5 Nutrient Uptake in Relation to Disease Development

Plant nutrients, in addition to their essential roles in plant growth and development, can be directly involved in plant defense pathways. For example, foliar application of phosphate salts can induce systemic protection against anthracnose in cucumbers (Gottstein and Kuć 1989), and phosphate-mediated resistance induction has been associated with localized cell death (Walters and Murray 1992). Nitrogen deficiency can compromise elicitor-induced resistance to pathogen infection (Dietrich et al. 2004), whereas excessive nitrogen increases disease incidence by promoting growth of large plant canopies (Simón et al. 2003). In addition, many other mineral nutrients such as potassium, calcium, sulfur, and micronutrients also play significant roles in plant defense mechanisms (Walters and Bingham 2007).

7.7.6 Grafting-Induced Systemic Defense

Defense-related enzymes are often induced by a variety of biotic stresses. Studies have shown that plants grafted on certain rootstocks generally

Table 7.6 Enzymes reported with higher activities in grafted vegetables compared with self-rooted plants

Enzymes	Scion	Rootstock	Plant tissue	Pathogen	Reference
Phenylalanine ammonia-lyase (PAL)	Eggplant	*Solanum torvum*	Roots and leaves	Inoculation of *V. dahliae*	Zhou et al. 1998
	Pepper	Pepper	Roots and leaves	Inoculation of *F. solani*	Jiang et al. 2010b

exhibit a higher activity of defense-related enzymes compared with self-rooted plants under certain stressful conditions (Table 7.6).

7.8 Conclusions

Grafting is a valuable disease management tactic for heirloom tomato growers. This practice originated as a way to ensure fruit quality while keeping disease resistance high for melon production systems with soilborne disease pressure. This same principle lends itself well to heirloom tomato production. Furthermore, research worldwide has demonstrated increased yields from grafted vegetable plants in comparison to nongrafted plants. As Southeastern growers realize its potential benefits and relative ease, grafting will become a more popular disease management technique in the USA.

Even though there are many problems associated with cultivating grafted vegetable seedlings, the need for successfully grafted seedlings is growing rapidly. Breeding multipurpose rootstock and developing efficient grafting machines and techniques will undoubtedly encourage increased use of grafted seedlings not only in Korea and Japan but also in many other countries. Large-scale commercial production of vegetable seedlings is expanding rapidly in many developed countries, and this will lead to an increased commercial supply and use of grafted vegetable seedlings throughout the world.

References

Ali M, Matsuo N, Okubo H, Fujieda K (1991) Resistance of non-tuberous *Solanum* to root-knot nematode. J Jpn Soc Hort Sci 60(4):921–926

Biles CL, Martyn RD, Wilson HD (1989) Isozymes and general proteins from various watermelon cultivars and tissue types. HortSci 24(5):810–812

Cohen R, Burger Y, Horev C, Koren A, Edelstein M (2007) Introducing grafted cucurbits to modern agriculture: the Israeli experience. Plant Dis 91:916–923

Cortada L, Sorribas FJ, Ornat C, Andrés MF, Verdejo-Lucas S (2009) Response of tomato rootstocks carrying the *Mi*-resistance gene to populations of *Meloidogyne arenaria, M. incognita* and *M. javanica*. Eur J Plant Pathol 124:337–343

Davis AR, Perkins-Veazie P, Sakata Y, LópezGalarza S, Maroto JV, Lee SG, Huh YC, Sun Z, Miguel A, King SR, Cohen R, Lee JM (2008) Cucurbit grafting. Crit Rev Plant Sci 27:50–74

Dietrich R, Plob K, Heil M (2004) Constitutive and induced resistance to pathogens in *Arabidopsis thaliana* depends on nitrogen supply. Plant Cell Environ 27:896–906

Dole JM, Wilkins HF (1992) In vivo characterization of a graft-transmissible, free-branching agent in poinsettia. J Am Soc Hort Sci 117(6):972–975

Dong L, Zuo Y, Li X, Wang Q (2010) Effects of grafting on cucumber soil biochemistry (in Chinese). J Chin Agric Univ 15:51–56

Doumbou CL, Hamby Salove MK, Crawford DL, Beaulieu C (2001) Actinomycetes, promising tools to control plant diseases and to promote plant growth. Phytoprotection 82:85–102

Edelstein M, Cohen R, Burger Y, Shriber S, Pivonia S, Shtienberg D (1999) Integrated management of sudden wilt in melons, caused by *Monosporascus cannonballus*, using grafting and reduced rates of methyl bromide. Plant Dis 83:1142–1145

FAO (1992) FAO production yearbook. FAO of the United Nations, Rome

Gottstein HD, Kuć JA (1989) Induction of systemic resistance to anthracnose in cucumber by phosphates. Phytopathology 79:176–179

Grimault V, Prior P (1994) Grafting tomato cultivars resistant or susceptible to bacterial wilt: analysis of resistance mechanisms. J Phytopathol 141:330–334

Gu X, Zhang S, Zhang S (2006) The screening of cucumber rootstocks resistant to southern root-knot nematode (in Chinese). Chin Veg 2:4–8

Hartmann HT, Kester DE (1975) Plant propagation: principles and practices, 3rd edn. PrenticeHall, Englewood Cliffs

Hasna MK, Ögren E, Persson P, Mårtensson A, Rämert B (2009) Management of corky root disease of tomato in participation with organic tomato growers. Crop Prot 28:155–161

Page: 23 Heo YC (1991) Effects of rootstock on exudation and mineral elements contents in different parts of oriental melon and cucumber (in Korean with English summary). MS thesis, Kyung Hee University, Seoul, p 53

Iouannou N (2001) Integrating soil solarization with grafting on resistant rootstocks for management of soil-borne pathogens of eggplant. J Hort Sci Biotech 76:396–401

Itagi T (1992) Status of transplant production systems in Japan and new grafting techniques (in Korean). Symp Protected Hort. Hort Expt Stn, Rural Development Admin, Suwon, pp 32–67

Ito T (1992) Present state of transplant production practices in Japanese horticultural industry. In: Kurata K, Kozai T (eds) Transplant production system. Kluwer, Yokohama, pp 65–82

Jang KU et al (1992) Utilization of sap and fruit juice of *Luffa cylindrica* L. res. rept. Korean Ginseng and Tobacco Inst, Taejon, p 116

Jiang F, Liu YX, Ai XZ, Zhen N, Wang HT (2010a) Study on relationship among microorganism, enzymes' activity in rhizosphere, soil and root rot resistance of grafted capsicum (in Chinese). Scientia Agr Sinica 43:3367–3374

Jiang F, Liu YX, Liu W, Zheng N, Wang HT, Ai XZ (2010b) Relationship between root rot resistance and phenylpropanoid metabolism in graft capsicum (in Chinese). Chin Veg 8:46–52

King SR, Davis AR, Liu W, Levi A (2008) Grafting for disease resistance. HortSci 43:1673–1676

Kousik CS, Thies JA (2010) Response of US bottle gourd (*Lagenaria siceraria*) plant introductions (PI) to crown rot caused by *Phytophthora capsici*. Phytopathology 100:S65

Kurata K (1992) Transplant production robots in Japan. In: Kurata K, Kozai T (eds) Transplant production system. Kluwer, Yokohama, pp 313–329

Kurata K (1994) Cultivation of grafted vegetables II. Development of grafting robots in Japan. HortSci 29:240–244

Lee JM (1989) On the cultivation of grafted plants of cucurbitaceous vegetables (in Korean with English summary). J Kor Soc Hort Sci 30(3):169–179

Lee WY (1992) Use of *Sicyos angulatus* cv. Andong as rootstock for cucurbits (in Korean). Rural Dev Admin Suwon, Korea Res Ext 33(4):69–73

Lee JM (1994) Cultivation of grafted vegetables I. Current status, grafting methods, and benefits. HortSci 29:235–239

Lee WY, Hwang GS, Shin CG, Ha HT, Hwang JM (1992) Yield performance test of watermelon (*Citrullus vulgaris* Schrad) grafted onto the rootstock, *Sicyos angulatus* L in the farm field (Abstr) (in Korean with tables and figures in English). Kor Soc Hort Sci 10(2):38–39

Louws FJ, Rivard CL, Kubota C (2010) Grafting fruiting vegetables to manage soil-borne pathogens, foliar pathogens, arthropods and weeds. Sci Hort 127:127–146

Matsuzoe N, Okubo H, Fujieda K (1993) Resistance of tomato plants grafted on *Solanum* rootstocks to bacterial wilt and root-knot nematode. J Jpn Soc Hort Sci 61:865–872

Mertely JC, Martyn RD, Miller ME, Bruton BD (1993) An expanded host range for the muskmelon pathogen *Monosporascus cannonballus*. Plant Dis 77:667–673

Mysore KS, Ryu CM (2004) Non-host resistance: how much do we know? Trends Plant Sci 9:97–104

Namiki F, Shiomi T, Kayamura T, Tsuge T (1994) Characterization of the *formae speciales* of *Fusarium oxysporum* causing wilts of cucurbits by DNA fingerprinting with nuclear repetitive DNA sequences. Appl Environ Microbiol 60:2684–2691

Oka Y, Offenbach R, Pivonia S (2004) Pepper rootstock graft compatibility and response to *Meloidogyne javanica* and *M. incognita*. J Nematol 36:137–141

Padgett M, Morrison JC (1990) Changes in grape berry exudates during fruit development and their effect on mycelial growth of *Botrytis cinerea*. J Amer Soc Hort Sci 115(2):256–257

Paplomatas EJ, Elena K, Tsagkarakou A, Perdikaris A (2002) Control of Verticillium wilt of tomato and cucurbits through grafting of commercial varieties on resistant rootstocks. Acta Hort 579:445–449

Park YD (1987) Effect of fertilizer levels, growth regulators and salt treatment on the growth and sex expression in cucumbers grafted to different rootstocks. MS thesis, Kyung Hee University, Suwon, Korea, 50 pp

Rivard CL, Louws FJ (2006) Grafting for disease resistance in heirloom tomatoes. North Carolina Coop Ext Serv Bul Ag-675

Sakata Y, Ohara T, Sugiyama M (2005) The history and present state of the grafting of cucurbitaceous vegetables in Japan. Acta Hort 731:159–170

Sigüenza C, Schochow M, Turini T, Ploeg A (2005) Use of *Cucumis metuliferus* as a rootstock for melon to manage *Meloidogyne incognita*. J Nematol 37:276–280

Simón MR, Cordo CA, Perelló AE, Struik PC (2003) Influence of nitrogen supply on the susceptibility of wheat to *Septoria tritici*. J Phytopathol 151:283–289

Tachibana S (1989) Respiratory response of detached roots to lower temperatures in cucumber and figleaf gourd grown at 20°C root temperature. J Jpn Soc Hort Sci 58(2):333–337

Thies JA, Ariss JJ, Hassell RL, Olson S, Kousik CS, Levi A (2010) Grafting for management of southern root-knot nematode, *Meloidogyne incognita*, in watermelon. Plant Dis 94:1195–1199

Trionfetti Nisini P, Colla G, Granati E, Temperini O, Crinò P, Saccardo F (2002) Rootstock resistance to Fusarium wilt and effect on fruit yield and quality of two muskmelon cultivars. Sci Hort 93:281–288

Walters DR, Bingham IJ (2007) Influence of nutrition on disease development caused by fungal pathogens: implications for plant disease control. Ann Appl Biol 151:307–324

Walters DR, Murray DC (1992) Induction of systemic resistance to rust in *Vicia faba* by phosphate and EDTA: effects of calcium. Plant Pathol 41:444–448

Wang J, Zhang DW, Fang Q (2002) Studies on antivirus disease mechanism of grafted seedless watermelon (in Chinese). J Anhui Agric Univ 29:336–339

Wang HR, Ru SJ, Wang LP, Fong ZM (2004) Study on the control of Fusarium wilt and Phytophthora blight in cucumber by grafting (in Chinese). Acta Agric Zhejiangensis 16:336–339

Yamakawa B (1983) Grafting. In: Nishi K (ed) Vegetable handbook (in Japanese). Yokendo Book, Tokyo, pp 141–153

Yao H, Wu F (2010) Soil microbial community structure in cucumber rhizosphere of different resistance cultivars to Fusarium wilt. FEMS Microbiol Ecol 72:456–463

Yetisir H, Kurt S, Sari N, Tok FM (2007) Rootstock potential of Turkish *Lagenaria siceraria* germplasm for watermelon: plant growth, graft compatibility, and resistance to Fusarium. Turk J Agric For 31:381–388

Yi H, Kim HJ, Kim CG, Harn CH, Kim HM, Park S (2009) Using T-RFLP to assess the impact on soil microbial communities by transgenic lines of watermelon rootstock resistant to cucumber green mottle mosaic virus (CGMMV). J Plant Biol 52:577–584

Yin YL, Zhou BL, Li YP, Fu YW (2008) Allelopathic effects of grafting on rhizosphere microorganisms population of eggplants (in Chinese). Acta Hort Sinica 35:1131–1136

Zhang S, Gun X, Wang Y (2006) Effect of bur cucumber (*Sicyos angulatus* L) as rootstock on growth physiology and stress resistance of cucumber plants (in Chinese). Acta Hort Sinica 33:1231–1236

Zhou B, Gao Y, Lin G, Fu Y (1998) Relationship between disease resistance and electrolytic leakage, proline content and PAL activity in grafted eggplant (in Chinese). Acta Hort Sinica 25:300–302

8 Biorational Pest Management

Abstract

Biorational pest control under protected cultivation is an important tool for pest control in an economically and ecologically sound way. The term biorational is derived from the two words biological and rational, referring to pesticides of natural origin that have limited or no adverse effects on the environment or beneficial organisms. Biorational pesticides, which include botanicals, minerals, minimum risk pesticides, hot pepper wax and garlic extract, and synthetics, are becoming popular owing to environmental awareness and consumer concern. Biorational insect and disease control products are discussed in this chapter.

Keywords

Biorational pesticides • Pests • Diseases • Botanicals • Minerals • Synthetics • Insect growth regulators

8.1 Introduction

Insects and diseases are the major challenge to protected cultivation (glasshouses and plastic houses and tunnels). It is estimated that, worldwide, the area of protected crops is 307,000 ha, with vegetables occupying 65 % (200,000 ha) of this area and ornamentals 35 % (107,000 ha). Year-round warm temperatures and relatively high humidity make the protected environment of greenhouses highly favorable to arthropod pests. In many countries, production from protected cultivation is export oriented, and there fore, the grower has to comply with the maximum residue levels set by market regulatory bodies.

Following the chemical revolution of the 1950s, insect control mainly relied on use of chemical insecticides, many of which are effective but simultaneously kill natural predators and parasitoids. Furthermore, cost, availability, and risks to health and the environment impose serious limitations on the use of chemical control by growers of protected crops. Therefore, an alternative approach is urgently needed. One possible solution is the use of biorational pest control. Biorational pesticides currently represent only a small fraction, 1.3 % of the world pesticide mar-

P.P. Reddy, *Sustainable Crop Protection under Protected Cultivation*,
DOI 10.1007/978-981-287-952-3_8

ket. However, owing to growing concern about the use of toxic chemical pesticides, the growth rate of the biorational pesticide sector over the next 10 years has been forecast at 10–15 % annually in comparison to 2 % for chemical pesticides (Menn 1996).

8.2 Biorational Insect Control Products

Biorational pest control is an important tool to control pests of protected cultivation. The main objective of this approach is to optimize pest control in an economically and ecologically sound way. The term biorational is derived from the two words biological and rational, referring to pesticides of natural origin that have limited or no adverse effects on the environment or beneficial organisms. Biorational pesticides are becoming popular owing to environmental awareness and consumer concern.

These pest management products (Table 8.1) generally fit well into an integrated pest management (IPM) strategy, which relies on monitoring for early detection of pests and emphasizes the use of selective products that protect crops while minimizing negative effects on water, air, and soil and on pollinators and beneficial insects.

8.2.1 Botanicals

Botanicals are plant-derived materials such as pyrethrin, azadirachtin, and extract of *Chenopodium ambrosioides*. Plant-derived oils such as neem oil, canola oil, and jojoba oil are also included in this group. Botanicals are generally short-lived in the environment, as they are broken down rapidly in the presence of light and air. Products derived from the seeds of the neem tree, including azadirachtin and neem oil, are selective and have low mammalian toxicity. Many botanicals are considered "minimum risk" pesticides and are exempt from registration by the EPA.

Neem, *Azadirachta indica* (syn. *Antelaea azadirachta*, *Melia azadirachta*), is a tree belonging to the Meliaceae (mahogany) family, which originates from Southern and Southeastern Asia. It is an evergreen, deciduous, fast-growing plant which may reach a height of 25 m. The fruit are the most important source of the ingredients of neem that affect insects in various ways. The main active ingredient of neem seed oil is azadirachtin which has proven to be effective against insects, mites, and phytopathogens (Schmutterer 1990). Neem oil is obtained by cold-pressing neem seeds, which is typically 0.2–0.6 % of total seed content (Isman 2006).

Azadirachtin is considered nontoxic to mammals (rat oral acute LD_{50} is >5000 mg/kg), fish, and pollinators. It is easily and rapidly degraded by sunlight. Russell IPM developed a commercial formulation of neem oil, Fytomax. It is broad-spectrum emulsified neem oil which contains 70 % enriched neem oil, 10 % plant-originated emulsifier, 10 % UV stabilizer, and 10 % antioxidant. Fytomax is effective against mites, whiteflies, aphids, cutworms, fungus gnats, shore flies, leaf miners, thrips, and other types of soft-bodied insects on contact. This product is also effective against rots, mildews, rusts, scab, leaf spot, and blights. The multiple modes of action make it unlikely to become insect and plant pathogen resistant.

Azadirachtin has numerous effects on insects. It is a powerful insect growth regulator, a feeding deterrent, and an oviposition deterrent. Insect growth regulators affect the hormonal system of insects, preventing them from developing normally into mature insects. However, this property does not cause the immediate death of an insect pest. Azadirachtin has two profound effects on insects. At the physiological level, it blocks the synthesis and release of molting hormones (ecdysteroids) from the prothoracic gland, leading to incomplete ecdysis in immature insects. In adult female insects, a similar mechanism of action leads to sterility. In addition, azadirachtin is a potent antifeedant to many insects.

The females of some lepidopterous insects are repelled by neem products and will not lay eggs on treated plant parts or other substrates under laboratory conditions. This has been observed in *Crocidolomia binotalis*, *Helicoverpa armigera*,

Table 8.1 Biorational insecticides and miticides

Insecticides	Trade name	Target pests	Comments
Azadirachtin	Agroneem Plus (M); Aza-Direct; Azatin XL; Azatrol EC; Azera (M); Ecozin Plus 1.2 ME; Molt-X; Neemix 4.5, 0.25	Aphids, caterpillars, leafhoppers, leaf miners, thrips, whiteflies, beetles, and other insects	An insect growth regulator extracted from the seeds of the neem tree. Works by contact or ingestion against immature stages and has antifeedant properties
Chlorantraniliprole	Coragen	Caterpillars, Colorado potato beetle, leaf miners	May be used as soil or transplant water treatment at planting, in drip or as foliar. Avoid runoff in surface waters. Nontoxic to bees
Clove oil, thyme oil, and cinnamon oil	Ecotrol G	Root maggots, wireworm, cutworm, symphylans	Soil-incorporated volatile plant oils on a corn cob carrier, applied before, during, or after planting. Exempt from registration and reporting under federal EPA standards
Cryolite	Kryocide Prokil Cryolite	Colorado potato beetle (larvae only), cucumber beetle, flea beetle, cutworms, some caterpillars	Stomach poison, which must be ingested. Sodium aluminofluoride mineral applied as spray or dust. Reapply to cover new growth. May cause extensive wear of nozzles
Neem oil extract	Trilogy	Primarily labeled for (but not limited to) mite control	Can be used to control mites. For best results, use when population levels are low to prevent buildup. Repeat applications are needed
Flubendiamide	Belt SC, Synapse WG	Caterpillars including armyworms, loopers, hornworm, corn earworm, European corn borer	A diamide, active by insect larval ingestion, leading to a rapid cessation of feeding followed by death. Labeled for use as a foliar spray on brassicas and turnip greens, leafy vegetables, sweet corn, cucurbits, fruiting vegetables, and legumes to control caterpillars
Insecticidal soap (potassium salts of fatty acids)	M-Pede, Des-X	Aphids, leaf miners, mites, thrips, whiteflies	Works on contact. Can be phytotoxic to some crops; test on small plot. Avoid treatment when plants are stressed. May also harm some beneficials. Also active against powdery mildews
Iron phosphate	Sluggo Snail and Slug Bait	Snails, slugs	Bait which causes feeding to cease. Death occurs over 3–5 days. Exempt from tolerance and has a zero hour reentry interval due to low toxicity to people and wildlife

(continued)

Table 8.1 (continued)

Insecticides	Trade name	Target pests	Comments
Kaolin	Surround WP	Flea beetle, striped cucumber beetle, leafhopper, thrips	Interferes with insects' ability to recognize their host; particles cling to cuticle. Plant leaves turn white but growth is not inhibited. Reapply after heavy rain. To avoid lumps in mixing, form a slurry and then dilute. Maintain agitation
Methoxyfenozide	Intrepid	Many species of caterpillars	Mimics molting hormone; causes premature molt and death
Petroleum oil (mineral, paraffinic oils)	JMS Stylet-Oil, SuffOil-X	Aphids, leaf miners, beetle larvae, mites, thrips, leafhoppers, whiteflies	A horticultural oil (80 % petroleum oil) insecticide, miticide, and fungicide. Kills eggs, larvae, and nymphs of insects and mites and adults of soft-bodied insects through suffocation. Registered for use as a foliar spray on a variety of crops. Also labeled for certain diseases
Pyrethrin	PyGanic EC 5.0, PyGanic EC 1.4, Pyrenone (includes piperonyl butoxide)	Asparagus beetle, blister beetle, cucumber beetle, flea beetles, Mexican bean beetle, potato leafhopper, many caterpillars, aphids, stink bugs	Botanical insecticide with broad-spectrum activity. Contact toxin with rapid knockdown but short period of activity (nonpersistent). Highly toxic to fish. Derived from a chrysanthemum species cultivated in Africa. Some formulations are approved for organic crops
Pyriproxyfen	Esteem, Distance IGR, Knack	Thrips, whiteflies, shore flies, fungus gnats	Label may include greenhouse or field use. An insect growth regulator that suppresses development of embryo within the egg, immature, and pupal stages of the insect. No activity against adult insects
Rosemary oil, peppermint oil, other essential plant oils	Ecotrol EC, Ecotec	Aphids, beetles, bugs, early stages of caterpillars, leaf miners, mites, thrips	A broad-spectrum, plant-based, horticultural oil that acts as an insecticide and miticide. Works on contact. Spray when pests first appear; repeat at 7–10 days. Greenhouse and field use. Thorough coverage is needed. Exempt from registration under federal EPA standards; label gives instructions for use

(continued)

Table 8.1 (continued)

Insecticides	Trade name	Target pests	Comments
Soybean oil	Golden Pest Spray Oil	Primarily soft-bodied insects	A soybean-derived horticultural oil that works as a contact insecticide, as well as a feeding and oviposition deterrent. Labeled for use in sweet corn for control of earworm and armyworm. Also can be applied as a low-volume application for mites on eggplant and tomatoes and other soft-bodied insects on a range of vegetable crops
Spinetoram	Radiant	See spinosad below	A second-generation spinosyn, similar to spinosad below
Spinosad	Entrust, Seduce Insect Bait, GF-120 Naturalyte	Caterpillars, Colorado potato beetle, asparagus beetle, flea beetle, leaf miners, thrips. Baits labeled for cutworms, earwigs (Seduce), or fruit flies (GF-120)	Acts both as a contact and stomach toxin. Derived from soil bacterium *Saccharopolyspora spinosa*. Nondisruptive to most predator insect species and some parasites. Rotate with other selective biorationals to prevent resistance
Spiromesifen	Oberon	Whiteflies, nymphs and pupae, mites	Contact insecticide and miticide

and *Spodoptera frugiperda*. The fly *Lucilia sericata* was also deterred from egg laying as were some beetles, *Callosobruchus* spp. (Isman 2006). In field experiments, it was found that neem oil could reduce egg hatch by destroying egg masses (Fig. 8.1).

The antifeedant effect of neem is well established. There are numerous records of the antifeedant effects of neem derivatives on insects of various orders. Studies on the feeding behavior of larvae of various lepidopterous insects (such as *Spodoptera littoralis*, *S. frugiperda*, *S. exempta*, *Heliothis virescens*, *Helicoverpa zea*, *H. armigera*, *Trichoplusia ni*, and *Mamestra brassicae*) revealed that azadirachtin reduced feeding of all species tested. Oligophagous species were more sensitive than polyphagous ones (Simmonds and Blaney 1984).

8.2.2 Minerals

Some biorational pesticides are minerals mined from the earth and minimally processed. Kaolin clay, insecticidal soap, copper, and iron phosphate are examples.

8.2.2.1 Insecticidal Soap

Insecticidal soap is another potential biorational product for the growers of protected cultivation. Insecticidal soap can kill soft-bodied insects and mites. Insecticidal soap is one of the safest insecticides: it is safe, eco-friendly, and nontoxic to animals or humans, and nontarget insects are mostly unaffected by it. Though the oil from which insecticidal soaps are made may be of plant origin, they are not generally considered as botanical insecticides. Insecticidal soap acts rapidly and does not leave residue on fruit or vegetables. Therefore, it can be used right up to the crop harvest.

In chemical terms, insecticidal soaps are made from the salts of fatty acids. Russell IPM developed an insecticidal soap, Fytoclean. Fytoclean is a potassium salt of fatty acids and is produced by adding potassium hydroxide to fatty acids from plant oils and animal fats, which are saponified by the potassium hydroxide. The soap salts are

Fig. 8.1 Effect of Fytomax (neem oil) as an antifeedant on insect larvae

derived from fatty acids with chains of 12–18 saturated carbons and 18 carbons with one or two double bonds (unsaturated). Potassium salt formation confers the property of water solubility on the fatty acids. The lipoprotein matrix of the insect cuticle is disrupted when the lipophilic carbons of Fytoclean (fatty acids) penetrate it. This results in evacuation of cellular contents, causing the cells to dehydrate and die.

Fytoclean disrupts the cuticle of soft-bodied insects (e.g., aphids, whiteflies, thrips) which makes them vulnerable to dehydration.

The efficacy of Fytoclean was compared with synthetic insecticides buprofezin and malathion and biological insecticide Mycotal. In a field experiment, different treatments were evaluated against whitefly, *Bemisia tabaci*, crawlers on runner beans. The results showed that Fytoclean significantly reduced the number of live pupae that subsequently developed compared to buprofezin and a control treatment, although the number of live pupae obtained with Fytoclean was not significantly different from numbers obtained with the malathion or Mycotal treatments (Fig. 8.2).

Fytoclean is only effective against those insects that come into direct contact with the spray before it has dried. The dried residues on plant surfaces are not insecticidal and they degrade rapidly. Fytoclean is not very effective against insects with thicker cuticles, such as adult beetles, bees, wasps, and grasshoppers. In addition, highly mobile insects may escape by flying away. Fytoclean should be applied for controlling soft-bodied insects at the rate of 0.5–1 % of total spray volume. A full plant cover spray in the morning is preferable because this allows the effect of the product to be maximized during the heat of the day. Fytoclean can be applied at any time right up to the crop harvest and has no preharvest interval. In conclusion, these various biorational crop protection materials give an opportunity to manage pest populations without toxic chemicals. A combination of biorational pest control products will be a better option for growers of protected crops and also help organic growers to manage their crops in a sustainable and effective way.

8.2.3 Synthetics

Minerals and other natural materials that are heated, chemically reacted, or mixed with surfactants may be considered synthetics. Synthetics also include insect growth regulators (IGRs), which interrupt or inhibit the life cycle of a pest. They may also work by strengthening plant defenses.

8.2.3.1 Insect Growth Regulators

Insect growth regulators (IGRs) are another least-toxic pesticide control option for pests.

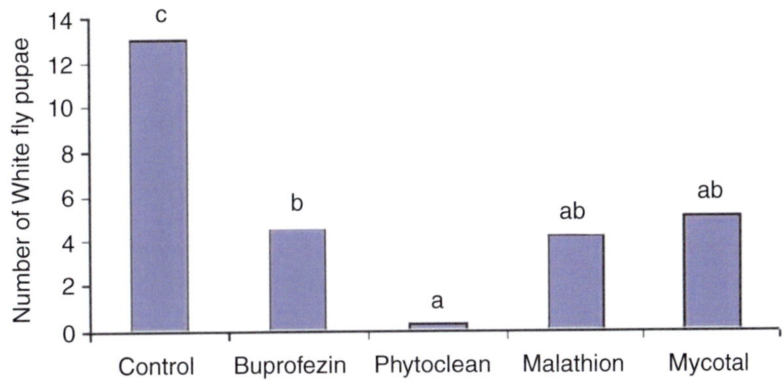

Fig. 8.2 Efficacy of Fytoclean compared with two synthetic insecticides and one biopesticide. Common letters indicate treatments that did not differ from each other at P≤5 % (ANOVA, LSD test)

IGRs typically kill insects by disrupting their development. They have a complex mode of action that precludes insects from rapidly developing resistance. IGRs can work in one of several ways:

- They can mimic juvenile hormones, so that insects never enter the reproductive stage of development.
- They can interfere with the production of chitin, which makes up the shell of most insects.
- They can interfere with the molting process.

IGRs usually work through ingestion, so good spray coverage is essential. They generally do not affect nontarget species, such as humans, birds, fish, or other vertebrates. For most IGRs, there are minimal reentry restrictions. IGRs typically take several days to have an effect on pest populations. Because IGRs do not affect mature insects, adult beneficials released into the greenhouse after an IGR application are not likely to be affected. Use of IGRs is generally prohibited by organic certification organizations because the products are synthesized.

IGRs can sometimes be used in conjunction with biological control efforts and may provide growers with a "safety net" should beneficial fail to keep the pests below economically damaging levels. The integration of biorational pesticides (also known as least toxic or biopesticides) in greenhouses is often necessary in addition to cultural and biological control measures. When the use of a pesticide is necessary, materials should be selected that are least harmful to the predators and parasites released into the greenhouse. Insecticidal soap, horticultural oils, and the bacterium *Bacillus thuringiensis* are examples of insecticides that can be safely integrated into a biological control program. The advantages of biopesticides over conventional chemicals are their selectivity to a targeted pest, lower toxicity to beneficial insects and greenhouse workers, and shorter reentry intervals (REIs).

Enhanced control with pesticides is achieved with thorough spray coverage. Wider plant spacing and removal of dead lower leaves improve pesticide coverage and pest control.

The table below lists some well-known insect growth regulators (Table 8.2).

8.2.4 Hot Pepper Wax and Garlic Extract

These products are actually insect repellents. Hot pepper wax contains paraffin (refined wax) and capsaicin (the "hot" in hot peppers). Another study at Ohio State showed that hot pepper wax did not control western flower thrips but was effective against two-spotted spider mite (Lindquist 1998). Products containing garlic also work to repel insects.

Table 8.2 Selected insect growth regulators

Brand name	Supplier	Active against
Adept	Uniroyal Chemicals	Fungus gnats
Azatin	Hydro-Gardens, Olympic Horticultural Products	White flies, leaf miners, thrips, mealybugs, fungus gnats, aphids, cabbage loopers, diamondback moths, armyworms
Citation	Novartis	Leaf miners, fungus gnats, shore flies
Confirm	Rohm and Haas	Caterpillars
Dimilin	Uniroyal Chemicals	Beet armyworm, whiteflies, fungus gnats, shore flies
Distance	Valen	Whiteflies, fungus gnats, shore flies
Enstar II	Wellmark Intl.	Whiteflies, fungus gnats, aphids, soft and armored scales, mealybugs
Insegar	Novartis	Caterpillars, psyllids, scales
Neemazad	Thermo Trilogy	Whiteflies, leaf miners, thrips, mealybugs, fungus gnats, aphids, cabbage loopers, diamondback moths, armyworms
Neemix	Thermo Trilogy	Whiteflies, leaf miners, thrips, mealybugs, fungus gnats, aphids, loopers, diamondback moths, armyworms, cabbage loopers
Precision	Novartis	Whiteflies, soft scales, fungus gnats, shore flies, leaf miners
Preclude	Whitmire Micro-Gen	Whiteflies, thrips, scales, aphids
Pyrigro	Whitmire Micro-Gen	Whiteflies, fungus gnats, scales
Trigard	Novartis	Leaf miners

8.2.5 Minimum Risk Pesticides

These are a special class of pesticides that are not subject to federal registration requirements because their ingredients, both active and inert, are demonstrably safe for the intended use. This exemption falls under section 25(b) of the Federal Insecticide, Fungicide, and Rodenticide Act (FIFRA). The list of 25(b) exempt materials includes the following: corn gluten meal; sodium chloride; corn, linseed, sesame, soybean, and cottonseed oil; garlic and garlic oil; and essential oils including mint, thyme, geranium, lemongrass, cinnamon, and rosemary. Some pest control products listed in this guide meet the criteria for exemption and do not have EPA pesticide registration or a pesticide label. More on this class of pesticides can be found.

8.3 Biorational Disease Control Products

Biorational disease control products (fungicides, bactericides, and nematicides) are listed in Table 8.3. Sulfur, potassium bicarbonate, phosphites, and copper compounds are examples of minerals or synthetics that can control fungal and bacterial diseases. Not all of these products are OMRI listed. Botanicals such as rosemary oil, soybean oil, or garlic extracts also appear in this table and are generally approved for use in organic production by OMRI. Products listed in the table require thorough coverage, application at the first signs of disease, and frequent repeated dosages to be effective.

Table 8.3 Biorational disease control materials

Active ingredient	Target diseases	Labeled crops	Comments
Acibenzolar-S-methyl (Actigard 50 WG)	Downy mildew, bacterial "speck, white rust, *Xanthomonas*	Crucifers, tomato, spinach	Plant defense activator
Calcium polysulfide (Rex Lime Sulfur Solution, Green Cypress Lime Sulfur Solution)	Anthracnose, brown rot (*Monilinia*), powdery mildew, *Phomopsis* cane blight, scab, rust	Apple, blueberries, caneberries, currants, gooseberries, quince	OMRI listed[a]
Citric acid, fish oil, and mint oil (Fungastop)	Broad-spectrum antifungal compound	All crops	
Cottonseed oil, corn oil, garlic oil (GC-3, Mildew Cure)	Powdery mildew	Most crops	OMRI listed
Copper sulfate (Basic Copper 53)	Bacterial blight, powdery mildew, downy mildew, leaf spots, anthracnose, *Cercospora*, *Alternaria*	Most crops	OMRI listed
Cupric hydroxide (Champ WG)	Bacterial blight, black rot (*Xanthomonas*), downy mildew, powdery mildew, anthracnose, *Phomopsis*, late blight, *Botrytis*, leaf spots	Most crops	OMRI listed
Cuprous oxide (Nordox 75 WG)	Bacterial blight, black rot (*Xanthomonas*), downy mildew, powdery mildew, anthracnose, *Phomopsis*, late blight, *Botrytis*, leaf spots	Most crops	OMRI listed
Hydrogen dioxide (OxiDate)	*Alternaria*, *Phytophthora*, *Pythium*, *Rhizoctonia*, *Fusarium* wilt, *Sclerotinia*, anthracnose, bacterial blight, *Botrytis*, powdery mildew, rust	Beans, cucurbits, celery, crucifers, leafy vegetables, tomato, pepper, onions, potato, herbs, root crops	OMRI listed. Contact activity only
Kaolin (Surround WP)	Powdery mildew, sunburn and heat stress	All crops	OMRI listed
Neem oil (Trilogy, NeemGard, Triact 70[OG])	Anthracnose, *Botrytis*, downy mildew, powdery mildew, scabs, rusts, leaf spots, and blights	Most crops	OMRI listed
Oils, petroleum based (Organic JMS Stylet-Oil, Civitas, SAF-T-SIDE, SuffOil-X)	*Alternaria*, gummy stem blight, powdery mildew, rust	Most crops	OMRI listed
Potassium bicarbonate (Kaligreen, Armicarb 100, MilStop)	*Alternaria*, anthracnose, *Botrytis*, downy mildew, *Fusarium*, leaf spots, *Phytophthora*, powdery mildew	Beans, crucifers, cucurbits, eggplant, lettuce, peppers, potato, tomato, spinach, wheat	Armicarb 100 is not OMRI listed, but MilStop and Kaligreen are OMRI listed
Potassium phosphite (Helena ProPhyt)	Downy mildews, *Phytophthora* species, *Pythium* species	Cucurbits, crucifers, onion, potato, tomato, lettuce, spinach	Systemic. Drench at transplant

(continued)

Table 8.3 (continued)

Active ingredient	Target diseases	Labeled crops	Comments
Mono- and dibasic sodium, potassium, and ammonium phosphites (Phostrol)	Downy mildews, *Phytophthora* species, *Pythium* species	Most crops	Use caution when using in a tank mix
Potassium silicate (Sil-Matrix)	Powdery mildew, leaf spots	Most crops	Broad-spectrum preventive fungicide
Reynoutria sachalinensis extract (Regalia Biofungicide)	Powdery mildew, *Botrytis*, leaf spots, bacterial spot, speck, and canker	Cucurbits, peppers, lettuce, tomato, strawberry, grapes	Plant defense activator. Use caution in tank mix. OMRI listed
Rosemary and clove oils (Phyta-Guard EC)	Powdery mildews, bacterial spot	Most crops	
Rosemary, clove, and thyme oils (Sporan EC)	Bacterial spot, early blight, gray mold, late blight, powdery mildew, downy mildew	Most crops	
Saponins from *Quillaja saponaria* (c)	Nematodes	*Brassica* crops, cucurbits, pepper, tomato, leafy vegetables	OMRI listed
Rosemary, thyme, and clove oil (Sporatec)	Anthracnose, *Botrytis*, downy mildew, powdery mildew, leaf spots, rusts, bacterial spot	Most crops	OMRI listed
Sesame oil (Organocide)	Powdery mildew, leaf spots	Most crops	OMRI listed
Sesame oil (Dragonfire-CPP)	Nematodes	All crops	OMRI listed
Sesame seed meal (Ontrol)	Nematodes	All crops	Ontrol is not OMRI listed, but many sesame seed formulations are OMRI listed
Streptomycin sulfate (Agri-Mycin 17)	Bacterial diseases	Celery, peppers, tomato, potato	
Sulfur (Micro Sulf, Sulfur DF, Kumulus DF, Britz Sulfur Dust)	Powdery mildew, leaf spots	Most crops	OMRI listed
Tetracycline (MycoShield)	Bacterial diseases	Tomato, peppers, tree fruit	
Thyme oil (Proud 3, ProMax)	*Pythium*, *Phytophthora* species, clubroot, *Sclerotium*, nematodes	Most crops	OMRI listed

[a]Organic Materials Review Institute (OMRI)

References

Isman MB (2006) Botanical insecticides, deterrents, and repellents in modern agriculture and an increasingly regulated world. Annu Rev Entomol 45:45–66

Lindquist RK (1998) Evaluations of non-conventional pesticides for insect and mite control on greenhouse ornamental plants. Greenhouse Product News, July, 52–55

Menn JJ (1996) Biopesticides: has their time come? J Environ Sci Health B 31:383–389

Schmutterer H (1990) Properties and potential of natural pesticides from the neem tree, *Azadirachta indica*. Annu Rev Entomol 35:271–297

Simmonds MSJ, Blaney WM (1984) Some neurophysiologic effects of azadirachtin on lepidopterous larvae and their feeding response. In: Natural pesticides from the neem tree and other tropical plants. Proceedings of the 2nd International Neem Conference, Germany, pp 163–180

9 Drip Chemigation for Insect Pest Management

Abstract

The application of pesticides through a drip irrigation system has been used successfully for the control of a variety of vegetable pests. Because many vegetable growers already use a drip irrigation system for water management, the injection of a pesticide can easily and inexpensively be achieved with the addition of an injection pump and the required safety equipment for the injection of soluble pesticides. The current availability of novel, systemic insecticides such as neonicotinoids or anthranilic diamides, as well as additional future pesticides under development, has advanced the opportunities for drip chemigation as an effective and environmentally sound pest management tactic. History, need for drip chemigation, selection of chemicals, chemigation system, and considerations for drip chemigation with insecticides in vegetable crops are discussed.

Keywords

Drip chemigation • Neonicotinoids • Anthranilic diamides • Vegetable crops

9.1 Introduction

"Chemigation" is defined as the application of agricultural chemicals, including herbicides, insecticides, fungicides, and fertilizers through a drip/trickle irrigation system (Chalfant and Young 1984). However, other terms have been used to denote the application of agricultural chemicals through an irrigation system including "pestigation" (Larsen 1982), "insectigation" (Owens 1981), "fertigation" (Hall 1982), "herbigation" (Johnson et al. 1987), "fungigation" (Johnson et al. 1987), "drip chemigation" (Wildman and Cone 1986), and "drip/trickle chemigation" (Certis USA 1999). It is an efficient and economical means of applying inputs necessary for crop, nursery, greenhouse, and landscape management. Chemigation can be applied only if the product label allows such application and only according to the label directions and restrictions. This is possible through application of a right amount of agricultural

P.P. Reddy, *Sustainable Crop Protection under Protected Cultivation*,
DOI 10.1007/978-981-287-952-3_9

chemicals at the right time, at the right place, and in the right manner:

- To achieve high yield and the quality produce
- To minimize groundwater/surface water pollution

The three basic crop irrigation systems include overhead/sprinkler, drip/trickle, and surface/gravity flow systems. The proper equipment, depending on the irrigation system, is essential for successful (as well as legal) chemigation. Although overhead irrigation and surface flow systems may be suitable for some vegetable crops, drip irrigation is widely used in vegetable production.

The application of pesticides through a drip irrigation system has been used successfully for the control of a variety of vegetable pests. Because many vegetable growers already use a drip irrigation system for water management, the injection of a pesticide can easily and inexpensively be achieved with the addition of an injection pump and the required safety equipment (switches, check valves, drain valves) for the injection of soluble pesticides (Fig. 9.1). The current availability of novel, systemic insecticides that are labeled under the EPA Reduced Risk Pesticide Program such as neonicotinoids (including imidacloprid, clothianidin, thiamethoxam, and dinotefuran) or anthranilic diamides (including chlorantraniliprole and cyantraniliprole), as well as additional future pesticides under development, has advanced the opportunities for drip chemigation as an effective and environmentally sound pest management tactic.

Chemigation via the drip system is a sound and economical option in place of traditional foliar sprays of pesticides for control of specific vegetable pests. It enables growers to precisely apply a pesticide under virtually any weather condition for control of many insect pests, including aphids, whiteflies, leafhoppers, beetles, and caterpillars while at the same time reducing the total pesticide, labor, and energy inputs as compared with foliar sprays. It enhances a pest management program because many of the new-chemistry pesticides labeled for drip chemigation are selective to specific pests, and because they are applied to the root zone, they are generally less toxic to beneficial and nontarget organisms. Over a dozen labels currently allow chemigation via the drip system for a wide range of pests, and several new-chemistry pesticides are being added to this list every year.

Fig. 9.1 Chemigation of staked tomatoes under black plastic with trickle irrigation

9.2 History of Drip Chemigation

Insecticides were first injected into a drip irrigation system in bell peppers in New Jersey for the control of European corn borer (*Ostrinia nubilalis*) in 1980 (Ghidiu and Smith 1980) and the following year in lima beans for control of the Mexican bean beetle (*Epilachna varivestis*) (Ghidiu 1981). However, the insecticides were not effective against these pests when applied systemically and neither trial resulted in insect pest reduction. The first successful drip application of an insecticide for insect pest reduction was conducted in 1985 (Wildman and Cone 1986), where asparagus aphid (*Brachycorynella asparagi*) numbers in asparagus were significantly reduced using disulfoton (Di-Syston 6E) as compared with the untreated. Successful insectigation trials were reported with entomopathogenic nematodes for the control of spotted cucumber beetles (*Diabrotica undecimpunctata*) in 1986 (Reed et al. 1986), followed by the effective control of aphid (Aphididae) populations by chemigating imidacloprid in vegetables in Arizona in the mid-1990s (Palumbo 1997) and effective control of spotted cucumber beetles in melons in Virginia with drip-applied imidacloprid and thiamethoxam (Kuhar and Speese 2002). In a 3-year field trial starting in 2004, chlorantraniliprole was shown to be highly effective against the European corn borer in bell pepper when applied through a drip irrigation system (Ghidiu et al. 2009). Further, chlorantraniliprole applied through a drip irrigation system significantly reduced armyworms (*Spodoptera* spp.) and fruit worms (*Helicoverpa zea*) in tomatoes in field tests in both Virginia (Kuhar et al. 2009) and Florida (Schuster et al. 2009). Ghidiu et al. (2009) reported that chlorantraniliprole and thiamethoxam injected via a drip irrigation system significantly reduced damage to eggplant foliage caused by flea beetles (*Epitrix* spp.) and leaf miners (*Liriomyza* spp.).

9.3 Why Drip Chemigation?

Many growers throughout the vegetable-growing regions are already using drip/trickle irrigation as a water management tactic. Chemigation via drip irrigation allows growers to apply pest control materials through a distribution system they already have in place (the drip system), thus saving significant time and money over ground-applied chemicals. Over the past 10–12 years, many field tests throughout the USA have demonstrated that application of pesticides via the drip system results in highly effective control of specific pests of vegetable crops using low rates of labeled insecticides.

9.3.1 Advantages

- The total pesticide input for the control of pests in most crops is reduced using drip chemigation when compared with that of traditional foliar applications.
- The entire plant is protected.
- In many trials with vegetable crops, 1–2 drip applications of an insecticide per season resulted in equivalent, or better, control of specific insect pests as compared with multiple foliar applications (Kuhar et al. 2010).
- With fewer applications needed, less total energy inputs are required (either by tractor or by large horsepower overhead irrigation system pumps), and no soil compaction occurs as a result of heavy spray equipment being operated within the crop.
- Certain plant diseases, such as *Phytophthora capsici*, a soilborne fungus, which produces spores that are spread via water splashing onto the foliage by rainfall, tractor wheels, and other factors, may be reduced because foliage is not wetted and the soil moisture can be better controlled (Ristaino and Johnston 1999).
- Drip-chemigated fields also reduce potential worker exposure to pesticides by eliminating the application of insecticides to the plant foliage.
- Weather is not a factor during or after application, as drip chemigation is not affected by wind and application can be made when fields are too muddy to operate ground equipment.
- Allows for flexibility in application timing when injections can be made virtually any time during the season from stand establishment until harvest.
- Drip chemigation can be completed without spray drift, eliminating "application visibility" that concerns many growers.
- Since many of the newer insecticides such as the neonicotinoids and anthranilic diamides are selective to certain insect pests, they are generally less disruptive to nontarget species and beneficial organisms.
- No need to spread the agricultural chemicals (so coverage is more even, and a lot of work is saved).
- Chemical placement at right place and better uniformity.
- Efficiency is high – saving of chemicals.
- Controlling the depth of application.
- Prevent chemical leaching to the groundwater.
- Mechanical damage to the crop by sprayers is reduced by chemigation.
- It may reduce environmental hazards associated with spray drift.

- Chemical application cost saved by 40 % or more.
- Reduces energy consumption by 90 %.
- The total insecticide input for control of targeted insect pests in most crops is significantly reduced when compared with that of traditional foliar applications, while at the same time essentially 100 % protection of the plant is obtained because these materials are root systemic and translocate throughout the plant, resulting in a more even distribution of the pesticide within the plant. For some vegetable crops, 1–2 drip/trickle irrigation applications of an insecticide during the season result in equivalent control, or better control, of insect pests than that of multiple foliar sprays. Kuhar et al. (2009) reported that a single injection of the high labeled rate of chlorantraniliprole (Coragen) into a drip system was as effective as four foliar applications of the pyrethroid lambda-cyhalothrin (Warrior II) for control of caterpillar damage in fresh market tomatoes, and Ghidiu et al. (2009) reported that two injections of chlorantraniliprole into a drip system were as effective as seven applications of a standard grower foliar spray program consisting of two applications of acephate (Orthene 97) followed by five applications of indoxacarb (Avaunt 30WDG) for control of European corn borer in bell peppers.
- Less energy is required to transport water and insecticide solutions at the lower pressures and velocities in drip/trickle systems as compared with other irrigation systems. Also, because fewer applications are needed, less energy input is required than by tractor or other application methods, and no soil compaction occurs that results from heavy tractors or spray equipment being operated over the field. The fewer times a tractor goes over the field, the less potential for plant damage caused by the tractor operation.
- Pathogen movement through the field via water flowing over the soil surface can be reduced through the use of a drip/trickle irrigation system if plastic mulch is used in combination with drip tubing. This is especially true for plant diseases such as *Phytophthora capsici*, a soilborne fungus, which produces spores that are spread via water splashing up onto the foliage by rainfall, operation of field equipment and sprayers, etc.
- Weather is not a factor during application, as injection of an insecticide into a drip/trickle irrigation system can be made in wind or rain or when fields are too muddy or soft to operate ground equipment without getting stuck.
- Applicator exposure, both in terms of physical contact and time of exposure, to insecticides during application is significantly reduced.
- Plant growth may be enhanced through the use of drip irrigation systems because with frequent drip waterings, it is possible to maintain a more optimum balance between soil water, plant needs, and aeration. Healthy plants are less susceptible to insect pest problems than unhealthy plants or plants under water stress.
- Many of these new-chemistry insecticides are selective to specific insect pests, and they are generally less toxic or disruptive to nontarget species and beneficial organisms, including insect predators, parasites, and pollinators. The injection of insecticides through a drip/trickle irrigation system thus fits well into an integrated pest management program.

9.3.2 Disadvantages

- Requires considerable management input and personnel training.
- Requires a change in management techniques.
- Some chemical may react with irrigation system components.
- Additional equipment may be required for chemigation.
- Also, drip irrigation systems generally require a high level of maintenance and regular monitoring of the entire system for pressure fluctuations, leaks in the system, plugged emitters, and other potential problems, all of which are even more important when insecticides are injected.

- Since some of the new compounds have exhibited long residual activity after chemigation, caution should be taken not to overapply the products to sustain insect susceptibility.
- The initial capital expenditures for a complete drip system and additional injection equipment can often times be greater than that for sprinkler, overhead, or other irrigation systems. However, the more the system is used, the smaller the costs per hour of operation. And most drip systems have many reusable components (pumps, filters, tubing, hoses, injectors, etc.).
- Water carries particulates that can clog the emitters if the filters malfunction or are not maintained properly (backwashing, cleansing, etc.). If plastic mulch is used in combination with the drip/trickle system, clogged emitters are sometimes difficult to locate, resulting in uneven distribution of the insecticide when pumped into the drip system. Also, if plastic drip lines and tubing are not properly rinsed after each use, it is possible that emitters may become clogged with residue.
- Drip line repairs can sometimes be time consuming and costly. Bright, direct sunlight may affect some plastic tubing used for drip irrigation, shortening their usable life span. Also, small rodents (mice, rabbits, chipmunks, etc.) and certain soil insects (crickets, wireworms, ground beetles, and others) may chew on drip hose, especially in droughts, causing small leaks that subsequently result in loss of pressure and uneven distribution of the insecticide. Oftentimes, such small leaks are difficult to detect under row covers such as black plastic until the row becomes saturated and wet spots appear.
- After the final harvest, cleanup costs of drip/trickle systems may be higher than with other irrigation systems. Costs may include removal of plastic row covers, all drip lines and tubing, and injection and safety equipment. The disposal cost of used plastics (plastic mulches, drip lines) continually increases.
- Top-dressed fertilizers and some herbicides may need additional sprinkler application for activation, especially if plastic mulch is not used over the row.
- Salts can accumulate as a result of inadequate flushing of the drip/trickle irrigation system (Dasberg and Or 1999), particularly at the perimeter of the wetted area.

9.4 Effectiveness of Insectigation

During the mid-1990s, researchers reported effective control of beetles, aphids, whiteflies, and several other insect pests using foliar applications of a newly developed class of insecticides, the neonicotinoids. Neonicotinoid insecticides (imidacloprid, thiamethoxam, dinotefuran, and clothianidin) have been widely used on vegetable crops to control leaf-feeding beetles and sucking insect pests such as leafhoppers, psyllids, aphids, thrips, and whiteflies. Neonicotinoids are neurotoxins that target the nicotinic acetylcholine receptor acting as agonists (Maienfisch et al. 2001). Although they are effective as contact insecticides, it is the ability of these chemicals to be taken up by the roots from the soil as systemic insecticides and transported to the foliage that has been one of the primary reasons for their success. These insecticides are particularly suitable for application through a drip irrigation system because they are effective against specific insect groups, they are soluble and have systemic properties, they are relatively nonphytotoxic, and they are considered reduced risk pesticides under the US Environmental Protection Agency (USEPA) Federal Reduced Risk Pesticide Program. Felsot et al. (1998) examined the distribution of imidacloprid in soil when applied through a drip irrigation system and concluded that imidacloprid was a good candidate for drip chemigation. Imidacloprid and other neonicotinoids applied through a drip system have been shown to provide effective control of whiteflies and aphids in desert vegetable crops (Palumbo et al. 2001) as well as cucumber beetles in cucurbit crops (Kuhar and Speese 2002).

A single application of these insecticides through a drip irrigation system provided similar control as multiple foliar sprays.

More recently, another new class of insecticides, the anthranilic diamides, has been shown to be highly toxic to numerous caterpillar pests (Lahm et al. 2005). One of these insecticides, chlorantraniliprole, is xylem mobile through root uptake and controls caterpillars and other leaf-feeding pests (Lahm et al. 2007). Like the neonicotinoid-class insecticides, chlorantraniliprole is also highly soluble, root systemic, and effective against specific insect pests, especially caterpillars, leaf miners, and beetles. Because both of these materials are selective against certain insect pests, they are ideal materials for a pest management program.

Schuster et al. (2009) reported that a single or double drip application of chlorantraniliprole effectively reduced leaf miner, *Liriomyza trifolii*, and armyworm, *Spodoptera* spp., damage in tomatoes. Ghidiu et al. (2009) injected chlorantraniliprole for European corn borer control in bell peppers over a 3-year period and reported significant reduction of damaged fruit. Further, they reported that two injections of chlorantraniliprole were as effective for European corn borer control as multiple (up to nine) foliar applications of a standard insecticide program by using a ground sprayer. Kuhar et al. (2008) achieved similar levels of European corn borer control with soil applications of chlorantraniliprole in pepper. Kuhar et al. (2010) further reported that two drip applications of chlorantraniliprole significantly reduced the percentage of tomatoes damaged by tomato fruit worm, *Helicoverpa zea*, as compared with that typically achieved using multiple foliar applications of insecticides. Palumbo (2008) reported that two drip applications of chlorantraniliprole during stand establishment provided excellent residual control of *Trichoplusia ni*, *Spodoptera exigua*, and *Liriomyza* spp. in romaine lettuce (*Lactuca sativa* var. *longifolia*), with no significant marketable yield loss. In addition to chlorantraniliprole, there are additional new-chemistry systemic insecticides such as cyantraniliprole (an anthranilic diamide) soon to be registered in vegetables for control of an even greater pest spectrum (Kuhar et al. 2011a, b; Palumbo 2011).

9.5 Selection of Chemicals

Application of insecticides through irrigation system depends on the:

- Nature of insecticides
- Special needs of the crop at various growth stages
- Efficacy of the insecticide:
- Insecticides can be injected undiluted or they can be diluted with water or oil to a volume necessary for calibrating within the range of the pump. For many insecticides, mixing with oil before injection can improve efficacy and residual control.
- Oil carriers should not contain emulsifiers.
- Studies have shown that insecticide formulation best suited for chemigation are those insoluble in water and soluble in oil.
- Insecticide label must specially state that the product can be applied through an irrigation system.

Currently, the USEPA has approved and labeled numerous insecticides of different classes for application through a drip irrigation system in fruits and vegetables for the control of a wide variety of insect pests (Table 9.1).

Application of nematicides through drip irrigation has been very successful.

9.6 Chemigation System

Efficiency shall not be less than 85 % where pesticides are applied through the system. Injectors (pesticides) and other automatic operating equipment shall be located adjacent to the pump and power unit, placed in accordance with manufacturer's recommendation, and include integrated backflow prevention protection. Chemigation shall be accomplished in the minimum length of

Table 9.1 USEPA-approved and labeled insecticides of different classes for application through a drip irrigation system in fruits and vegetables

Common name	US brand name	Insecticide class
Azadirachtin	Aza-Direct	Limonoid insect growth regulator (neem)
Chlorantraniliprole	Coragen	Anthranilic diamide
Clothianidin	Belay	Neonicotinoid
Dimethoate	Dimate	Organic phosphate
Diazinon	Diazinon	Organic phosphate
Dinotefuran	Venom	Neonicotinoid
Imidacloprid	Admire PRO	Neonicotinoid
Malathion	Malathion 8 Aquamul	Organic phosphate
Methomyl	Lannate	Carbamate
Oxamyl	Vydate	Carbamate
Rosemary + peppermint oils	Ecotec	Botanical
Thiamethoxam	Platinum	Neonicotinoid
Thiamethoxam + chlorantraniliprole	Durivo	Neonicotinoid

time needed to deliver the chemicals and flush the pipelines. Application amounts shall be limited to the minimum amount necessary, as recommended by the chemical label.

Drip chemigation systems that will have insecticides injected into them must be properly engineered, installed, and maintained over the season to ensure a uniform distribution of outflow (Ross 2004). Chemigation requires that two separate hardware systems be joined together, operating as a single system. The components of the first system, a typical drip irrigation system, include:

- Main water source (well, irrigation pond, etc.). Public water supplies cannot be used.
- Main water pump.
- Water filter system (with sand screens, screen filters, flush valve, or drain).
- Backflow prevention valve/backwash controller.
- Pressure gauge.
- Low-pressure shutoff valve.
- Low-pressure sensor/shutoff switch.
- Pressure relief valve.
- Various diameter hoses and polytubing carry water to the plant roots (main, lateral, drip lines, etc). Drip lines are available with a wide range of emitter spacings and can be found with spacings of 4", 8", 12", 18", or even 24", with flow rates of 12–64 gallons or more per hour.

Most growers that currently use some form of drip/trickle irrigation as a water management tool can easily, and inexpensively, add the necessary equipment to properly inject agrichemicals.

The components of the second system, in addition to the typical drip irrigation system equipment listed above, include the following pieces of equipment that are needed before injecting any agricultural chemical into the system:

- Chemical mix tank, such as a 19-L plastic jug, preferably with an agitator and an outlet filter to prevent clogging of emitters by the chemical solution
- Containment tray or pan to catch any chemical solution leakage or spillage
- Positive displacement pump or other reliable solution metering devices which provide a consistent flow rate at low pressures
- Backflow prevention valve to prevent backflow of solution into the mixing container
- Additional safety equipment such as a solenoid-operated valve on the intake side of the injection pump to prevent fluid from being sucked out of the supply tank when the system is not operating
- Low-pressure shutoff valve to shut the injection system off should a loss of pressure occurs

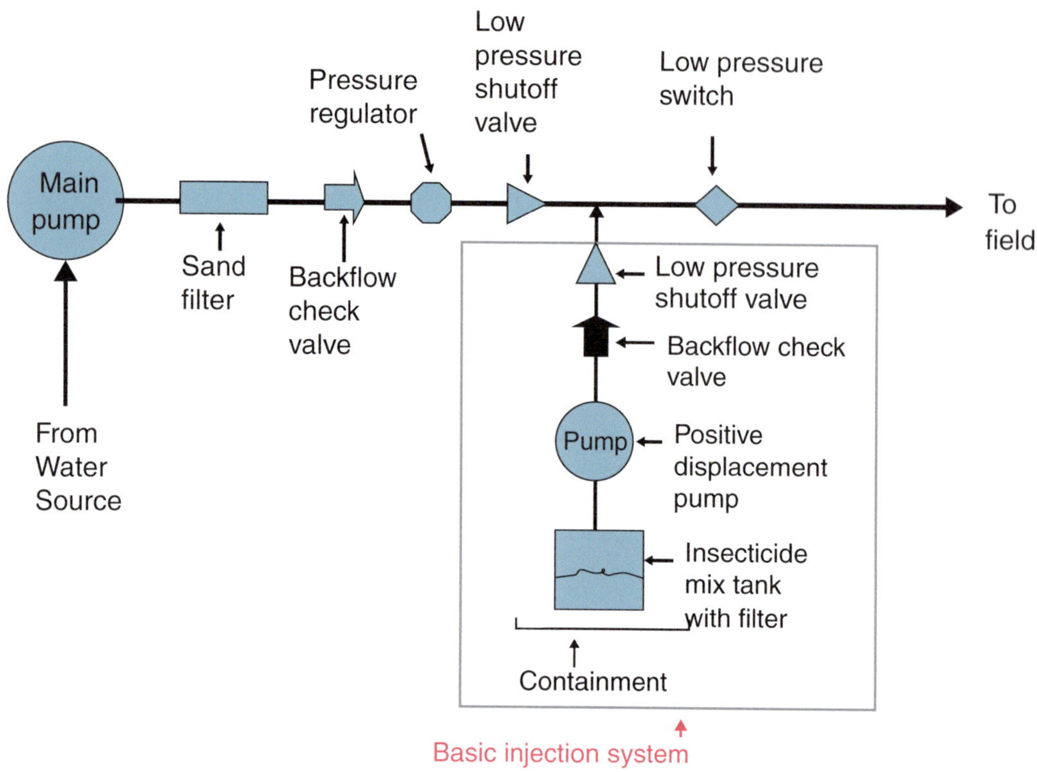

Fig. 9.2 A basic drip/trickle irrigation system with chemigation capabilities using a positive displacement injection metering pump

A basic drip/trickle irrigation system with an additional chemigation system using a positive displacement metering pump for injection of an insecticide is shown in Fig. 9.2.

9.7 Considerations for Drip Chemigation with Insecticides in Vegetable Crops

Optimal application timing of a root systemic pesticide through a drip irrigation system depends on the ability to efficiently deliver the compound to the actively growing roots. Several agronomic and operational factors should be considered before pesticides are applied through any drip/trickle irrigation system.

9.7.1 Crop Growth Stages and Root Types

Plant size at the time of chemigation may determine how a pesticide is applied (Syngenta 2009). Early in crop establishment, seedling plants and transplants generally occupy a small root zone as compared with the well-developed and expanded root zone of larger, more mature plants. In either case, insecticide efficacy will occur only if the soil in the root system has adequate soil moisture for plant growth and insecticide uptake. Root type also will influence the results of chemigation. For example, broccoli tends to have a relatively shallow and fibrous root system and will have different irrigation requirements than would a deep, taproot crop like watermelon (*Citrullus lanatus*).

9.7.2 Soil Properties

Movement of insecticides through the soil during chemigation will depend partially on the soil texture and the amount of organic matter in the soil. Irrigation water moves at higher flow rates in coarse-textured sandy soils than in fine-textured clay soils, which tend to hold water longer. Consequently, soil texture will determine drip irrigation frequency and duration as well as when a pesticide is applied via drip chemigation. Furthermore, soils with high organic matter may actually reduce the availability of a pesticide for root uptake, depending on the chemical properties of the pesticide.

9.7.3 Physiochemical Properties of the Pesticide

The water solubility and partition coefficient of a pesticide will influence movement of the pesticide in soil and determine how the pesticide is best applied via drip chemigation. For example, because of the low water solubility and low partition coefficients of imidacloprid, clothianidin, and chlorantraniliprole insecticides, they have little mobility in soil. Thus, when applying a compound such as chlorantraniliprole through a drip irrigation system, it should be injected at the beginning of the irrigation event followed by a long period of irrigation to push the compound into the root zone (DuPont 2008). In contrast, insecticides with a higher water solubility and high mobility in soil (e.g., oxamyl, thiamethoxam) should be applied near the middle third of the irrigation cycle to avoid leaching the compound out of the effective root zone (DuPont 2008; Syngenta 2009). Currently, many pesticide labels that permit drip chemigation recommend an optimal injection time during the irrigation cycle for that particular pesticide.

9.7.4 Type of Drip/Trickle Irrigation System

The type of drip tape used can influence the performance of a drip chemigation treatment. The flow rate and emitter spacing on the drip tape, coupled with the crop irrigation requirements, largely will determine irrigation frequency and how the pesticide is delivered to the root system. Ideally, emitter spacing should match up with plant spacing whenever possible to maximize delivery to plant roots. Drip tape placement relative to the roots may also influence application timing. Drip application of pesticides with low soil mobility (e.g., chlorantraniliprole) in subsurface drip tape buried 12.7 cm below the plants should be applied after the root system has become established near the tape. Drip chemigation in crops where the drip tape is placed on the soil surface at the base of plants is more flexible.

Another important factor for optimal pesticide activity with soil systemic compounds is to ensure that the correct amount of irrigation water is applied through the drip system during and after the pesticide injection. Insufficient amounts of water can prevent the pesticide from being adequately available to the roots for uptake, and, conversely, too much water may result in the compound being moved or leached out of the root zone. Either of these events can delay pesticide uptake by the plant and reduce efficacy. It is critical that the water–pesticide mixture be applied as uniformly as possible across the entire field during drip chemigation. Depending on the irrigation system, the amount of time to uniformly distribute the pesticide will depend on type of drip tape used, the pesticide pumping system, the length of time of the pesticide injection, and the length of time the irrigation system is allowed to operate after the injection. Ghidiu (2012) discusses and lists the benefits and disadvantages of various drip chemigation pumping equipment. Growers should be aware of these factors before making pesticide applications through drip irrigation systems.

9.8 Conclusions

The precision farming and high-tech horticulture for the improved input use efficiency, higher yield, and quality produce in sustainable manner is incomplete without efficient irrigation and pesticide application techniques. Chemigation offers growers a sound option in place of traditional foliar sprays of pesticides for control of specific pests of vegetables produced using a drip/trickle irrigation system. Use of the drip/trickle irrigation system for application of pesticides allows for precise placement of systemic pesticides into the root zone of vegetable crops, eliminating the need for multiple foliar sprays of pesticides. Many growers currently use drip irrigation systems for water management, and the addition of an agrichemical injection system is a cost-effective method of pesticide application. It enables growers to apply a pesticide under virtually any weather condition for control of a wide range of pests, including aphids, whiteflies, leafhoppers, leaf miners, beetles, caterpillars, and others while at the same time reducing the total pesticide inputs as compared with foliar sprays. The overall benefits of using chemigation include less application labor, less energy inputs, less time needed for application, less pesticide inputs, less worker and applicator exposure to the pesticide, less potential of soilborne disease problems, a more even distribution of the pesticide, and less soil compaction. It suits a pest management program well because many of the new-chemistry pesticides labeled for drip/trickle irrigation system application are selective to specific pests and, because they are applied to the plant root zone, are generally less toxic to beneficial and nontarget organisms. And in an urban state such as New Jersey, where urban populations border rural populations, chemigation can be conducted with no spray drift or misapplications, eliminating the ever-important "application visibility" that concerns both growers and the public.

References

Certis USA (1999) Chemigation Bulletin. Certis USA, LLC, Columbia

Chalfant RB, Young JR (1984) Management of insect pests of broccoli, cowpeas, spinach, tomatoes and peanuts with chemigation by insecticides in oils, and reduction of watermelon virus 2 by chemigated oil. J Econ Entomol 77:1323–1326

Dasberg S, Or D (1999) Drip irrigation. Springer-Verlag Berlin and Heidelberg GmbH & Co., New York, p 171

DuPont (2008) Drip chemigation: best management practices, DuPont Crop Protection Bulletin K-14594. EI du Pont de Nemours and Company, Wilmington

Felsot AS, Cone W, Yu J, Ruppert JR (1998) Distribution of imidacloprid in soil following subsurface drip chemigation. Bull Environ Contam Toxicol 60:363–370

Ghidiu GM (1981) Vydate injected through a trickle irrigation system to control Mexican bean beetle in pole lima beans. Rutgers Res Dev Cent/NJAES Rep 2:1

Ghidiu GM (2012) Insectigation in vegetable crops: the application of insecticides through a drip or trickle irrigation system. In: Larramendy ML, Soloneski S (eds) Integrated pest management and pest control: current and future tactics. InTech Press, Rijeka, pp 173–190

Ghidiu GM, Smith NL (1980) Trickle irrigation system injected insecticides to control the European corn borer in bell pepper. Results of pest control studies, Rutgers University Cooperative Extension Service Publication report 1:5–6

Ghidiu GM, Ward DL, Rogers GS (2009) Control of European corn borer in bell peppers with chlorantraniliprole applied through a drip irrigation system. Int J Veg Sci 15:93–201

Hall BJ (1982) Row crop fertigation. Am Veg Grow 2:73

Johnson AW, Young JR, Threadgill ED, Dowler CC, Sumner DR (1987) Chemigation's strong future. Agrichemical Age 30:8–9

Kuhar TP, Speese J (2002) Evaluation of drip line injected and foliar insecticides for controlling cucumber fly in melons, 2001. Entomol Soc Am Arthrop Manag Tests 27:E46

Kuhar TP, Doughty H, Hitchner E, Cassell M (2008) Evaluation of insecticide treatments for the control of lepidopteran pests in bell peppers in Virginia, 2007. Arthropod Manage Tests 33:7

Kuhar TP, Doughty HB, Cassell M, Wallingford A, Andrews H (2009) Control of Lepidopteran larvae in fall tomatoes through drip irrigation systems. In: Arthropod pest management research on vegetables in Virginia for 2009. VPI&SU Eastern Shore AREC report #308:7–28

Kuhar TP, Walgenbach JF, Doughty HB (2010) Control of *Helicoverpa zea* in tomatoes with chlorantraniliprole applied through drip chemigation. Online. Plant Health Prog. doi:10.1094/PHP-2009-0407- 01-RS

Kuhar TP, Schultz P, Doughty H, Wimer A, Wallingford A, Andrews H, Philips C, Cassell M, Jenrette J (2011a) Evaluation of foliar insecticides for the control of green peach aphids in broccoli in Virginia, 2010. Arthropod Manage Tests 36:E4

Kuhar TP, Schultz P, Doughty H, Wimer A, Wallingford A, Andrews H, Philips C, Cassell M, Jenrette J (2011b) Evaluation of soil and foliar insecticides for the control of lepidopteran larvae in bell pepper in Virginia, 2010. Arthropod Manage Tests 36:E55

Lahm GP, Selby TP, Freudenberger JH, Stevenson TM, Myers BJ, Seburyamo G, Smith BK, Flexner L, Clark CE, Cordova D (2005) Insecticidal anthranilic diamides: a new class of potent ryanodine receptor activators. Bioorg Med Chem Lett 15:4898–4906

Lahm GP, Stevenson TM, Selby TP, Freudenberger JH, Cordova D, Flexner L, Bellin CA, Dubas CM, Smith BK, Hughes KA et al (2007) Rynaxypyr: a new insecticidal anthranilic diamide that acts as a potent and selective ryanodine receptor activator. Bioorg Med Chem Lett 17:274–6279

Larsen R (1982) Local need for pestigation will continue to be the rule. Irrig Age 1982:6

Maienfisch P, Angst M, Brandl F, Fischer W, Hofer D, Kayser H, Kobel W, Rindlisbacher A, Senn R, Steinemann A et al (2001) Chemistry and biology of thiamethoxam: a second generation neonicotinoid. Pest Manag Sci 57:906–913

Owens JC (1981) Insectigation proved effective against southwestern corn borer. Irrig Age 68:7–58

Palumbo JC (1997) Evaluation of aphid control in lettuce with Admire applied through drip irrigation. Arthropod Manage Tests 22:61

Palumbo JC (2008) Systemic efficacy of Coragen applied through drip irrigation on romaine lettuce, Fall 2007. Arthropod Manage Tests 33:24

Palumbo JC (2011) Cross-spectrum insect control with novel insecticides in cantaloupes 2010. Arthropod Manage Tests 36:22

Reed DK, Reed GL, Creighton CS (1986) Introduction of entomogenous nematodes into trickle irrigation systems to control striped cucumber beetle (Coleoptera: Chrysomelidae). J Econ Ent 79:330–1333

Ristaino JB, Johnston SA (1999) Ecologically based approaches to management of Phytopthora blight on bell peppers. Plant Dis 83:1080–1089

Ross DS (2004) Drip irrigation and water management. In: Lamont WJ Jr (ed) Production of vegetables, strawberries, and cut flowers using plasticulture. Natural Resource, Agriculture and Engineering Service Cooperative Extension Publication NRAES-133, Ithaca, pp 14–35

Schuster DJ, Shurtleff A, Kalb S (2009) Management of armyworms and leaf miners on fresh market tomatoes, Fall 2007. Arthropod Manage Tests 34:E79

Syngenta (2009) Best use guidelines for drip application of crop protection products, Syngenta Bulletin G&S 409.6026. Syngenta Crop Protection Inc, Greensboro

Wildman TE, Cone WW (1986) Drip chemigation of asparagus with disulfoton: *Brachycorynella asparagi* (Homoptera: Aphididae) control and disulfoton degradation. J Econ Ent 76:1617–1620

Selective Pesticides in IPM

10

Abstract

For control of pest and diseases, the pesticides still act as first line of defense. The use of pesticides will remain, at least in the short to medium term, an important strategy, allowing the grower to continue to produce economically a quality crop. Guided chemical control aims to reduce pesticide use by determining whether a control for a certain pest is necessary and, if so, when it should be applied (for optimum effect). In guided control, chemical control is only deemed necessary when the economic benefits counterbalance the costs; it implies the use of curative rather than preventive pesticides. It is essential that the use of chemicals in IPM be based on economic threshold and only after considering the impact of the factors regulating the populations of pests and pathogens, making sure that there are no other effective management tools. Importance of selective pesticides in IPM programs, effects of chemical pesticides on beneficial organisms, influence of pesticide application on the selectivity of a pesticide, and pesticide resistance and anti-resistance strategies in IPM are discussed.

Keywords

Guided chemical control • Pesticides • IPM • Pests • Diseases

10.1 Importance of Selective Pesticides in IPM Programs

The success of released or naturally occurring biological control agents in preventing pest outbreaks in protected crops has led the greenhouse industry to be particularly conscious of the necessity of applying selective pesticides. The activity of a selective pesticide is confined to a narrow range of specific pests (Heitefub 1975). In IPM, the process of developing the selectivity of a pesticide aims to maximize its specific effect against pests and diseases and minimize its effect on nontarget organisms (Hull and Beers 1985). Thus, the selectivity of a pesticide is often used to express its harmlessness for beneficial organisms. The selectivity of the action and of the toxicity of a pesticide is dependent on its physiological

P.P. Reddy, *Sustainable Crop Protection under Protected Cultivation*,
DOI 10.1007/978-981-287-952-3_10

selectivity and/or on the application procedures (Poehling 1989). Physiological selectivity is expressed by reduced sensitivity of an organism to the pesticide due to pesticide metabolism and to the availability of the appropriate enzymes in the target organisms (Hassall 1982). Application procedures comprise the dose rate, mode of action, method, and timing.

The use of chemical pesticides that cause undesired side effects on nontarget beneficial organisms may lead to pest outbreaks. In tomatoes, multiple application of the broad-spectrum carbamate (methomyl) for the control of leaf miner infestation (*Liriomyza sativae*) eliminated the naturally occurring beneficial parasitoid complex, which, without chemical treatment, reduced the pest population to 50 % of the level found in pesticide-treated plots (Oatman and Kennedy 1976). To avoid these consequences, the harmful effects of pesticides on the natural enemies of target pests must be avoided or minimized for successful implementation of biological control agents within IPM strategies. Some pests and pathogens have developed resistance toward certain chemical pesticides, and this must also be considered in order to prevent misuse of pesticides.

The selectivity of pesticides in relation to their effects on beneficial organisms that can be used in greenhouses, the potential for improving applications for better performance and selectivity, and the problems of resistance of the pests or diseases to the chemicals used in greenhouses need to be worked out.

10.2 Effects of Chemical Pesticides on Beneficial Organisms

Pesticides can exhibit primary or secondary effects on predators, parasitoids, and pathogens of target pests. Primary effects are direct or indirect, depending on their exposure and on the biological parameter influenced. Direct mortality of beneficial organisms may be caused by direct contact during application, pesticide residues, taking up contaminated prey, intoxication by fumigants, and contact or contamination with soil disinfectants.

Indirect or sublethal effects on beneficial arthropods include decreases in reproduction, oviposition, parasitization, predation, longevity, and egg viability and a delay in the development and shifting of the sex ratio. Morphological and behavioral changes may also occur (Elzen 1989).

Secondary effects due to pesticides include killing the prey/host of a beneficial organism or of species which produce alternative food like honeydew (Huffaker 1990), taking up contaminated food (Celli et al. 1997), and directly stimulating the pest; for example, some pyrethroids enhance reproduction in *Tetranychus urticae*.

Pesticides directly affect entomopathogenic fungal biocontrol agents by inhibition of spore germination and vegetative development (mycelial growth), and they also reduce the viability of conidia (McCoy et al. 1988) and their survival and activity on plant surfaces. Viability and infectivity of the infective juveniles (J_3) of entomopathogenic nematodes are also adversely affected (Rovesti et al. 1988).

Side effects of pesticides on natural enemies may vary between and within taxonomic groups. From their comprehensive data on the side effects of pesticides, Theiling and Croft (1988) concluded that predators were more tolerant to pesticide treatment than parasitoids, except for fungicides, toward which susceptibility was not greatly affected. The tolerance of aphid natural enemies decreases from Coccinellids > Chrysopids > Syrphids > Hemiptera > Hymenoptera (Hodek 1973). Evaluation of effects within taxonomic groups revealed that the classification of the effects of 74 compounds tested against the parasitoids *Encarsia formosa*, *Aphidius matricariae*, and *Leptomastix dactylopii* corresponded by more than 78 % (Hassan et al. 1994). In a comparison of trial results with 81 test compounds for predatory mite species occurring in orchards and vineyards with *Phytoseiulus persimilis*, the same level was reached in 64 % of the test compounds.

Differences in susceptibility have been recorded between taxonomically close species and even between strains within the same species.

Eretmocerus mundus adults were less susceptible to residues of amitraz, thiodicarb, and cypermethrin than *E. formosa* or *Encarsia pergandiella* (Jones et al. 1995). Among *Aphidius* species, *A. matricariae* was more tolerant to dimethoate than *A. rhopalosiphi* or *A. colemani* (Maise et al. 1997). *Paecilomyces farinosus* was more tolerant than *Verticillium lecanii*, *Beauveria bassiana*, and *Metarhizium anisopliae* (Baath 1991). The entomopathogenic nematodes *Steinernema carpocapsae*, *S. feltiae*, and *Heterorhabditis* HP88 exhibited different tolerance levels to nine tested pesticides (Zimmerman and Cranshaw 1990). Repeated exposure of local strains to chemicals may cause natural enemies to develop tolerance to pesticides. This is the case of *P. persimilis* and organophosphorus compounds (OPs) (Goodwin and Wellham 1992) and of *Aphidoletes aphidimyza* and azinphos-methyl (Warner and Croft 1982). Developmental stage may greatly influence the response of natural enemies to pesticides. The susceptibility of *A. aphidimyza* and *Chrysoperla carnea* to pesticides with contact mode of action increased from the egg stage to the adults (Bartlett 1964). In contrast, pesticide susceptibility was lowest in treated adults of *Coccinella septempunctata* (Zeleny et al. 1988) and in eggs of *P. persimilis*, while in the coccinellid the egg stage and in the predatory mite the larvae or protonymph stage were the least tolerant (Blümel and Stolz 1993). However, compounds with modes of action that regulate or inhibit insect growth resulted in high mortality of *Chrysoperla carnea* larvae but not of the adults, whose fertility was only slightly affected (Vogt 1992).

The host may offer parasitoids different degrees of protection against pesticides; unprotected stages of parasitoids (e.g., adult Hymenoptera) and protected stages (e.g., different developmental stages in aphid mummies) show different levels of mortality after the same pesticide treatment. Avermectin B killed 50 % of *E. formosa* protected in the whitefly scales in a direct contact test but 79 % of the adult wasps after contact with the dried residue (Zchori-Fein et al. 1994). *Leptomastix dactylopii* protected in *Planococcus citri* were barely affected by topical treatment of endosulfan, while the adults were severely damaged in residual tests (Reddy and Bhat 1993). Even sexes of the same species may present different susceptibility against pesticides. In five different populations of *Diglyphus begini* (Rathman et al. 1992) and in predatory mites, males are less tolerant than females.

Generally herbicides, acaricides, and fungicides have less effect than insecticides, although mycopesticides are highly susceptible to fungicides.

10.2.1 Effect on Beneficial Predators

For predatory mites most pyrethroids and carbamates were harmful, both in initial toxicity and in reproduction and persistence trials with the susceptible juvenile predators. *Aphidoletes aphidimyza* showed a similar susceptibility to insecticide/acaricide treatments and was also affected by OPs. OPs caused varying levels of mortality in predatory mites. In coccinellids, high mortality rates were caused by nearly all tested compound groups, except the microorganisms and soap. Chrysopids were not harmed by acaricides, most pyrethroids, soap, or microorganisms, but were affected by most of the insect growth regulators (IGRs) and most of the OPs. For predatory bugs, pyrethroids, carbamates, most OPs, and few of the IGRs proved to be harmful. Fungicides and herbicides were relatively harmless for coccinellids, chrysopids, and predatory bugs, but partly harmful to predatory mites.

10.2.2 Effects on Beneficial Parasitoids

Synthetic pyrethroids and pyrethrin were very harmful to adults, regardless of the test species. In tests with the protected stages, several pyrethroids were only slightly harmful, but in combination with persistence of more than 1 week, this advantage was neutralized. OPs were very harmful to the unprotected stages and with few exceptions also to the protected life stages and showed high persistence as residues. Carbamates were harmful in both types of laboratory tests, but some had persistence shorter than 3 days. IGRs and most of the acaricides were harmless to both the susceptible and the protected developmental stage of the parasitoids. Plant extracts (except pyrethrin), soap, and microorganisms were harmless. Fungicides belonging mainly to the group with a broad-spectrum and protective mode of action were harmful to adult parasitoids and revealed detrimental effects which persisted over 1 week. In tests with the protected life stage, however, all fungicides were considered harmless. Very few herbicides were harmful to adult wasps, but not for other developmental stages.

10.2.3 Entomopathogens

Only a small number of carbamates out of the tested insecticides/acaricides affected entomopathogenic nematodes, while fungicides proved to be mainly harmless. Insecticides, acaricides, and herbicides in most cases did not adversely influence the mycelial growth or the sporulation of the fungal species *V. lecanii*, *B. bassiana*, and *M. anisopliae* in laboratory tests or during infectivity tests in the greenhouse. Half of the fungicides examined in all types of tests affected at least one of the three test fungi, whereas one fourth of the fungicides were harmless for all of them. Effects could not generally be attributed to the mode of action of the fungicides. *Verticillium lecanii* was slightly more affected than *B. bassiana.*

10.2.4 Sublethal Effects on Natural Enemies

Besides direct toxicity caused by a number of "classical" insecticides, sublethal effects were also demonstrated in several investigations. Among sublethal effects of pesticide application on natural enemies which are reported in the literature are development prolongation, reduced egg production or its total inhibition, decrease in prey consumption, changes in searching or foraging behavior, alteration of pathogenicity in entomopathogens, and increased tendency to escape from treated surfaces. The importance of repellence of pesticide compounds for beneficials is difficult to classify. On the one hand, repellence may negatively influence natural enemies by expelling them from their host or prey which they need for further population development; on the other hand, beneficials can be protected from possibly hazardous contact with contaminated plant surfaces or prey/hosts. Both effects are undesirable, especially in greenhouses, where mass-reared arthropods are intentionally introduced as biological control agents and because the natural enemies would cease to be effective as control agents, particularly when untreated refuges are scarce.

Insect growth regulators, like diflubenzuron, chlorfluazuron, fenoxycarb, flufenoxuron, and teflubenzuron, which are incorrectly considered as harmless to many beneficials, in fact interfere with the viability of eggs, the molting process, and the reproduction of several predators.

The influence of different formulations of pesticides on their effects on natural enemies was shown for endosulfan, which as an emulsifiable concentrate (EC) formulation resulted in up to 17 % less mortality of *P. persimilis* than the wettable powder (WP) formulation in a residual laboratory test (Blümel et al. 1993). For *E. formosa* the EC formulation of tebufenpyrad was more toxic than the WP formulation (van de Veire 1995).

10.3 Influence of Pesticide Application on the Selectivity of a Pesticide

The relatively small areas in greenhouses – compared to arable agriculture – and high plant density dictate in many cases the use of manually operated spraying equipment. In an enclosed structure, good ambiental conditions can exist for applying very small particles and using artificial air movement to improve pesticide distribution and pest control. Conversely, improved chemical control can adversely affect bioagents such as bumblebees, antagonistic fungi, and beneficial arthropods, factor which has to be considered when choosing a pesticide. Pesticide application in enclosed areas also imposes the risk of breathing air that contains small particles of pesticides. Personal protective clothing is often hot and uncomfortable, and farmers tend to spray unprotected.

Unfortunately, many growers continue to use high-volume (HV) spraying (>1000 l/ha of spraying solution). HV spraying to run off leads to wastage to the order of 70–90 % of the chemical dripping to the ground (Matthews 1992). The low concentration of a.i. with HV applications reduces the hazard to the operator, who is often heavily contaminated by the pesticide, but may not give adequate control, and growers are thus forced to repeat sprays at frequent intervals. The whole area becomes contaminated with pesticides, making it impossible to integrate biological control with chemicals. The volume of spray and wastage due to runoff can be reduced significantly by changing nozzles to produce small droplets which do not coalesce on the target (Matthews 1992). A widely used piece of equipment is the knapsack mist blower.

As an alternative to HV spraying, the use of thermal or cold foggers gives the grower clear savings in time and labor, although they are only suitable in totally enclosed greenhouses. Deposition is improved with cold fogging, but persistence is less. The shorter persistence obtained with cold foggers allows the introduction of natural enemies quicker after treatment than when a thermal fogger is used, and a greenhouse can be treated when parasitoids are protected inside the infested host stages (Lingappa et al. 1972). Additionally, cold fogging allows the use of a wider range of pesticides, e.g., insecticides perhaps with higher selectivity, such as *Bacillus thuringiensis* which has been used successfully by cold fogging.

Another technique, vaporization, is suitable for small areas (approximately 100 m^2). The pesticide (e.g., sulfur) is placed on a small heater installed inside a wide pipe. After evaporation or sublimation, the pesticide condenses to small particles (e.g., 2–8 μm) and is carried up by the heated air directed by the pipe. The dispersion and settling of particles of this size is influenced by the inside air circulation systems, and they fall mainly on the upper side of the leaves, rendering minimal residual effect.

Alternatives to spray treatments include application of granules or drenches and chemigation by drip irrigation to the soil, when systemic pesticides can be used. Specific treatments can be combined with a pesticide or other types of lure, e.g., yellow cards in a "lure and kill" method. Thrips have been controlled with a polybutene sticky surface combined with an insecticide (Thripstick). Specific baits cause only minimal damage to nontarget organisms, as their chance of exposure is very low.

The timing of the pesticide treatment is crucial in order to avoid the susceptible life stage of the nontarget organism. Where chemical pesticides adversely affect the entomopathogenic fungus *V. lecanii*, they should not be applied at the same time, but after a delay (Schuler 1991). Similarly, the alternation of chemical fungicides with the fungal biocontrol agent *Trichoderma harzianum* T39 is preferred to the use of a tank mix of this biocontrol agent with chemicals for the control of foliar pathogens (Shtienberg and Elad 1997). Selective application can also be carried out by considering spatial factors and using the systemic pesticides as granules or seed treatment to preserve plant-inhabiting beneficials. Limited areas can be treated with handheld air-assisted spinning disk sprayers. Multiple applications of a pesticide may cause a severe reduction in the number of natural enemies, without achieving a satisfactory control of the target pest. In contrast, a single, better timed application of the same pesticide can control the pest to the same extent, without seriously damaging the natural enemies, thus improving overall control. Keeping the pest below the economic threshold has been achieved with different use of oxamyl and methamidophos against *L. sativae* and its parasitoid complex in tomatoes (Schuster et al. 1979).

Systemic fungicides, which were harmful to *V. lecanii* when applied as sprays, did not affect the fungus pathogenicity against *Aphis gossypii* on cucumber when applied as a soil drench (Wilding 1972). Another possibility for the partial preservation of natural enemies is the treatment of selected strata of the plants, e.g., flowers, and leaving the lower part of the canopy untreated, thus maintaining a significant population of natural enemies (Scopes and Biggerstaff 1974). These localized treatments are gaining acceptance where insects are used to pollinate crops and growers release natural enemies such as *E. formosa.* In one study, the application of pyriproxyfen to the upper parts of tomato plants infested with greenhouse whitefly effectively reduced the pest, but did not damage the parasitoid *E. formosa*, which, though susceptible to this compound, was situated in the whitefly pupae on the lower parts of the plants (van de Veire 1995).

10.4 Pesticide Resistance and Anti-resistance Strategies in IPM

Pests and pathogens may overcome the toxic effect of pesticides by metabolizing the active ingredient into less toxic components, developing a change in the target site, reducing the absorption of the chemical, or avoiding exposure to the compound. Resistance development is the most severe challenge to pesticide. In greenhouses, pesticide-resistant strains of fungi and pests have appeared frequently. This phenomenon occurs because the greenhouse is a closed system in which the population of selected strains is not diluted by the outdoor wild population. Usually, the existence of epidemic conditions in greenhouses is a prerequisite for the development of resistant populations of pathogens and pests. Moreover, the optimal conditions for their development in greenhouses prevail for long periods. The number of life cycles is increased due to the optimal conditions or the extended time they prevail, and control necessitates frequent pesticide applications which results in high selection pressure toward resistance to pesticides. The main pathogens which are known to develop resistance to fungicides in greenhouses are *Botrytis cinerea*, *Pseudoperonospora cubensis* (downy mildew of cucurbits), *Didymella bryoniae* (gummy stem blight of cucurbits), *Sphaerotheca fusca* (*Sphaerotheca fuliginea*) (powdery mildew of cucurbits), *Puccinia horiana*, *Uromyces dianthi* (*Uromyces caryophyllinus*), and *Fusarium oxysporum* f sp *gladioli.*

The benzimidazole fungicides (benomyl, carbendazim, thiophanates) have a high resistance potential against pathogens because they have a specific mode of action. The resistance is usually not associated with a significant loss of fitness of the pathogen. It occurs in populations of *B. cinerea*, *D. bryoniae*, *Fusarium*, and powdery mildews. Mixtures and alternations with multisite contact fungicides may delay this selection, before resistance becomes apparent.

Acute problems of resistance to dicarboximide fungicides (e.g., iprodione, procymidone, vinclozolin) have arisen when fungicides are used intensively and exclusively over many seasons (Gullino et al. 1989). Isolates are moderately resistant and tend to be almost as fit as sensitive strains in the absence of fungicides. It is recommended to restrict the number of dicarboximide treatments to not more than three per crop in greenhouses where resistance is found and even in the absence of detectable resistant strains. When infection pressure is high, it is usually recommended to alternate or mix these fungicides with protectants such as chlorothalonil, captan, and TMTD or with biocontrol which do not usually select for resistance. However, TMTD may interfere with natural enemies.

Ergosterol biosynthesis inhibitors (EBIs) are a group of fungicides which include triazole, imidazole, and pyrimidine fungicides which inhibit C14 demethylation and morpholines. Unlike the sharp, significant nature of resistance toward benzimidazoles and dicarboximides mentioned above, the resistance toward EBIs develops in the form of slow shifts in the pathogen population. For instance, powdery mildews in greenhouses were controlled for several years by benzimidazoles, hydroxypyrimidines, pyrazophos, and EBIs. Resistance is known in populations of *S. fusca*, but the alternation of fungicides, which is practiced in many countries, is helping to deal with the problem. It is generally recommended to rotate or mix EBI fungicides with fungicides from other groups as well as with biocontrol.

The failure of disease control in greenhouses is exemplified by the history of gray mold epidemics. Multiple resistant isolates occur in greenhouses that bear the resistance toward benzimidazole, diethofencarb, dicarboximides, and ergosterol biosynthesis inhibitors (Elad et al. 1992). The extreme summer conditions do not interfere with the survival of fungicide-resistant isolates (Yunis and Elad 1989).

Phenylamide fungicides that inhibit RNA synthesis were introduced in the late 1970s for Phycomycetes control. During the 1970s, *P. cubensis* was controlled mainly with protective applications of dithiocarbamates and chlorothalonil. In the early 1980s the phenylamide metalaxyl was released and soon afterward resistant strains were selected. Metalaxyl-resistant strains seem to be more competitive than wild-type strains (Cohen et al. 1983). Resistance was found also in *Phytophthora infestans* on tomato and *Bremia lactucae* on lettuce. Anti-resistance mixtures of metalaxyl with protectant fungicides were developed to cope with phenylamide resistance. In order to reduce the pressure toward the development of resistance in pathogen populations, it is usually better to limit the exposure of the pathogen to a group of fungicides. The number of applications of fungicides of the same mode of action has to be limited, especially against fungi with many cycles during the growing season. Moreover, the application of nonchemical methods is also recommended.

Insecticide and acaricide resistance of nearly all important arthropod greenhouse pests is well documented (Georghiou and Mellon 1983). Besides genetic and operational factors that influence the selection of resistant individuals, biotic reasons such as generation turnover, number of offspring per generation, and type of reproduction have a major impact on resistance development. Most of the pest species on greenhouse crops favor resistance selection with regard to these biological parameters.

Recently *Bemisia tabaci* and *B. argentifolii* have developed resistance against a range of conventional insecticides as well as against IGRs and juvenile hormone analogues (Cahill et al. 1994; Horowitz el al. 1994), and *Frankliniella occidentalis* developed resistance against most pesticide groups (Anonymous 1988), resulting in severe economic losses in the affected crops. Pesticide resistance can also develop in natural enemies and has been found in all taxonomic groups (Croft and Strickler 1983). The differences in the occurrence and the level of pesticide resistance in predators and parasitoids can be explained by the influence of the factors such as food limitation and differential susceptibility to the chemical.

Chemical resistance management strategies for pests comprise different approaches classified as management by moderation (low dosages, reduced number of applications), management by saturation (suppressing detoxification), and management by multiple attack (application of mixtures) (Georghiou 1983). For IPM programs, additionally nontarget effects on natural enemies have to be considered, which might not always correspond with the aforementioned strategies.

Chemical control of insect pests under protected cultivation using selected pesticides is presented in Table 10.1.

10.5 Future Prospects and Conclusions

Modern techniques used in greenhouses for pesticide application allow a low input of chemicals while achieving good coverage of the right part of the plant. Selective application can also direct the active ingredient to the right target, with lowered effect on beneficial organisms. However, it is important to know the undesired side effects of chemical use in greenhouses.

Uniform labeling of the nontarget effects of plant protection products during the process of authorization as proposed in the European Plant Protection Legislation (EU-Directive 414/91, including all annexes) is desirable. In view of this joint initiative by governmental research centers and industry, commercial test laboratories, and contributions from the European and Mediterranean Plant Protection Organization (EPPO), a harmonized labeling of plant protection products concerning the nontarget effects is expected.

Table 10.1 Chemical control of insect pests in protected cultivation

Major insect pest	Crop affected	Symptoms	Management
Cutworm, *Agrotis* spp.	Nursery of tomato and capsicum	Larvae cut young plants at ground level	Spray the crop with chlorpyrifos 20 EC at 2 ml/L
Aphid	Tomato, capsicum, cucumber, cole crops	Nymph and adult suck sap from plants. Cause stunted growth with curled, twisted leaves and deformed fruits. Secrete honey dew which allows the growth of black sooty molds	Neem oil, 5 ml/L; imidacloprid 17.8SL, 0.4 ml/L; quinalphos 25 EC, 2 ml/L
Whitefly, *Bemisia tabaci*	Capsicum, chillies, cabbage, cucumber	Nymph and adults suck sap from leaves, and vitality of plant is lower through loss of cell sap. Secrete honey dew which attracts fungus (sooty mold in capsicum). Transmit virus diseases in tomato, capsicum, and cucurbits	Imidacloprid 17.8SL, 0.4 ml/L; quinalphos 25 EC, 2 ml/L
Cabbage butterfly, *Pieris brassicae*	Cole crops	First instar larvae scrape leaf surface; subsequent instars eat up leaves from margin inward, leaving intact the main veins	Quinalphos 25 EC, 2 ml/L
Leaf miner	Tomato, capsicum, cucumber	Larvae mine into tender leaves in zigzag manner and scrape chlorophyll. Photosynthesis is adversely affected, vitality is reduced, and there is appreciable reduction in yield	Methyldemeton 25 EC, 1 ml/L
Thrips	Capsicum, cucumber, cole crops	Affected leaves become curled and wrinkled and gradually dry up. Feeding causes flower or leaf buds to abort or emerging leaves to become distorted	Neem oil, 5 ml/L; imidacloprid 17.8SL, 0.4 ml/L; quinalphos 25 EC, 2 ml/L
Mites	Tomato, capsicum, cucumber	Small brown or yellow dots on leaves are most easily observed sign of mite infestation. Severely infested plant leaves become discolored, producing an unthrifty gray or bronze look to plant	Dicofol, 2.5 ml/L; penconazole, 0.4 ml/L; neem oil, 5 ml/L; wettable sulfur, 2 g/L

Other topics for the implementation of side-effect data into IPM practice still need to be addressed. Most of the data about side effects of pesticides on beneficials is derived from laboratory tests or even higher test levels with only one application of the product. However, in practice, even when natural enemies are used against arthropod pests, chemical treatment can be necessary against fungal diseases. Often these fungicides have to be applied not once but several times at certain intervals. These applications can lead to an accumulation of the product on the plants, affecting the beneficial organisms. This situation becomes more complicated when mixtures of different active ingredients are used.

Very few chemical pesticides are selective for natural enemies. Improvements in the compatibility of beneficial organisms with pesticide application by selecting beneficial with some resistance toward chemical pesticides have been attempted, but this is often a cumbersome procedure as the pesticides used may change quickly.

Besides the degree of resistance, its stability and its possible influence on the fitness of the tolerant organisms are features that must be assessed before the selected organisms can be used in pest or disease control. For phytoseiids development of pesticide resistance against several insecticide groups, acaricides and fungicides, and even against sulfur has been extensively described (Croft and van de Baan 1988). Alternatively, pesticides are applied spatially to selected areas or in frequencies which reduce the target pest to a sufficient extent, but minimize harm to natural enemies and thus allow a combination or synergized effect of both the chemical and the biological controls (Zhang and Sanderson 1990).

Another important topic in the assessment of side effects is examining whether natural pesticides or natural enemies themselves affect beneficial organisms, as reported in studies of the impact of entomopathogenic nematodes on non-target organisms (Bathon 1996). Fransen and van Lenteren (1993) could not find detrimental effects of the entomopathogenic fungi *Aschersonia aleyrodis* on the parasitoid *E. formosa*, while Sterk et al. (1995) observed no effect of a commercial strain of *Paecilomyces fumosoroseus* on *P. persimilis*, *E. formosa*, and *Onus insidiosus*. However, Pavlyushin (1996) detected direct and sublethal effects of entomopathogenic fungi on chrysopids in the laboratory.

The present status of resistance of pests or pathogens in greenhouses is often unknown; growers tend to apply excess amounts of chemical, and control is not achieved. The development of tools for monitoring resistance should facilitate the assessment of different management options.

Research efforts are needed for developing disease management technology under protected environment with emphasis on avoidance and selective use of pesticides. Safe waiting intervals based on harvest time pesticide residues needed to be established for the crops under protected environment as this information is lacking completely. Apart from this, emphasis to improve the awareness level of the growers for timely diagnosis and judicious use of pesticides needs to be taken up on priority.

References

Anonymous (1988) Meeting-report Adhoc panel-meeting on *Frankliniella occidentalis* – its biology and control. EPPO Panel Information, Paris

Baath E (1991) Tolerance of copper by entomogenous fungi and the use of copper-amended media for isolation of entomogenous fungi from soil. Mycol Res 95(9):1140–1142

Bartlett BR (1964) Toxicity of some pesticides to eggs, larvae and adults of the green lacewing, *Chrysopa carnea*. J Econ Entomol 57:366–369

Bathon H (1996) Impact of entomopathogenic nematodes on non-target hosts. Biocontrol Sci Technol 6:421–434

Blümel S, Stolz M (1993) Investigations on the effect of insect growth regulators and inhibitors on the predatory mite *Phytoseiulus persimilis* AH. with particular emphasis on cyromazine. J Plant Dis Prot 100(2):150–154

Blümel S, Bakker F, Grove A (1993) Evaluation of different methods to assess the side effects of pesticides on *Phytoseiulus persimilis* AH. Exp Appl Acarol 17(3):161–169

Cahill M, Byrne FJ, Denholm I, Devonshire AL, Gorman KJ (1994) Insecticide resistance in *Bemisia tabaci*. Pestic Sci 42:135–142

Celli S, Bortolotti L, Nanni C, Porrini C, Sbrenna G (1997) Effects of the IGR fenoxycarb on eggs and larvae of *Chrysoperla carnea* (Neuroptera, Chrysopidae). Laboratory test. In: Haskell PT, McEwen PK (eds) New studies in ecotoxicology. The Welsh Pest Management Forum, Cardiff, pp 45–49

Cohen Y, Reuveni M, Samoucha Y (1983) Competition between metalaxyl-resistant and sensitive strains of *Pseudoperonospora cubensis* on cucumber plants. Phytopathology 73:1516–1520

Croft BA, Strickler K (1983) Natural enemy resistance to pesticides: documentation, characterization, theory and application. In: Georghiou GP, Saito T (eds) Pest resistance to pesticides. Plenum Press, New York, pp 669–702

Croft BA, van de Baan HE (1988) Ecological and genetic factors influencing evolution of pesticide resistance in Tetranychid and Phytoseiid mites. Exp Appl Acarol 4:277–300

Elad Y, Yunis H, Katan T (1992) Multiple fungicide resistance to benzimidazoles, dicarboximides and diethofencarb in field isolates of *Botrytis cinerea* in Israel. Plant Pathol 41:41–46

Elzen GW (1989) Sublethal effects of pesticides on beneficial parasitoids. In: Jepson P (ed) Pesticides and non-target invertebrates. Incept, Wimborne Dorset, pp 129–150

Fransen JJ, van Lenteren JC (1993) Host selection and survival of the parasitoid *Encarsia formosa* on greenhouse whitefly *Trialeurodes vaporariorum* in the presence of hosts infected with the fungus *Aschersonia aleyrodis*. Entomol Exp Appl 69:239–249

Georghiou GP (1983) Management of resistance in arthropods. In: Georghiou GP, Saito T (eds) Pest resistance to pesticides. Plenum Press, New York, pp 769–792

Georghiou GP, Mellon RB (1983) Pesticide resistance in time and space. In: Georghiou GP, Saito T (eds) Pest resistance to pesticides. Plenum Press, New York, pp 1–47

Goodwin S, Wellham TM (1992) Comparison of dimethoate and methidathion tolerance in four strains of *Phytoseiulus persimilis* (Athias-Henriot) (Acarina: Phytoseiidae) in Australia. Exp Appl Acarol 16(3):255–261

Gullino ML, Aloi C, Garibaldi A (1989) Influence of spray schedules on fungicide resistant populations of *Botrytis cinerea* Pers on grapevine. Neth J Plant Pathol 95(Suppl.1):87–94

Hassall KA (1982) The chemistry of pesticides. Verlag Chemie, Weinheim

Hassan SA, Bigler F, Bogenschütz H, Boller E, Brun J, Calis JNM, Chiverton P, Coremans-Pelseneer J, Duso C, Grove A, Helyer N, Heimbach U, Hokkanen GB, Lewis H, Mansour F, Moreth L, Polgar L, Samsoe-Petersen L, Sauphanor B, Stäubli A, Sterk G, van de Veire M, Viggiani G, Vogt H (1994) Results of the sixth joint pesticide testing programme carried out by the IOBC working group "Pesticides and Beneficial Organisms". Entomophaga 39:107–119

Heitefub R (1975) Pflanzcnschutz. Georg Thieme Verlag, Stuttgart

Hodek I (1973) Biology of Coccinellidae. Dr W Junk, The Hague [cited in Horn DJ, Wadleigh RW (1993) Resistance of aphid natural enemies to insecticides. In: Minks AK, Hanewijn P (eds) World crop pests: aphids, their biology, natural enemies and control, vol 2B. Elsevier, Amsterdam, pp 337–347]

Horowitz AR, Forer G, Ishaaya I (1994) Managing resistance in *Bemisia tabaci* in Israel with emphasis on cotton. Pestic Sci 42:113–122

Huffaker CB (1990) Effects of environmental factors on natural enemies of armoured scale insects. In: Rosen D (ed) World crop pests: armoured scale insects, their biology, natural enemies and control, vol 4B. Elseviers, Amsterdam, pp 205–220

Hull LA, Beers EH (1985) Ecological selectivity: modifying chemical control practices to preserve natural enemies. In: Hoy MA, Herzog DC (eds) Biological control in agricultural IPM systems. Academic, Orlando, pp 103–121

Jones WA, Wolfenbarger DA, Kirk AA (1995) Response of adult parasitoids of *Bemisia tabaci* (Hom: Aleyrodidae) to leaf residues of selected cotton insecticides. Entomophaga 40(2):153–162

Lingappa SS, Starks KJ, Eikenbary RD (1972) Insecticidal effect on *Lysiphlebus testaceipes,* a parasite of greenbug at all three developmental stages. Environ Entomol 1:520–521

Maise S, Candolfi MP, Neumann C, Vickus P, Mader P (1997) A species comparative study: sensitivity of *Aphidius rhopalosiphi, A. matricariae* and *A. colemani* (Hymenoptera: Aphidiidae) to Dimethoate 40EC under worst case laboratory conditions. In: Haskell PT, McEwen PK (eds) New studies in ecotoxicology. The Welsh Pest Management Forum, Cardiff, pp 45–49

Matthews GA (1992) Pesticide application methods, 2nd edn. Longman, London

McCoy CW, Samson RA, Boucias DC (1988) Entomogenous fungi. In: CM Ignoffo (ed) CRC handbook of natural pesticides, vol V, Microbial insecticides, Part A. CRC Press, Boca Raton, pp 151–237

Oatman ER, Kennedy GG (1976) Methomyl induced outbreak of *Liriomyza sativae* on tomato. J Econ Entomol 69(5):667–668

Pavlyushin VA (1996) Effect of entomopathogenic fungi on entomophagous arthropods. IOBC/WPRS Bull 19(9):247–249

Poehling HM (1989) Selective application strategies for insecticides in agricultural crops. In: Jepson PC (ed) Pesticides and non-target invertebrates. Incept, Wimborne Dorset, pp 151–176

Rathman RJ, Johnson MW, Rosenheim JA, Tabashnik BE, Purcell M (1992) Sexual differences in insecticide susceptibility and synergism with piperonyl butoxide in the leaf miner parasitoid *Diglyphus begini* (Hymenoptera: Eulophidae). J Econ Entomol 85(1):15–20

Reddy KB, Bhat PK (1993) Effect of endosulfan on the mealybug parasitoid *Leptomastix dactylopii* How. J Coffee Res 23(1):19–23

Rovesti L, Heinzpeter EW, Tagliente F, Deseo KV (1988) Compatibility of pesticides with the entomopathogenic nematode *Heterorhabditis bacteriophora* Poinar (Nematoda: Heterorhabditidae). Nematologica 34:462–476

Schuler T (1991) *Verticillium lecanii* (Zimmermann) Viegas (Hyphomycetales: Moniliaceae): Geschichte, Systematik, Verbreitung. In: Schuler T, Hommes M, Plate HP, Zimmermann G (eds) Biologie und Anwendung im Pflanzenschutz, Mitteilungen aus der Biologischen Bundesanstalt für Land- und Forstwirtschaft Berlin-Dahlem 269:118–125

Schuster DJ, Musgrave CA, Jones JP (1979) Vegetable leaf miner and parasite emergence from tomato foliage sprayed with oxamyl. J Econ Entomol 72:208–210

Scopes NEA, Biggerstaff SM (1974) Progress towards integrated pest control on year round chrysanthemums. In: Proceedings of the 7th British insecticide and fungicide conference, Brighton, 19–22 Nov 1973, BCPC, London, pp 227–234

Shtienberg D, Elad Y (1997) Incorporation of weather forecasting in integrated, biological-chemical management of *Botrytis cinerea.* Phytopathology 87:332–340

Sterk G, Bolckmans K, van de Veire M, Sels B, Stepman W (1995) Side-effects of the microbial insecticide PreFeRal (*Paecilomyces fumosoroseus,* strain Apopka 97) on different species of beneficial arthropods.

Mededelingen van de Faculteit Landbouwkundige en Toegepaste Biologische Wetenschappen, Universiteit Gent 60(3a):719–724

Theiling KM, Croft BA (1988) Pesticide side-effects on arthropod natural enemies: a data-base summary. Agric Ecosyst Environ 21:191–218

van de Veire M (1995) Integrated pest management in glasshouse tomatoes, sweet peppers and cucumbers in Belgium. PhD thesis, University of Gent

Vogt H (1992) Investigations on the side effects of insecticides and acaricides on *Chrysoperla carnea* Stph (Neuroptera, Chrysopidae). Mededelingen van de Faculteit Landbouwwetenschappen, Rijksuniversiteit Gent 57(2b):559–567

Warner LA, Croft BA (1982) Toxicity of azinphosmethyl and selected orchard aphid predator, *Aphidoletes aphidomyza*. J Econ Entomol 75(3):410–415

Wilding N (1972) The effect of systemic fungicides on the aphid pathogen *Cephalosporium aphidicola*. Plant Pathol 21:137–139

Yunis H, Elad Y (1989) Survival of dicarboximide-resistant strains of *Botrytis cinerea* in plant debris during summer in Israel. Phytoparasitica 17:13–21

Zchori-Fein E, Roush RT, Sanderson JP (1994) Potential for integration of biological and chemical control of greenhouse whitefly using *Encarsia formosa* and abamectin. Environ Entomol 23(5):1277–1282

Zeleny J, Vostrel J, Ruzicka Z, Kalushkov PK (1988) Impact of various pesticides on aphidophagous Coccinellidae. In: Niemczyk E, Dixon AFG (eds) Ecology and effectiveness of Aphidophaga. SPB Academic Publishing, The Hague, pp 327–332

Zhang ZQ, Sanderson JP (1990) Relative toxicity of abamectin to the predatory mite *Phytoseiulus persimilis* (Acari: Phytoseiidae) and two spotted spider mite (Acari: Tetranychidae). J Econ Entomol 83(5):1783–1790

Zimmerman RJ, Cranshaw WS (1990) Compatibility of three entomogenous nematodes (Rhabditia) in aqueous solutions of pesticides used in turf grass maintenance. J Econ Entomol 83(1):97–100

11 Plant Diseases and Their Management

Abstract

Some of the major diseases in greenhouse crops are caused by fungi, bacteria, viruses, and nematodes. Greenhouses are designed to protect crops from many adverse conditions, but most pathogens and several pests are impossible to exclude. Windblown spores and aerosols containing bacteria enter doorways and ventilators; soilborne pathogens enter in windblown dust and adhere to footwear and machinery. Aquatic fungi can be present in irrigation water; insects that enter the greenhouse can transmit viruses and can carry bacteria and fungi as well. Once inside a greenhouse, pathogens and pests are difficult to eradicate. Factors favorable to disease development; fungal, bacterial, and viral problems in protected cultivation; and disease management strategy are discussed.

Keywords

Fungal diseases • Bacterial diseases • Viral diseases • Disease management

11.1 Introduction

Insects and diseases are a major challenge to greenhouse production. Pest and disease control is probably the greenhouse crop practice with the greatest impact not only on the environment and human health but also on public opinion in recent years. As agriculture under protected cultivation develops, plants become more susceptible to pests for several reasons, including monoculture cultivation and the use of selected, high-yielding varieties which sometimes stimulate pest and disease development. IPM is an important tool in the management of these pests. The primary goal of IPM is to optimize pest control in an economically and ecologically sound way. IPM involves the integration of cultural, physical, biological, and chemical practices to grow crops with minimal use of pesticides.

P.P. Reddy, *Sustainable Crop Protection under Protected Cultivation*,
DOI 10.1007/978-981-287-952-3_11

11.1.1 Factors Favorable to Disease Development

Well-grown and productive crops are generally less susceptible to diseases, but in many cases compromises have to be made between optimum conditions for economic productivity and conditions for disease prevention. High host plant densities and the resulting microclimate are favorable to disease spread. Air exchange with the outside is restricted, so water vapor transpired by the plants and evaporated from warm soil tends to accumulate, creating a low vapor pressure deficit (high humidity). Therefore, the environment is generally warm, humid, and wind-free inside the greenhouse. Such an environment promotes the fast growth of most crops, but it is also ideal for the development of bacterial and fungal diseases (Jarvis 1992) and of insects vectoring viruses.

11.1.2 Problems in Protected Cultivation

Greenhouses are designed to protect crops from many adverse conditions, but most pathogens are impossible to exclude as:

- Windblown spores and aerosols containing bacteria enter doorways and ventilators.
- Soilborne pathogens enter in windblown dust and adhere to footwear and machinery.
- Aquatic fungi can be present in irrigation water.
- Insects that enter the greenhouse can transmit viruses and can carry bacteria and fungi as well.
- Once inside a greenhouse, pathogens are difficult to eradicate.

11.2 Diseases

11.2.1 Fungal Diseases

Fungal diseases constitute one of the biggest groups of foliar pathogens causing immense damage under protected environment. It was found that the incidence and severity of diseases vary considerably under protected environment when compared to open field. As observed in tomato, *Phytophthora infestans*, *Pseudocercospora fuligena*, and *Fulvia fulva* causing late blight, black leaf mold, and leaf mold, respectively, were observed to be of higher significance under polyhouse condition. Further, the notion of early and late blight was found to be obscure under polyhouse condition, since it was observed that late blight appears early in the crop season, whereas early blight appears late during crop season. This may be attributed to the fact that temperature and humidity are nearly balanced inside protected structure, even when outside field temperature is comparatively low and so on. *Alternaria alternata* is found to be a major disease affecting fruits. Capsicum is found to be infected primarily by *Colletotrichum capsici*; *Cercospora capsici*; *Pythium*, *Fusarium*, and *Phytophthora* spp; *Stemphylium solani* and *S. lycopersici*; *Sclerotium rolfsii*; *Verticillium albo-atrum* and *V. dahliae*; *Phytophthora capsici*; *Leveillula taurica*; and *Botrytis cinerea* causing anthracnose, leaf spot, damping-off, gray leaf spot, stem rot, wilt disease, *Phytophthora* blight, powdery mildew, and gray mold, respectively. Cucumber also attracts a considerable quantum of fungal pathogens. Out of which downy mildew (*Pseudoperonospora cubensis*), powdery mildew (*Erysiphe cichoracearum* and *Sphacelotheca fuliginea*), *Alternaria* leaf spot (*Alternaria cucumerina*), anthracnose (*Colletotrichum lagenarium*), and damping-off (*Pythium* spp) are important. Survival of pathogen is also enhanced inside polyhouse due to availability of host because of longer growing season.

11.2.2 Bacterial Diseases

Bacterial diseases are less frequent but under high moisture and poor irrigated conditions may cause huge damage. *Erwinia carotovora* ssp *carotovora* (bacterial soft rot), *Xanthomonas campestris* pv *vesicatoria* (bacterial spot), *Ralstonia solanacearum* (bacterial wilt), *Pseudomonas syringae* pv *lachrymans* (angular leaf spot), *Erwinia tracheiphila*, and *Ralstonia solanacearum* (bacterial wilt) are pronounced to name some.

11.2.3 Viral Diseases

Tomato, cucumber, and capsicum are very sensitive to viral diseases under protected environment. It often spreads in the plantation by insect vectors such as whitefly, thrips, and aphids. The damage caused by the virus is usually much greater than the mechanical injury caused by the insect vector. Plant tissue damaged by a viral disease does not die immediately. The most important symptom of viral infections is the light (white or yellow) color of the leaves or a mosaic pattern of light and darker shades of green on the leaves. In many cases, viral disease leads to dwarfed growth, rosette formation, or other strange stem, fruit, and leaf deformations. The symptoms of viral infections are often not found everywhere in a cultivated field but rather in patches and also sometimes without symptoms. Viruses prevalent among greenhouse crops include tobacco mosaic virus or tomato mosaic virus (TMV or ToMV), cucumber mosaic virus (CMV), tobacco etch virus (TEV), potato virus-Y (PVY), potato leaf roll virus (PLRV), tomato spotted wilt virus (TSWV), alfalfa mosaic virus, pepper veinal mottle virus (PVMV), pepper mild mottle virus (PMMY), chili veinal mottle virus (CVMV Or ChiVMV), tomato yellow leaf curl virus (TYLCV), and tomato big-bud mycoplasma (TBB).

Most viruses are spread by insect vectors, such as whitefly, aphid, and thrips; others are seedborne and mechanically transmitted (Table 11.1).

11.3 Disease Management Strategy

Proper field sanitation is one of the most important management strategies, since once the buildup of inoculum occurs inside the polyhouse, it is very difficult to manage it. So prevention is always better than cure. Use disease-resistant varieties. Reduced incidence of leaf wetness by staking plants, providing ample spacing between plants to allow for good air movement, and avoiding overhead irrigation also helps in disease control. Judicious use of chemicals with least toxicity recommended specially for polyhouse cultivation should be sprayed. Chemicals like chlorothalonil, cymoxanil, and azoxystrobin are prohibited in polyhouse-grown tomato and thus should be avoided. As for viruses, scouting of fields for the first occurrence of virus disease is a prerequisite. Where feasible, pull up and destroy infected plants, but only after spraying them thoroughly with an insecticide to kill any insects they may be harboring. Use reflective mulches to repel insects, thereby reducing the rate of spread of insect-borne viruses. Monitor vector population early in the season and apply insecticide treatments when needed. Minimize plant handling to reduce the amount of virus spread mechanically.

Polyhouse disease control may be classified into two approaches: (i) those aimed at the root environment and (ii) those aimed at the aerial environment.

Table 11.1 Main viral diseases of vegetable crops and means of transmission

	Crop		
Transmission	Tomato	Pepper	Cucumber
Seedborne	TMV	TMV	CGMMV
Mechanical	TMV	TMV	CGMMV
Aphids	CMV, PVY	CMV, PVY	CMV, ZYMV
Whiteflies	TYLCV	–	CVYV, CYCDV
Thrips	TSWV	TSWV	–

11.3.1 The Root Environment

Soil disinfection (i.e., sterilization) is an important part of soilborne disease control when raising vegetables by the ground culture method or when soil-based potting mixes are used. Soilborne diseases include damping-off (*Pythium* and *Rhizoctonia*), black root rot (*Thielaviopsis*), and several other root rots and wilts caused by *Fusarium* and *Phytophthora*. Potting mixes based on compost, peat moss, vermiculite, perlite, and bark are typically pathogen-free and do not require prior sterilization.

Sterilization involves heating the soil to 212 °F for 30 min, a process that kills most organisms in the soil. Chemical changes also occur in the soil and may affect plant nutrient uptake to the following crop. Pasteurization involves heating the soil to only 160 °F for 30 min. While most harmful pathogens are killed at this temperature, many beneficial soil organisms survive. Thus, pasteurization is the preferred method in organic programs.

Chemical biocides, electrical heat, steam heat, and soil solarization are the primary methods of soil disinfection in greenhouse production. A fifth method of disease suppression is biological control. Soil fumigants such as methyl bromide are, of course, restricted in organic production.

Electrical heat treatment, which is done inside a steel chamber surrounded by heating coils, is limited to treating about a cubic yard of soil at a time. Its primary use is in the pasteurization of small batches of sand and soil for potting mixes.

Accordingly, steam pasteurization and soil solarization are the two most viable options for sterilizing greenhouse soils or large volumes of soil-based mixes. Biological control is complementary to these two methods.

11.3.1.1 Steaming

Steam was the primary method of soil sterilization in the greenhouse industry prior to the emergence of soil fumigants. Steam heat is highly effective and environmentally safe. Equipment and fuel costs are expensive, however, and treatment between crops is labor and time consuming. These are the chief reasons many growers shifted to soilless mixes.

There are three methods of steaming in common use today (Klassen 1993):

- Tarping an area and piping in steam for 6–8 h, heating and sterilizing the top 8 in. of soil
- Pumping steam into subsurface drainage pipes, sterilizing the top 60 cm of soil
- Negative pressure steaming, where pumps pull steam applied at the surface through pipes buried 60 cm deep and 300 cm apart

11.3.1.2 Solarization

Solarization in greenhouses is a proven means of soil sterilization and is practiced worldwide in Japan, Israel, Greece, France, Italy, Belgium, Portugal, and Spain (Mahrer 1991; Garibaldi and Bullino 1991; Horiuchi 1991), in addition to the USA and Canada.

Soil solarization is the process of tarping moist soils with clear polyethylene to trap solar radiation and raise soil temperatures to levels lethal to most pathogens and weed seeds. Solarization is most effective when applied for at least 30 days in midsummer. Two layers of polyethylene, separated by fillers (i.e., PVC pipes of 60- to 120-cm length) spaced every few feet to create an air space, increase the efficiency of solarization.

Solarization can also be used to pasteurize sand or soil intended for potting media, or to treat used media (Gamliel et al. No date). A simple technique developed in Florida (Giblin and Verkade 1987) is to fill a black plastic trash bag with media, which is then sealed in a transparent plastic bag. The double-wrapped media is placed in the open on asphalt or concrete and spread to a uniform depth of 7.5 cm. As long as the trash bag received one full day of sunshine from April through August, the temperatures reached or exceeded 113 °F for more than 2 h. This temperature, which was monitored with a thermometer inserted into the media, was considered the minimum treatment period for plant-parasitic nematodes. This method yields about 24 L of media.

11.3.1.3 Biological Control

Biological control of pathogens in greenhouse soils and potting mixes is accomplished through natural control via cultural practices and through applied biocontrols. The introduction of biocontrol agents, or antagonists, is the next form of biological soil disinfection. Biological fungicides are a promising alternative to synthetic fungicides.

There are several new biocontrols available for suppressing root diseases. One product is RootShield (contains the fungus *Trichoderma harzianum*), a biofungicide that controls root diseases caused by *Pythium*, *Rhizoctonia*, *Fusarium*, and *Sclerotinia*. SoilGard 12G (contains the fungus *Gliocladium virens*) is another biofungicide that controls *Pythium*, *Fusarium*, *Rhizoctonia*, *Phytophthora*, and *Thielaviopsis* (when co-applied with sulfur). Another product is Mycostop (contains the actinomycete *Streptomyces griseoviridis*) which controls *Fusarium*, although it may control other root diseases. Deny contains the bacterium *Burkholderia* (*Pseudomonas*) *cepacia* and controls diseases caused by *Rhizoctonia*, *Fusarium*, and *Pythium*. It also provides some nematode control. Intercept also uses *B. cepacia* to control disease.

11.3.1.4 Cultural Practices

Cultural practices that promote soil health include crop rotation and use of tilled-in crop residues and green manures or organic amendments brought in from off site.

The beneficial effect of crop rotation on plant health and yield is probably due to changes in soil microflora. These shifts result in increasing numbers of beneficial microorganisms (including nitrifying and other useful bacteria, antagonists of pathogens, etc.) and in decreasing numbers of noxious ones (pathogens and antagonists of beneficial microorganisms) (Gindrat 1979).

In some greenhouses, a short-term break crop – such as annual ryegrass, oats, buckwheat, or rapeseed – may fit into the rotation and be used to increase organic matter. Another option is to rely on organic amendments or green leaf manures brought in from off site. Amendments such as fresh residues, composts, and manures result in soil improvement through a proliferation of microflora.

Green leaf manures are field-grown cover crops that are chopped and harvested green and then transported inside the greenhouse to be incorporated into the soil. Many studies have shown that crucifer residues from rapeseed and oilseed radish cover crops produce toxic gases that suppress soilborne pathogens. Additionally, soil solarization, following incorporation of cruciferous residues, increases the efficiency of soilborne pathogen control.

The role of composts as a slow-release nutrient source in organic production is well established. Recently, the role of composts as amendments for the control of soilborne plant pathogens has increased interest in similar use in conventional agriculture.

In container production, disease-suppressive composts are commercially available in both peat and soil-based media. The disease-suppressive properties come from a certain composting technique that yields antibiotic (antagonistic) and competitive (saprophytic) responses (Roberts 1992). Natural disease-suppressive potting mixes – based on composts – are available through several commercial sources. The potting mixes can also be mixed with mycorrhizae which colonize plant roots and protect them against certain pathogenic fungi, including various *Pythium* and *Fusarium* species (Anon 1992).

11.3.2 The Aerial Environment

Foliar and stem diseases include gray mold (*Botrytis*), powdery mildew (*Erysiphe* spp), early blight (*Alternaria* spp), soft rot (*Erwinia* spp), and several other fungal and bacterial diseases caused by *Xanthomonas*, *Fusarium*, and *Pseudomonas*.

Greenhouse climates are warm, humid, and wind-free – an ideal environment for the development of many foliar and stem diseases. For the majority of pathogenic fungi and bacteria, infection usually occurs when a film or drop of water on the plant surface persists. Unless temperature, humidity, and ventilation are well regulated, this surface water can remain in the greenhouse until infection becomes assured (Jarvis 1992).

Integrated disease management, therefore, is based on climate control for disease infection and optimum crop yield and quality. It eliminates inoculum through high standards of hygiene (sterilizing soil or using soilless media, obtaining disease-free planting material, chlorine bleach rinses of footwear and equipment, vegetative-free floors, etc.), cultural practices for limiting disease spread, biological and pesticidal control, and, most important, when available, resistant cultivars (Jarvis 1992).

11.3.2.1 Environmental Control

Temperature regulation and humidity regulation – functions of heating and cooling, ventilation, vapor pressure, and structure – are increasingly becoming computerized. Expert software that reduces disease-infective conditions while promoting crop growth is available through commercial vendors.

11.3.2.2 Foliar Products

Greenhouse growers have fewer alternative control products for diseases than for insects. Copper- and sulfur-based fungicides are the only commercial fungicides acceptable in certified organic programs. Coppers exhibit both fungicidal and bactericidal properties. Sulfurs are noted for control of mildews. Alternative disease control strategies, although based more on grower's experience and limited research, include compost watery extracts, biodynamic herbal extracts, and foliar feeding below.

Fungicides are potentially very effective with some diseases, but may be ineffective with others (Table 11.2).

Biofungicides for foliar disease control are also available. AQ10 contains the fungus *Ampelomyces quisqualis* and controls powdery mildew. Trichodex controls *Botrytis*. TopShield contains *Trichoderma harzianum* and controls *Botrytis* and powdery mildew (Table 11.3).

11.3.2.3 Foliar Feeding

Foliar feeding is used by many organic growers to induce resistance to foliar diseases. Seaweed and fish emulsion are the two products most commonly applied. Although the mechanism for resistance is not clear, modification of the leaf surface and enhanced plant nutrition is suspected. It has been stated that foliar fertilization indirectly protects against plant pathogens by increasing natural plant immunity through improvement of the plant's nutritional status (Nowosielski et al. 1988).

Evidence is also strong that foliar feeding can have a dramatic effect on the rhizosphere microflora via changes in root exudates (Gindrat 1979). Such changes in root exudates may lead to an increase of antagonists and thus enhance biological control of pathogens in the root environment as well (Gindrat 1979).

11.3.2.4 Baking Soda

Baking soda has for many years been used in Europe as a mild fungicide by rose growers and organic gardeners. Recently, plant pathologists at Cornell University confirmed through research that baking soda exhibits fungicidal activity against powdery mildew and several other diseases on ornamentals (Anon 1990). They determined that a mixture of 0.5 % baking soda (about 5 level teaspoons per gallon of water) and 0.5 % horticultural oil showed the greatest activity.

A new product containing baking soda, called Remedy, is now available for use as a pesticide. Remedy is effective against black spot, powdery mildew, leaf spots, anthracnose, *Phoma*, *Phytophthora*, scab, and *Botrytis*. Remedy should be applied at the first sign of disease and repeated at one- or two-week intervals until the problem subsides.

11.3.2.5 Polyethylene Films

Reuveni et al. (1989) observed a reduction in the number of infection sites of *B. cinerea* on tomato and cucumber when a UV-absorbing material was added to polyethylene film to increase the ratio of blue light to transmitted UV light. Blue photoselective polyethylene sheets have been suggested for their ability to reduce gray mold on tomato (Reuveni and Raviv 1992) and downy mildew on cucumber (Reuveni and Raviv 1997). Green-pigmented polyethylene reduced the

Table 11.2 Selected fungicides and bactericides labeled for vegetable plants

Fungicide	Targeted pest	Labeled crops	Comments
Basic copper sulfate (Cuprofix Ultra 40 Disperss), 12 h REI, Group M1	Many diseases – angular leaf spot, downy mildew, *Alternaria* blight, anthracnose, bacterial blight, bacterial spot (depending on crop)	Many – cucumbers, eggplant, peppers, tomatoes	Crops grown in greenhouse may be more sensitive to copper injury. Observe for 7–10 days for symptoms of injury
Bacillus pumilus (Sonata), 4 h REI, Group 44, OMRI listed	Downy mildew, powdery mildew	Many – cole crops, cucurbits, fruiting, leafy vegetables	Begin applications when greenhouse conditions favor disease development
Bacillus subtilis (Serenade), 4 h REI, Group 44, OMRI listed	Many diseases – downy mildew, powdery mildew, bacterial spot, early blight	Many vegetables – broccoli, leafy vegetables, cucurbits, peppers, tomatoes	Preventive biofungicide. Thorough coverage essential
Bacillus subtilis (Cease), 4 h REI, Group 44, OMRI listed	Many diseases – leaf spots, powdery mildew, *Botrytis* blight, downy mildew	Many including cole crops, cucurbits, fruiting vegetables, leafy vegetables	Begin applications when greenhouse conditions favor disease development. Thorough coverage essential
Copper hydroxide (Champ DP Dry Prill, Champ Formula 2 Flowable, Champion WP, Champ WG [OMRI listed], Kocide 101, Kocide 2000, Kocide 4.5LF, Kocide DF), 24 h REI, Group M1	Leaf spots, anthracnose, bacterial spots, and other diseases	See labels for specific crops	See labels for specific usage instructions
Copper salts of fatty and rosin acids (Camelot), 12 h REI, Group M1	Many – bacterial leaf spots, leaf spots and blights, downy mildew, powdery mildew	Greenhouse vegetables	See label for specific usage instructions
Cuprous oxide (Nordox 75 WG), 24 h REI, Group M1	Anthracnose, *Phomopsis*, *Botrytis*, various leaf spots and blights	Tomato, pepper, eggplant	Begin applications when disease first threatens
Dichloran (Botran 75-W), 12 h REI, Group 14	*Botrytis*, white mold (*Sclerotinia*)	Cucumber, leaf lettuce, tomato	Seedlings or newly set transplants of tomato may be injured by drenching
Fenhexamid (Decree 50WDG), 12 h REI, Group 17	*Botrytis*	Fruiting vegetables, tomatoes, cucumber, leafy greens (except spinach)	Thorough coverage needed. Do not make more than two consecutive applications
Horticultural oil, paraffinic oil (Ultra-Pure Oils), 4 h REI, NC Saf-T-Side (spray oil emulsion fungicide, insecticide, and miticide), 12 h REI, NC, OMRI listed (Organic JMS Stylet Oil)	Powdery mildew	Cucurbits, melons, squash, and others	Contact fungicide. Application should be made when disease is first noticed. Applications should be preceded by a phytotoxicity check to ensure that material is safe
Hydrogen dioxide (OxiDate), 0 h REI (nonspray), 1 h REI (spray), OMRI listed	Many – mildews, leaf spots, and blights	Tomato, pepper, leafy and cole crops, cucurbits, and others	Strong oxidizing agent. Contact, oxidizing sanitizer

(continued)

Table 11.2 (continued)

Fungicide	Targeted pest	Labeled crops	Comments
Insecticidal soap, potassium salts of fatty acids (M-Pede), 12 h REI, OMRI listed	Powdery mildew	Greenhouse cucumber	Works by contact
Kaolin (Surround WP), 4 h REI, Group NC, OMRI listed	Powdery mildew	Cucurbit vegetables	Forms a mineral-based particle film resulting in a dry, white film. May be unsightly for retail sales. Uniform coverage important for effectiveness

Table 11.3 Commercial biopesticides

Biopesticide	Supplier	Pests controlled	REI	Application/comments
AQ10 *Ampelomyces quisqualis*	Plant Health Care	Powdery mildew	4 h	Begin application as soon as host tissue emerges. Apply at least two sequential sprays 7–14 days apart. Works best under conditions of high humidity
SoilGard *Gliocladium virens*	Thermo Trilogy	*Rhizoctonia, Fusarium, Pythium, Phytophthora, Thielaviopsis*	4 h	Incorporate into soil before planting
Mycostop *Gliocladium virens*	Green Spot	*Fusarium, Alternaria, Phomopsis*	4 h	Can be incorporated into medium or applied to seed
Deny *Pseudomonas (Burkholderia) cepacia*	Stine Seed	*Rhizoctonia, Fusarium, Pythium*		
Intercept *Pseudomonas (Burkholderia) cepacia*	Soil Technologies	*Rhizoctonia, Fusarium, Pythium,* nematodes		
Actinovate *Streptomyces lydicus*	Green Spot	*Pythium, Fusarium, Phytophthora, Sclerotinia*		Can be incorporated into medium or applied to seed
Bio-Trek HB *Trichoderma harzianum*	Wilbur-Ellis	*Fusarium, Pythium, Rhizoctonia*	12 h	Apply to seed
Bio-Trek Nursery Drench *Trichoderma harzianum*	Wilbur-Ellis	*Fusarium, Pythium, Rhizoctonia*	12 h	
RootShield *Trichoderma harzianum*	BioWorks	*Fusarium, Pythium, Rhizoctonia, Sclerotinia*		
Trichodex *Trichoderma harzianum*	Abbott	*Botrytis*		
TopShield *Trichoderma harzianum*	BioWorks	*Botrytis*, powdery mildew		

conidial load and gray mold in commercial tomato and cucumber greenhouses by 35–75 %. *Sclerotinia sclerotiorum* on cucumber, *Fulvia fulva* (*Cladosporium fulvum*) on tomato, and cucumber powdery mildew were also reduced (Elad 1997).

11.3.2.6 Greenhouse Structures and Covers

The influence of greenhouse structures and covers on greenhouse climatic regimes may have strong consequences for diseases. In high-tech greenhouses, regulation of temperature and water pressure deficit enables the creation of conditions less favorable to pathogens and, in some cases, more favorable to biocontrol agents. The use of heating to limit the development of a number of pathogens is well known (Jarvis 1992). The use of high root temperatures in winter-grown tomatoes in rock wool offers a nonchemical method of controlling root rot caused by *Phytophthora cryptogea*. The high temperature was shown to enhance root growth while simultaneously suppressing inoculum potential and infection and, consequently, reducing or preventing aerial symptoms (Kennedy and Pegg 1990). Careful control of the temperature also proved important in the case of hydroponically grown spinach and lettuce, in which it prevented or reduced attack by both *Pythium dissotocum* and *Pythium aphanidermatum* (Bates and Stanghellini 1984). Recently, attacks of *P. aphanidermatum* on nutrient film technique (NFT)-grown lettuce in Italy were related to the high temperature (>29 °C) of the nutrient solution. Root rot was inhibited by reducing the temperature below 24 °C (Carrai 1993).

References

Anon (1990) Baking soda can ward off fungus. Greenhouse Manager, June, p 24

Anon (1992) Mycori-Mix contains beneficial fungi, suppresses *Pythium*. Greenhouse Manager, September, p 68

Bates ML, Stanghellini ME (1984) Root rot of hydroponically grown spinach caused by *Pythium aphanidermatum* and *P. dissotocum*. Plant Dis 68:989–991

Carrai C (1993) Marciume radicale su lattuga allevata in impianti NFT. Colture Protette 22(6):77–81

Elad Y (1997) Effect of filtration of solar light on the production of conidia by field isolates of *Botrytis cinerea* and on several diseases of greenhouse-grown vegetables. Crop Prot 16:635–642

Gamliel A et al. No date. Solarization for the recycling of container media. The Hebrew University of Jerusalem, Rehovot. Unpublished manuscript, 8 pp

Garibaldi A, Bullino ML (1991) Soil solarization in Southern European countries, with emphasis on soilborne disease control of protected crops. In: Katan J, DeVay JE (eds) Soil solarization. CRC Press, Boca Raton, pp 227–235

Giblin RM, Verkade SD (1987) Solarization of small volumes of potting soil for disinfection of plant-parasitic nematodes. Proc Fla State Hortic Soc 100:174–176

Gindrat D (1979) Biological soil disinfection. In: Mulder D (ed) Soil disinfection. Elsevier Scientific Publishing Co, New York, pp 253–287

Horiuchi S (1991) Soil solarization in Japan. In: Katan J, DeVay JE (eds) Soil solarization. CRC Press, Boca Raton, pp 215, 218–223, 225

Jarvis WR (1992) Managing diseases in greenhouse crops. American Phytopathological Society, St Paul, pp 3, 5, 220–221

Kennedy R, Pegg GF (1990) *Phytophthora cryptogea* root rot of tomato in rock wool nutrient culture. III. Effect of root zone temperature on infection, sporulation and symptom development. Ann Appl Biol 117:537–551

Klassen P (1993) Mulling over methyl bromide. Greenhouse Grower, August, pp 118, 120

Mahrer Y (1991) Physical properties of solar heating of soils by plastic mulching in the field and in glasshouses and simulation models. In: Katan J, DeVay JE (eds) Soil solarization. CRC Press, Boca Raton, pp 75, 81–86

Nowosielski O, Dziennik W, Kotlinska D, Narkiewicz J, Dobrzanska J (1988) A biological basis for the efficiency of plant protecting foliar fertilizers in vegetable production. Acta Hortic 222:105–116

Reuveni R, Raviv M (1992) The effect of spectrally-modified polyethylene films on the development of *Botrytis cinerea* in greenhouse grown tomato plants. Biol Agric Hortic 9:77–86

Reuveni R, Raviv M (1997) Control of downy mildew in greenhouse-grown cucumbers using blue photoselective polyethylene sheets. Plant Dis 81:999–1004

Reuveni R, Raviv M, Bar R (1989) Sporulation of *Botrytis cinerea* as affected by photoselective sheets and filters. Ann Appl Biol 115:417–424

Roberts DR (1992) Insect-, disease-suppressive mixers help growers minimize crop losses. Greenhouse Manager, September, pp 68, 70–71

12 Fungal Diseases and Their Management

Abstract

Fungal diseases constitute one of the biggest groups of foliar pathogens causing immense damage under protected environment. The important fungal diseases include damping-off, crown and root rots, wilts, powdery mildews, downy mildews, *Botrytis* diseases, *Sclerotinia* rot, *Alternaria* diseases, and rust diseases. Proper field sanitation is one of the most important management strategies, since once the buildup of inoculum occurs inside polyhouse, it is very difficult to manage it. The use of disease-resistant varieties, reduced incidence of leaf wetness by staking plants, providing ample spacing between plants to allow for good air movement, and avoiding overhead irrigation also help in disease control. For judicious use, chemicals with least toxicity recommended specially for polyhouse cultivation should be sprayed.

Keywords

Fungal diseases • Damping-off • Wilts • Powdery mildews • Downy mildews • *Botrytis* diseases • *Alternaria* diseases

12.1 Introduction

Fungal diseases constitute one of the biggest groups of foliar and soilborne pathogens causing immense damage under protected environment. It was found that the incidence and severity of diseases vary considerably under protected environment when compared to open field. As observed in tomato, *Phytophthora infestans*, *Pseudocercospora fuliginea*, and *Fulvia fulva* causing late blight, black leaf mold, and leaf mold, respectively, were observed to be of higher significance under polyhouse condition. Further, the notion of early and late blight was found to be obscure under polyhouse condition since it was observed that late blight appears early in the crop season, whereas early blight appears late during crop season. This may be attributed to the fact that temperature and humidity are nearly balanced inside a protected structure, even when outside field temperature is comparatively low and so on. *Alternaria alternata* is found to be a

P.P. Reddy, *Sustainable Crop Protection under Protected Cultivation*,
DOI 10.1007/978-981-287-952-3_12

major disease affecting fruits. Capsicum on the other hand is found to be infected primarily by anthracnose (*Colletotrichum capsici*), leaf spot (*Cercospora capsici*), damping-off (*Pythium*, *Fusarium*, and *Phytophthora* spp.), gray leaf spot (*Stemphylium solani* and *S. lycopersici*), stem rot (*Sclerotium rolfsii*, *Verticillium albo-atrum*, and *V. dahliae*), *Phytophthora* blight (*Phytophthora capsici*), powdery mildew (*Leveillula taurica*), and gray mold (*Botrytis cinerea*). Cucumber also attracts a considerable quantum of fungal pathogens, out of which downy mildew (*Pseudoperonospora cubensis*), powdery mildew (*Erysiphe cichoracearum* and *Sphacelotheca fuliginea*), leaf spot (*Alternaria cucumerina*), anthracnose (*Colletotrichum lagenarium*), and damping-off (*Pythium* spp.) are important. Survival of pathogen is also enhanced inside polyhouse due to availability of a host because of longer growing season.

12.2 Damping-Off, Crown and Root Rots

Plants in seedbeds may be diseased, either before or after their emergence from the soil, and the disease is called pre- or postemergence damping-off, respectively. In the first case, seedlings do not emerge in patches of the seedbeds. In the second case, plants rot quickly and drop down on the soil. Low temperatures and very wet soils, which delay the growth of the plants, favor infection. A large number of fungi may cause damping-off, but *Pythium* spp., *Phytophthora* spp., *Fusarium* spp., and *Rhizoctonia solani* are the most common. Nowadays, due to the use of improved technology, damping-off is no longer a severe disease in greenhouses (Blancard et al. 1991). However, root rots and crown rots are still destructive in soil, though not in soilless cultures (Davies 1980).

All *Pythium* and *Phytophthora* species as well as *R. solani* are common soil inhabitants. They survive in the soil. Infection usually takes place at the time of planting and symptoms appear very soon in *Pythium* and *Phytophthora* rot or several weeks later in *Rhizoctonia* stem rot. *Rhizoctonia solani* may infect at moderate soil moisture levels, but *Pythium* spp. and *Phytophthora* spp. infect only in water-saturated soils (Strider 1985).

12.2.1 Symptoms

12.2.1.1 *Pythium* and *Phytophthora* Rots

Various *Pythium* spp. and *Phytophthora* spp. may damage the lower part of tomato, pepper, cucumber, carnation, poinsettia, gerbera, etc., both in soil and soilless cultures.

In tomato, a root and crown rot extending to a considerable height above the soil level may occur. The infected area has a dark discoloration and the pith is usually destroyed. *Phytophthora nicotianae* var. *parasitica* is the most common pathogen. In pepper, a similar disease caused by *P. capsici* is very common. Collar, stem, and fruit rot as well as leaf spots may occur. In cucumber, a soft rot of the young plants at the soil level may occur soon after transplanting. Infected tissues shrink and in wet weather a white mycelium develops. Infected plants wilt and die quickly. Poinsettia grown in pots also suffers from *Pythium* rot. Severe root rot, extending above ground in succulent plants, and quick death are the main symptoms. In cucumber and poinsettia, *Pythium ultimum*, *P. irregulare*, *P. debaryanum*, and *P. aphanidermatum* are mostly involved (Tompkins and Middleton 1950). Carnations infected by *Pythium* and *Phytophthora* species develop soft rot at the collar and in the root system, resembling *Rhizoctonia* stem rot.

12.2.1.2 *Rhizoctonia* Stem Rot (*R. solani*)

This infects a large number of plants, such as tomato, carnation, poinsettia, etc., causing symptoms resembling *Pythium* or *Phytophthora* rots. *Rhizoctonia* stem rot is mainly confined to the collar. Carnation is very susceptible. Infected plants show pale brown dry lesions, with circular rings, at soil level. Growth is stunted and leaves become dull green. Complete wilting soon follows. Strands of the pathogen develop on the lesions and stems break easily at the infected area (Parmeter 1970).

12.2.2 Management

Prevention is the only effective approach, because damping-off is difficult to stop once symptoms occur. There are several strategies to prevent damping-off:

- Use only certified disease-free seed from reputed seed companies.
- Use fungicide-treated seed: certain fungicides are labeled for damping-off for selected vegetable crops.
- Use pasteurized soil or compost-based or soilless mixes.
- Avoid overwatering, excessive fertilizer, overcrowding, poor air circulation, careless handling, and planting too deeply.
- Provide adequate light for rapid growth.

Effective control of the abovementioned diseases may be obtained selectively by the following means: (i) use of naturally or artificially suppressive substrates, (ii) early drenching by effective fungicides, (iii) soil disinfestation, (iv) use of resistant cultivars, (v) grafting on resistant rootstocks, and (vi) biological control (Tjamos 1992).

Hygiene is of utmost importance. Since the fungi concerned are soilborne, it is important to use disease-free soil, substrates, pots, etc. This is preferably done by using new or steam-sterilized material, but cleaning and sterilization of pots, glass, etc., is also possible with chemical disinfectants, such as copper sulfate, quaternary ammonium compounds, formaldehyde, and sodium hypochlorite. Moist conditions often favor these fungi and should be avoided. In systems where recirculated water is used, the diseases can spread rapidly. For these systems, the water should therefore be disinfected. This can be done by heating, with sand filters, or by UV radiation. Fungicides can be used, by application to soil in pots or by spray. They are generally used preventively, in cases where the risk is considered to be high. Seed treatment against these fungi is also recommended. Cultivars have different degrees of resistance toward these diseases.

12.2.3 Main Fungicides

Against All the Fungi Concerned Etridiazole, furalaxyl, and propamocarb (by pot soil treatment) and etridiazole, fosetyl-Al, furalaxyl, and propamocarb (by spray).

Against *Pythium* spp. and *Phytophthora* spp. Furalaxyl, metalaxyl, and oxadixyl.

Against *Rhizoctonia solani* Iprodione, pencycuron, and tolclofos-methyl.

12.3 Wilts

All major greenhouse crops suffer from one or more wilts. In several crops, wilts are the main diseases due to the damage they cause and the difficulty of controlling them.

12.3.1 Symptoms

12.3.1.1 *Fusarium Wilt (F. oxysporum)*

The most common *Fusarium* wilts in greenhouses appear on tomato (*F. oxysporum* f. sp. *lycopersici*), cucumber (*F. oxysporum* f. sp. *cucumerinum*), melon (*F. oxysporum* f. sp. *melonis*), carnation (*F. oxysporum* f. sp. *dianthi*), gladiolus (*F. oxysporum* f. sp. *gladioli*), cyclamen (*F. oxysporum* f sp. *cyclaminis*), and chrysanthemum (*F. oxysporum* f. sp. *chrysanthemi*).

Wilting, yellowing, chlorosis, drooping (mostly of the lower leaves), stunting, and brown discoloration of the vascular bands up to the top of the stem are the dominant symptoms. Wilting of the lateral shoots and large lesions on the lower part of the stem are also common in *Fusarium* wilt of carnation and melon. All the above *F. oxysporum formae speciales* have more than one race. Each of them infects cultivars of one host but may colonize the root system of other plants as well. They survive in the soil for several years, due to the production of thick-walled chlamydospores, but inoculum is reduced over the years. *Fusarium* wilt in tomato, watermelon, carnation, cyclamen, chrysanthemum, and gladiolus is favored by higher temperatures than *Fusarium* wilt in melon (Sherf and Macnab 1986).

12.3.1.2 *Verticillium–Phialophora* Wilt, *Verticillium dahliae*, *V. albo-atrum*, *Phialophora cinerescens* (*Verticillium cinerescens*)

This infects a huge number of plants and among them the majority of the plants grown in greenhouses. It is more severe in Solanaceae such as tomato, eggplant, and pepper. Of the floral crops, chrysanthemum seems to be more susceptible. Symptoms are very similar to those of *Fusarium* wilt. *Verticillium* wilt is favored by moderate temperatures. *Verticillium dahliae*, which is more common, survives in the soil for many years due to the abundant production of black resistant microsclerotia, while *V. albo-atrum* survives by producing dark dormant mycelium. A similar wilt caused by *P. cinerescens* damages carnations in several areas (Sherf and Macnab 1986).

12.3.2 Management

Hygiene solves many problems and all possible precautions and actions associated with proper sanitation and management should be taken. If there have been previous problems with wilting, the soil should be steam-sterilized before planting or use as pot soil and again after harvest. Infected plants should be removed. After harvest, the pallets, water hoses, glass walls, etc., should be cleaned thoroughly. The water table should be checked and should be kept sufficiently low. High sodium and nitrogen concentrations should be avoided and only healthy planting material should be used. Fungicides can be used prophylactically (by soil treatment before planting or added to water for pouring after planting) or curatively (by spraying or pouring from a watering can). The sensitivity of the different pathogens to individual fungicides varies considerably. Therefore, in case of doubt, the problem should be thoroughly investigated, the pathogen identified, and expert advice obtained on suitable fungicides.

12.3.3 Main Fungicides

Prophylactic Furalaxyl, metalaxyl, prochloraz, propamocarb, tolclofos-methyl, and zineb.

Curative Carbendazim, furalaxyl, iprodione, metalaxyl, prochloraz, and tolclofos-methyl.

12.4 Powdery Mildews

Powdery mildews are very destructive of several greenhouse crops. The following are some of the powdery mildew fungi which mostly attack greenhouse-grown plants:

- Cucurbits – *Sphaerotheca fusca* (*Sphaerotheca fuliginea*), *E. cichoracearum*, and *L. taurica* (only on cucumber)
- Solanaceous plants – *L. taurica* and *Oidium lycopersicum* (only on tomato)
- Roses – *Sphaerotheca pannosa*
- Begonia – *Microsphaera begonia*
- Gerbera – *Erysiphe cichoracearum*

12.4.1 Symptoms

Powdery mildew fungi (except for *L. taurica*) may attack all green tissues. The dominant symptoms include initial white powdery spots, which enlarge and coalesce to cover large areas. *Leveillula taurica* infects only leaves. Light yellow or yellow-green spots on the upper leaf surface, which later become brown, and scarce white mold on the lower surface are the main characteristics. Powdery mildew-infected plant parts may be chlorotic and distorted. Premature defoliation and poor growth are common features of severely infected plants (Braun 1995).

Infections take place by conidia. Under favorable conditions, powdery mildew progresses rapidly. By the end of the season, some powdery mildew fungi, such as *S. fuliginea*, *E. cichoracearum*, etc., may develop cleistothecia with ascospores, but these do not play an important role in the epidemiology of the disease (Braun 1995).

Conidia are mostly discharged and transferred by wind currents. Animal pests may also disseminate conidia in greenhouse crops. Young conidia readily germinate on plant surfaces depleted of nutrients. The relative humidity (RH) favoring infection by powdery mildew fungi and development of the disease differs from species to species. For instance, high RH is more favorable for *S. fuliginea* than for *E. cichoracearum*. Therefore, the first fungus is more frequent on greenhouse cucurbits than the second. High RH may favor spore germination of powdery mildew fungi, but free water may be deleterious. RH at 97–99 % is optimal for spore germination of *S. pannosa* and *S. fuliginea*. At RH below 75 %, spores of *S. pannosa* do not germinate, but mycelium development and sporulation may occur at RH as low as 21–22 %. Powdery mildew fungi overwinter on cultivated plants or weeds, which survive in or outside the greenhouse (Coyier 1985a).

12.4.2 Management

Dense stand of the plants, high nitrogen concentrations, and temperature fluctuations should be avoided. The plants should be watered at a proper time (allowing the soil to dry before night) and frequently ventilated. Affected plants or their parts (leaves, stems, and flowers) should be removed. Contrary to most fungal diseases, the release of conidia of these fungi is stimulated by low relative humidity. If under these conditions an attack develops, the application of a fungicide spray becomes necessary. The use of resistant cultivars can, however, reduce fungicide treatments:

- Maintain proper plant spacing to reduce relative humidity levels within the plant canopy and improve spray coverage.
- Keep relative humidity levels below 90 % in the greenhouse. Heat and ventilate in the late afternoon and early morning to reduce high relative humidity at night.
- Clean the greenhouse thoroughly between crops, removing all weeds that could be potential hosts.

12.4.3 Main Fungicides

Benomyl, bitertanol, bupirimate, carbendazim, dichlofluanid, dinocap, dodemorph, fenarimol, imazalil, penconazole, propiconazole, thiophanate-methyl, triadimenol, and triforine

Chemicals such as demethylation inhibitors (DMIs) (triadimefon, fenarimol, etc.), pyrimidines (ethirimol, bupirimate, etc.), pyrazophos, and dinocap remain the main means of controlling powdery mildews in greenhouses. Biological control agents have also been effectively tested against *S. fuliginea* and *S. pannosa*. Finally, fully resistant cvs of melon and partially resistant cvs of long-type cucumber are available (Coyier 1985b; Molot and Lecoq 1986).

12.5 Downy Mildews

Downy mildews of tomato (*P. infestans*), cucurbit (*Pseudoperonospora cubensis*), lettuce (*Bremia lactucae*), rose (*Peronospora sparsa*), and snapdragon (*Peronospora antirrhini*) are the most destructive in greenhouse-grown plants.

12.5.1 Symptoms

In tomato, leaves and young shoots are infected first. Fruit infection starts mostly near the stalk and spreads very quickly to the whole fruit. Infected tissues of fruits and shoots are firm and brown (Sherf and Macnab 1986).

In cucurbits, downy mildew appears as yellow, angular, or circular spots on the upper surface of the mature leaves of the plant. Soon the tissues at the center of the spots die and become light brown. Cucumber and melon are more susceptible than watermelon.

Downy mildew of lettuce causes scattered light-green to yellow spots on the upper leaf surface. Old spots become brown and dry up.

Downy mildew of rose damages all green plant parts, but leaves are more susceptible. Leaf infection resembles the effect of toxins. Infected leaves have purplish red to dark-brown irregular spots and shed readily (Strider 1985). Snapdragon plants infected by *P. antirrhini* are stunted and the top internodes of the young plants are short. The borders of the lower leaves curl down and then dry. Eventually the entire plant dies (Garibaldi and Rapetti 1981). A white fungal growth (brown for cucurbit downy mildew) on the infected tissues under moist conditions is typical of all downy mildews.

Plant infection takes place through stomata and mycelium develops intercellularly. Soon branched conidiophores are produced and protrude through the stomata. Infection progresses in the periphery of the spot which gradually enlarges. Conidiospores of downy mildews are ovoid and hyaline, except for *P. cubensis* which are brown. They are discharged by hygroscopic changes and disseminate in greenhouses by wind currents and water splashes. Initial infection may take place by spores that are transferred over long distances by the wind. Abundant oospores of *P. antirrhini* develop on dead plant stems. Oospores of *P. sparsa* also very often develop on infected roses, whereas *P. cubensis* and *P. infestans* oospores are rare. *Phytophthora infestans* survives on seed potato tubers and spreads to young potato plants after they have been planted. Inoculum is disseminated from potatoes to neighboring tomato crops. Cucurbit downy mildew can infect several species of cucurbits, grown either in greenhouses or open fields all year round. There is evidence that *P. sparsa* survives as a dormant mycelium on the infected stems of roses. *Peronospora antirrhini* perennates as dormant thick-walled oospores in dead plant parts and soil (Sherf and Macnab 1986).

Free water on plant tissues is necessary for downy mildew fungi to cause infection. High RH is also required for good sporulation. *Peronospora antirrhini* is favored by low temperature and high RH. Free water or high relative humidity is not often a factor limiting downy mildew development in plastic greenhouses. It seems that temperature is more critical. For instance, *P. cubensis*, with a high maximum temperature for development and infection, may, under certain conditions, infect all year round, whereas *P. infestans* and *P. sparsa* do not infect during the hot period of the year. Downy mildews complete a cycle within about 6–8 days. Thus, under favorable weather conditions, they may have several cycles and spread rapidly (Strider 1985).

12.5.2 Management

These fungi like moist conditions, wet leaves, and a high relative humidity. Wet crops should therefore be dried as quickly as possible by heating or ventilation and temperature fluctuations should also be avoided. In many crops, a prophylactic treatment with fungicides is necessary. In case of an infestation, a treatment with a curative fungicide should be done, followed by a series of prophylactic sprayings.

Dithiocarbamates, chlorothalonil, and the systemic phenylamides (metalaxyl, Ridomil, etc.) are the most commonly used in greenhouses. There are some tomato cvs fully resistant to downy mildew and some partially resistant cucumber cvs suitable for greenhouses, but all rose cvs grown for cut flowers are susceptible to downy mildew. Ventilation of the greenhouses may also effectively prevent infection (Strider 1985).

12.5.3 Main Fungicides

Prophylactic Chlorothalonil, fosetyl-Al, mancozeb, and propamocarb.

Curative Furalaxyl and metalaxyl.

12.6 *Botrytis* Diseases

Botrytis cinerea, *B. tulipae*, and *B. gladiolorum* are *Botrytis* spp. that cause most damage in greenhouse crops.

12.6.1 Symptoms

Botrytis cinerea causes gray mold on a large range of hosts, including nearly all the major greenhouse plants. All plant parts at different growth stages may be damaged. Due to the diversity of the infected plant parts, several types of symptoms appear on one or on various hosts. On young stems, leaves, flowers, and fruits, initially water-soaked spots occur, which rapidly enlarge under favorable weather conditions. In tomato fruits, green-white circular spots called "ghost spots" also appear. On hard plant parts, such as stems and collars, *B. cinerea* causes cankers and parts above them may die. These symptoms are very common on vegetables such as tomato, eggplant, pepper, and cucumber. Infected tissues die soon and a gray mold which consists of conidiophores with clusters of spores develops on their surface. In plants, like tomato, black sclerotia develop inside the infected stems. *Botrytis cinerea* also causes very characteristic collar rot in lettuce. The infected plants usually develop large brown necrotic lesions on the stem near the soil surface and the lower leaves. The infection gradually progresses upward. Infected plants may wither and die in a short time (Sherf and Macnab 1986).

Botrytis tulipae causes tulip fire blight. Spots of various types on leaves and flowers, lesions on the stem, blossom blight, and bulb rot are the dominant characteristics. *Botrytis gladiolorum* damages gladiolus and some other Iridaceae. Large spots on leaves and the stem, pinpoint spots on the flowers, neck rot, and soft rot of corms are the most common symptoms. *Botrytis* spp. also infect all types of propagating material, which are either destroyed before planting out or become weak plants which may die before or after emergence. Finally, *Botrytis* spp. may cause severe postharvest losses in plant products during storage or transportation (Trolinger and Strider 1985).

Botrytis cinerea develops and sporulates profusely on any organic material. Spores are disseminated by wind over long distances or by water splashes. Healthy plants are infected through wounds, through senescent tissues, directly through the epidermis, and rarely through stomata. Symptoms may appear very quickly or infection may remain quiescent and symptoms appear later when tissues age or during storage. In greenhouses, initial infection depends on spores transferred from outdoors. Later, the inoculum established in the greenhouse is the main source of infection. In plants grown in non-heated greenhouses, low temperature, high RH, and low light intensity, prevalent from late November till late March, create good conditions for infection by *B. cinerea* (Elad et al. 1992; Jarvis 1992).

12.6.2 Management

Hygiene is very important. Detritus and infected leaves, stems, and flowers should be removed. Soil or growing medium should be well drained and a dense stand should be avoided. The relative humidity should be generally low, without too much water sprinkled over the plants. Plants should not be allowed to become wet (e.g., by condensation) or, if they do, should be dried as soon as practicable. Fungicides should be applied as sprays, as soon as symptoms are seen.

Management of environmental conditions, such as temperature, humidity, and duration of leaf wetness, together with sound cultural practices and use of fungicides, will help prevent disease development:

- Remove plant debris between crop cycles and during production.
- Dispose of diseased plants and debris in plastic bags; keep bags closed to prevent the

spread of spores to uninfected plants as the bag is removed from the greenhouse.
- Reduce humidity and leaf wetness duration to prevent spore germination, provide good air circulation, and reduce humidity within the canopy.
- Pay attention to proper planting dates, fertility, watering, and height management to prevent plants from becoming overgrown, with the aim of reducing humidity within the canopy.

Botrytis-incited diseases are prevented by ventilation and heating of greenhouses. Fungicides, mostly benzimidazoles and dicarboximides, are also used extensively. Nowadays, due to the predominance of resistant strains of the pathogen, they are only marginally effective and growers are advised to combine dicarboximides with other means of control such as biocontrol preparations. New fungicides have recently been released, but in greenhouses, they are used on a limited scale. Formulations of biological control agents such as Trichodex (*Trichoderma harzianum* T39) are also available (Elad et al. 1992; Gullino 1992).

12.6.3 Main Fungicides

Sprays Carbendazim, chlorothalonil, dichlofluanid, iprodione, procymidone, thiram, tolylfluanid, and vinclozolin.

Space Treatments Chlorothalonil, dicloran, iprodione, and vinclozolin.

12.7 Sclerotinia Rot, *Sclerotinia sclerotiorum*, *S. minor*

12.7.1 Symptoms

This is a common greenhouse disease that damages lettuce, eggplant, tomato, cucumber, pepper, etc. Infection on lettuce begins close to the soil, where a water-soaked area appears. Infection may spread downward to the roots or upward to the heart of the plants. Infected leaves fall onto the soil and dry up. The other plants are infected along the stem, leaves, flowers and fruits. Infected areas become water-soaked. Stem infection is more severe. Leaves above the infected area become yellow, then wither, and die. In wet weather, a white mass of mycelia appears on the infected areas, which gradually develops into black sclerotia. *Sclerotinia sclerotiorum*, which is the most common pathogen, produces sclerotia up to the size of bean seeds, whereas *S. minor* produces smaller sclerotia. Sclerotia fall onto the soil where they can survive for several years. When weather conditions are favorable, they germinate to produce apothecia which release ascospores and cause new infection. High RH and moderate temperature are required for infection (Fletcher 1984).

12.7.2 Management

The spaces between plants should be wide, moist conditions should be avoided, and affected plants should be removed. If infestation is to be expected, the soil should be steam-sterilized or the soil surface should be covered with mulch. Fungicides should be applied as sprays, as soon as symptoms are seen.

The elimination of sclerotia and the control measures recommended against gray mold are effective against *Sclerotinia* rot as well:

- Practice strict sanitation, removing debris from the greenhouse to prevent renewed infection.
- Control the quantities and the duration of irrigation to reduce soil wetness.
- Adopt sound cultural practices and fungicides to help prevent disease development.

12.7.3 Main Fungicides

Iprodione, procymidone, thiophanate-methyl, thiram, and vinclozolin

12.8 *Alternaria* Diseases

12.8.1 Symptoms

Strains of *Alternaria alternata* have been recorded: (i) causing cankers in tomato crops, (ii) causing leaf spotting in cucumber, and (iii) causing mostly postharvest rotting on tomato fruits. Also *A. cucumerina* may on occasion infect cucumber, melon, watermelon, and squash (Vakalounakis and Malathrakis 1987). At present, none of them has any economic impact on greenhouse crops.

All *Alternaria* species are facultative parasites mostly infecting weak plants. They survive in the soil on plant debris, but their black spores may also survive on several surfaces in greenhouses. *Alternaria solani* may survive on potato, which is an alternative host. Spores growing on dead material or on host plants are easily disseminated by wind or by splashed water. Plant infection takes place through stomata or directly through leaf surface. Spore germination and subsequent infection take place under a wide range of temperature. RH needs to be higher than 97 % for rapid germination, but germination may take place in some cases at RH >75 %. Senescent tissues are preferentially infected. The optimal temperature reported for *A. solani* is 18–25 °C and for *A. cucumerina* 20–32 °C. However, temperatures prevailing during the growing period of the respective hosts are not a factor limiting infection.

12.8.2 Management

Alternaria diseases can be prevented by dithiocarbamates, chlorothalonil, iprodione, etc. Hygienic measures and use of healthy propagating material are very important, especially when crops are grown in the soil. Inoculum surviving on plant debris in the soil and spores remaining on the greenhouse frames should be eradicated.

12.9 *Didymella* Diseases

12.9.1 Symptoms

Two very severe diseases of greenhouse crops are caused by *Didymella* spp.: *Didymella* stem rot or canker in tomato and eggplant {*Didymella lycopersici* [teleomorph of *Phoma lycopersici* (*Diplodina lycopersici*)]} and gummy stem blight in cucurbits [*Didymella bryoniae* (anamorph of *Phoma cucurbitacearum*)]. Both diseases damage all aerial plant parts of their hosts in greenhouses when weather is cool and RH high. They may infect the collar and root system causing yellowing and withering of the plants, which may later die. Cankers along the stem and the petioles are also very common. Plant parts above cankers may die. Both diseases cause large spots on the leaves which may cover the entire leaf surface. Tomato fruits are infected at the stem end. Initially, the infected area is light brown but it soon turns pink due to the large amount of pycnidiospores released. Infected parts may cover one-third of the fruit surface. Infection of cucumber and melon fruits by *D. bryoniae* appears mostly at the blossom end. Infection may occur only inside the fruit without being visible on the surface. Soon after infection, a lot of pycnidia appear on the infected areas and their color turns dark brown. Dark perithecia also appear a little later than pycnidia produced by *D. bryoniae*, while those of *D. lycopersici* are rare (Blancard et al. 1991).

The inoculum remains in plant residues inside and outside greenhouses. In the first case, infection starts through the collar. There is good evidence that infection of the aerial parts by *D. bryoniae* is initiated by ascospores released from infected plant material left outside greenhouses. In greenhouses, the two diseases are rapidly spread by water splashes and cultural practices.

12.9.2 Management

Soil disinfestation, destruction of plant residues, and strict hygienic conditions delay the outbreak of the disease. However, disinfested soil is readily reinfested. The fungicides commonly used in greenhouse against other fungal diseases are also effective. Moreover, the reduction of the RH and of free water on the leaf surfaces is very effective (Sherf and Macnab 1986).

12.10 Rust Diseases

12.10.1 Symptoms

The rust fungi (*Uredinales*) form orange or brown spots containing spores on stems or leaves. The spores are widely dispersed in the air. Examples are *Coleosporium tussilaginis* on cineraria, *Puccinia horiana* on chrysanthemum, *Puccinia pelargonii-zonalis* on pelargonium, *Pucciniastrum epilobii* on fuchsia, and *Uromyces dianthi* on carnation.

12.10.2 Management

Healthy plant material and resistant cultivars should be used. If plants are infected or seem to be infected, they should be removed. Overhead irrigation should be avoided because high humidity stimulates the disease. Fungicides should be applied as sprays, as soon as symptoms are seen.

12.10.3 Main Fungicides

Bitertanol, oxycarboxin, tolylfluanid, and triforine

References

Blancard D, Lecoq H, Pitrat M (1991) Maladies des cucurbitacées. Revue Horticole, INRA, Paris

Braun W (1995) The powdery mildews (Erysiphales) of Europe. Gustav Fischer, New York

Coyier DL (1985a) Powdery mildews. In: Strider DL (ed) Diseases of floral crops, vol 1. Praeger, New York, pp 103–140

Coyier DL (1985b) Roses. In: Strider DL (ed) Diseases of floral crop, vol 2. Praeger, New York, pp 405–417

Davies JML (1980) Diseases in NFT. Acta Horticult 98:299–305

Elad Y, Shtienberg D, Yunis H, Mahrer Y (1992) Epidemiology of grey mould, caused by *Botrytis cinerea* in vegetable greenhouses. In: Verhoeff K, Malathrakis NE, Williamson B (eds) Recent advances in Botrytis research. Pudoc Scientific Publishers, Wagenigen, pp 147–158

Fletcher JT (1984) Diseases of greenhouse plants. Longman, New York

Garibaldi A, Rapetti S (1981) Grave epidemia di peronospora su antirrino. Colture Protette 9:35–38

Gullino ML (1992) Chemical control *of Botrytis* spp. In: Verhoeff K, Malathrakis NE, Williamson B (eds) Recent advances in Botrytis research. Pudoc Scientific Publishers, Wagenigen, pp 217–222

Jarvis WN (1992) Managing diseases in greenhouse crops. APS Press, St Paul

Molot PM, Lecoq H (1986) Les oidiums des cucurbitacées. I. Données bibliographiques. Travaux preliminaries. Agronomie 6(4):355–362

Parmeter JR Jr (1970) Rhizoctonia solani: biology and pathology. University of California Press, Berkeley

Sherf AF, MacNab AA (1986) Vegetable diseases and their control. John Wiley and Sons, New York

Strider DL (ed) (1985) Diseases of floral crops, vols 1 and 2. Praeger Special Studies, New York

Tjamos EC (1992) Selective elimination of soil-borne plant pathogens and enhancement of antagonists by steaming, sub-lethal fumigation and solarization. In: Tjamos EC, Papavizas GC, Cook RJ (eds) Biological control of plant diseases. Plenum Press, London

Tompkins CM, Middleton JT (1950) Etiology and control of poinsettia root and stem rot caused by *Pythium* spp. and *Rhizoctonia solani*. Hilgardia 20:171–182

Trolinger JC, Strider DL (1985) Botrytis diseases. In: Strider DL (ed) Diseases of floral crops, vol 1. Praeger Special Studies, New York, pp 17–101

Vakalounakis DJ, Malathrakis NE (1987) A cucumber disease caused by the fungus *Alternaria alternata*. J Phytopathol 121:325–336

13 Bacterial Diseases and Their Management

Abstract

Some diseases affecting greenhouse-cultivated crops are caused by one or more bacterial infestations. Some bacteria primarily damage the root system (such as *Ralstonia solanacearum*, *Erwinia tracheiphila*, *R. solanacearum*), while others primarily manifest aboveground plant parts (like *Xanthomonas campestris* pv. *vesicatoria*, *Pseudomonas syringae* pv. *lachrymans*). Bacteria cannot be combated using crop protection agents, so solutions have to be found in prevention and in competition with other bacteria. As a result, it is necessary to find new methods of crop protection and disease combating. Greenhouse horticulture researchers are working on alternative methods and applications based on prevention (hygiene), antagonism and competition (other bacteria or fungi), climate control (ventilation), or physical solutions (light) for management of bacterial diseases.

Keywords

Bacterial diseases • Bacterial wilt • Leaf spot • Disease management

13.1 Introduction

Several bacterial diseases damage all types of greenhouse crops. Bacterial diseases are less frequent but under high moisture and poor irrigated condition may cause huge damage. The most common bacterial diseases are bacterial soft rot (*Erwinia carotovora* ssp. *carotovora*), bacterial leaf spot (*Xanthomonas campestris* pv. *vesicatoria*), angular leaf spot (*Pseudomonas syringae* pv. *lachrymans*), and bacterial wilt (*Ralstonia solanacearum*, *Erwinia tracheiphila*).

P.P. Reddy, *Sustainable Crop Protection under Protected Cultivation*,
DOI 10.1007/978-981-287-952-3_13

13.2 Bacterial Leaf Spot, *Xanthomonas campestris* pv. *vesicatoria*

13.2.1 Symptoms

Bacterial leaf spot occurs on tomato and pepper. The symptoms are water-soaked circular brown spots during rainy periods. On green fruits, lesions become slightly raised and surrounded by greenish white halos that eventually turn irregular and brown, with a sunken scabby surface. While it occurs primarily on peppers, all aboveground parts of tomatoes are also susceptible. The spots on the leaves are chocolate brown and irregularly shaped with areas of dead leaf tissue; initially they are less than 0.6 cm in diameter (Fig. 13.1). Severely spotted leaves appear scorched and defoliation may occur. It is most prevalent during moderately high temperatures and long periods of leaf wetness.

13.2.2 Survival and Spread

The bacterium cannot live in soil for more than a few weeks. *Xanthomonas campestris* pv. *vesicatoria* survives on volunteer tomato, plant debris, soil, and seeds. It is spread by heavy rainfall, wind, and seeds. Disease development is favored by warm temperatures (24–30 °C) and high relative humidity.

Fig. 13.1 Bacterial leaf spot on bell pepper

13.2.3 Management

Xanthomonas campestris pv. *vesicatoria* does not survive well in decayed plant material. Therefore, removal of crop debris and crop rotation are effective for preventing the spread of this bacterium. Another important management method is to eliminate volunteer plants (plants that grow without being specifically planted or managed by farmers) as wild pepper and tomato plants can harbor the disease in warmer climates (The American Phytopathological Society website 2012).

Diseased seeds may be treated with sodium hypochlorite, calcium hypochlorite, or trisodium phosphate to kill bacteria on the surface of the seed (The American Phytopathological Society website 2012). Infected fields are treated with multiple application approach involving spraying a mixture of copper and mancozeb, which prevents further infection and kills a variety of pathogens present at the time of spraying (Obradovic et al. 2004).

13.3 Bacterial Wilt, *Ralstonia solanacearum*

Bacterial wilt caused by *R. solanacearum* is highly challenging and one of the most destructive diseases of solanaceous crops including tomato, potato, chili, and brinjal worldwide. The disease is predominant in warm humid tropical and temperate regions of the world (Hayward 2005). In India, it occurs across the country mainly in states of coastal areas, foot hills, and lower altitude of hills (Singh et al. 2010). It has a wide host range of about 200 different groups of plants with 50 families, which accounts severe damage to the crops. The damage caused by this disease to the crop is>60 % depending on environmental conditions and variety of crops. Bringing about severe crop losses worldwide, the disease is now receiving global profile (Allen et al. 2005). It is disturbing agriculture and even trade negotiations in the developing as well as developed world. It is listed as one of the top ten microorganisms affecting agriculture in the US Agro-terrorism Protection Act of USDA (2002) and subjected to strict quarantine regulations throughout Europe and America.

13.3.1 Symptoms

The wilting symptom on solanaceous crops (tomato, chili, and brinjal) can be induced by bacterial and fungal pathogen, root-knot nematode, and deficit or excess of soil moisture. The most characteristic symptoms of tomato, chili, and brinjal are very rapid wilting, especially where the plants are young and succulent. The flowering stage is the most critical stage where a plant shows sudden wilting (Fig. 13.2).

The first visible symptom is the flaccid appearance of the youngest leaves in the field. Other primary indications of infection are stunting and downward curling of leaflets and petioles. Usually infected plants collapse quickly, but where this fails to happen, there is a development of blackening of the vascular system at the junction between the stem and leaf. Further down the stem, the whole vascular system may be completely blackened and when cut oozes creamy bacterial slime.

Under favorable environmental conditions for the pathogen (soil temperatures of approximately 25 °C; saturated humidity), epinasty and wilting of one side or of the whole plant follows within a few days leading to total plant collapse. Under less favorable conditions (soil temperature below 21 °C), less wilting occurs, but large numbers of adventitious roots may develop on the stem. It is possible to observe water-soaked streaks from the base of the stem, which is evidence of necrosis in the vascular system. When the stem is cut crosswise, discolored brown vascular tissues exude white or yellowish bacterial ooze.

Fig. 13.2 Bacterial wilt symptoms on tomato

13.3.2 Bacterial Ooze Test

Ralstonia solanacearum is a limited xylem-invading pathogen and plants wilted by this pathogen have $>10^8$ cfu/g of tissue. A common sign of bacterial wilt of tomato observed at the surface of freshly cut sections from severely infected stems is a sticky, milky-white exudate, which indicates the presence of dense masses of bacterial cells in infected vascular bundles and particularly in the xylem (Fig. 13.3), which is responsible for transportation of raw sap (water and nutrients) from roots to aerial parts of the plant. Ooze also may accumulate on the cut surface on the infected tissue. Allen et al. (2001) reported that, even if ooze does not form spontaneously, a streaming test may be positive. Other wilt-inducing pathogens do not produce comparable ooze. The ooze is usually an almost pure culture of *R. solanacearum,* which can be cultured on standard, low-ionic-strength bacteriological media. This water streaming test (ooze test) is of presumptive diagnostic value in the field.

13.3.3 Survival and Spread

Bacterial wilt caused by *R. solanacearum* can survive in weeds, in seeds, and in soil for 2–10 years. It is disseminated by running water, soil movement, or movement of infected transplants. Disease development is favored by high temperatures (30–35 °C) and high relative humidity.

Fig. 13.3 Exudates of dense masses of bacterial cells from infected vascular bundles

13.3.4 Management

- Use disease-free seeds.
- Treat seeds with streptocycline at 500-ppm concentration.
- Grow bacterial wilt-resistant varieties of the following crops:
  - Brinjal: Arka Keshav (long type), Arka Nidhi (long type), Arka Neelkanth (long type), Pusa Anupam (round type), Hisar Shyamal (round type), and Singnath
  - Tomato: Palam Pink, Palam Pride, and F1 7711, Arka Rakshak, Arka Abha, and Swarna Sampada
  - Chili: Surajmukhi

- Disinfecting the cutting knives with suitable bactericide.
- Avoidance of surface water for irrigation.
- The incidence of wilt diseases in tomato was reduced by applying panchagavya which contains 20 ml of cow ghee, 50 ml each of cow milk and curd, 40 ml of cow urine, 400 g of cow dung, 20 g of common salt, and 10 g of baker's yeast.
- Biocontrol agents *Pseudomonas fluorescens* and *Bacillus subtilis* were found effective to control the disease.
- Crop rotation is not effective as the pathogen can survive for a long period (several years) in the soil and also attack a wide range of crops and solanaceous weeds. Crop rotation 2–3 years with wheat and *Crotalaria juncea* is helpful.
- Do not grow crops in soil where bacterial wilt has occurred.
- Rogue out wilted plants from the field to reduce spread of the disease from plant to plant.
- Control root-knot nematodes since they could facilitate infection and spread of bacterial wilt.
- Where feasible, extended flooding (for at least 6 months) of the infested fields which can reduce disease levels in the soil.
- Soil amendments (organic manures) can suppress bacterial wilt pathogen in the soil. Good results have been encountered by using biofumigation as soil treatment for bacterial wilt.
- Application of bleaching powder 12–15 kg/ha in the furrows during transplanting in the field.
- Actigard (acibenzolar-s-methyl) has shown to be effective, but only when inoculum densities are low.
- In India, CRA 66 rootstocks were used in grafting of tomato to reduce bacterial wilt in tomatoes (Tikoo 1979). Several Hawaiian lines (Hawaii 7996–7998) have been identified as suitable candidates for resistance to bacterial wilt. This technique could be a very valuable tool for eliminating bacterial wilt in tomato, pepper, and eggplant production systems.

13.4 Bacterial Soft Rot, *Erwinia carotovora* ssp. *carotovora*

13.4.1 Symptoms

Bacterial soft rot occurs on potato, tomato, cucumber, lettuce, and iris. The disease causes the plant to wilt rapidly; the pith usually disintegrates, becoming wet and slimy and leaving a hollow stem (Fig. 13.4). Whole fruit may become a soft, watery, colorless, decayed mass in 3–5 days. Infected plant parts produce a foul odor that is very characteristic.

13.4.2 Survival and Spread

Erwinia carotovora ssp. *carotovora* survives in soil, plant debris, and weeds. It is spread by surface irrigation of soil, by wind and seeds, and through cultural practices. Disease development is favored by warm temperatures (25–30 °C) and high relative humidity.

13.4.3 Management

Avoid reuse of irrigation tail water and overhead irrigation. Prevent bruising and wounding of plants during field operations. Copper bactericides provide some bacterial control when applied before disease is observed.

Fig. 13.4 Soft rot on lettuce

13.5 Angular Leaf Spot, *Pseudomonas syringae* pv. *lachrymans*

Angular leaf spot is more or less prevalent in almost every cucumber growing area of the world causing heavy losses not only to cucumber but to other cucurbits as well.

13.5.1 Symptoms

Under natural conditions, the disease initially appears as minute water-soaked spots on leaves, which became and turn light tan in color (Fig. 13.5). With the passage of time, spots turn necrotic and many of these slough-off giving leaves a tattered appearance. The spots on infected fruits lead to rot or malformation of fruit in case of severe infection.

13.5.2 Survival and Spread

Both naturally infected as well as the artificially inoculated seed can harbor the pathogen beyond two sowing seasons. Pathogen survives on diseased crop debris buried or kept on soil surface from harvesting to next sowing time.

Fig. 13.5 Angular leaf spot on cucumber

13.5.3 Management

The disease is controlled by hot water treatment (hot water at 50 °C for 30 min as seed treatments proved effective), chemical seed treatment (immersion of seeds in streptomycin at 80–100 ppm for 15 min gave good disease control), foliar sprays, and use of systemic resistance inducers. Soak treatment of infected seeds in sodium hypochlorite solution at 1: 20 for 20 min or calcium hypochlorite at 0.25–0.5 % for 60 min was recommended as effective seed treatments against the disease (Watanabe and Ohuchi 1983). Klimach et al. (2001) reported that, under greenhouse conditions, application of bronopol at 0.01 and 0.02 % significantly decreased the intensity of angular leaf spot of cucumbers.

Raupach and Kloepper (1998) reported that level of systemic resistance induced by seed treatments with PGPR *Bacillus pumilus*, *B. subtilis*, and *Curtobacterium flaccumfaciens* in cucumber against angular leaf spot. Mixtures of PGPR were able to induce systemic resistance and offer significant protection with or without use of soil fumigant methyl bromide also (Raupach and Kloepper 2000).

13.6 Bacterial Wilt, *Erwinia tracheiphila*

Bacterial wilt is a cucurbit disease, caused by the Gram-negative bacterium in the family *Enterobacteriaceae*. Cucumber and muskmelon plants are most susceptible, but squash, pumpkins, and gourds may also become infected. Watermelons are immune to the disease.

13.6.1 Symptoms

Bacterial wilt is a disease of the vascular tissue. When a plant is infected, *E. tracheiphila* multiplies within the xylem, eventually causing mechanical blockage of the water transport system. The first sign of infection, which appears about 5 days after acquisition, is the wilting of individual leaves on a single stem (Fig. 13.6). However, the disease will soon spread down the runner and then infect the whole plant, causing it to shrivel and die. There is a diagnostic test for bacterial wilt that can be done in the field. The presence of the *E. tracheiphila* causes the sap to become a milky color and acquire a sticky consistency. If the stem is cut near the crown and the ends are slowly pulled apart, the sap should form a viscous string (Fig. 13.6).

13.6.2 Transmission

Erwinia tracheiphila is spread between plants by two species of insect vectors, striped cucumber beetles (*Acalymma vittatum*) and spotted cucumber beetles (*Diabrotica undecimpunctata*). The beetles acquire *E. tracheiphila* by feeding on infected plants and carry the bacteria in their digestive tracts. The disease may be spread to susceptible plants through feeding wounds, by way of infected mouthparts or frass. The bacterium is capable of overwintering in the gut of its insect vectors.

Fig. 13.6 Symptoms of bacterial wilt on cucurbit

13.6.3 Management

The most effective preventative measure is to keep beetle populations as low as possible through careful field monitoring and insecticidal sprays.

13.7 Management of Bacterial Diseases

A similar management approach can be adopted for all the above bacterial diseases:

- Use certified seed from a reputed source.
- If you do not grow your own transplants, ensure that the transplant grower follows a disease management program (to start a bacterial disease management program at transplanting time is far too late).
- Promptly remove infected plants and adjacent plants to prevent further infection.
- Avoid unnecessary handling of plant material.
- Reduce extended leaf wetness: never allow dew to form on the plants; minimize the length of time the leaves are wet; carry out irrigation early in the day, under conditions in which the foliage can dry in 1–2 h; and maintain good ventilation (one of the most important management factors in controlling bacterial diseases in the greenhouse).
- Monitor the plants in the greenhouse: at the first sign of bacterial disease symptoms, remove the affected plants and as many surrounding trays as possible (plants in surrounding trays may be infected and could spread the pathogen even when not yet showing symptoms).
- Disinfect all benches, equipment, flats, and stakes.
- Include copper in the disease management program: there are no products available to cure bacterial diseases, but copper fungicides can slow bacterial disease development, and if used correctly (preventively), they can avert an outbreak, even if the pathogen is present, by keeping bacterial populations low. A number of copper products are registered for bacterial disease control.

References

Allen C, Kelman A, French ER (2001) Brown rot of potatoes. In: Stevenson WR, Loria R, Franc GD, Weingartner DP (eds) Compendium of potato diseases, 2nd edn. The American Phytopathological Society Press, St Paul, pp 11–13

Allen C, Prior P, Hayward AC (2005) *Bacterial wilt disease and the* Ralstonia solanacearum *species complex*. The American Phytopathological Society, St Paul

Hayward AC (2005) Research on bacterial wilt: a perspective on international links and access to literature. In: Allen C, Prior P, Hayward AC (eds) *Bacterial wilt disease and the* Ralstonia solanacearum *species complex*. The American Phytopathological Society, St Paul, pp 1–8

Klimach A, Kubica S, Owska K, Wieczorek W (2001) Evaluation of bronopol antibacterial activity against some bacterial diseases of plants. Pestycydy 3–4:39–46

Obradovic A, Jones JB, Momol MT, Balogh B, Olson SM (2004) Management of tomato bacterial spot in the field by foliar applications of bacteriophages and SAR inducers. Plant Dis 88:736–740

Raupach GS, Kloepper JW (1998) Mixtures of plant growth-promoting rhizobacteria enhance biological control of multiple cucumber pathogens. Phytopathology 88:1158–1164

Raupach GS, Kloepper JW (2000) Biocontrol of cucumber diseases in the field by plant growth promoting rhizobacteria with and without methyl bromide fumigation. Plant Dis 84:1073–1075

Singh D, Sinha S, Yadav DK, Sharma JP, Srivastava DK, Lal HC, Mondal KK, Jaiswal RK (2010) Characterization of biovar/ races of *Ralstonia solanacearum*, the incitant of bacterial wilt in solanaceous crops. Indian Phytopath 63(3):261–265

The American Phytopathological Society website (2012) Bacterial spot of tomato and pepper. http://www.apsnet.org/edcenter/intropp/lessons/prokaryotes/Pi

Tikoo S (1979) Successful graft culture of tomato in bacterial wilt sick soils. Curr Sci 48:259–260

Watanabe Y, Ohuchi A (1983) Angular leaf spot of cucumber in Japan. J Agric Res Q 17:112–119

14 Viral Diseases and Their Management

Abstract

Viruses are very small, submicroscopic particles that cause plant disease. There is no control for plants infected with a virus. Some of the more common viruses that may infect greenhouse crops include (but are not limited to): impatiens necrotic spot virus (INSV) and tomato spotted wilt virus (TSWV) which are also known as tospoviruses, cucumber mosaic virus (CMV), and tobacco mosaic virus (TMV). Viruses can be transmitted by insects, primarily aphids, thrips, and whiteflies. Major virus diseases are transmitted by aphids, whiteflies, thrips, beetles, and fungi and mechanically. Management of viral diseases using production of virus-free vegetables by using insect-proof nets, integrated management, and GAP recommendations are discussed.

Keywords

Viral diseases • Virus transmission • Management strategy

14.1 Introduction

Viruses are a major problem in greenhouse crops especially in temperate regions. However, most efforts in programs for integrated pest and disease management are focused on pest and fungal or bacterial disease control and few recommendations are given for viral diseases. In general, viruses are not considered at all or are treated in a very simplistic manner. The main reason for this is the lack of information about viral disorders to give recommendations to deal with plant virus problems. In addition, in contrast to pests, fungi, or bacteria, no direct control methods can be used against viruses. Nevertheless, in recent years, a significant progress in knowledge on plant viruses has occurred, and valuable information has been obtained that will facilitate the development of control strategies. Because of difficulties and costs of reducing the spread of viruses by controlling their vectors and sources of infection, the introduction of resistance to a particular virus into commercially useful cultivars is the best control method but, unfortunately, the exception. Most virus management programs involve the integration of indirect measures to avoid or reduce the sources of infection and dispersal of the virus or the minimization of the effect of infections on crop yield. When confronted with a virus problem, the understanding of the ecology

P.P. Reddy, *Sustainable Crop Protection under Protected Cultivation*,
DOI 10.1007/978-981-287-952-3_14

and epidemiology of the disease will provide the information needed to make strategic decisions for virus disease control.

Tomato, cucumber, and capsicum are very sensitive to virus diseases under a protected environment. Virus often spreads in the plantation by insect vectors such as whitefly, thrips, and aphids. The damage caused by the virus is usually much greater than the mechanical injury caused by the insect vector. Plant tissue damaged by a viral disease does not die immediately. The most important symptom of viral infections is the light (white or yellow) color of the leaves or a mosaic pattern of light and darker shades of green on the leaves. In many cases, viral disease leads to dwarfed growth, rosette formation, or other strange stem, fruit, and leaf deformations. The symptoms of viral infections are often not found everywhere in a cultivated field but rather in patches and also sometimes without symptoms. Viruses prevalent among greenhouse crops include tobacco mosaic virus or tomato mosaic virus (TMV or ToMV), cucumber mosaic virus (CMV), tobacco etch virus (TEV), potato virus Y (PVY), potato leaf roll virus (PLRV), tomato spotted wilt virus (TSWV), alfalfa mosaic virus (AMV), pepper veinal mottle virus (PVMV), pepper mild mottle virus (PMMY), chili veinal mottle virus (CVMV or ChiVMV), tomato yellow leaf curl virus (TYLCV), and tomato bigbud mycoplasma (TBB).

Ornamentals are affected by a great number of viruses, of which some examples are given below (with their means of transmission in brackets): on rose, *Prunus necrotic ringspot ilarvirus* (PNRSV) (vegetatively propagated through rootstocks); on chrysanthemum, *tomato aspermy cucumovirus* (TAV) (aphids); on pelargonium, *pelargonium leaf curl tombusvirus* (PLCV) (mechanically transmitted) and others; on carnation, *carnation vein mottle potyvirus* (CVMV) (aphids), *carnation ringspot dianthovirus* (CRSV) (mechanically transmitted), *carnation latent carlavirus* (CLV) (aphids), *carnation etched ring caulimovirus* (CERV) (aphids), and *carnation mottle carmovirus* (CarMV) (mechanically transmitted); on tulip and other liliaceous bulbs, *tulip breaking potyvirus* (TBV) (aphids) and *tobacco necrosis necrovirus* (TNV) (*Olpidium* spp); on achimenes, alstroemeria, eremurus, gerbera, ornithogalum, and phlox, *tobacco rattle tobravirus* (TRV) (free-living nematodes); and on alstroemeria, anthurium, begonia, chrysanthemum, impatiens, kalanchoë, pelargonium, and phalaenopsis, *tomato spotted wilt tospovirus* (TSWV) (thrips).

Virus diseases cause major losses in the quality of ornamental plants. Infection by viruses may cause various symptoms such as yellowing, leaf roll, leaf curl, chlorotic spots and rings, necrosis, and leaf and flower malformation.

In many circumstances, control strategies are based on the dispersal procedures used by viruses in nature, and similar control measures are recommended for viruses with equivalent dispersal manners. Therefore, virus dispersal mechanisms and the deduced control methods will be briefly reviewed in before major diseases caused by plant viruses in protected crops are described.

14.2 Plant Virus Dispersal Mechanisms

The ability of a virus to be disseminated and perpetuated in time and space depends upon which methods are used for dispersal. Figure 14.1 summarizes the main transmission mechanisms of plant viruses; one or several of them can be exploited by a specific virus. The knowledge about the main dispersal procedures of a virus in nature will provide a means to prevent and control viral diseases: to minimize sources of infection, to reduce dissemination during growing practices, and/or to limit spread by vectors. Some aspects of virus dispersal and their importance in virus control are analyzed below (Fig. 14.1).

14.2.1 Sources of Infection

As a general rule, virus-infected plants are sources for secondary spread by mechanical or biological vector means and, therefore, should be eliminated as soon as possible. Mechanical transmission is one of the most dangerous dispersal methods for viruses in protected crops due to the frequent handling of plants during the intensive

- **Transmission by biological vectors**
  - **Aerial**
    - **Mites**
    - **Insects**
      - **Chewing mouthparts (beetles)**
      - **Sap sucking mouthparts (aphids, thrips, leafhoppers, whiteflies, etc.)**
  - **Soil inhabiting**
    - **Fungi: *Olpidium, Polymyxa, Spongospora***
    - **Nematodes: *Xiphinema, Longidorus, Trichodorus***
  - **By dodder (*Cuscuta* sp.)**
- **Transmission without biological vectors**
  - **Mechanical transmission: growing practices, plant contact, infected plant debris (roots, aerial parts)**
  - **Through seed and pollen**
  - **Through vegetative propagation: cuttings, tubers, bulbs, corms, rootstocks**
  - **Graft transmission**

Fig. 14.1 Main transmission strategies used by plant viruses

cropping practices. Some viruses are extremely important in protected crops because of their efficient transmission by mechanical inoculation during cultural operations. If plants infected with some of these viruses are suspected to be present in a crop, secondary spread can be reduced by adequate treatment of hands and implements during plant handling. In these cases, plant debris in soil and greenhouse structures is important source for primary infections in subsequent sensitive crops and, therefore, as long as possible, they should be eliminated and soil and structures disinfected. The propagation material used for planting can be a very effective means of introducing viruses into a crop at an early stage, giving randomized foci of infection within the planting. If other transmission methods (e.g., mechanical, insects) are coupled, which may rapidly spread the virus within the crop, then infected seeds, plantlets, etc., can be of significant importance in the epidemic of the disease. In these cases, certified virus-free material should be used as the basis to control the virus. Approximately 18 % of the known plant viruses are seed transmitted in one or more hosts (Mink 1993; Johansen et al. 1994). The rate of seed transmission is very variable depending on the virus/host combination and is not necessarily a good indicator of the epidemiological importance; low transmission rates combined with efficient secondary spread can be very important epidemically. Tolerance levels in a seed certification program will depend, therefore, on the kind of secondary spread. For example, only very low infection levels are permitted in lettuce seed lots for an effective control of the lettuce mosaic virus (LMV) because of its efficient secondary spread by aphids; good control was obtained in California if less of one seed in 30,000 was infected (Dinant and Lot 1992).

For many vegetatively propagated crops like ornamentals (carnation, tulip, etc.), the main virus sources are infected plants themselves and their vegetative derivatives (cuttings, tubers, bulbs, corms, rootstocks). In these cases, control may be done by using virus-free stocks and certification schemes to produce propagation material free of virus.

Soil may be another source of virus infection. Soilborne viruses can be transmitted by fungi or nematodes or can have no biological vector like tobamoviruses that are very stable and are maintained in infected plant debris mixed with the soil. Control usually is through soil disinfection if no resistant cultivars are available. The maintenance of virus-sensitive crops continuously throughout the year will ensure the permanent presence of significant levels of inoculum and, then, of virus infection. Therefore, crop rotations should incorporate nonsensitive species. However, although a rupture of the infection cycle is done, the presence of alternative hosts for the virus in the surroundings of the protected crop can be of special relevance to perpetuate the virus. The management of these hosts will help to the control of the virus.

14.2.2 Vector Transmission

Viral diseases may infect hundreds of plant species including tomatoes, peppers, eggplant, and cucumber. Most viruses are spread by insect vectors, such as whitefly, aphid, and thrips; others are seedborne and mechanically transmitted (Table 14.1).

Many important viruses in protected crops are transmitted from plant to plant by invertebrates. Sap-sucking insects are the main vectors, mostly Homoptera, and among them, aphids are the most important, transmitting 43 % of known viruses.

Control of insect-transmitted viruses has been traditionally done by spraying insecticides to reduce the vector populations. However, the effectiveness of treatments in controlling the virus depends on virus/vector transmission relationships. Table 14.2 summarizes the main properties of the different kinds of relationships based on the feeding times needed by the vector to acquire (*acquisition time*) and inoculate (*inoculation time*) the virus, on the *latent period* from acquisition until the vector is able to transmit the virus, and on the *retention time* during which the vector remains infective following inoculative feeding without further access to the virus. This classification is mainly based on aphid-transmitted viruses. No evidence for virus in the hemocoel or salivary system exists in the *noncirculative* transmission. In the *circulative transmission*, virus is acquired by feeding, enters the hemocoel via the hindgut, circulates in the hemolymph, and enters the salivary gland. Inoculation results from transport of virus into the salivary duct and introduction of saliva into the plant during feeding. If virus multiplies in the insect cells, then the transmission is called *propagative.*

Insecticide treatments may be ineffective in controlling nonpersistently transmitted viruses (short acquisition and inoculation times, no latent period) (Table 14.2), because acquisition, latent, and inoculation times are so short that the virus is acquired and transmitted before the vector can be affected by most insecticides. However, especially in protected crops, chemical treatments can help to reduce the overall vector populations and therefore secondary spread of the disease. For nonpersistently transmitted viruses, oils or tensioactive film-forming products have been reported to be effective in controlling virus acquisition and inoculation in outdoor crops. Insecticidal treatments used to control semipersistently (long acquisition and inoculation times, no latent period), or circulatively (long acquisition and inoculation times, latent period), transmitted viruses can be effective in controlling the virus because longer acquisition, inoculation, and/or latent times are needed and the vector may die before the virus can be transmitted. In any case, it should be noted that the small percentages of insects that usually survive the treatments are enough to cause important infections if virus sources are present. Accurate knowledge of disease epidemiology in a certain region will provide information about the critical periods of infection, which will facilitate decisions on when treatments should be done, or the adjustment of

Table 14.1 Main viral diseases of vegetable crops by host and means of transmission

	Crop		
Transmission	Tomato	Pepper	Cucumber
Seedborne	TMV	TMV	CGMMV
Mechanical	TMV	TMV	CGMMV
Aphids	CMV, PVY	CMV, PVY	CMV, ZYMV
Whiteflies	TYLCV	–	CVYV, CYCDV
Thrips	TSWV	TSWV	–

Table 14.2 Classification of virus/vector relationships based on time needed for acquisition and inoculation of the virus, latent period, and time of retention

	Noncirculative		Circulative (persistent)	
	Nonpersistent	Semipersistent	Non-propagative	Propagative
Acquisition time	Seconds	Minutes/hours	Minutes/hours	Minutes/hours
Latent period	0	0	Hours/days	Hours/days
Inoculation time	Seconds	Minutes/hours	Hours	Hours
Retention time	Minutes	Hours	Days	Entire life

planting dates to avoid high vector populations in young plantings (Zitter and Simons 1980).

14.3 Major Virus Diseases

14.3.1 Aphid-Transmitted Viruses

14.3.1.1 Cucumber Mosaic Virus (CMV)

CMV is the type species of the genus *Cucumovirus* of the family *Bromoviridae* of plant viruses. CMV virions are 29-nm icosahedral particles. A great variability among CMV isolates has been reported. According to biological properties of symptomatology, thermosensitivity in vivo, and molecular and serological characteristics, most CMV isolates have been assigned to two main groups.

14.3.1.1.1 Distribution and Economic Importance

CMV is distributed worldwide, predominantly in temperate regions but with increasing importance in tropical countries. It causes serious diseases in many important crops grown in the open but also in protected conditions (tomato, pepper, cucurbits, etc.) (Jordá et al. 1992). Yield reductions are mainly due to decreased fruit set and production of nonmarketable fruits because of reduced size or presence of symptoms like mosaics, malformations, or necrosis. Control of CMV is difficult because of the wide host range and its rapid natural transmission by aphids.

14.3.1.1.2 Symptoms

Symptoms are extremely variable depending on the CMV isolate, host species or cultivar, plant age at infection time, and environmental conditions. Early infected plants can show marked stunting. Symptoms in leaves are mosaic, mottle, and/or distortion. Necrosis is induced by certain isolates. Flower abortion and fruit discoloration and malformations are caused.

14.3.1.1.3 Host Range

CMV has an extremely wide host range that comprises more than 1000 species of dicotyledons and monocotyledons. Host range includes many important vegetable crops like melon, cucumber, zucchini squash, watermelon, tomato, pepper, eggplant, lettuce, carrot, celery, spinach, pea, etc.; ornamentals like anemone, aster, dahlia, delphinium, geranium, lily, periwinkle, primula, petunia, viola, zinnia, etc.; and woody and semi-woody plants like banana, ixora, passion fruit, etc.

14.3.1.1.4 Transmission

In nature, CMV is transmitted in a nonpersistent manner by more than 60 aphid species including *Aphis gossypii*, *Macrosiphum euphorbiae*, and *Myzus persicae*. Variable rates of seed transmission have been described in 20 species including some vegetable crops like bean or spinach or weeds like *Stellaria media*. There is no evidence of seed transmission in cucurbits. CMV can be mechanically transmitted in experimental conditions.

14.3.1.1.5 Management

Integrated control measures are recommended in protected crops to reduce CMV incidence:

- Elimination of infected plants
- Avoidance of aphid entrance in the greenhouse by covering entrances with aphid-proof nets
- Reduction of aphid populations by using insecticides
- Use of virus-free seeds (e.g., in bean and spinach)
- Elimination of alternative spontaneous hosts present in and around the crop

Resistance to CMV is available in cucumber and programs are in course in melon using Korean and Chinese varieties. Sources of resistance or tolerance have been found in most cultivated or related species. However, in most cases, resistance or tolerance is not absolute and is overcome by some CMV species. Aphid vector tolerance or resistance incorporated in the plant can be combined with other control methods. Transgenic melon, cucumber, and squash plants expressing the coat protein gene of CMV offer a good level of resistance to several strains of the virus.

14.3.1.2 *Potyvirus* Genus

The *Potyvirus* genus of the family *Potyviridae* is by far the largest of the plant virus groups. Many members cause important economic losses in protected crops and can be a major limiting factor for production. Virus particles are elongated and flexuous (680–900 × 11 nm) rods. Virus infections are associated with characteristic cytoplasmic and nuclear inclusions, pinwheels, bundles, and laminated aggregates (Shukla et al. 1994).

14.3.1.2.1 Distribution and Economic Importance

The *Potyvirus* genus is the most devastating among plant viruses. Damaging members like BCMV, bean yellow mosaic virus (BYMV), ZYMV, WMV2, PRSV-W, and PVY are spread worldwide and cause economically important problems wherever present. Several authors reported losses up to 100 % in squash, cucumber, and watermelon caused by ZYMV. Potyviruses are mainly a problem in outdoor crops; however, they can also be a severe threat in protected crops.

14.3.1.2.2 Symptoms

Potyviruses can induce severe diseases in important crops. Symptoms may vary depending on host species, virus strain, environmental conditions, and plant age at infection time. Potyviruses like ZYMV, WMV2, and papaya ringspot virus-W strain (PRSV-W) can cause severe diseases in zucchini, squash, melon, cucumber, and watermelon, inducing stunting, chlorosis, mosaic, leaf malformation, flower abortion, and fruit and seed malformation. Vein clearing, mosaic, yellow mottling, and growth reduction are often observed in LMV infections of lettuce, endive, and spinach. Legume-infecting potyviruses like BCMV cause abnormal formation of seeds that are smaller, discolored, and/or distorted.

14.3.1.2.3 Host Range

In nature, most potyviruses have relatively narrow host ranges, few species within one genus or closely related genera; for example, BCMV is restricted to *Phaseolus* species; potato virus Y (PVY) to members of the Solanaceae; watermelon mosaic virus 2 (WMV2), zucchini yellow mosaic virus (ZYMV), and zucchini yellow fleck virus (ZYFV) mainly to species of the Cucurbitaceae; and LMV to species mainly in Compositae.

14.3.1.2.4 Transmission

Potyviruses are transmitted in nature by aphids in a nonpersistent manner. Some aphid species (especially those of the genera *Myzus*, *Aphis*, and *Macrosiphum*) are associated with high virus incidences in crops. Seed transmission is important epidemiologically in certain potyviruses, like bean common mosaic virus (BCMV) in French bean or LMV in lettuce. Potyviruses can be transmitted experimentally by mechanical inoculation.

14.3.1.2.5 Management

Control should be done by an adequate management of the crop, integrating different control measures: if seed transmitted, the use of certified virus-free seeds is the basis for effective control; the use of virus-free plantlets will avoid primary infections; because transmitted in a nonpersistent manner, spraying insecticides is not effective for preventing virus spread; however, effective control has been obtained in some cases by spraying with light mineral oils in outdoor crops. Successful breeding programs for resistant cultivars have been done in lettuce to LMV (resistance-breaking strains have recently been described), in French bean to BCMV, and in melon to PRSV-W. Transgenic approaches have also been explored, overcoming difficulties associated with conventional breeding methods. Cross protection using an attenuate poorly aphid transmissible strain of ZYMV (ZYMV WK) have been successfully used to control CMV.

14.3.1.3 *Luteovirus* Genus

There are a number of yellowing diseases transmitted in nature by aphids that are caused by viruses in the *Luteovirus* genus. This is the case of beet western yellows (BWYV) and cucurbit aphid-borne yellows (CABYV) viruses. Viral particles are 25–30 nm and icosahedral in shape.

14.3.1.3.1 Distribution and Economic Importance

It was first described in France in outdoor and protected crops and has been found throughout the Mediterranean area, Asia, Africa, and California.

It has been reported in North America, Europe, and Asia and is probably distributed worldwide.

In melon and cucumber, important yield losses are reported, due to reduced number of fruits per plant caused by flower abortion but not by altering fruit shape or quality.

14.3.1.3.2 Symptoms

BWYV infects lettuce, cucumber, watermelon, squash, sugar beet, carrot, spinach, pepper, and tomato, symptoms being mild chlorotic spotting, yellowing, thickening, and brittleness of older leaves. CABYV causes a yellowing disease of melon, cucumber, and zucchini squash; symptoms are initial chlorotic patches, leaf thickening, and general bright yellowing of leaves.

14.3.1.3.3 Transmission

Transmission in nature is by aphids in a circulative, non-propagative, persistent manner.

14.3.1.3.4 Management

Disease management should be by integrating measures to reduce aphid populations within the greenhouse via avoidance of insect entrance (nets in windows) and chemical spraying, with measures to reduce infection foci (virus-free planting material, elimination of infected plants). Sources of resistance have been found for CABYV in melon germplasm and for BWYV in lettuce (Dogimont et al. 1996).

14.3.2 Whitefly-Transmitted Viruses

14.3.2.1 *Tomato Yellow Leaf Curl Virus* (TYLCV)

TYLCV is a member of the *Geminivirus* genus of plant viruses whose virions have twin isometric particle morphology. Based on its transmission by the whitefly *Bemisia tabaci* to dicotyledons, TYLCV belongs to the subgroup III of geminiviruses.

14.3.2.1.1 Distribution and Economic Importance

TYLCV causes devastating damages in tomato crops of the Mediterranean basin, subtropical Africa, and Central America. Losses are caused by reduced fruit yield and by the limitation of the economically feasible growing areas and periods.

14.3.2.1.2 Symptoms

In nature, TYLCV-caused diseases mainly affect to tomato crops. Symptoms in tomato consist in stunting, curling of leaflet margins with or without yellowing, reduction in leaf size, and flower abortion.

14.3.2.1.3 Host Range

TYLCV has a very narrow host range that covers some Solanaceae species like tomato, *Datura stramonium*, and different *Nicotiana* spp and has also been described in French bean and *Malva parviflora* (Mansour and Al-Musa 1992; Cohen and Antignus 1994).

14.3.2.1.4 Transmission

TYLCV is transmitted from plant to plant by *B. tabaci* in a circulative manner (Mehta et al. 1994); propagation in insect cells is still under discussion.

14.3.2.1.5 Management

Effective control through crop management measures to avoid the vector and inoculum sources is possible in greenhouse crops. In the semi-protected crops typical of the Mediterranean regions, chemical control of vectors is ineffective to limit the spread of TYLCV. In these cases, control should be based on crop management following recommendations derived from the epidemiological knowledge of the disease and/or the use of the resistant/tolerant cultivars commercially available.

14.3.2.2 *Closterovirus* and *Clostero*-like Viruses

In recent years there is an emerging threat in worldwide agriculture, particularly in temperate regions, that is caused by a number of viruses that are transmitted by whiteflies and induce yellowing symptoms in plants. This is probably related to the increasing importance of whitefly populations worldwide and to changes in the relative predominance of existing species. These viruses are not generally well characterized; however, most of them seem to be members of the *Closterovirus* genus of plant viruses. Whitefly-transmitted closteroviruses have flexuous particles of variable length depending on species (900 × 12 nm).

14.3.2.2.1 Distribution and Economic Importance

Most of these viruses have been first described in the USA and cause important diseases in outdoor and protected crops. Tomato infectious chlorosis virus (TICV) was found infecting field and greenhouse tomato crops in California. Cucumber yellow stunting disorder virus (CYSDV) is present in the Mediterranean area, causing disease in cucurbits, and has not been described in America. Yield losses of up to 50–75 % occur in affected lettuce crops.

14.3.2.2.2 Symptoms

Symptoms usually consist in interveinal yellowing of the leaves, stunting, and/or necrosis.

14.3.2.2.3 Host Range

Lettuce infectious yellows virus (LIYV) infects lettuce, sugar beet, melon, squash, watermelon, and carrot. Lettuce chlorosis virus (LCV) does not infect cucurbits.

14.3.2.2.4 Transmission

Beet pseudoyellows virus (BPYV) and tomato infectious chlorosis virus (TICV) are transmitted by *Trialeurodes vaporariorum*, and cucumber yellow stunting disorder virus (CYSDV), lettuce infectious yellows virus (LIYV), and lettuce chlorosis virus (LCV) are transmitted by *Bemisia tabaci* and *B. argentifolii* (Célix *et al.* 1996; Duffus et al. 1996a, b). A semipersistent transmission manner has been demonstrated in certain cases.

14.3.2.2.5 Management

Integrated management of the disease in protected crops should be based on the early elimination of primary infected plants, avoidance of entrance of whiteflies, and rational insecticide treatments to reduce overall vector populations in the greenhouse. In melon, resistance to BPYV has been described few years ago and, recently, to CYSDV (Gómez-Guillamón et al. 1995).

14.3.3 Thrips-Transmitted Viruses

14.3.3.1 *Tomato Spotted Wilt Virus* (TSWV)

TSWV is the type species of the *Tospovirus* genus of the family *Bunyaviridae*. TSWV has isometric, membrane-bound particles of approximately 80 nm in diameter.

14.3.3.1.1 Distribution and Economic Importance

TSWV causes serious diseases worldwide in both outdoor and protected economically important crops. Significant yield losses are caused in vegetable crops like tomato, pepper, or lettuce and in different ornamental species.

14.3.3.1.2 Symptoms

The symptomatology vary from no symptoms to chlorotic or necrotic local lesions, ring spots, line patterns, mosaic, mottling, bronzing, chlorosis, necrosis, leaf or stem malformation, and stunting. Flower abortion is observed and fruits can exhibit malformation, necrosis, and abnormal coloration. Symptoms vary depending on virus/host isolate combination, plant age at infection time, and environmental conditions.

14.3.3.1.3 Host Range

TSWV has a wide host range, infecting more than 250 species in 70 different families of both monocotyledons and dicotyledons including important cultivated species (Edwardson and Christie 1986).

14.3.3.1.4 Transmission

TSWV is transmitted by several species of thrips of which *Frankliniella occidentalis* is the most important worldwide. Transmission is circulative and propagative and is unique in that the virus is only acquired by first-stage larvae and is transmitted by second-stage larvae and adults. Adults are the most important epidemiologically because they are more mobile and remain viruliferous for their entire life (Aramburu et al. 1997).

14.3.3.1.5 Management

Control of TSWV is difficult because of the wide host ranges of both the virus and the vector and the efficient natural transmission by thrips. The use of insecticides to reduce virus incidence by controlling the vector is ineffective and crop management practices are difficult to implement. In this situation, the use of resistant cultivars is the best solution. Genetic resistance to TSWV has been difficult to identify, characterize, and incorporate into commercial cultivars. Some important progress has been done in this field in tomato, where resistant cultivars are available, and in pepper and lettuce. However, the durability of resistance depends upon the biological variability that seems to exist among TSWV isolates (Roca et al. 1997). The development of genetically engineered virus-resistant plants is also under investigation. While efforts to produce resistant crops are going on, control in protected crops should be done integrating measures to limit the spread of the disease using certified virus-free planting material, rouging infected plants, and by biological or chemical control of thrips.

14.3.4 Beetle-Transmitted Viruses

14.3.4.1 *Squash Mosaic Virus* (SqMV)

SqMV is a member of the *Comovirus* genus. Virions are 30-nm isometric particles. SqMV has several pathogenically different strains. Isolates could be grouped into 2 serological groups that differ in seed transmissibility and, to a certain extent, in host range and symptomatology (Campbell 1971).

14.3.4.1.1 Distribution

SqMV is widely distributed in the Western hemisphere and also occurs in other countries throughout the world, probably introduced through seed lots.

14.3.4.1.2 Symptoms

In cucurbits, SqMV cause symptomless infection or may induce ringspots, systemic mosaic, malformation, and vein banding, depending on virus strain, host, and environmental conditions. Symptoms in fruits vary from small chlorotic areas to severe malformation with dark green areas. Isolates in subgroup 1 cause severe symptoms in melon and mild ones in pumpkin; some strains infect watermelon. Subgroup 2 isolates do not infect watermelon and cause mild symptoms in melon and severe in pumpkin.

14.3.4.1.3 Host Range

Natural host range is narrow, restricted to the Cucurbitaceae, in which most species are susceptible. Experimentally, it also infects plants in other families.

14.3.4.1.4 Transmission

SqMV is naturally transmitted by chewing insects, especially chrysomelid beetles, in a non-persistent manner, and, like all comoviruses, is seedborne (embryoborne). Subgroup 1 isolates are seed transmitted in pumpkin, squash, melon, and watermelon, and subgroup 2 isolates in pumpkin and squash. Mechanical transmission easily occurs by plant contact and during cultural operations. Commercial and experimental seed lots generally yield about 1–10 % infected seedlings but up to 94 % transmission has been reported in melon.

14.3.4.1.5 Management

Control is achieved by testing seed lots to prevent seed transmission (Nolan and Campbell 1984). If present, mechanical transmission should be avoided by elimination of symptomatic plants and reducing handling and pruning transmission possibilities.

14.3.5 Fungal-Transmitted Viruses

14.3.5.1 *Melon Necrotic Spot Virus* (MNSV)

MNSV belongs to the genus *Carmovirus* of the family *Tombusviridae*. Virions are 30-nm icosahedral particles.

14.3.5.1.1 Distribution and Economic Importance

MNSV has been found as a natural pathogen in melon, cucumber, and watermelon protected crops in Japan, the USA, and Europe in which it causes significant yield losses.

14.3.5.1.2 Symptoms

In melon, cucumber, and watermelon, MNSV causes small chlorotic spots in young leaves that turn into necrotic spots and large necrotic lesions. In melon and watermelon, necrotic streaks appear along the stems and petioles and sometimes are the only visible symptoms. In fruits, discoloration, necrosis, and malformation both externally and internally are observed.

14.3.5.1.3 Host Range

MNSV isolates have a narrow experimental host range mainly restricted to cucurbits and differ in the systemic infection of certain hosts: watermelon isolates failed to infect melon and cucumber plants systemically, melon isolates systemically infect melon plants but not watermelon and cucumber, and cucumber isolates infect melon and cucumber plants systemically and inoculated but not uninoculated leaves of watermelon plants.

14.3.5.1.4 Transmission

MNSV is naturally transmitted by the zoospores of the fungal vector *Olpidium bornovanus* (*Olpidium radicale*). Seed transmission is reported: 10–40 % of the seedlings from seeds of muskmelon were affected and plants became infected when grown in the presence of *Olpidium-contaminated* soil. Mechanical transmission is possible experimentally and has been reported during cultural operations.

14.3.5.1.5 Management

Apart from recommended control methods for soil-, seed-, and mechanically transmitted viruses (soil, seeds, and tool disinfection, etc.), grafting on immune *Cucurbita ficifolia* rootstocks has been used in cucumber to control MNSV. Melon cultivars resistant to this virus are commercially available.

14.3.6 Mechanically Transmitted Viruses

14.3.6.1 *Tobamovirus* Genus

The genus *Tobamovirus* of plant viruses includes species that cause devastating diseases in protected crops. Virions are elongated rigid rod-shaped particles of about 300 × 18 nm. The type member is tobacco mosaic virus (TMV). Tobamoviruses cause severe diseases in susceptible species especially in protected crops because of the intensive production that implies high density of plants and frequent cultural operations which favor mechanical transmission.

14.3.6.1.1 Distribution and Economic Importance

Tobamoviruses are a first-order problem in protected crops. Most tobamoviruses are easily distributed worldwide via infected seeds. PMMV is one of the most destructive pathogens of protected pepper crops; infections may reach 100 % of the plants and the yield of marketable fruit be drastically reduced. Tomato mosaic virus (ToMV) has been for years a virus of great economic importance in protected tomato crops; however, the development of resistant cultivars has reduced considerably the incidence of the disease, but it is still a serious threat where resistant cultivars are not grown. In pepper, ToMV can also cause severe losses on susceptible cultivars (Brunt 1986).

14.3.6.1.2 Symptoms

PMMV induces a faint mosaic in pepper leaves, whereas fruits are severely malformed with distorted coloration and often exhibit depressed necrotic areas. ToMV causes a wide range of symptoms on tomato depending on virus strain, cultivar, plant age at infection time, and environmental conditions: mottle or mosaics are observed in leaves that are malformed, plants are stunted, and fruits show external mottling and, sometimes, internal browning. In pepper, symptoms vary with cultivar and can be mosaics, systemic chlorosis, necrotic local lesions, leaf abscission, and/or systemic leaf and stem necrosis. In cucurbits, cucumber green mottle mosaic virus (CGMMV) causes more or less prominent leaf symptoms (mosaic, mottling, and malformation), stunting, flower abortion, and fruit mottling, distortion, and/or internal discoloration.

14.3.6.1.3 Host Range

Natural host range is very narrow, usually restricted to specific hosts; however, experimentally can be transmitted to numerous species of different families. For example, pepper mild mottle virus (PMMV) naturally infects pepper, ToMV infects tomato and pepper (Brunt 1986), and cucumber green mottle mosaic virus (CGMMV) infects some cucurbits like cucumber, watermelon, and melon and spontaneous perennial hosts like *Lagenaria siceraria* (Okada 1986).

14.3.6.1.4 Transmission

In nature, tobamoviruses are the most infectious and persistent disease agents; they are transmitted and easily spread between plants by contact and during cultural operations, through contaminated implements. The viruses can survive over years in plant debris that may be source for new infections via the roots or aerial parts if infected remains are present in the greenhouse structures. In certain cases, these viruses are seed transmitted: the virus is carried in the external seed surface, testa, and sometimes in the endosperm (Johansen et al. 1994). Seed samples with endosperm infection can remain infected for years. No natural vectors are known; presence in irrigation water has been reported for tomato mosaic virus (ToMV). Tobamoviruses are easily transmitted experimentally by mechanical inoculation.

14.3.6.1.5 Management

In first term, control methods are addressed to eliminate or reduce primary inoculum sources. Virus-free seeds should be used: sanitation of seeds can be done by soaking seeds in different solutions of active reagents (trisodium phosphate, hydrochloric acid, sodium hypochlorite) or by dry heat treatment (Rast and Stijger 1987). Removal of plant debris from previous susceptible crops and steam treatment of the soil and greenhouse structures will aid to avoid primary infections. Secondary spread can be reduced by washing hands and implements with soap and water before and during plant handling and/or frequent dipping into skim milk solutions. Cross protection has been largely used in greenhouse tomato crops to control ToMV by inoculation of tomato seedlings with an attenuated strain obtained by Rast (1972) in the Netherlands, thus avoiding ulterior infection with virulent ToMV strains. Other solanaceous crops that are susceptible to the mild strain (like pepper) must not be grown in proximity. Resistant genes have been described and incorporated in commercial tomato against ToMV and pepper against different tobamoviruses [TMV, ToMV, PMMV, and paprika mild mottle virus (PaMMV)]. However, resistance-breaking strains can be detected (Tenllado et al. 1997).

Table 14.3 summarizes the characteristics of the main virus species that cause diseases in protected crops, for which comprehensive reviews are available (Shukla et al. 1994; Murphy et al. 1995; Brunt et al. 1996).

14.4 Management Strategy for Virus Diseases

Proper field sanitation is one of the most important management strategies, since once the buildup of inoculum occurs inside the polyhouse, it is very difficult to manage it. So prevention is always better than cure. Use disease-resistant varieties. Leaf wetness is reduced by staking plants providing ample spacing between plants to allow for good air movement and avoiding overhead irrigation. For judicious use, chemicals with least toxicity recommended specially for polyhouse cultivation should be used. Chemicals like chlorothalonil, cymoxanil, and azoxystrobin are prohibited in polyhouse-grown tomato and thus should be avoided. As for viruses, scout fields for the first occurrence of virus disease. Where feasible, pull up and destroy infected plants, but only after spraying them thoroughly with an insecticide to kill any insects they may be harboring. Use reflective mulches to repel insects, thereby reducing the rate of spread of insect-borne viruses. Monitor vector population early in the season and apply insecticide treatments when needed. Minimize plant handling to reduce the amount of virus spread mechanically.

It is essential to start with virus-tested, vector-free planting material and to ensure continuous growth. General hygiene is equally important and means for the removal of affected plants and debris, sterilizing working tools, and the use of sterilized soil and resistant cultivars. The vectors should be controlled.

Most viruses infecting vegetables are transmitted by sap-sucking insects. Weeds and other hosts are crucial in the life cycle of many viruses and their vectors. Infected plants cannot be cured: control involves the prevention or delay of infection. No single method is likely to provide perfect control, but if combinations of the management options are utilized, it can be possible to successfully implement disease control.

14.4.1 Production of Virus-Free Vegetables by Using Insect-Proof Nets

Continuous cropping of vegetables in a particular area leads to buildup of diseases and their vectors. To reduce the problem, one of the methods is to give a rest period of 60 days during dry season so that the vector's life cycle is discontinued. This also requires clean cultivation so that all the plant and leaves are killed or removed so that the virus cycle is discontinued. But in many areas this practice is not possible. First farmers do not agree to stop cropping. In many areas dry season is the production time, so another alternative is to grow the crop under nets. Normally a 40-mesh net is used which prevents most of the sucking insects to get inside.

Net houses must have double-entry sliding gates so that with the movement of the farmer, the pest does not get an easy chance to get inside.

Nurseries should also be raised inside the net house to prevent the infection. Use of plastic mulching sheet also reduces the population of sucking pests and thus reduces the chances of viral diseases in the crop.

14.4.2 Integrated Management

Greenhouse crops represent a singular case for disease management. They are closed systems where external exchanges are reduced to the minimum, although the intermediate situation present in the protected crops grown under the simple and less hermetic structures typical of the Mediterranean area should also be considered. The most damaging viruses in protected crops are soilborne viruses [MNSV, PMMV, ToMV, tomato bushy stunt virus (TBSV)] or those imported via contaminated seed (TMV, ToMV, PMMV, CGMMV, SqMV, MNSV, BCMV, LMV,

Table 14.3 Major virus species affecting protected crops

Natural transmission[a]	Family	Genus	Species[b]	Main infected crops[c]
Aphids (np)	*Bromoviridae*	*Cucumovirus*	CMV	cu, le, me, pe, to, wa, zu
Aphids (np), seed (be)	*Potyviridae*	*Potyvirus*	BCMV	be, fa
Aphids (np)	*Potyviridae*	*Potyvirus*	BYMV	be, fa
Aphids (np), seed (be)	*Potyviridae*	*Potyvirus*	LMV	le
Aphids (np)	*Potyviridae*	*Potyvirus*	PVY	pe, to
Aphids (np)	*Potyviridae*	*Potyvirus*	PRSV-W	cu, me, wa, zu
Aphids (np)	*Potyviridae*	*Potyvirus*	WMV2	cu, me, wa, zu
Aphids (np)	*Potyviridae*	*Potyvirus*	ZYMV	cu, me, wa, zu
Aphids (np)	*Potyviridae*	*Potyvirus*	ZYFV	cu, me, wa, zu
Aphids (p)	—	*Luteovirus*	BWYV	lu, ce, pe, to, wa, zu
Aphids (p)	—	*Luteovirus*	CABYV	cu, me, zu
Aphids (p)	*Bunyaviridae*	*Tospovirus*	TSWV	fa, le, pe, pea, to
Bemisia tabaci (p)	*Geminiviridae*	*Geminivirus*	TYLCV	to
Bemisia tabaci (sp)	—	—	CVYV	cu, me
Bemisia tabaci	—	*Closterovirus*	LCV	le
Bemisia tabaci (sp)	—	*Closterovirus*	LIYV	le, me, wa, zu
Bemisia tabaci (sp)	—	*Closterovirus*	CYSDV	cu, me
Trialeurodes vaporariorum (sp)	—	*Closterovirus*	BPYV	cu, me, le
Trialeurodes vaporariorum (sp)	—	*Closterovirus*	TICV	to
Trialeurodes vaporariorum	—	*Closterovirus*	ToCV	to
Beetles (np)	*Comoviridae*	*Comovirus*	SMV	me, wa, zu
Olpidium bornovanus	*Tombusviridae*	*Comovirus*	MNSV	cu, me, wa
Olpidium brassicae	—		LBVV	le
Olpidium brassicae	—		LRNV	le
Fungus	—		PYVD	pe
Mechanical	—	*Tobamovirus*	TMV	pe, to
Mechanical, seed (lo)	—	*Tobamovirus*	ToMV	pe, to
Mechanical, seed (pe)	—	*Tobamovirus*	PMMV	pe
Mechanical, seed (cu)	—	*Tobamovirus*	CGMMV	cu, me, wa
Mechanical, fungus	—	*Potexvirus*	PVX	to
Unknown, soil	*Tombusviridae*	*Tobamovirus*	TBSV	eg, pe, to
Unknown, soil, seed (cu)	*Tombusviridae*	*Tobamovirus*	CLSV	cu

[a]*np* nonpersistent, sp semipersistent, *p* persistent

[b]*CMV* cucumber mosaic virus, *BCMV* beam common mosaic virus, *BYMV* bean yellow mosaic virus, *LMV* lettuce mosaic virus, *PVY* potato virus Y, *PRSV-W* papaya ringspot virus-W, *WMV2* watermelon mosaic virus 2, *ZYMV* zucchini yellow mosaic virus, *ZYFV* zucchini yellow fleck virus, *BWYV* beet western yellow virus, *CABYV* cucurbit aphid-borne

(continued)

Table 14.3 (continued)

yellow virus, *TSWV* tomato spotted wilt virus, *TYLCV* tomato yellow leaf curl virus, *CVYV* cucumber vein yellowing virus, *LCV* lettuce chlorosis virus, *LIYV* lettuce infectious yellows virus, *CYSDV* cucurbit yellow stunting disorder virus, *BPYV* beet pseudoyellows virus, *TICV* tomato infectious chlorosis virus, *ToCV* tomato chlorosis virus, *SMV* squash mosaic virus, *MNSV* melon necrotic spot virus, *LBVV* lettuce big vein virus, *LRNV* lettuce ring necrosis virus, *PYVD* pepper yellows vein disease, *TMV* tobacco mosaic virus, *ToMV* tomato mosaic virus, *PMMV* pepper mild mottle virus, *CGMMV* cucumber green mottle mosaic virus, *PVX* potato virus X, *TBSV* tomato bushy stunt virus, *CLSV* cucumber leaf spot virus

[c]*be* French bean, *cu* cucumber, *eg* eggplant, *fa* faba bean, *le* lettuce, *me* melon, *pe* pepper, *to* tomato, *wa* watermelon, *zu* zucchini squash

etc.) or contaminated plantlets. The precise knowledge about which virus problems are affecting in a specific crop, the dispersal mechanisms, and the epidemiology of the disease induced will help to make strategic management decisions within an integrated management strategy.

The means to prevent and control viral diseases based on the knowledge of their dispersal mechanisms have been discussed earlier. Other strategies for virus control are focused to the minimization of the impact of the infection on crop yield; breeding for resistance and cross protection are two of these strategies. When possible, the best control method against plant viruses would be the development of resistant cultivars (Sherf and Macnab 1986). However, experience has shown that breeding for resistance or the development of transgenic plants is unlikely to give permanent solutions for any particular virus and crop. Variable virus populations may be present (Luis-Arteaga et al. 1996; Tenllado et al. 1997) and/or virus can mutate (Aranda et al. 1997) in the field with respect to virulence and the range of crops and cultivars they can infect. Cross protection is based in that mild virus strains can be used to protect plants against infection by severe strain(s) of the same virus. Basic criteria for selection of cross protection as a disease control strategy are well known (Gonsalves and Gamsey 1989). Mildness of a strain is usually relative to a certain target crop, and this should be taken into account if cross protection is to be used in greenhouses where other crops that may be sensitive to the protective virus strain are grown simultaneously. The same applies for precautions to be taken to avoid dispersal of the mild strain to sensitive crops grown in the vicinity of the protected greenhouse crop. Due to possible virus mutations, the reversion of the mild strain used in the cross protection program to a severe one must be continuously verified. When using cross protection, the risk of coinfection with other virus(es) that may have synergistic effects with the protective strain should also be evaluated. Cross protection alone is not enough to give a high level of control of the disease because protection depends on the homology of the severe strain and on challenge pressure (Gonsalves and Gamsey 1989). Therefore, the combination of various virus management practices compatible with an integrated management of the greenhouse is often desirable. Indirect measures for virus control have been discussed, e.g., (i) adjustment of planting dates to avoid high vector populations in young plantings if epidemiological data of the disease are available, (ii) use of virus-free propagation material, (iii) disinfection of soil and greenhouse structures, (iv) minimization of external entrance of insects, (v) rapid elimination of virus-infected plants, (vi) adequate plant handling, and (vii) avoidance of overlapping or continuous cultivation of sensitive species in the rotation.

14.4.3 GAP Recommendations

14.4.3.1 Exclusion/Avoidance

- Plant virus-free seeds and seedling transplants.

- Grow crops in areas where the disease seldom occurs or during periods when the virus or its vector is at a low level.

14.4.3.2 Reduction in Virus Inoculum Level

- Control weeds and other virus hosts and insect vectors.
- Destroy old crops promptly.
- Separate new crops from maturing crops and avoid overlapping crops, especially continuous year-round cropping.

14.4.3.3 Protection of the Host

- Plant virus-resistant or virus-tolerant varieties.
- Use highly reflective mulches and oil sprays to deter insects.
- Use insecticides strategically to protect plants from insects.

References

Aramburu J, Riudavets J, Arnó J, Laviña A, Moriones E (1997) The proportion of viruliferous individuals in field populations of *Frankliniella occidentalis:* implications for tomato spotted wilt virus epidemics in tomato. Eur J Plant Pathol 103:623–629

Aranda MA, Fraile A, Dopazo J, Malpica JM, García-Arenal F (1997) Contribution of mutation and RNA recombination to the evolution of a plant pathogenic RNA. J Mol Evol 44:81–88

Brunt AA (1986) Tomato mosaic virus. In: van Regenmortel MHV, Fraenkel-Conrat H (eds) The plant viruses. Vol. II: the rod-shaped plant viruses. Plenum Press, New York, pp 180–204

Brunt AA, Crabtree K, Dallwitz MJ, Gibbs AJ, Watson L (1996) Viruses of plants. Descriptions and lists from the VIDE database. CAB International, Wallingford

Campbell RN (1971) Squash mosaic virus, CMI/AAB descriptions of plant viruses no 43, CMI, Kew

Célix A, López-Sesé A, Almarza N, Gómez-Guillamón ML, Rodríguez-Cerezo E (1996) Characterization of cucurbit yellow stunting disorder virus, a *Bemisia tabaci*-transmitted closterovirus. Phytopathology 86:1370–1376

Cohen S, Antignus Y (1994) Tomato yellow leaf curl virus, a whitefly-borne geminivirus. Adv Dis Vector Res 10:259–288

Dinant S, Lot H (1992) Lettuce mosaic virus. Plant Pathol 41:528–542

Dogimont C, Slama S, Martin J, Lecoq H, Pitrat M (1996) Sources of resistance to cucurbit aphid-borne yellows luteovirus in melon germplasm collection. Plant Dis 80:1379–1382

Duffus JE, Lin HY, Wisler GC (1996a) Tomato infectious chlorosis virus – a new clostero-like virus transmitted by *Trialeurodes vaporariorum.* Eur J Plant Pathol 102:219–226

Duffus JE, Lin HY, Wider GC (1996b) Lettuce chlorosis virus – a newly whitefly-transmitted closterovirus. Eur J Plant Pathol 102:591–596

Edwardson JR, Christie RG, Edwardson JR, Christie RG (1986) Tomato spotted wilt virus. In: Edwardson JR, Christie RG (eds) Viruses infecting forage legumes, vol III, Monograph series no. 14. Florida Agricultural Experiment Stations, Gainesville, pp 563–580

Gómez-Guillamón ML, Torés JA, Soria C, Sesé AIL (1995) Searching for resistance to *Sphaerotheca fuliginea* and two yellowing diseases in *Cucumis melo* and related *Cucumis* species. In: Lester GE, Dunlap JR (eds) Proceedings cucurbitaceae '94: evaluation and enhancement of cucurbit germplasm. Gateway Printing, South Padre Island, Tex, pp 205–208

Gonsalves D, Gamsey SM (1989) Cross-protection techniques for control of plant virus diseases in the tropics. Plant Dis 73:592–597

Johansen E, Edwards MC, Hampton RO (1994) Seed transmission of viruses: current perspectives. Annu Rev Phytopathol 32:363–386

Jordá C, Alfaro A, Aranda MA, Moriones E, García-Arenal F (1992) An epidemic of cucumber mosaic virus plus satellite RNA in tomatoes in Eastern Spain. Plant Dis 76:363–366

Luis-Arteaga M, Rodríguez-Cerezo E, Fraile A, Sáez E, García-Arenal F (1996) Different tomato bushy stunt virus strains cause disease outbreaks on solanaceous crops in Spain. Phytopathology 86:535–542

Mansour A, Al-Musa A (1992) Tomato yellow leaf curl virus: host range and virus-vector relationships. Plant Pathol 41:122–125

Mehta P, Wyman JA, Nakhla MK, Maxwell DP (1994) Transmission of tomato yellow leaf curl geminivirus by *Bemisia tabaci* (Homoptera: Aleyrodidae). J Econ Entomol 87:1291–1297

Mink GI (1993) Pollen- and seed-transmitted viruses and viroids. Annu Rev Phytopathol 31:375–402

Murphy FA, Fauquet CM, Bishop DHL, Ghabrial SA, Jarvis AW, Martelli GP, Mayo MA, Summers MD (eds) (1995) Virus taxonomy: classification and nomenclature of viruses. Archiv Virol 10(Suppl):268–274

Nolan PA, Campbell RN (1984) Squash mosaic virus detection in individual seeds and seed lots of cucurbits by enzyme-linked immunosorbent assay. Plant Dis 68:971–975

Okada Y (1986) Cucumber green mottle mosaic virus. In: van Regenmortel MHV, Fraenkel-Conrat H (eds) The plant viruses, vol II, The rod-shaped plant viruses. Plenum Press, New York, pp 267–281

Rast ATB (1972) MH-16, an artificial symptomless mutant of tobacco mosaic virus for seedling inoculation of tomato crops. Neth J Plant Pathol 78:110–112

Rast ATB, Stijger CCMM (1987) Disinfection of pepper seed infected with different strains of capsicum mosaic virus by trisodium phosphate and dry heat treatment. Plant Pathol 36:583–588

Roca E, Aramburu J, Moriones E (1997) Comparative host reactions and *Frankliniella occidentalis* transmission of different isolates of tomato spotted wilt tospovirus from Spain. Plant Pathol 56:407–415

Sherf AF, Macnab AA (1986) Vegetable diseases and their control. John Wiley & Sons Inc, New York

Shukla DD, Ward CW, Brunt AA (1994) The potyviridae. CAB International, Wallingford

Tenllado F, García-Luque I, Serra MT, Díaz-Ruiz JR (1997) Pepper resistance-breaking tobamoviruses: can they co-exist in single pepper plants? Eur J Plant Pathol 103:235–243

Zitter TA, Simons JN (1980) Management of viruses by alteration of vector efficiency and by cultural practices. Annu Rev Phytopathol 18:289–310

15 Nematode Diseases and Their Management

Abstract

The soil conditions are highly conducive for nematode infestations of the polyhouse crops. Nematode problems arise from contaminated soil or soil mixture used as a component of the growing medium, monocropping, and infested planting materials. Nematode management here must be considered primarily as exclusion or avoidance. Once nematodes are introduced, it is difficult to manage them. Nematode management using exclusion, sanitation, physical approaches, host plant resistance, chemical nematicides, biological control, and integrated methods is discussed.

Keywords

Nematode diseases • Exclusion • Sanitation • Host resistance • Nematicides • Biocontrol • Integrated management

15.1 Introduction

The most economically important nematode genera of greenhouse production include *Meloidogyne* (root knot), *Pratylenchus* (lesion), *Radopholus* (burrowing), *Rotylenchulus* (reniform), and *Aphelenchoides* or *Ditylenchus* (foliar or stem). They feed on the roots devitalizing the plants ultimately affecting both qualitative and quantitative production. In most cases, problems with these nematodes arise from planting of infected seed or planting stocks, in systems utilizing bare ground or raised ground beds, or where contaminated soil is used as a component of the growing medium. Subsequent spread of nematodes within the greenhouse usually occurs via movement of infested soil or plant material by workers, water, or greenhouse equipment. In some greenhouse systems, nematodes can be introduced into and recirculated quickly via drainage and irrigation water to inoculate most other healthy plants within the greenhouse. The latter scenario clearly defines how nematode problems can be amplified quickly once introduced. Commonly followed nematode management practices such as summer plowing, fallowing, and crop rotation which are meant for nematode suppression in field crops are not practicable in polyhouses. For greenhouse production, nematode management must, therefore, be

P.P. Reddy, *Sustainable Crop Protection under Protected Cultivation*,
DOI 10.1007/978-981-287-952-3_15

viewed first as one of exclusion or avoidance. Once nematodes are introduced and damage becomes visible, it is currently not possible to chemically or nonchemically resolve the problem to avoid potentially significant yield losses thereafter. Some of the most important sources of contamination and hence major factors to consider include (i) seed and planting stocks, (ii) irrigation water, (iii) soil and potting media, and (iv) general cleanliness.

The ideal conditions provided by protected cultivation and continuous availability of the host plant round the year often result in high population buildup of soilborne pathogens including plant-parasitic nematodes. Indiscriminate use of agrochemicals in crop ecosystems has affected the beneficial microflora and has adversely altered the natural biological suppression which has hitherto been operating in Indian soils. This serious situation has led to emergence of nematode problems and nematode-induced disease complexes caused by several nematodes and pathogenic fungi or bacteria in various crops. There are numerous reports mentioning the difficulties faced by the polyhouse growers due to severe infestation of various nematodes.

The root-knot nematode has been a major pest of vegetable and floricultural crops under protected cultivation. The yield losses to the extent of 25–27% have been recorded due to the infestation of root-knot nematodes in various vegetables and ornamentals. The population buildup is very rapid in the polyhouses, and nematode population reaches 5–6 times of threshold levels within 18–24 months, making the polyhouses useless. Crops such as capsicum, tomato, okra, gherkins, muskmelon, watermelon, carnations, and gerbera are being grown under protected cultivation (in polyhouses/greenhouses/shade nets). These crops grown in all the states of India are seriously infested with nematodes such as *Meloidogyne incognita*, *M. javanica* (root-knot nematodes), *Rotylenchulus reniformis* (reniform nematode), and *Pratylenchus* spp. (lesion nematodes). Nematode problems on all these crops under protected conditions have assumed alarming proportions leading to huge crop losses of up to 80% in some cases.

Fig. 15.1 Root-knot nematode and *Fusarium* wilt disease complex on tomato plants

Generally, not a single pathogen but a community of pathogens, which might interact each other, is the cause of crop failure (Fig. 15.1). Therefore, control means must be effective against all soilborne pathogens occurring in the soil to be cropped.

15.2 Nematode Management

Greenhouse nematode management strategies include exclusion, sanitation, protection, and eradication (Table 15.1).

15.2.1 Exclusion

15.2.1.1 Site Selection

Many different factors should be considered in selecting a suitable site for greenhouse placement, such as proximity to water, labor, road systems, and agricultural markets. There are also important pest management considerations. Whenever practical, the site should be removed from major open-field areas of commercial fruit and vegetable production. This will serve to minimize opportunity for pest introduction into greenhouse facilities. If ground-based production systems will be used, then soils free of economically important plant-parasitic nematodes should only be selected for greenhouse placement. Once nematodes are introduced, it is currently not

Table 15.1 Greenhouse nematode management tactics

Management tactic	Management tactic description
Exclusion	Tactics that prevent the introduction and spread of nematodes into the greenhouse Nematode-free planting materials – use plug transplants produced from soilless culture devoid of nematodes
Sanitation	Removal and disposal of nematode-infected plant materials
Protection	Chemical control – no fumigant or non-fumigant nematicides currently registered for greenhouse use Solarization – trapping of solar energy under clear plastic panels to heat soil to lethal levels of nematodes Soilless substrates – use of pathogen-free artificial substrates for crop production (e.g., perlite, rock wool) Host plant resistance – crop varieties which reduce nematode reproduction, typically with reduced crop damage
Eradication	Elimination of nematodes from soil or plant materials within greenhouse structure. Generally not feasible from natural soil-based greenhouse systems Steam pasteurization – 60 °C for 30 min Steam sterilization – 100 °C (not recommended) Hot water dips – plants emerged in hot water to kill internal nematode parasites of roots or foliar tissue

possible to eliminate or eradicate the problem. Greenhouses structures should also be situated to avoid introduction of nematodes and other soil-borne pests via downstream movement of drainage or runoff water and soil.

15.2.1.2 Irrigation

Another primary source of plant-parasitic nematode spread and introduction into greenhouse structures has been demonstrated via irrigation water from streams, ditches, and ponds. Nematodes have even been found in irrigation waters from shallow wells also indicating that they can be from those sources as well. Since nematodes can be carried through irrigation water that has drained from an infested field, growers should avoid use of ditch or pond waters for greenhouse irrigation or for spray mixtures. Recycling of greenhouse irrigation water should also be avoided since nematodes can be recirculated quickly via the reused irrigation water to inoculate most other healthy plants within the greenhouse. If greenhouse water must be reused, it should be decontaminated by heating to 90 °C or greater. Given the small, microscopic size of nematodes, irrigation screens and filters should not be relied upon as means of nematode exclusion or control.

15.2.2 Sanitation

15.2.2.1 Nematode-Free Planting Materials

In most cases, greenhouse problems with both soilborne and foliar nematodes arise from planting of infected seed or planting stocks. At present no chemical or nonchemical management tactics are available that can resolve nematode problems within the greenhouse once introduced into the crop. As a result, nematode-free transplants or plug plants that rely upon soilless substrates from production are increasingly used to exclude foliar and soilborne species of nematodes but also to expedite plant establishment and crop production. In addition, hot water dips have been developed as a control strategy to eliminate plant-parasitic nematodes form both roots and foliar tissues of infested plant materials to be used as transplants (e.g., strawberry). The dip treatment consists of submerging plants in hot water for a specified time and then immediately transferring the treated plants into cold water for an additional time period. The temperature at which different species of nematode are killed is variable and depends upon temperature and time of exposure (a dosage concept of concentration x time). Plants that have been treated with hot water generally require additional care and water after planting than non-treated plants, but the hot water treatment has been used to successfully eliminate nematodes from planting stock. Rouging infected plants when early symptoms are first recognized may also help minimize any plant introduced-nematode problem within the greenhouse.

15.2.2.2 Other Sanitation Practices

Other cultural measures that reduce nematode problems in the greenhouses include rapid destruction of infested crop root systems following harvest. Discarding infected potted plants will help prevent spread of nematode. Entry points for greenhouse structures should also contain sanitizing stations for hands, shoes, boots, tools, and other equipment.

15.2.3 Physical Approaches

Because of the relatively low thermal sensitivities of nematodes and other soilborne pests, physical approaches which utilize heat have generally proved to be the most effective and widely used management strategies within greenhouses. In general terms, the thermal sensitivities of a wide range of soilborne crop pests that can be encountered are relatively broad (Fig. 15.2). On a relative temperature scale, nematodes and water molds would be considered relatively intolerant of high temperature, being effectively killed by 30-min exposure at temperatures as low as 120 °F. Within the range of 120–140 °F, most plant-pathogenic fungi and bacteria are killed. Certain weeds and viral pathogens compose the most tolerant group, requiring 30-min exposures at temperatures of 160 °F or greater for pest elimination. A 30-min exposure of 180 °F or greater is generally recognized as the recommended soil pasteurization temperature and the period of exposure to provide a broad spectrum measure of soilborne pest and disease control. The following section defines specific pest control systems that rely on heat to control nematodes in greenhouse production.

15.2.3.1 Steam

Steam is a well-established and effective technique for soilborne pest and disease control and is extensively used for bulk soil, substrate, or even small-scale field treatments within greenhouses and nursery plant operations. Steaming, as defined here, is the introduction of water vapor (>212 °F:100 °C) into soil to elevate soil temperatures to levels lethal (160–180 °F) to soilborne pests. Soil temperature and treatment duration determine whether complete elimination (sterilization) or only partial removal of soil microflora (pasteurization) occurs.

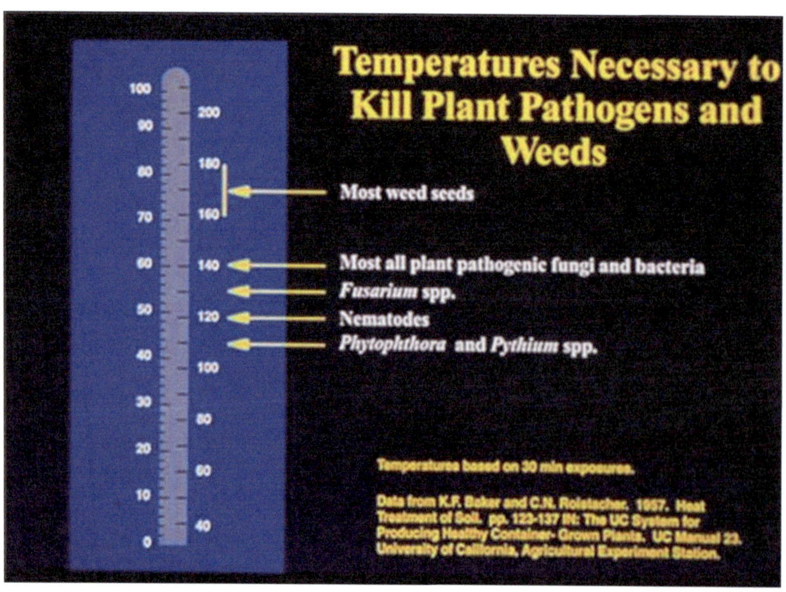

Fig. 15.2 Generalized relationship of soil temperatures for 30-min exposures to kill various soilborne pests and pathogens (After Baker and Rolstacher 1957)

Since its inception in 1888, various methods have been developed to deliver and apply steam for soilborne pest and disease control. The most common methods include systems in which soil is brought to the boiler where steam was introduced into bins or chambers for bulk soil treatment. For other systems, a small portable steam boiler is brought to the treatment site where the steam is introduced directly, via a conduit, into buried perforated pipe or tile or winch-drawn plows or directly under an inverted pan, tray, or tarpaulin cover. For the latter systems, the soil needs to be covered with a pan, tray, or canvas sheet to conserve energy and minimize heat loss from soil. As a general rule, it is recommended that steam treatment is maintained so that the coldest spot in the ground bed or substrate is held at 180 °F for 30 min. A negative pressure steaming system has also been developed in which steam is blown under a tarpaulin cover. The steam is pulled into the soil by negative pressure created by an exhaust fan that pulls air out of the soil through buried perforated pipes. This process has not only proved to be effective in greenhouses but has expedited the steaming treatment process. Self-propelled soil steaming systems are also commercially available for small-scale field treatments.

Due to the high costs, heating inefficiencies, pest control inconsistencies, and other treatment impracticalities, steam is not currently used for large-scale, open-field greenhouse uses. In most cases, where comparisons have been made, greenhouse uses of steam have proven to be as effective as chemical soil sterilant approaches (e.g., methyl bromide). Problems have developed, however, when soils have been sterilized, practically when the complete elimination of all soil microorganisms occurs after prolonged exposures to temperatures in excess of 200 °F. These problems include increased soil aggregation and destruction of soil structure and release of phototoxic substances from soil and organic matter including soluble salts, ammonia, and heavy metals such as manganese. Fortunately, some of these adverse side effects can be remedied when the soil is only partially sterilized, generously irrigated afterward, or pasteurized with steam at reduced temperatures of 160–180 °F.

Other disadvantages of steam include (1) the large capital investments in steam equipment, (2) permanently installed duct/drain systems, and (3) time-consuming treatment involving moving, burial, or repositioning of pans or tarpaulin covers each time the system needs to be relocated to a new steaming site. Steam also cannot be effectively used in heavier soils with slow vapor and water percolation, soils with high organic matter content, or those with high water soil-water table. Steam also does not uniformly heat the soil in open-field uses nor always heat ground beds (Fig. 15.3) deeply enough to control major wilt diseases such as *Verticillium* or *Fusarium*. Steam has proven effective when a limited amount of substrate is treated but not ground soil. This is due to the depth at which harmful organisms can be found in the soil, which too often is either out of the reach of steam or can be reached only at extremely high costs. Heating the soil to depths of more than 12 in. requires much longer use of the boiler, more hand labor, and fuel quantities.

To improve the temperature range in deeper soil layers, a permanent steam sterilization system has been developed in which steam is introduced into drain pipes buried at a depth of 60 cm, after which the steam moves to the soil surface. This method is an improvement over sheet steaming, especially on sand and loam soils. As mentioned earlier, the most recent development in

Fig. 15.3 Use of steam applied under tarpaulin cover via buried perforated pipe for nematode control in a raised bed, greenhouse production system

steam sterilization is a combination of sheet steaming and steaming via drain pipes, a so-called negative pressure steaming. With this method, steam is introduced under the steaming sheet and pulled into the soil by negative pressure (created in the soil through buried perforated pipes). The temperatures achieved with negative pressure steaming are considerably better than those obtained with sheet steaming.

15.2.3.2 Soil Solarization

Soil solarization is a nonchemical technique in which *transparent* polyethylene plastic mulch is laid over moist soil for a 4–6-week period to heat non-cropped, greenhouse soils to temperatures lethal to nematodes and other soilborne pathogens. Soil temperatures are magnified because of the trapping of incoming solar radiation under the clear polyethylene plastic. To be effective, soils must be maintained at high soil moisture content to increase the susceptibility (thermal sensitivity) of soilborne pests and thermal conductivity of soil. Moist plastic mulched soils increase soil temperatures due primarily to the elimination of heat loss by evaporation and upper heat convection, in addition to a greenhouse effect by prohibiting dissipation of radiation from soil.

On a global scale, the most successful use of soil solarization has been reported in ground beds of greenhouses within sunny Mediterranean countries with heavier (loamy to clay soils) rather than sandy soils, as well as in closed plastic houses in cooler climates. Solarization of bags of nursery potting mixes was also recently certified in California as an alternative to steam or fumigation with methyl bromide. This technique has not been extensively tested in Florida. It is known that soils with poor water holding capacity and rapid drainage can significantly inhibit heat transfer to deeper soil horizons. As a result, loss of pest control is oftentimes directly correlated to soil depth. The depth to which lethal temperature can be achieved (15–20 cm) is also dependent on the intensity and duration of sunlight and ambient temperature. At present, the only time to consider soil solarization for pest control is during hot summer months.

15.2.3.2.1 Effect of Soil Solarization on the Nematode Population Densities (Tomato)

In tomato, soil solarization treatment markedly reduced the root-knot nematode, *M. incognita*, larval density per 100 ml of soil compared with the untreated control (Fig. 15.4). The soil solarization method had a significant influence to reduce the severity of root damage. Root galling index is dramatically reduced in both solarized greenhouses compared with untreated control where the highest root galling index was found (Fig. 15.5).

The growth of tomato plants, as indicated by visual assessment and plant height measurements, was significantly improved by the soil solarization. Tomato fruit yields in solarized greenhouses were fourfold higher than those obtained in untreated control (Fig. 15.6). Such

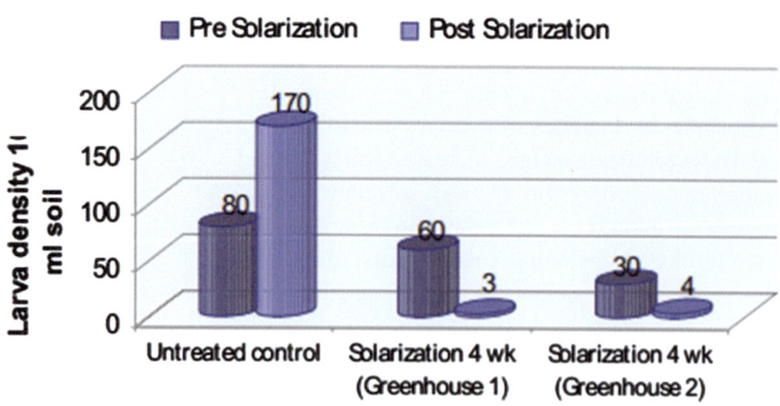

Fig. 15.4 Effect of soil solarization on nematode population densities in 100-ml soil

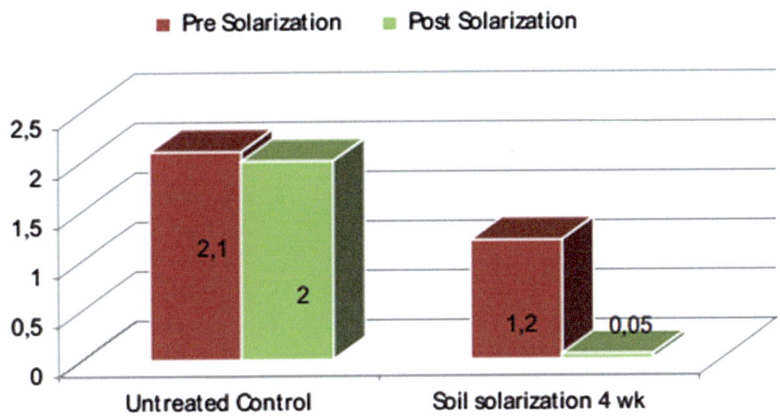

Fig. 15.5 Effect of soil solarization on root gall index due to root-knot nematode

significant increases were consistent with effective control of nematode provided by solarization.

Solarization is also known to improve plant growth and yield through the release of nutrient induced by high soil temperatures. The soil solarization treatment reduced nematode populations more than 90% also in lettuce. The yield of lettuce was about twofold higher in solarized soil as compared with untreated control in several experimental plots.

Soil solarization should be done during the hottest period of the year. Soil solarization offers a satisfactory and environmentally friendly solution for the control of root-knot nematode. This method is easily to be used into organic, conventional, and integrated control growing system.

This technique at the same time shows good results to reduce the infection of a severe tomato disease like corky root disease (*Pyrenochaeta lycopersici*).

15.2.4 Host Plant Resistance

Use of nematode-resistant crop varieties is often viewed as the foundation of a successful integrated nematode management program. In a resistant variety, nematodes fail to develop and reproduce normally within root tissues. Commercially available nematode-resistant vegetable varieties are currently available for tomato, pepper, southern pea, and sweet potato. However, new nematode-resistant crop varieties are

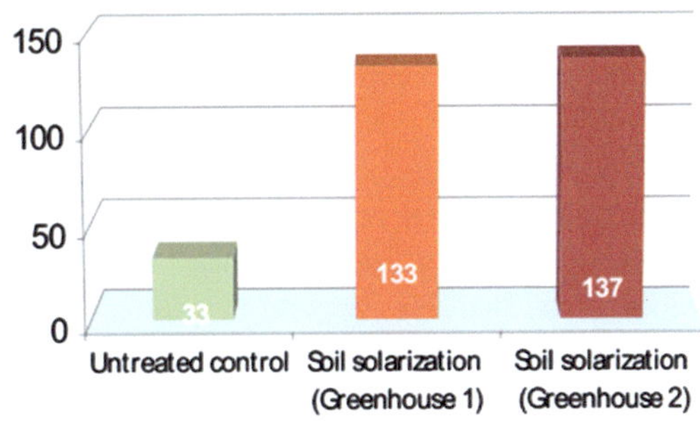

Fig. 15.6 Effect of soil solarization on tomato yield

increasingly becoming available as emphasis increases on nonchemical forms of nematode management. In many European and Mediterranean countries, grafting of susceptible plants onto nematode-resistant rootstocks is being used in protected culture systems for a number of annual greenhouse crops including tomato, eggplant, and various cucurbits. This technology has yet to be fully developed for field or greenhouse production systems of the US greenhouse growers who should consult their local county Cooperative Extension Service agent for new nematode-resistant crop varieties suitable for greenhouse production that may have become available.

Resistant rootstocks may also provide good control of soilborne pathogens of vegetables. In Italy, satisfactory control of root-knot nematodes was obtained by grafting tomato on resistant rootstocks (Renzoni and Lamberti 1974). Later, because of the availability of resistant cultivars of this crop, the use of resistant rootstocks did not receive further attention. However, because of other useful traits (Morra 1998), rootstocks are stimulating interest and some of tomato, eggplant, and melon may also possess resistance to nematodes.

15.2.5 Chemical Nematicides

15.2.5.1 Fumigants

Nematicides containing or developing methyl isothiocyanate after being incorporated into the soil against nematodes also exhibit herbicide and fungicide activities.

For nematode control dazomet is the most effective of these compounds. Being a granular product it is relatively easy to apply. However, it is effective only at high rates (500–700 kg/ha) of application, requires appropriate soil moisture content to hydrolyze, and releases methyl isothiocyanate. Therefore, several irrigations might be necessary during a 4–6-week period prior to planting to improve efficacy and facilitate degradation of by-products into non-phytotoxic substances. Usually, this nematicide does not cause severe residue problems.

Metam sodium is a liquid product which also develops methyl isothiocyanate. However, it has a rather slow nematicide activity and is effective only at very high application rates (1000–2000 L/ha). It must be injected into the soil or diluted in water and drenched or, better, used in chemigation.

15.2.5.2 Non-fumigants

These nematicides are generally less effective than the best fumigants. They kill only active stages of the nematodes but not eggs of which they may only delay hatching. Nevertheless, they are not phytotoxic and, therefore, can be used on established crops. Their use is suggested when nematode soil population densities at planting are from low to medium. They possess also insecticide activity. In general, these nematicides are formulated as granules containing 5–10% active ingredients and, therefore, their application is rather simple. The suggested rates are 10–20-kg active ingredient/ha.

Aldicarb, a carbamate compound with systemic activity, is one of these nematicides. Although aldicarb seems more effective than other non-fumigant nematicides against cyst-forming nematodes, its efficacy against a wide range of nematodes and insects is well documented. It might show phytotoxicity to some crops even at the suggested rates.

Oxamyl is also a carbamate which shows good nematicidal activity in acid soils, while in those with pH ≤ 7 it is degradated in a few days in compounds with no nematicidal activity. For this reason this chemical is not popular in Italy. Oxamyl is also systemic and effective against soil insects. Usually, accumulation of its residues in plant tissues is low if properly applied.

Carbofuran is also a carbamate compound with systemic activity, although used as a soil insecticide, and also shows nematicidal activity, but it may be phytotoxic to some crops. Its nematicidal activity lasts shortly but, usually, accumulation of its residues in plants is not a problem.

Cadusafos is among the last compounds with nematicidal activity that has become available. It is a phosphorganic compound which acts only as contact pesticide, has low mobility in the soil, and does not seem to accumulate in soil and plants. According to preliminary investigations, cadusafos appears as effective as the previously known nematicides. However, further investigations are necessary to obtain more insights on its efficacy as a nematicide and its fate in the environment.

Fenamiphos is probably the most widely used non-fumigant nematicide. It is a phosphorganic compound possessing also systemic activity and is effective against nematodes, especially root-knot nematodes, and insects. It has medium persistence in the soil and, therefore, provides good protection of the crops for 1–2 months after planting. It is not phytotoxic at the suggested rates.

Prophos (ethoprophos) is another phosphorganic which acts as a contact nematicide and, therefore, is effective only against nematodes living free in the soil. It has also insecticidal activity and sometime may show phytotoxicity.

15.2.6 Biological Control

Many of the fungi that parasitize nematodes are common soil inhabitants such as *Purpureocillium lilacinum* (syn. *Paecilomyces lilacinus*), *Pochonia chlamydosporia* (syn. *Verticillium chlamydosporium*), and trapping fungi. Bacteria in the genus *Pasteuria* are also regularly found parasitizing nematodes in soil. The use of the biocontrol agent *Pasteuria penetrans* was found to be effective against root-knot nematode species *Meloidogyne javanica* and *M. incognita*.

15.2.7 Integrated Methods

The following integrated methods can be utilized to manage plant nematodes under greenhouse conditions:

- Treat the bed with neem or *Pongamia* cake at 50 g + *T. harzianum* or *Pochonia chlamydosporia* both at 25 g + *Paecilomyces lilacinus* at 25 g/m^2 at an interval of 2 months.
- Apply 2 t of farmyard manure enriched with biopesticide *Trichoderma harzianum* per acre before planting, along with 100–200 kg of neem or *Pongamia* cake.
- Treat the seeds with biopesticide *Pseudomonas fluorescens* at 10 g/kg seed.
- Raise the seedling in the coco peat or any substrate by treating it with *P. fluorescens* at 1 kg + *T. harzianum* at 1 kg + neem or *Pongamia* cake at 50 kg/t.
- Apply *P. fluorescens/Trichoderma viride* at 2.5 kg/ha mixed with 50-kg farmyard manure 10 days before sowing/planting and use of chemical pesticide like carbofuran 3G at 1 kg a.i./ha.

Preplant treatment of beds with dazomet followed by the application of neem cake (1 kg/m^2, 15 days later) along with *P. lilacinus* or *P. chlamydosporia* significantly reduced populations of *M. incognita* and the mortality of plants and suppressed the nematode infection for nearly 2 years. On a long-term basis, soil management with preplant treatment of dazomet, followed by the application of oil cakes plus antagonistic fungi, was more effective against *M. incognita* than post-plant treatment with carbofuran, carbosulfan, and chlorpyrifos in polyhouses.

References

Baker KF, Rolstacher CN (1957) Heat treatment of soil. In: The UC system for producing healthy container-grown plants. UC Manual 23, University of California, Agric Expt Stn

Morra L (1998) Potenzialità e limiti dell'innesto in orticoltura. Inf Agrar 54(49):43–46

Renzoni G, Lamberti F (1974) Innesti erbacei e nematocidi nella lotta contro I nematodi galligeni (*Meloidogyne* spp.) su pomodoro. Nematol Mediterr 2:83–90

16 Insects Pests and Their Management

Abstract
Well-fertilized and irrigated greenhouse crops are, however, often more sensitive to pests, like aphids, whiteflies, and leaf miners. The environment is generally warm, humid, and wind-free inside the greenhouse which promotes the fast growth of most crops, but it is also ideal for the development of herbivorous insects. Some of the important pest groups affecting greenhouse crops are aphids (*Myzus persicae, Aphis gossypii*), caterpillars (*Spodoptera* spp., *Helicoverpa armigera*), leaf miner (*Liriomyza trifolii*), mites (*Tetranychus urticae, Polyphagotarsonemus latus*), thrips (*Thrips tabaci, Scirtothrips dorsalis*), and whiteflies (*Trialeurodes vaporariorum, Bemisia tabaci*). Insects that enter the greenhouse can transmit virus diseases (like tomato spotted wilt virus) and can carry bacteria and fungi as well. Management strategies involving avoidance, early detection, and curative measures are discussed.

Keywords
Insect pests • Aphids • Thrips • Whiteflies • Mites • Integrated pest management

16.1 Introduction

The warm, humid conditions and abundant food under protected conditions provide an excellent, stable environment for pest development. Often, the natural enemies that keep pests under control outside are not present under protected environment. For these reasons, pest situations often develop in the indoor environment more rapidly and with greater severity than outdoors. The damage inflicted by arthropod pests on greenhouse crops varies with the pest and season. The level of damage that can be tolerated is greatly dependent on the type of crop. Producers of vegetable crops generally can accept a higher level of damage than those of ornamental crops that are produced for their aesthetic value.

Integrated pest management (IPM) is a systematic approach to manage pests that combines a variety of techniques and strategies to either

P.P. Reddy, *Sustainable Crop Protection under Protected Cultivation*,
DOI 10.1007/978-981-287-952-3_16

reduce pest populations or lessen their economic impact. It is a site-specific strategy for managing pests that relies on correct pest identification and understanding the pest biology. With a long-term perspective, it is easier to see that an investment in IPM can pay for itself in a higher-quality crop and a cleaner environment.

The growing conditions within the protected environment of greenhouse/polyhouse are highly favorable to arthropod pests. In India, about 20 insect species have been recorded to be associated with the crops under a protected environment. Some of the important pest groups are aphids (*Myzus persicae, Aphis gossypii*), caterpillars (*Spodoptera* spp., *Helicoverpa armigera*), leaf miner (*Liriomyza trifolii*), mites (*Tetranychus urticae*, *Polyphagotarsonemus latus*), thrips (*Thrips tabaci, Scirtothrips dorsalis*), and whiteflies (*Trialeurodes vaporariorum, Bemisia tabaci*). In addition to damaging the crop, some insects vector diseases like tomato spotted wilt virus. The detailed account of the insect pests associated with crops under a protected environment is being presented in Table 16.1.

Each crop has a set of typical pests that attack under certain conditions. This information should be available with the grower to aid in diagnosis and to help understand the life cycle of the pest. Correct identification is critical to selection of appropriate management tactics, because individual species can display different susceptibilities to chemical and other measures.

16.2 Aphids

Aphids or plant lice are small, soft-bodied, sedentary insects that form colonies on the leaves and stems of the host plants. They suck the cell sap from leaf or apical growing parts and prefer to feed on tender, young growth. Aphids multiply at a rapid pace completing one generation in 7–10 days. They give birth to six to ten young ones per day with 50–100 in total life span. Thus, enormous populations can build up in a relatively short period.

Feeding by aphids can cause leaves or stems to curl or pucker. This leaf distortion often protects the aphids from contact insecticides. They also secrete honeydew on which black sooty mold develop in the periods of high atmospheric relative humidity which interferes with photosynthetic activity. They also act as vectors of many viral diseases. Common aphid species on ornamentals include the polyphagous *Aulacorthum circumflexum*, *Macrosiphum euphorbiae*, *Myzus ornatus*, *M. persicae*, *A. gossypii*, *Macrosiphoniella sanborni* on chrysanthemum, *Aulacorthum solani* on pelargonium, *Macrosiphum rosae* on *rose*, and *Dysaphis tulipae* on bulbs.

16.2.1 Monitoring

Examine the foliage, stems, and new growth of key plants such as peppers, eggplants, cole crops, and leafy greens for early detection of aphid infestation. Signs of aphid activity include shedding of white skins, shiny honeydew, curled new leaves, distorted growth, and the presence of ants. Yellow sticky cards help detect the entrance of winged aphids into the greenhouse from outdoors. Yellow cards will not, however, permit the monitoring of aphids within the crop, as most of the aphids will be wingless.

16.2.2 Management

Starting with plant material free from aphids at least delays the development of the population. So does netting of openings of the glasshouse. Yellow traps and regular inspection of the crop provide information concerning presence, population increase, and the need for treatment.

Chemical treatment, by spray or trickle irrigation, is widely used against aphids. Repeated treatments may be needed to keep populations at low levels. Treatment should be started as soon as aphids are detected. There is considerable resistance of aphid populations (especially *A. gossypii* and *M. persicae*) to pirimicarb and organophosphorus compounds. In these cases, active substances with a different mode of action, such as imidacloprid or nicotine, can be used.

Table 16.1 Insect pest scenario under a protected environment in India

Group	Insect and mite pests	Host	Distribution
Aphids	*A. gossypii*	Capsicum	Punjab, Delhi
	M. sanborni	Chrysanthemum	Karnataka, Himachal Pradesh
	Macrosiphum luteum	Orchid	Sikkim
	Myzus escalonicus	Strawberry	New Delhi
	M. persicae	Capsicum, gerbera	Punjab, Maharashtra
	Toxoptera aurantii	Orchid	Sikkim
Caterpillars	*H. armigera*	Capsicum, tomato, carnation	Punjab, Uttarakhand, Himachal Pradesh
	S. litura	Rose, tomato, capsicum, cucumber	Karnataka, Punjab, HP
Leaf miner	*L. trifolii*	Tomato, cucumber, chrysanthemum, gerbera, and many ornamentals	Karnataka, Himachal Pradesh
Thrips	*S. dorsalis*	Rose	Karnataka
	Thrips palmi	Gerbera	Karnataka
	T. tabaci	Gerbera	Maharashtra
Whiteflies	*B. tabaci*	Gerbera, capsicum	Karnataka, Punjab
	T. vaporariorum	Tomato, cucumber, capsicum, beans, gerbera, and more than 30 hosts	Himachal Pradesh, Nilgiri hills (TN)

Biological control agents like *Aphidoletes aphidimyza*, *Aphidius matricariae*, *A. colemani*, and *Verticillium lecanii* are available. They are generally used preventively but *V. lecanii* can also be used curatively.

16.2.3 Main Insecticides

Acephate, bifenthrin, cypermethrin, deltamethrin, diazinon, dichlorvos, dimethoate, formothion, heptenophos, imidacloprid, malathion, methomyl, mevinphos, oxamyl, permethrin, phosalone, pirimicarb, propoxur, resmethrin, and thiometon

Space Treatments Diazinon, dichlorvos, nicotine, pirimicarb, propoxur, or sulfotep

16.3 Caterpillars

Two species of caterpillars are associated with the crops under protected environment. These are tobacco caterpillar (*Spodoptera litura*) and tomato fruit borer (*H. armigera*).

16.3.1 Tobacco Caterpillar, *S. litura*

They are mainly foliage feeders but occasionally damage fruits. Eggs are deposited by female moths that have made their way into the greenhouse and lay eggs in large masses. A single female lays about 300 eggs in clusters covered by brown hair. They pupate in soil. Heavy infestations may occur in isolated areas within the greenhouse, and timely spot treatments can often provide effective control.

Spodoptera exigua caterpillars feed on young leaves, shoots, and flowers. Young larvae may skeletonize leaves near the soil. Moths are brownish gray and hide during the day.

16.3.2 Tomato Fruit Borer, *H. armigera*

Tomato fruit borer is one of the most significant insect pests of field-grown crops. They are less common in greenhouses, but when infestations do occur, they can cause serious losses. This is because, as their name implies, this caterpillar feeds directly on the fruit. This insect attacks many other crops.

Tomato fruit borer feed on blooms, as well as immature and mature fruits. The individually laid eggs are most often deposited on leaves near upper bloom clusters. If not controlled, heavy infestations of these caterpillars can destroy the crop. Because large larvae that have already bored into fruit are very difficult to control, it is important to detect infestations and begin treatment when caterpillars are small.

Eggs hatch within 3–7 days, depending on species and temperature. The small caterpillars begin feeding on leaf tissue. The caterpillar stage lasts about 11–17 days with larvae eating more and causing proportionally more damage as they become larger. Upon completion of its larval development, the caterpillar transforms to a pupa on the foliage or in the soil, depending on species. Moths may emerge from these pupae in 7–11 days.

16.3.3 Management

Infestation may develop very rapidly because of the short generation time. Frequent inspection of the crop is thus necessary. The glasshouses should be kept closed at night to prevent the moths from entering. Application of insecticides may be necessary, but only young larvae are susceptible. *Spodoptera exigua* nucleopolyhedrovirus can be used as a biological insecticide, but also in that case only larvae are susceptible. After ingesting the insecticide, they will die within 3–6 days.

16.3.4 Main Insecticides

Against Young Larvae *Spodoptera exigua* nucleopolyhedrovirus, diflubenzuron, lufenuron, methomyl, and teflubenzuron

Against Adult Moths Space treatment with dichlorvos or pyrethroids

16.4 Leaf Miners

The adults of *L. trifolii* and *L. huidobrensis* cause small white feeding spots on leaves, normally of little significance. The larvae of both *L. huidobrensis* and *L. trifolii* leave the mines to pupate in the soil; however, the pupae of *L. trifolii* can sometimes also be found on the leaves. These organisms are quarantine pests. Therefore, plants for export have to be free from these insects.

Leaf miners are larvae of small flies. They damage plants by feeding between the upper and lower surface of the leaf by forming mines. The mines increase in width as the larva grows. When fully grown, the larva emerges from the leaf and fall to the ground to pupate. Each female fly lays 50–100 eggs by making punctures in the leaf tissue.

Damage is caused by larvae mining the leaves. If the infestation is severe, photosynthetic activity is reduced affecting plant vigor. Because the damaging stages of the insects occur entirely inside the leaf, control with contact insecticides is ineffective once the damage appears. Infestations can be avoided through the use of good cultural practices, hand removal and disposal of infested leaves, and use of chemical controls when necessary.

16.4.1 Management

Infestation in the glasshouse is caused by infested planting material and by the entering of flies through doors and windows. The use of different traps (yellow, sticky, water) or frequent inspection of the plants can indicate the presence of the pest. The use of healthy seedlings is important. Insect nets are effective in keeping the flies out and the crop clean.

Chemical treatment, by spraying against the larvae or space treatment against the adults, is widely used against leaf miners. Treatment should start as soon as damage due to adults (little white dots) is detected. Repeated treatments may be needed to keep populations at low levels. However, chemical control of *L. huidobrensis* larvae in a crop is not very effective. The same applies to killing the pupae in the soil. *L. trifolii* is easier to handle. The parasitic wasps *Dacnusa sibirica*, *Diglyphus isae*, and *Opius pallipa* are being developed as biological control agents in ornamental crops in glasshouses. Where possible and when available, these parasites should be used.

16.4.2 Main Insecticides

Abamectin, cyromazine, methamidophos, methomyl, oxamyl, pyrazophos, thiocyclam hydrogen oxalate, and triazophos

Against Adult Moths Space treatment with dichlorvos or pyrethroids is effective.

Against *L. huidobrensis* Larvae Abamectin, cyromazine, oxamyl, and thiocyclam hydrogen oxalate

16.5 Thrips

Thrips are tiny, slender insects about 1–1.5 mm long. Female thrips insert eggs into slits in the tissue. Eggs hatch in 2–7 days. Nymphs feed much like adults and molt four times during development. They pupate in debris or soil. Winged adults are carried into the greenhouse on contaminated plant material, or they fly in.

Thrips infest the leaves, flowers, buds, and young fruits of a crop. They feed by rasping the plant surface and sucking up the exuding sap. Heavily infested leaves have a mottled or silvery appearance.

16.5.1 Monitoring

Early detection of thrips infestation is critical for effective management because populations are lower and it is easier to obtain good coverage when plant canopies are small. Symptoms of feeding are often not noticed until the damage has occurred. Eggplant, tomatoes, and peppers are especially prone to thrips infestations. Blue sticky cards, key plants, and indicator plants are all effective for the detection of onset of an infestation. Yellow sticky cards should be placed just above the crop canopy, near doors and vents, and over thrip-sensitive cultivars to monitor the movement of the thrips. Recent research has shown that light- to medium-blue sticky cards catch more thrips than yellow ones. However, it is more practical to use yellow cards for general pest monitoring to attract fungus gnats, whiteflies, and winged aphids. The number of thrips per card should be recorded and graphed weekly to monitor population levels and movement in or out of the greenhouse and thus help make control decisions.

16.5.2 Management

It is important to start with thrips-free planting material. Yellow or blue (especially for *Frankliniella occidentalis*) sticky traps can indicate the presence of thrips. The yellow traps also indicate the presence of other pests (e.g., whiteflies, aphids, and leaf miners).

The predatory mites *Neoseiulus cucumeris* and *Amblyseius degenerans* and the predatory bugs *Orius* spp. should be used for preventive biological control where possible. The fungus *V. lecanii* has a side effect on thrips. If population densities increase, single or repeated spray with an insecticide is necessary. If only chemical control is used, the first spray is applied as soon as thrips are detected. In the case of *F. occidentalis,* this is not easy because of resistance to many products. Because of the hidden way of life of many thrips, including *F. occidentalis*, space treatment may be necessary, particularly if the thrips have infested the flower buds. Although pyrethroids are effective, they can destroy all biological control efforts for weeks (e.g., *N. cucumeris*).

16.5.3 Main Insecticides

Abamectin, acephate, acrinathrin, chlorpyrifos-ethyl, cypermethrin, deltamethrin, fenpropathrin, malathion, methamidophos, methomyl, mevinphos, oxamyl, pirimiphos-methyl, pyrazophos, and thiocyclam hydrogen oxalate

Space Treatments Dichlorvos or sulfotep

For Thrips Other Than F. occidentalis Imidacloprid and several organophosphates and pyrethroids may be used.

16.6 Whiteflies

Whiteflies are serious pests in the greenhouse. Two species of whiteflies are prevalent under protected environment, namely, greenhouse whitefly (*T. vaporariorum)* and cotton whitefly (*B. tabaci*). The former is prevalent in temperate regions, whereas the other is a serious pest in tropical and subtropical parts of the country.

Both the nymphs and adults of whitefly feed on phloem cell sap and causes chlorotic spots. The leaves dry up prematurely and plant growth is affected. Nymphs also secrete a sticky substance known as honeydew, which covers leaf surfaces and flowers on which the sooty growth takes place under humid weather conditions.

The lower surface of the leaves is infested with all life stages of whiteflies. The female of these sap-sucking insects may lay up to 150 eggs at the rate of 25 per day. The entire life cycle takes 21–36 days, depending on the greenhouse environment. They can complete more than 12 generations in a year.

16.6.1 Monitoring

To monitor whiteflies, check susceptible plants, such as tomatoes, at ten locations in an area of 100 m^2, examining plants on each side of the aisle. Look on the undersides of one or two leaves per plant, for nymphs, pupa, and adults. Yellow sticky traps can also be used to detect adult whiteflies once populations have reached higher densities. Begin treatments as soon as the first sign of infestation is noted.

16.6.2 Management

Cultural methods or resistant cultivars are of no importance for the control of these pests in ornamental plants. The use of traps (yellow, sticky, or water) or frequent inspection of the plants can indicate their presence. Control is essential as soon as populations are detected and can be assured by single or repeated sprays or space treatments with insecticides or biologically. *Encarsia formosa*, a parasitic hymenopteran well known for its effective control of whiteflies in vegetables, is now also used successfully in protected ornamentals. *E. formosa* should be used when available and effective and, in that case, any insecticides used should be known to be harmless to this natural enemy. The fungus *V. lecanii* or the predatory ladybird *Delphastus pusillus* can also be used for the biological control of whiteflies. *V. lecanii* has a side effect on thrips.

16.6.3 Main Insecticides

Sprays Abamectin, bifenthrin, buprofezin, cyfluthrin, cypermethrin, deltamethrin, diazinon, dichlorvos, fenazaquin, fenpropathrin, imidacloprid, malathion, methomyl, mevinphos, nicotine, oxamyl, permethrin, pirimiphos-methyl, and propoxur

Space Treatments Buprofezin, diazinon, dichlorvos, pyrethrins, and teflubenzuron

16.7 Leaf Rollers (Tortricidae)

They are mainly brown or green caterpillars of tortricid leaf rollers, like *Adoxophyes orana*, *Cacoecimorpha pronubana*, *Clepsis spectrana*, and *Epichoristodes acerbella*, which live between leaves spun together. They feed on leaves and buds.

16.7.1 Management

Tortricids normally enter the glasshouse through openings in spring and summer during cultivation or are imported with planting material. Where pheromone traps are available, monitoring of adult moths by trapping can provide information on the appropriate timing of application, when the sensitive stages of the pests are exposed (i.e., larvae before leaf rolling or flying adults). Chemical control can be necessary, by spraying against larvae or by use of space treatment against adults. *Trichogramma evanescens* can be used as a biological control agent.

16.7.2 Main Insecticides

Against Larvae Acephate, bifenthrin, cypermethrin, cyfluthrin, deltamethrin, diflubenzuron, fenpropathrin, lufenuron, methomyl, permethrin, teflubenzuron, and trichlorfon

Against Adult Moths Space treatment with dichlorvos or pyrethroids

16.8 Scale Insects and Mealybugs

A large number of species belong to this group. Larvae and adults feed on stems and leaves by puncturing the vascular bundles of woody or annual plants. Many of them produce a waxy secretion and/or honeydew, thus reducing the ornamental value of the plants. Examples include the polyphagous mealybugs *Planococcus citri*, *Pseudococcus affinis*, and *P. longispinus* and the polyphagous scale *Coccus hesperidum* and *Diaspis boisduvalii* on orchids.

16.8.1 Management

It is important to use healthy planting material and to avoid growing plant batches of different growth stages and direct contact between them. Multiplication of these insects is rather slow. Direct movement of the larvae is limited and slow, and adults are (almost) sedentary. Their way of life is often hidden, and a thorough and regular, but not necessarily frequent, inspection is necessary to be sure to detect them. If scale insects or mealybugs are detected, a chemical treatment is required. Biological control is possible, e.g., with *Cryptolaemus montrouzieri* against mealybugs. The parasitic wasps *Leptomastix dactylopii* and *Metaphycus helvolus* can be used against *Planococcus* spp.

16.8.2 Main Insecticides

Acephate, buprofezin, bifenthrin, chlorpyrifos-ethyl, deltamethrin, etrimfos, methidathion, mevinphos, mineral oil, oxamyl, pirimiphos-methyl, and propoxur

16.9 Weevils, *Otiorhynchus* spp

These are gray- to black-colored weevils of about 1 cm that feed on leaves and stems. More important are the whitish larvae that live in the soil and feed on roots.

16.9.1 Management

Sanitation and general hygiene are important. Larvae are spread with soil, e.g., potted plants. Detritus should be removed and hiding places should be kept free from adult weevils. The soil should be sterilized/treated before planting or potting or should be drenched after potting. If in spite of these precautions infestation occurs, a chemical or biological treatment may be necessary. Larvae can successfully be controlled by the entomopathogenic nematodes *Steinernema carpocapsae* and *Heterorhabditis* spp. The nematodes are most effective at soil temperatures higher than 13 °C.

16.9.2 Main Insecticides

Against Larvae Carbofuran, chlorpyrifos-ethyl, and fonofos

Against Weevils Acephate, cypermethrin, deltamethrin, diflubenzuron, methamidophos, methomyl, and pirimiphos-methyl

16.10 Sciaridae

The larvae live in the soil and feed on young plants.

16.10.1 Management

Larvae can successfully be controlled by entomopathogenic nematodes of the genus *Steinernema* spp. The nematodes are most effective at soil temperatures higher than 13 °C. The predatory mite *Hypoaspis* can also be used for biological control. Chemical control is also possible, by spraying against the larvae as soon as detected or space treatment against the adults.

16.10.2 Main Insecticides

Permethrin and teflubenzuron

16.11 Slugs

Slugs (e.g., *Arion* spp., *Deroceras* spp., and *Lehmannia* spp.) may inflict considerable damage to young plants and occasionally the leaves of older plants by feeding and by forming slimy tracks.

16.11.1 Management

General hygiene is important. All plant debris, leftover plastic, growing medium, etc., that may serve as hiding places or oviposition sites should be removed. The soil should be free from weeds and also under the tables and along the walls. Moist conditions favor the development of a slug population. Biological control is possible with *Phasmarhabditis hermaphrodita*.

16.11.2 Main Molluscicides

Mercaptodimethur pellets and metaldehyde pellets

16.12 Management Strategies

Since greenhouse conditions allow rapid development of pest populations, successful control of insect pests on greenhouse vegetables and ornamentals depends on several factors. The IPM programs for protected cultivation can be described as a pyramid constructed having three key components, namely, avoidance of problem, sampling and early detection, and curative measures.

With "avoidance" as the foundation, virtually all the non-curative preventive management components (e.g., physical, mechanical, and cultural means) can be fitted. Confronted with a pest crisis, short-term function depends on the upper two levels of the pyramid. However, sustainable, long-term strategies must depend on the development of this solid foundation. At the same time, a pyramid strategy developed for one pest must be compatible with like strategies in place for all pests of a system.

Although physical and cultural measures do not have a curative effect after heavy infestations of greenhouse pests, their preventive and delaying effect on the rapid increase of pests is of great importance in integrated pest management.

16.12.1 Avoidance

16.12.1.1 Use of Physical Barriers

16.12.1.1.1 Use of Insect-Proof Screens

One strategy particularly suited to greenhouse IPM programs is the use of physical barriers to exclude insect pests. Screening removes an important variable from a grower's pest management program – the movement of pests into the greenhouse from outside. This includes common greenhouse pests such as thrips, aphids, leaf miners, and whiteflies but also some less common pests such as fruit borers. Screening vents and doorways can greatly limit the movement of insect pests into the greenhouse. However, selection of proper screen size mesh and assuring adequate airflow are more important.

Mesh size depends upon the targeted insect (Table 16.2). Mesh with holes less than 200 μm is required for complete exclusion of thrips; however, screening with holes as large as 600 μm is sufficient for excluding leaf miners.

For the exclusion of thrips, a screen mesh of 76 is required. Obviously, screening that keeps out thrips also keeps out larger pests, such as aphids and whiteflies. However, selection of screen with particular mesh size is site specific and depends on the prevalent pests of the crop in the locality. The smaller the holes in the mesh, the more reduction there will be in airflow. This reduction can be counteracted by increasing the surface area through which air flows. Screening can be stretched from gutter to gutter to increase surface area.

16.12.1.1.2 Provision of Double Door

Limited access to screened areas is beneficial since insects may come in the protected structure on clothing or be swept in with the wind. Building a screened foyer to create a double-door entry partially solves the problem of wind-carried insects. Special efforts must be put in for repairing holes or tears immediately and cleaning the screens to maintain airflow.

16.12.1.1.3 Ultraviolet Radiation-Absorbing Sheets

Altering the visual behavior of insects has been used successfully as a tool in IPM programs directed to protect crops from insects and insect-borne viral diseases. The first evidence that UV-absorbing films may reduce insect invasion of greenhouses came from Japan.

Table 16.2 Screen mesh sizes needed to exclude major greenhouse pest species

Insect pest	Hole size (micron)	Mesh (number of threads per linear inch)
Leaf miner (*L. trifolii*)	610	34
Cotton whitefly (*Bemisia tabaci*)	462	42
Aphid (*M. persicae*)	340	52
Greenhouse whitefly (*T. vaporariorum*)	290	58
Thrips (*Thrips* spp.)	192	76

Insects perceive light signals through their compound eyes. The anatomy and physiology of the compound eye is adapted to sense UV wavelengths alone or a mixture of UV and visible radiation. The UV part of the solar spectrum plays an important role in the ecological behavior of insects, including orientation, navigation, feeding, and interaction between the sexes.

Spectrally modified sheets are produced commercially by the introduction of a UV-absorbing additive into the raw material which blocks the transmission of most wavelengths in the UV range below 370–380 nm without interfering with the transmission of photosynthetically active radiation (400–700 nm).

The manipulation of the UV vision of insects by using UV-blocking greenhouse cladding materials has been shown to be effective in preventing the immigration of a wide range of insect pests (whiteflies, aphids, thrips, and leaf miners) from the external environment into the protected crop. It was found that populations of aphid (*A. gossypii*), greenhouse whitefly (*T. vaporariorum*), thrips (*Frankliniella occidentalis* and *Scirtothrips dorsalis*), and leaf miner (*Liriomyza* sp.) were lower on tomatoes grown in a plastic house made of polyethylene treated to exclude UV wavelengths than on crops grown in an ordinary plastic house. The number of whiteflies, aphids, and thrips trapped on sticky yellow cards under a UV-absorbing film were 10–100 times lower than the number trapped under regular films. The use of UVA films also helped in reducing the number of insecticide applications by 50–80 % for the management of *S. littoralis*.

16.12.1.2 Sanitation and Cultural Practices

Many of the methods incorporated in an IPM strategy are logical greenhouse operating procedures and basic horticultural practices. The following methods produce a healthier crop, prevent many pest problems, and isolate pests to smaller areas in the greenhouse. Some of the important ones are:

16.12.1.2.1 Preseason Cleanup

A basic component of cultural practices is sanitation. Infestations are easier to prevent than to cure. It has been observed that insects can enter the protected structure in the following three ways:

- Introduction of infested seedlings/planting material
- Infestation from other plants within the protected structure
- Infestation from host plants outside but near the protected structures

Before introducing a new crop into the greenhouse, it is extremely important to eliminate pests from the previous crop. Remove all plant debris and weeds from the greenhouse. Many pests also occur on other crops or broadleaf weeds. For this reason, it is important to avoid growing other crops next to the greenhouse and to prevent heavy growths of broadleaf weeds around the outside edges of the greenhouse. Under a protected environment, monoculture is suggested; however, if one has to go for polyculture, then avoid staggered planting.

A fallow period of 2–4 weeks reduces the pest load considerably. To determine the presence of thrips, whiteflies, leaf miners, or other insects, set up yellow sticky cards and indicator plants after watering. Observe for any insects that are trapped on the cards after 2 days and continue till the activity is ceased and only thereafter that the decision regarding plantation of new crop be made.

16.12.1.2.2 Inspection on Arrival

One of the most important points in protected cultivation is to begin with insect-free planting material. When new plants arrive at the greenhouse, examine them closely for signs of pest infestation. If necessary, remove lower or damaged leaves to avoid spread of pests. Make the decision whether treatment is needed from the first sign of symptoms of insects or mites. It is much easier to manage a pest infestation by treating a group of small plants (in seedling stage) rather than larger plants where the dense canopy prevents thorough coverage.

16.12.1.2.3 Balanced Use of Fertilizers

Fertilization schedules based on balanced use of nutrients should be followed. Nitrogen should be applied only as needed for optimal growth. Periodic heavy applications set up nitrogen surpluses that cause excessive growth, which favor the population growth of aphids and other pests. Application of potassium at desired levels has been found to reduce the incidence of insect pests.

16.12.1.2.4 Pinching and Pruning

Pinching off damaged plant parts, flowers, and spotted leaves (and those with insect larvae or egg deposits) can be a very effective way of reducing the spread of pests in the greenhouse. The plant debris should be placed immediately in a covered container before being disposed off. This practice can be helpful in reducing the pest population of all the targeted pests.

Pruning lower leaves after harvesting lower fruit clusters is a helpful measure in removing large numbers of developing leaf miners and whiteflies.

16.12.1.2.5 Plant Quarantine

Workers in the greenhouse are frequently the mechanism for dispersal of insects and mite pests. One should try to avoid moving plants with mites or thrips and they should not be touched or moved immediately before handling clean, healthy plants.

16.12.1.2.6 Trap Crop/Indicator Plants

For early detection and trapping of the target pests, some of the preferred hosts of the target pests can be used. Planting border rows of *Portulaca oleracea* in rose can be used as a trap crop for tobacco caterpillar under a protected environment.

16.12.2 Early Detection

Scouting and early detection are critical to manage the insect infestation successfully. Monitoring or scouting is the regular, systematic inspection of the plants and exteriors to identify and assess pest problems. It includes inspection of foliage and flowers and the use of insect traps. Many insect infestations begin in isolated spots within the greenhouse. Timely crop monitoring identifies situations where pests are absent or are at levels well below those necessary to cause damage, thus preventing unnecessary control applications and expenditures.

16.12.2.1 Scouting

Scouting procedures for most greenhouse-grown crops are based on visual observations and are used to provide estimates of the pest population in protected environment. The common pests that attack greenhouse crops do not distribute themselves evenly throughout the crop. Therefore, it is imperative to scout the entire greenhouse in a consistent, uniform pattern. Inspect the entire plant, including the soil surface, for the presence of arthropod pests. Look at the plant systematically each time. Begin at the bottom and work up. Look at the older leaves, the young, tender leaves, and the flush growth. When the crop is young, it is important to check all the leaves on the plant. Because a majority of arthropod pests associated prefer the underside of a leaf, it is important to turn the leaves over to check for pests. The detailed account of observations to be recorded is presented in Table 16.3.

Scout the crop on a regular basis and at least 1–2 % of the total plants should be inspected at weekly interval. A thorough greenhouse inspection reveals the location and severity of any current pest problems. One should use a field data sheet to record the identification, location, and severity of all pests present and record the effectiveness of any treatments.

16.12.2.2 Monitoring

It is a relative method of insect population estimation where no direct observations on the plants for the presence of insect pests are needed. However, the pest population is estimated with the help of attractant traps. For whitefly, aphid, thrips, and leaf miner adults, yellow sticky cards (10 × 20 or 20 × 30 cm) are an excellent supplement to pest observation in the protected environment. Additionally, for thrips, blue-colored sticky traps can also be used. Though the sticky traps are available commercially, however, they can also be fabricated indigenously at growers' level by using pieces of fluorescent yellow chart paper and pasting castor oil/mustard oil or the commercial grade adhesive, the polybutene, on both the surfaces. The traps are placed in a grid pattern and 1–2 yellow sticky cards per 100 square meter of floor area are used. If the target is mass trapping, then the number of traps can be increased to five or more. Hang the yellow sticky cards/traps in the crop with the help of strings about 10–15 cm above the plant canopy. As the crop grows, cards can be moved up. Designate the location of each sticky card on a map of the greenhouse. Check the sticky cards every scouting visit (twice a week if possible) and record the total number of whiteflies, thrips, winged aphids, and leaf miner flies from each card on the field data sheet. Change the cards when more than 60–70 % of the area is covered by trapped insect.

Apart from this, sex pheromone-baited traps can also be used for detecting moths of tobacco caterpillar and tomato fruit borer. These traps allow visualizing population trends and can be used to time the application of pesticides or release of bioagents. The detail of the scouting and monitoring plan is given in Table 16.3.

Table 16.3 Monitoring and scouting technique for pests under a protected environment

	Aphids	Leaf miners	Yellow mites	Spider mites	Thrips	Whiteflies	Caterpillars
Scouting							
Inspect underside of leaf	x		x	x	x	x	x
Inspect upper surface for stippling/small white spots				x	x		
Inspect leaves for mines within tissue		x					
Inspect new growth or terminal for feeding	x		x		x		
Tap flowers over white surface and look for movement	x			x	x		
Press lower leaf surface on white card and look for spots on cards				x		x	
Observe for							
Honeydew or sooty mold	x						
Small black spots on leaves (fecal drops)					x		
Holes/skeletonized spot in/on leaves							x
Frass on leaves							x
Small black spots on leaves (fecal drops)					x		
Chlorotic (yellow) spots on upper leaf surface	x				x	x	
Casted-off skins on leaf	x						
Curling of leaves	x		x				
Distortion of new growth	x		x			x	
Buds fail to open or uneven opening of flowers					x		
Webbing on leaves/ flowers				x			
Monitoring							
Yellow sticky card at top of plant canopy	x	x			x	x	
Blue sticky card at top of plant canopy					x		
Pheromone traps							x

16.12.3 Curative Measures

16.12.3.1 Biological Control

Practically every crop pest has its natural enemies in the form of parasitoids or predators. The biological control involves a large-scale multiplication and release of such agents or creating conditions under which the naturally occurring agents can act effectively. In field situations, some very outstanding successes have been achieved by using biocontrol agents, but the method suffered a setback owing to the large-scale and indiscriminate use of insecticides.

The basics of biological control to manage pests include use of specific predators and parasitoids in greenhouse ecosystems. But to be effective, biological control must be well planned and begun when the target populations are low. In Western Europe and North America, the bioagents are commercially available and being used successfully for the management of pest problems under protected situations (Tables 16.4 and 16.5).

However, under Indian conditions, this technique though is having potential but have the following limitations:

- Temperature extremes and the effect of chemical pesticides affecting the capabilities of natural enemies
- Low tolerance of pest damage among consumers especially in ornamentals
- Lack of a system for supplying natural enemies of good quality

16.12.3.2 Chemical Control

The insecticides are curative in action and are one of the important tools to check the flaring pest populations. Insecticides belonging to different groups, namely, botanical pesticides, microbial pesticides, insect growth regulators (IGRs), synthetic chemical pesticides, etc., are in use. Some of the serious limitations of the insecticides have been highlighted in recent years. Out of these, the problems of insecticides residues on crops have been agitating the minds of people in developed and developing countries of the world due to indiscriminate use of pesticides. However, the sensible approach suggested and generally accepted is for need-based judicious and safe use of insecticides.

Under a protected environment, in order to avoid the contamination of produce from pesticide residues, the use of pesticides having less persistence and adopting the suggested waiting period between last insecticidal application and harvesting can be followed.

Another challenge with pest managers is to minimize the development of resistance in pest species to the pesticides. This can be checked or delayed by avoiding repeated application of the same insecticide and also from the same group. Incorporating botanicals, microbials, and others in the management schedule proves better in the insecticide resistance management program (IRMP).

Apart from this, the safety of the pesticide applicator should also be taken into consideration as the chemical pesticides become more volatile under high-temperature regimes of the protected environment. One should apply the pesticides by taking all safety measures including the protective kit to avoid the direct effect of pesticides to the applicator. A reentry period of at least 12 h should be observed after application of pesticides. Insecticides with fumigant action need to be avoided under protected structures.

The following are some of the insecticides labeled for the management of insect and mite pests (Table 16.6).

16.13 Future Thrusts

Research efforts are needed for developing pest management technology under a protected environment with emphasis on avoidance and selective use of pesticides. Safe waiting intervals based on harvest time pesticide residues need to be established for the crops under a protected environment as this information is lacking completely. Apart from this, emphasis to improve the awareness level of the growers for timely diagnosis and judicious use of insecticides needs to be taken up on priority.

Table 16.4 Predators and parasitoids of insect pests

Organism	Supplier	Pests controlled	Application/comments
A. degenerans or *Iphiseius degenerans* (predatory mite)	Intl. Technology Services, IPM Labs., Green Spot	Aphids	Release when pest levels are low. Prefer high humidity
Amblyseius fallacis or *Neoseiulus fallacis* (predatory mite)	IPM Labs., Rincon-Vitova, Green Spot	Mites	
Anagyrus pseudococci (parasitic wasp)	Praxis	Mealybugs	
Aphidius colemani (parasitic wasp)	Florikan, IPM Labs., Harmony Farm Supply, Praxis, Rincon-Vitova, Green Spot	Aphids	Release 5–2/m^2; humidity should be 70–85 %, temp. 65–77 °F. Release at first sign of pests and for three consecutive weeks thereafter. Sensitive to pesticides
A. matricariae (parasitic wasp)	Arbico, Hydro-Gardens, Green Spot	Aphids	1250–7500/ha. Can be used in greenhouses all year long
A. aphidimyza (predatory midge)	Nature's Control, Intl. Technology Services, IPM Laboratories, Hydro-Gardens, Arbico, Praxis, Rincon-Vitova, Harmony Farm Supply, Green Spot	Aphids	1–3/3 m^2; humidity should be 50–90 %, temp 60–80 °F. Release when aphids are first observed; release lower numbers for preventive control; apply every 2 weeks. Active at night; sensitive to day length
Aphytis melinus (parasitic wasp)	Arbico, Sespe Creek Insectary, Harmony, Hydro-Gardens, IPM Labs., Natural Pest Controls, Praxis, Green Spot	Many scales	25,000–250,000/ha; humidity should be 20–80 %, temp. 65–95 °F
Chilocorus spp. (beetle)	Praxis	Scales	
Chrysoperla carnea (predator)	Natural Pest Controls, Beneficial Insectary, Caltec, Arbico, A-1 Unique Insect Control, Praxis, Rincon-Vitova, Hydro-Gardens	Aphids, caterpillars, mealybugs, scales, spider mites, thrips, whiteflies	1 lacewing/5–30 aphids; 1000 eggs/200 m^2. Apply every 1–3 weeks as needed. May arrive as eggs, immatures, or adults
Chrysoperla rufilabris (predator)	Arbico, Beneficial Insectary, IPM Labs., A-1 Unique Insect Control, Nature's Control, Praxis, Rincon-Vitova	See above	
Chrysoperla spp. (predator)	M&R Durango, Florikan, Green Spot	See above	
Coccophagus lycinnia (parasite wasp)	Praxis	Scale	
Coleomegilla maculata (pink ladybird beetle)	Arbico	Aphids, mites, caterpillars, scales, thrips, whiteflies	1/m^2; shipped as larvae and eggs

(continued)

Table 16.4 (continued)

Organism	Supplier	Pests controlled	Application/comments
C. montrouzieri (predatory beetle)	Arbico, Caltec, Intl. Technology Services, IPM Laboratories, Natural Pest Controls, Nature's Control, Florikan, Harmony Farm Supply, Hydro-Gardens, Praxis, Rincon-Vitova, Green Spot	Aphids, mealybugs, soft scales	2–5/infested plant; humidity should be 70–80 %, temp. 70–80 °F. Larvae are cannibalistic; repeat as necessary for control; do not wear white while distributing
Dacnusa sibirica (parasitoid)	Arbico, Intl. Technology Services, Natural Pest Controls, Praxis, Koppert, Green Spot	Leaf miners	1250–2500/ha; humidity should be 50–90 %; temp. 60–85 °F. Apply at first appearance of pests
Deraeocoris brevis (predator)	Green Spot	Aphids, whiteflies, thrips	
Delphastus pusillius (predatory beetle)	Arbico, IPM Laboratories, Nature's Control, Harmony Farm Supply, Hydro-Gardens, Rincon-Vitova, Praxis, Green Spot	Greenhouse whitefly, sweet potato whitefly	2000/3000 m^2; temperature should be 60–85 °F. Will feed on spider mites if no whiteflies are available Should be used along with *E. formosa* and traps
Diaeretiella rapae (parasitoid)	Arbico, Praxis	Aphids	Release rates vary
Diglyphus isaea (parasitoid)	Arbico, Harmony Farm Supply, Natural Pest Controls, Praxis, Intl. Technology Services, Green Spot	Leaf miners	Temps. should be 75–90 °F and humidity around 80 %
E. formosa (parasitic wasp)	Arbico, Nature's Control, IPM Laboratories, Intl. Technology Services, Florikan, Harmony Farm Supply, Hydro-Gardens, Natural Pest Controls, A-! Unique Insect Control, Praxis, Rincon-Vitova, Green Spot	Greenhouse whitefly, sweet potato whitefly, silver leaf whitefly	Release 1/m^2 weekly for 3 weeks when pest numbers are low. Release 2–4/m^2 when pest numbers are high. Apply when pests are first observed. Should be used in conjunction with traps. May be used along with other beneficials. *E. formosa* is very susceptible to chemicals. Temps should be at least 64 °F. Reapply every 2 weeks
Encarsia luteola or *E. deserti*	Beneficial Insectary, Hydro-Gardens, IPM Laboratories, Arbico, Praxis, Green Spot	Whiteflies	

(continued)

Table 16.4 (continued)

Organism	Supplier	Pests controlled	Application/comments
Eretmocerus californicus or *E. eremicus* (parasitic wasp)	Hydro-Gardens	Greenhouse whitefly, silver leaf whitefly, sweet potato whitefly	Capable of handling hot, dry temperatures. Introduce when whiteflies are first observed. May be used in combination with other beneficials such as green lacewings. *Eretmocerus* is more tolerant of pesticides than *E. formosa*
Euseius spp. (predatory mite)	Praxis	Scale, mites	
Harmonia axyridis (Asian lady beetle)	Green Spot	Scale, whiteflies, mealybugs, aphids	Temperatures should be 70–85 °F; humidity around 70 %
Heterorhabditis bacteriophora (beneficial nematode)	M&R Durango, Arbico, BioLogic, Hydro-Gardens, Harmony Farm Supply, Plant Health Care, Green Spot	Fungus gnats, crown borers, thrips, grubs, cutworms, Japanese beetles, black fine weevil	Application rate varies; 1 million/3000 m^2 is suggested. Nematodes need a moist environment to survive and move through soil. Apply in evening directly into growing medium
Hippodamia convergens (lady beetle) (predator)	A-1 Unique Insect Control. Arbico, Caltec, IPM Laboratories, Natural Pest Controls, Nature's Control, Harmony Farm Supply, Hydro-Gardens, Praxis, Green Spot	Aphids, mites, whiteflies	Release at dusk near an immediate food source. Spray plants with water prior to release
Hypoaspis miles (predator)	Florikan, Harmony Farm Supply, Green Spot	Thrips	100–300/m^2
Iphiseius degenerans or *A. degenerans* (predatory mite)	Intl. Technology Services, IPM Labs., Green Spot	Aphids	
Leptomastida enormis (parasitic wasp)	Praxis	Mealybug	
Leptomastix dactylopii	Praxis, Arbico, IPM Labs.	Mealybug	
Lindorus lophanthae or *Rhyzobius lophanthae* (black lady beetle)	Harmony Farm Supply, Arbico, Green Spot	Armored scales	1–2/m^2. Release as soon as scales are detected. Temps. should be 59–77 °F, humidity 20–90 %
Lysiphlebus testaceipes (parasitic wasp)	Praxis	Aphids	
Mesoseiulus longipes or *Phytoseiulus longipes* (predator)	Arbico, Nature's Control, Harmony Farm Supply	Spider mites	Introduce at first sign of pests. Can tolerate hot, dry conditions (humidity 40 %, temp. 70–90 °F). Tolerates extremes of temp. and humidity better than other mite predators

(continued)

Table 16.4 (continued)

Organism	Supplier	Pests controlled	Application/comments
M. helvolus (parasitoid)	Arbico, IPM Labs., Harmony Farm Supply, Natural Pest Controls, Praxis, Sespe Creek Insectary, Green Spot	Scale	Temps. should be 73–87 °F; RH 50 %. Do not overwinter in cold climates. They are attracted to lights and sticky traps
Neoseiulus spp. or *Amblyseius* spp. (predatory mites) *N. barkeri*	IPM Laboratories, Arbico	Broad mites, thrips	10–30/plant per week
N. californicus	Arbico, Beneficial Insectary, Caltec, IPM Labs., Nature's Control, Harmony Farm Supply, Hydro-Gardens, Rincon-Vitova	Spider mites	Can tolerate hot conditions (humidity 60 % min.; temp. 70–90 °F). Introduce at first sign of spider mites. Can survive absence of prey longer than other predatory mites
N. cucumeris	Arbico, Natural Pest Controls, Nature's Control, Intl. Technology Services, Florikan, IPM Labs., Harmony Farm Supply, Hydro-Gardens, Rincon-Vitova, Green Spot	Thrips, mites	Humidity should be 70–90 %, temp 50–85 °F. Introduce at first sign of pests
N. cucumeris and *N. barkeri*	Hydro-Gardens	Thrips, aphids, mites	1 predator/m^2; humidity should be moderate, temp. 70 °F. Establish population early. Repeat every month during periods of warm, dry weather
Neoseiulus fallacis or *Amblyseius fallacis* (predatory mite)	IPM Labs., Rincon-Vitova, Green Spot	Mites	Release when pest levels are low. Prefer high humidity
Orius insidiosus (minute pirate bug) (predator)	Florikan, IPM Labs., Harmony Farm Supply, Arbico, Hydro-Gardens, Praxis, Koppert, Intl. Tech. Services, Green Spot	Aphids, caterpillars, thrips, whiteflies, mites	1/10 m^2 (preventive), 1 every 2 m^2 when pests are present. Temperature should be 70–90 °F. *Orius* are dormant during September–April. Reapply every 2–3 weeks. Very susceptible to pesticides. Works well in combination with *N. cucumeris*
Phytoseiulus longipes or *Mesoseiulus longipes* (predator)	Arbico, Nature's Control, Harmony Farm Supply	Spider mites	Introduce at first sign of pests. Can tolerate hot, dry conditions (humidity 40 %, temp. 70–90 °F). Tolerates extremes of temp. and humidity better than other mite predators

(continued)

Table 16.4 (continued)

Organism	Supplier	Pests controlled	Application/comments
Phytoseiulus persimilis (predatory mite)	Arbico, Beneficial Insectary, Hydro-Gardens, Intl. Technology Services, Natural Pest Controls, Nature's Control, Caltec, Florikan, IPM Labs., Harmony Farm Supply, Praxis, Green Spot	Spider mites	2000 mites/3000 m^2 or 1/5 m^2. Apply at first sign of spider mites. If pests persist, reapply every 3–5 weeks. Can be used in combination with either *Galendromus occidentalis* or *Neoseiulus californicus*. May be used in combination with *Bt* or Enstar II. Also available in *P. persimilis* Ht for higher temperatures. Humidity should be 60–90 %, temps. 65–80 °F
Propylea quatuordecimpunctata (predatory beetle)	Praxis	Aphids	
Rhyzobius lophanthae or *Lindorus lophanthae* (black lady beetle)	Harmony Farm Supply, Arbico, Green Spot	Armored scales	1–2/m^2. Release as soon as scales are detected. Temps. should be 59–77 °F, RH of 20–90 %
Stethorus punctillum (spider mite destroyer)	Nature's Control, Green Spot	Spider mites	100 spider mite destroyers start a colony. These lady beetles live 4–5 weeks
Steinernema spp. (beneficial nematodes)	Hydro-Gardens, Nature's Control	Cutworms, thrips, white grubs, shore flies, fungus gnats, etc.	1 million/3000 m^2; temperature should be above 55 °F. Apply in evening, directly to growing media. Water in after application. Needs moist environment to thrive
Steinernema spp. and *Heterorhabditis* spp. (beneficial nematodes)	Florikan, Green Spot	See above	See above
S. carpocapsae (beneficial nematodes)	BioLogic, Thermo Trilogy, Arbico, Harmony Farm Supply, M&R Durango, Hydro-Gardens, IPM Labs., Praxis, Green Spot	Moths, borers, earworms, rootworms, beetles, cutworms, fungus gnats, crickets, grubs, shore flies, weevils, maggots, etc.	11,000–35,000/m^2. Work in moist soil at temps. from 50 °F to 85 °F. For best results, apply at dusk
Steinernema feltiae (beneficial nematodes)	BioLogic, Geiger, M&R Durango, IPM Laboratories	See above	Apply when media temps. are 55–90 °F
Stethorus punctillum (lady beetles)	Praxis	Spider mites	
Thripobius semiluteus (parasitoid)	Arbico, Nature's Control	Thrips	

Source: Greer, Lane. 1999. ATTRA Technical Specialist. Compiled from magazine articles, extension bulletins, and product literature

Table 16.5 Biopesticides for key greenhouse pests

Insecticide	Target pests	Labeled crops	Comments
Bacillus thuringiensis	Leaf miner, caterpillars	Greenhouse vegetables, e.g., tomatoes, cole crops, peppers	Stomach poison that must be eaten by target insect to be effective. Most effective against small, newly hatched larvae. Insects stop feeding and die 1–5 days later
Beauveria bassiana	Aphids, thrips, whitefly	Many vegetables incl. eggplant, peppers, squash	Contains a fungus that must contact target pest. Thorough spray coverage needed for contact material to be effective. Treat when insect population is low. Repeated applications may be needed
V. lecanii	White fly, mealybug, leafhopper	See above	See above
Nucleopolyhedroviruses (NPVs)	Tobacco caterpillar, tomato fruit borer	Many vegetables	*Ha*NPV for tomato fruit borer, *Sl*NPV for tobacco caterpillar

Table 16.6 Selected insecticides labeled for insects and mites on vegetable plants

Insecticide	Target pests	Labeled crops	Comments
Azadirachtin	Many incl. aphids, thrips, caterpillars, leafhopper, leaf miner, whitefly, mites	Many vegetables incl. cucurbit, eggplant, tomato, peppers	Insect growth regulator for immature stages of insects. Repeat applications needed. Repels some insects and can be used as an antifeedant
Fenazate	Spider mites	Many vegetables	Compatible with beneficial predatory mites. Rapidly degraded in high-temperature alkaline water. Use solutions promptly or add a commercial buffering agent
Buprofezin	Whitefly, leafhopper	Many vegetables	Active against nymph stages. Chitin synthesis inhibitor and suppresses oviposition of adults and reduces viability of eggs. Treated pests may remain alive for 3–7 days, but feeding damage is low. Apply no more than two applications per season

(continued)

Table 16.6 (continued)

Insecticide	Target pests	Labeled crops	Comments
Chlorfenapyr	Caterpillars, spider mites, broad mites, western flower thrips	Many vegetables – do not use on tomato varieties <2.5 cm in diameter when mature	Do not apply more than three times during crop cycle. Do not make more than two consecutive applications before rotating to a chemical in a different class
Dinotefuran	Aphids, leaf miner, thrips, whitefly	Vegetable transplants	Do not make more than one application per crop
Imidacloprid	Aphids, leafhopper, leaf miner, thrips, whitefly	Vegetable bedding	–
Pyrethrins	Many incl. aphids, caterpillars, thrips, leafhopper, whitefly	Many vegetables	Flushes insects from hiding with knockdown effects
Pyriproxyfen (distance insect growth regulator)	Whitefly, aphids	See supplemental label for indoor-grown fruiting vegetables	Do not apply to tomato varieties <2.5 cm in diameter or to non-bell peppers Do not make more than two applications per season
Spinosad	Leaf miner, caterpillars, thrips	Many vegetables	See label for resistance management guideline
Spirotetramat	Aphids, spider mites, whitefly	Vegetable transplants	Will not control heavy population of spider mites
Sucrose octanoate esters (SucraShield)	Aphids, caterpillars, leafhopper, mites, thrips, whitefly	Vegetables	Contact insecticide with limited residual activity

17 Aphids and Their Management

Abstract

This chapter summarizes IPM for greenhouse aphids on both vegetable and ornamental crops. Focus is on monitoring, sanitation, biological controls, biorational pesticides, and insect growth regulators. Supplemental tables include information on the newest biopesticides and biological control organisms.

Keywords

Aphids • Scouting • Monitoring • Biocontrol • Biorational pesticides

17.1 Introduction

There are approximately 4000 aphid species in the world. Life cycles and preferred hosts vary with each type of aphid. Common aphid pests of greenhouse crops include the green peach aphid (*Myzus persicae*), the melon/cotton aphid (*Aphis gossypii*), the chrysanthemum aphid (*Macrosiphoniella sanborni*), the rose aphid (*Macrosiphum rosae*), the potato aphid (*Macrosiphum euphorbiae*), and the foxglove aphid (*Aulacorthum solani*) (Lindquist 1991).

The green peach aphid is probably the most notorious aphid pest of greenhouse crops because of its wide host range, worldwide distribution, and difficulty of control and the number of viral diseases it vectors (Sunderland et al. 1992).

Aphid management relies on understanding that the females of many aphid species do not have to mate in order to reproduce, and they typically produce live young rather than eggs. These characteristics contribute to the tendency of aphid populations to "explode."

17.2 Crop Scouting and Trapping

Plants should be visually inspected for signs of an aphid infestation. Look especially carefully at plants prone to aphid problems and at plant parts like the undersides of leaves, stems, and new growth. Choose plants randomly throughout the greenhouse and inspect undersides of leaves, buds, or tip growth and watch for honeydew and cast skins. Since aphids are difficult

P.P. Reddy, *Sustainable Crop Protection under Protected Cultivation*,
DOI 10.1007/978-981-287-952-3_17

to see on plants with fine foliage, hold such plants over a white piece of paper and gently tap to dislodge any aphids. Avoid moving infested plants to new areas where susceptible plants are growing. Locations where aphids are found should be flagged, so that population development and control efforts can be evaluated.

Different aphid species tend to infest different parts of their host plants. Green peach aphids tend to cluster on the succulent young growth, whereas melon aphids are usually evenly distributed along the plant stems. Melon aphid populations also have fewer winged adults than do green peach aphids. Knowing which species is infesting the crop is very important in successful detection and monitoring of aphid populations.

Yellow sticky cards placed horizontally at the top of the pot or container (if you are growing containerized plants) can be used for monitoring winged aphids. However, since winged aphids caught during the summer months may have blown in from the outdoors; sticky cards are not as reliable as visual inspections. Sticky cards are more useful in the winter months when aphids caught on the cards are not likely to have come in from the outside. It is better to rely primarily on visual inspections for aphid detection, and use sticky cards as a backup method (Table 17.1).

Signs of an aphid infestation include honeydew or sooty mold on leaves, yellow spots on upper leaf surfaces, cast skins on leaves, curling of leaves, and distortion of new growth.

Table 17.1 Monitoring and scouting techniques for aphids

	Aphids
Scouting	
Inspect underside of leaf	x
Inspect new growth or terminal for feeding	x
Tap flowers over white surface and look for movement	x
Observe for	
Chlorotic (yellow) spots on upper leaf surface	x

17.3 Biological Control

There are several biological control options for greenhouse aphid pests. Some common biological control agents (BCAs) include green lacewings (*Chrysoperla carnea, C. rufilabris, Chrysopa* spp.), aphid midges (*Aphidoletes aphidimyza*), parasitic wasps (*Aphidius colemani* and *A. matricariae*), and lady beetles (*Hippodamia convergens*) (Table 17.2).

A 1998 study showed that green lacewing larvae did not disperse as well as the parasitic wasp *Aphidius colemani* (Heinz 1998). To achieve equal aphid suppression, more of the slower-moving species need to be introduced and from more points (e.g., lacewings have to be released on each bench because they cannot move to adjacent benches). A study performed at Colorado State University compared the effectiveness of parasitic wasps, aphid midges, lady beetle larvae, and green lacewing larvae (Anon 1999). The researchers found that lacewings performed better in hot temperatures, while aphid midges and lady beetles were better in cooler temperatures. At all temperatures, *Aphidoletes* was the best of the four at controlling aphids.

Strains of the fungus *Beauveria bassiana* provide good control of aphids, including green peach aphids. The fungus works by attaching to the outside of the pest and then penetrating into the body and killing it. The fungus is available commercially for greenhouse ornamentals as Naturalis-O and for vegetables as BotaniGard (Table 17.3). Another fungus, *Verticillium lecanii*, can also provide good biological control of aphids. Formulations of this fungus are currently being sold in some European countries under the names of Vertalec and Mycotal, but neither of these products is yet registered for use in the USA.

Verticillium lecanii often occurs naturally in greenhouses, so it may be possible to encourage its growth and distribution in the greenhouse without the benefit of a commercially available product. *V. lecanii* spores require at least 93 % relative humidity at temperatures between 59 °F and 81 °F to germinate and grow. High humidity must be present for at

Table 17.2 Predators and parasitoids for biocontrol of aphids in greenhouses

Organism	Supplier	Application/comments
Chrysoperla carnea	Natural Pest Controls, Beneficial Insectary, Caltec, Arbico, A-1 Unique Insect Control, Praxis, Rincon-Vitova, Hydro-Gardens	1 lacewing/5–30 aphids; 1000 eggs/200 m^2. Apply every 1–3 weeks as needed. May arrive as eggs, immatures, or adults
Chrysoperla rufilabris	Arbico, Beneficial Insectary, IPM Labs., A-1 Unique Insect Control, Nature's Control, Praxis, Rincon-Vitova	
Chrysoperla spp. (predator)	M&R Durango, Florikan, Green Spot	
Coleomegilla maculata	Arbico	1/m^2; shipped as larvae and eggs
Cryptolaemus montrouzieri (predator beetle)	Arbico, Caltec, Intl. Technology Services, IPM Laboratories, Natural Pest Controls, Nature's Control, Florikan, Harmony Farm Supply, Hydro-Gardens, Praxis, Rincon-Vitova, Green Spot	2–5/infested plant; humidity should be 70–80 %, temp. 70–80 °F
Deraeocoris brevis (predator)	Green Spot	
Diaeretiella rapae (parasite)	Arbico, Praxis	Release rates vary
Harmonia axyridis (Asian lady beetle)	Green Spot	Temperatures should be 70–85 °F; humidity around 70 %
Hippodamia convergens (lady beetle) (predator)	A-1 Unique Insect Control, Arbico, Caltec, IPM Laboratories, Natural Pest Controls, Nature's Control, Harmony Farm Supply, Hydro-Gardens, Praxis, Green Spot	Release at dusk near an immediate food source. Spray plants with water prior to release
Iphiseius degenerans or *Amblyseius degenerans* (predatory mite)	Intl. Technology Services, IPM Labs., Green Spot	
Lysiphlebus testaceipes (parasitic wasp)	Praxis	
Neoseiulus cucumeris and *N. barkeri*	Hydro-Gardens	1 predator/m^2; humidity should be moderate, temp. 70 °F. Establish population early. Repeat every month during periods of warm, dry weather
Orius insidiosus (minute pirate bug) (predator)	Florikan, IPM Labs., Harmony Farm Supply, Arbico, Hydro-Gardens, Praxis, Koppert, Intl. Tech. Services, Green Spot	1/10 m^2 (preventive), 1 every 2 m^2 when pests are present. Temperature should be 70–90 °F. *Orius* are dormant during September–April. Reapply every 2–3 weeks. Very susceptible to pesticides. Works well in combination with *Neoseiulus cucumeris*
Propylea quatuordecimpunctata	Praxis Harmony Farm Supply, Praxis, Rincon-Vitova, Green Spot	

Table 17.3 Biopesticides based on *Beauveria bassiana*

Brand name	Supplier	REI	Application/comments
Naturalis-O	SePro	4 h	Apply when insects first appear and repeat every 7–10 days. Good spray coverage is needed. Not compatible with other fungicides
BotaniGard	Mycotech	12 h	See above

least 10–12 h/day. Unfortunately, most plant disease-causing fungi also grow best at these same temperature and humidity ranges. Fungicides used to control the plant disease-causing fungi would probably also kill any beneficial fungi present. Insecticides may also be harmful to *V. lecanii*.

17.4 Biorational Pesticides

Least-toxic pesticides used against aphids in greenhouses include insecticidal soap (M-Pede), horticultural oil (SunSpray Ultra-Fine Spray Oil), and botanical insecticides such as neem (Azatin, Neemazad, and Neemix) or natural pyrethrums (Table 17.4).

17.5 Insect Growth Regulators

Insect growth regulators (IGRs) are another least-toxic pesticide control option for pests. IGRs typically kill insects by disrupting their development. They have a complex mode of action that precludes insects from rapidly developing resistance. IGRs can work in one of several ways:

- They can mimic juvenile hormones, so that insects never enter the reproductive stage of development.
- They can interfere with the production of chitin, which makes up the shell of most insects.
- They can interfere with the molting process.

IGRs usually work through ingestion, so good spray coverage is essential. They generally don't

Table 17.4 Biorational pesticides

Brand name	Supplier	REI	Application/comments
Azadirachtin – extract of neem seed; IGR that works through contact or ingestion			
Azatin	Green Spot	4 h	Apply when pests first appear
Neemazad	Thermo Trilogy	12 h	Cannot be applied through irrigation. Low rate can be used as a preventative
Garlic extracts			
Garlic Gard	Soil Technologies		
Garlic Barrier	Green Spot	4 h	Use late in the day. Can be mixed with fish oil or horticultural oil. Do not use in combination with bumble bees or honeybees
Horticultural oil – includes dormant and summer superior oils			
All Seasons	Green Spot	4 h	Use on sunny days to promote rapid drying and decrease chance of phytotoxicity. Not compatible
Hot pepper wax – contains capsaicin, paraffin, and mineral oil			
Hot Pepper Wax	Green Spot	4 h	Also contains herbal essential oils. Not compatible with beneficials
Hot Pepper Wax	Hot Pepper Wax, Inc.	0 h	
Insecticidal soap – contains potassium salts of fatty acids			
M-Pede	Mycogen	12 h	Phytotoxicity is often a concern, esp. after repeated applications
Safer	Green Spot	4 h	See above
Insecticidal Soap	Olympic		
Neem oil – multipurpose organic insecticide/fungicide/miticide; kills eggs and larval and adult stages of insects			
Trilogy 90EC	Thermo Trilogy	4 h	Apply at first signs of damage. Repeat every 7–10 days as needed
Triact 90EC	Thermo Trilogy	4 h	For ornamental crops only
Soybean oil			
Golden Natural Spray Oil	Stoller	12 h	

affect nontarget species, such as humans, birds, fish, or other vertebrates. For most IGRs there are minimal reentry restrictions. IGRs typically take several days to have an effect on pest populations. Because IGRs do not affect mature insects, adult beneficials released into the greenhouse after an IGR application are not likely to be affected. Use of IGRs is generally prohibited by organic certification organizations because the products are synthesized.

IGRs can sometimes be used in conjunction with biological control efforts and may provide growers with a "safety net" should beneficial fail to keep the pests below economically damaging levels. The table below lists some well-known insect growth regulators (Table 17.5).

Table 17.5 Selected insect growth regulators

Brand name	Supplier
Azatin	Hydro-Gardens, Olympic Horticultural Products
Enstar II	Wellmark Intl.
Neemazad	Thermo Trilogy
Neemix	Thermo Trilogy
Preclude	Whitmire Micro-Gen

17.6 Conclusions

Greenhouse aphids are tiny insects, but they demand serious attention on the part of the greenhouse grower. Integrated pest management offers a sustainable approach for dealing with greenhouse aphids, and safer pest control products facilitate the adaptation of least-toxic control measures that dovetail very nicely with the IPM philosophy.

References

Anon (1999) Efficacy of four biocontrol agents on the green peach aphid, *Myzus persicae*, in greenhouse peppers. Midwest Biological Control News, Jan–Feb, p 7

Heinz KM (1998) Dispersal and dispersion of aphids and selected natural enemies in spatially subdivided greenhouse environments. Environ Entomol 27(4):1029–1038

Lindquist R (1991) A guide to aphid control. Grower Talks, Oct, p 75

Sunderland K et al (1992) Integrated pest management of greenhouse crops in Northern Europe: aphids. In: Janick J (ed) Horticultural reviews, 13th edn. Wiley, New York, pp 23–30

18 Thrips and Their Management

Abstract

This chapter summarizes IPM for greenhouse thrips on both vegetable and ornamental crops. Monitoring, biology and identification, sanitation, insect screening, cultural controls, biological controls, biorational pesticides, and insect growth regulators are discussed. Supplemental tables include information on the newest biopesticides and biological control organisms.

Keywords

Thrips • Biology • Scouting • Sanitation • Insect screening • Biocontrol • Biorational pesticides

18.1 Introduction

Thrips are one of the most difficult pests to control in greenhouses. They are tiny insects that reproduce rapidly, congregate in tight places that make pesticide coverage difficult, and feed with rasping–piercing–sucking mouth parts, resulting in deformation of flowers and leaves.

Tolerance to thrips on floriculture crops is particularly low. In addition, when the western flower thrips (WFT) feed on plants infected with the tomato spotted wilt virus (TSWV) or impatiens necrotic spot virus (INSV), the insect vectors these diseases to other plants in the greenhouse. Once plants are infected, it is too late to do anything except dispose of diseased plants. Thus, the best way to prevent virus infection is to control thrips.

There are at least 6000 species of thrips in the world. A few of the main species of thrips known to injure greenhouse crops are summarized in Table 18.1.

18.2 Biology and Identification

It is important to understand the life cycle and behavior of thrips in order to develop an effective control strategy. Knowledge about the weak link in a pest's life cycle can help in choosing the most appropriate control strategy. Table 18.2 and Fig. 18.1 summarize the life cycle of thrips.

As an aid to managing thrips species infesting greenhouse, positive identification needs to be obtained. Because they are extremely small, the best way thrips can be identified is by a trained entomologist with the aid of a microscope.

P.P. Reddy, *Sustainable Crop Protection under Protected Cultivation*,
DOI 10.1007/978-981-287-952-3_18

Table 18.1 Main species of thrips in greenhouses

Common name	Species name
Greenhouse thrips	*Heliothrips haemorrhoidalis*
Banded greenhouse thrips	*Hercinothrips femoralis*
Flower thrips	*Frankliniella tritici*
Western flower thrips (WFT)	*Frankliniella occidentalis*
Onion or tobacco thrips	*Thrips tabaci*

Table 18.2 Life cycle of thrips

Stage	Duration at temperatures between 68 °F and 98 °F
Egg	2–4 days
1st instar	1–2 days
2nd instar	2–4 days
Prepupal	1–2 days
Pupal	1–3 days
Adult	30–45 days

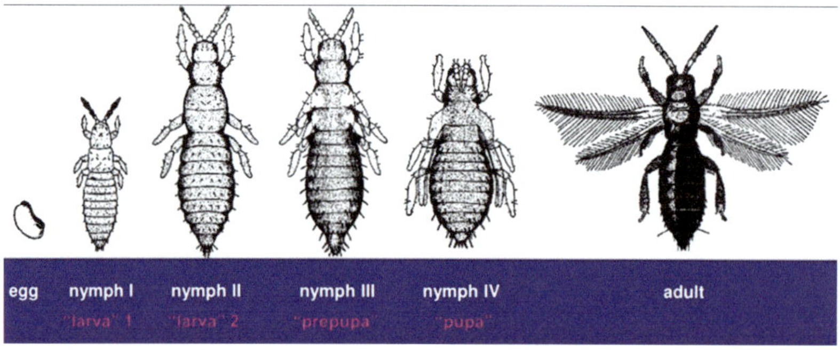

Fig. 18.1 Thrips life cycle

18.3 Crop Scouting and Trapping

To detect early infestations, a crop scouting program that includes both sticky trap cards and visual inspection is critical. Scouting should be done once a week and more often when an infestation is detected. Regular scouting is also necessary to monitor the efficacy of control measures. A hand lens is a useful tool to detect live thrips as well as signs of thrips activity – e.g., black feces and silvery, flecked areas on leaves. Lightly blowing on blossom and growing point aids in visual inspection as it causes thrips to become mobile, apparently because of the carbon dioxide contained in exhalation. Hot-pink sticky cards have been found to be the most attractive color for trapping thrips, though blue is often still used. Sticky traps should be placed 2.5–5.0 cm above the crop canopy so that the bottoms of the traps are just above the crop, at the rate of one or two per 400 m^2. The economic threshold, or number of thrips found on each trap in relation to crop injury, is not yet available for every greenhouse crop. A threshold level of 20 WFT/trap/week was figured to be appropriate for chrysanthemums in Switzerland (Parella 1995). In contrast, a greenhouse in New Mexico initiates pesticide application when 5 thrips/trap/week are detected, and chrysanthemums make up 80 % of its crop mix (Aylsworth 1994). When biological control organisms are used to control thrips, they should be released as soon as the first adult thrips have been trapped. Certain plants are especially appealing to thrips and can be used as "biological indicators" to detect the presence of thrips in the greenhouse. Majestic, Blue Magic, and Calypso petunias are recognized as thrips indicators because they will exhibit viral symptoms of TSWV within just a few days after feeding. Fortunately, the virus remains localized in the plants and does not become systemic, so that spread of this disease from infected petunias to other plants in the greenhouse is not a problem. Symptoms of thrips-vectored TSWV on petunias include a distinct brown rim at the feeding site within 3–4 days after feeding. This is followed by a localized circular lesion in about a week. Flowers should be removed from the petunias to discourage adult thrips from feeding on the flowers instead of the foliage, because viral lesions will not show up on flowers (Anon 1993). The following suggestions on effective integration of sticky cards and indicator plants are provided (Pundt et al. 1992):

- Place sticky cards throughout your operation to detect where thrips are located and to monitor their numbers.
- Place indicator plants among crops at bench or floor level. One plant every 20–30 ft seems to work well.
- Place incoming plant material with indicator plants and isolate for at least 3–4 days to allow thrips scars to develop and show viral lesions.

The following additional points are provided on use of sticky cards and indicator plants (Allen 1991):

- Indicator plants are not intended as a substitute for sticky cards, but rather they are an additional component to an effective monitoring system.
- While sticky cards monitor only flight activity and relative population of thrips, indicator plants monitor feeding activity and the presence of virus-carrying thrips near those plants.
- Studies show that attraction to indicator plants can be significantly increased if hot-pink or blue insect-monitoring cards, without adhesive, are attached to the petunia pots. In addition to increased attractiveness to thrips, the colored cards serve as location markers and reminders that the plants should be examined frequently.

A summary of monitoring and scouting techniques for greenhouse insects is provided in Table 18.3.

Table 18.3 Monitoring and scouting techniques for thrips

	Thrips
Scouting:	
Inspect underside of leaf	x
Inspect upper surface for stippling/small white spots	x
Inspect new growth or terminal for feeding	x
Tap flowers over white surface and look for movement	x
White or brown spots on flowers	x
Dark area on buds that are just opening	x
Observe for:	
Small black spots on leaves (fecal drops)	x
Chlorotic (yellow) spots on upper leaf surface	x
Buds fail to open or uneven opening of flowers	x

18.4 Sanitation

Sanitation is the key for controlling pests in greenhouses. The goal of sanitation is to eliminate all possible sources of the pest. Weeds inside and near the outside of the greenhouse can harbor pests. It is best to pull weeds inside the greenhouse rather than spray them, since insects may survive the spray and migrate onto crops. Bag all weeds and dispose them off outside the greenhouse.

In addition, a 3–10-m vegetation-free zone around the outside perimeter of the greenhouse especially near vents and openings can provide a dramatic decrease in pests. A heavy-duty geotextile weed barrier (e.g., DeWitt Sunbelt weed barrier) covered with bark mulch or gravel can provide a pleasant vegetation-free zone and eliminate the need for herbicides.

Plant debris from previous crops can also be a source of both immature and adult pests. Clean up all debris from previous crops and dispose off infested plants or any infested growth. Ideally, the greenhouse should be thoroughly cleaned and left empty for 1 week prior to beginning the next crop. This enables removal of all pest stages and starves any remaining adults. Closing up the greenhouse when it is empty in summer will increase the temperature and help eradicate pests.

Inside the greenhouse, a clean stock program should be in place. This includes temporary quarantine and inspection of all plants upon arrival from other greenhouses and regular monitoring of stock plants used for propagation. If a separate section of the greenhouse cannot be dedicated to this purpose, flag all incoming plants. All new plant material should be thoroughly inspected (with a 10× hand lens) for the presence of pests to ensure that no infested plants are introduced into the greenhouse.

Workers in the greenhouse should avoid wearing yellow clothing, since many pests are attracted to this color and may hitch a ride on the fabric from one greenhouse to the next.

18.5 Insect Screening

In the spring, adult thrips migrate from host plants and relocate to new host plants. Though thrips are not strong fliers, they are easily borne by the wind, which facilitates their movement. Prevention of airborne entry of thrips into greenhouses can be greatly aided with the use of an insect screen. Growers in New York, North Carolina, and Florida estimated a 30–70 % reduction in pest problems after installation of insect screening (Gill and Ross 1994).

Thrips are so tiny and elusive that the brand of insect screening you choose can make a big difference in how well it excludes thrips. In 1995, researchers from North Carolina reported that 27 types of insect screens were evaluated for thrips exclusion. Only three brands (FlyBarre, BugBed, No-Thrips) were effective in preventing the entry of thrips into greenhouses (Grossman 1996).

18.6 Cultural Controls

One option to consider is a fallow period in summer. To perform this operation successfully, the grower should first remove all plants and then all weeds and then heat the greenhouse (either artificially or naturally) until soil temperatures reach 60 °F. This temperature should be maintained for 3 weeks. During this time, any thrips eggs will hatch and the nymphs will starve for lack of food.

This strategy has worked for western flower thrips. Researchers found that an air temperature of 104 °F with a relative humidity of 10 % was sufficient to kill WFT (an environment that was fatal to plants within 4 days). However, when sufficient water was made available to keep plants alive in the greenhouse (whether weeds or crop plants), mortality of WFT was only 50 % (Anon 1992). Another recommended strategy for controlling thrips is to remove all flowers and buds, if not crucial to the crop. The flowers and buds should be put into plastic bags, sealed, and disposed off outside the greenhouse.

A large greenhouse grower in the Netherlands found that garlic plants are an effective way to repel thrips. He uses three potted garlic plants for every 30 ft^2 of bench area (Shipp and Gillespie 1993).

18.7 Biological Control

Biological control of greenhouse thrips can be achieved through release of biocontrol agents such as predators (predatory mites, lady beetles, soil-dwelling mites), parasitoids, and pathogens (Tables 18.4 and 18.5).

Certain crops are better adapted to biological control of thrips than others. Greenhouse vegetables, bedding plants, and tropical house plants are better adapted to biological control than are floriculture crops. Thrips preferentially feed on flowers and leaves and cause little or no damage to fruits themselves. Thus, while fruit-bearing vegetable crops like tomatoes, cucumbers, and peppers are well suited to low populations of prey food (i.e., thrips), tolerance in floriculture crops is greatly reduced because feeding results in cosmetic injury on the marketable portion of the crop. When TSWV or INSV are present, tolerance is nil in either case.

A recent study showed that beneficial insects can control western flower thrips on bedding plants, even though bedding plants have such a short turnaround time (Gill et al. 2000). Treatments consisted of BotaniGard (a *Beauveria bassiana*-based product) applied at 1 kg/ha, *Neoseiulus cucumeris* mites applied weekly at a rate of 12,000 mites per 1000 m^2 of growing area, and a combination of the two. Plants treated with *B. bassiana* had between 0.16 and 0.5 thrips/plant, while plants treated with predatory mites had a maximum of 0.15 thrips/plant. The untreated control plants had 0.75–1.25 thrips/plant.

18.7.1 Notes on Biological Control of Thrips

- Release of biocontrol agents should begin as soon as thrips are detected on sticky traps, or even before detection, as part of a planned biocontrol release program on sensitive crops.
- Adult female predatory mites (*Neoseiulus*) consume from 1 to 10 young thrips per day and have a 30-day life span. They can also survive on pollen and spider mites in the absence of

Table 18.4 Predators for the management of insect pests

Organism	Supplier	Application/comments
Chrysopa carnea	Natural Pest Controls, Beneficial Insectary, Caltec, Arbico, A-1 Unique Insect Control, Praxis, Rincon-Vitova, Hydro-Gardens	1 lacewing/5–30 aphids; 1000 eggs/67 m^2. Apply every 1–3 weeks as needed. May arrive as eggs, immatures, or adults
Chrysoperla rufilabris	Arbico, Beneficial Insectary, IPM Labs, A-1 Unique Insect Control, Nature's Control, Praxis, Rincon-Vitova	
Chrysoperla spp. (predator)	M&R Durango, Florikan, Green Spot	
Coleomegilla maculata	Arbico	3 predators/m^2; shipped as larvae and eggs
Deraeocoris brevis (predator)	Green Spot	
Neoseiulus cucumeris and *N. barkeri*	Hydro-Gardens	3 predators/m^2; humidity should be moderate, temp. 70 °F. Establish population early. Repeat every month during periods of warm, dry weather
Orius insidiosus (minute pirate bug) (predator)	Florikan, IPM Labs., Harmony Farm Supply, Arbico, Hydro-Gardens, Praxis, Koppert, Intl. Tech. Services, Green Spot	1/3 m^2 (preventive), 2 every m^2 when pests are present. Temperature should be 70–90 °F. *Orius* are dormant during September–April. Reapply every 2–3 weeks. Very susceptible to pesticides. Works well in combination with *Neoseiulus cucumeris*

Table 18.5 Pathogens for biological control of insect pests

Brand name	Supplier	REI	Application/comments
Beauveria bassiana Naturalis-O	SePro	4 h	Apply when insects first appear and repeat every 7–10 days. Good spray coverage is needed. Not compatible with other fungicides
Beauveria bassiana BotaniGard	Mycotech	12 h	See above

thrips. Repeat applications of predators must be made to establish a predator to prey of 1:2 ratio. *Neoseiulus* attacks first instar (very young) thrips only and does not move long distances from where it is first placed. They are most often applied in small piles at the base of plants or in paper bags. Usually, a small hole is made in the bag, and mites move out of the bag slowly.

- Adult pirate bugs (*Orius*) consume 5–20 thrips (all stages) per day. They can survive on pollen in the absence of prey. Both adults and nymphs are predacious. *Orius* is the only predator that attacks thrips in tight places like flower buds. Since *Orius* is a strong flyer, it moves easily throughout the greenhouse.
- Soil-dwelling predacious mites (*Hypoaspis*) attack thrips in their prepupal and pupal stages when they inhabit the soil or growing medium. *Hypoaspis* mites are usually applied only once per crop or season.
- *Thripobius semiluteus* is a parasitoid of greenhouse thrips nymphs.
- In laboratory trials and caged rose trials, *Beauveria bassiana* sprays killed up to 82 % of the thrips on rose foliage (Murphy et al. 1998). Mortality increased along with humidity. Oil formulations of *B. bassiana* worked more quickly than wettable powders.
- BotaniGard and Naturalis-O, both of which use *Beauveria bassiana*, have been effective on a schedule of three to five applications at 3–5-day intervals. The addition of Azatin (a neem product) may increase effectiveness (Sanderson 1999).
- Dr. Richard Lindquist at Ohio State University found that four applications of Naturalis-O

over a period of 15 days controlled thrips well on gerbera (Gilrein 1999a).

- In another study, Lindquist found that Conserve (a new biopesticide formulated from the soil-inhabiting actinomycete *Saccharopolyspora spinosa*) was significantly better than Orthene at controlling western flower thrips (Lindquist 1999). Plants treated with Conserve also had higher numbers of beneficial insects and mites – such as minute pirate bugs, predatory mites, and green lacewing larvae.

18.8 Biorational Pesticides

While the practices of inspection, sanitation, physical exclusion, and biological control will go a long way toward managing thrips, there may still be instances when pesticides are necessary. Biorational pesticides – also known as least-toxic or "soft" pesticides – are emphasized in biologically intensive IPM programs and by growers depending on organic pest management (OPM) as part of certified organic production (Table 18.6). Thrips control can still be difficult even with the use of biorational pesticides. During much of their life cycle, thrips exist as eggs, as pupae in the soil, or as extremely mobile adults. Once thrips infest a crop, the adult females begin feeding and laying their eggs. Thrips usually concentrate on rapidly growing tissues such as young leaves, flowers, and terminal buds. This affinity for tight places makes thorough coverage with a pesticide difficult. Just prior to pupation, the larvae move down the plant to pupate in the soil or leaf litter. They are most vulnerable just after hatching and before pupation.

Table 18.6 Biorational pesticides

Brand name	Supplier	REI	Application/comments
Azadirachtin – extract of neem seed; IGR that works through contact or ingestion			
Azatin	Green Spot	4 h	Apply when pests first appear
Neemazad	Thermo Trilogy	12 h	Cannot be applied through irrigation. Low rate can be used as a preventative
Garlic extracts			
Garlic Gard	Soil Technologies		
Garlic Barrier	Green Spot	4 h	Use late in the day. Can be mixed with fish oil or horticultural oil. Do not use in combination with bumblebees or honeybees
Horticultural oil – includes dormant and summer superior oils			
All Seasons	Green Spot	4 h	Use on sunny days to promote rapid drying and decrease chance of phytotoxicity. Not compatible
Hot pepper wax – contains capsaicin, paraffin, and mineral oil			
Hot Pepper Wax	Green Spot	4 h	Also contains herbal essential oils. Not compatible with beneficials
Hot Pepper Wax	Hot Pepper Wax, Inc.	0 h	
Insecticidal soap – contains potassium salts of fatty acids			
M-Pede	Mycogen	12 h	Phytotoxicity is often a concern, esp. after repeated applications
Safer	Green Spot	4 h	See above
Insecticidal soap	Olympic		
Neem oil – multipurpose organic insecticide/fungicide/miticide; kills eggs, larval, and adult stages of insects			
Trilogy 90EC	Thermo Trilogy	4 h	Apply at first signs of damage. Repeat every 7–10 days as needed
Triact 90EC	Thermo Trilogy	4 h	For ornamental crops only

18.8.1 Notes on Biorational Control of Thrips

- Neem extracts (trade names Azatin, Neemazad, and Neemix) prevent development of flower thrips in the early larval stages but have no effect on adults. Repeat applications are most effective.
- Biorational control of flower thrips was achieved with either (i) M-Pede insecticidal soap mixed with an emulsified crop oil or (ii) weekly applications of neem seed oil (NSO) for 4 straight weeks.

18.9 Insect Growth Regulators

Insect growth regulators (IGRs) are another least-toxic pesticide control option for pests (Table 18.7). IGRs typically kill insects by disrupting their development. They have a complex mode of action that precludes insects from rapidly developing resistance.

IGRs can work in one of several ways:

- They can mimic juvenile hormones, so that insects never enter the reproductive stage of development.
- They can interfere with the production of chitin, which makes up the shell of most insects.
- They can interfere with the molting process.

IGRs usually work through ingestion, so good spray coverage is essential. They generally do not affect nontarget species, such as humans, birds, fish, or other vertebrates. For most IGRs there are minimal reentry restrictions. IGRs typically take several days to have an effect on pest populations. Because IGRs do not affect mature insects, adult beneficials released into the greenhouse after an IGR application are not likely to be affected. Use of IGRs is generally prohibited by organic certification organizations because the products are synthesized.

IGRs can sometimes be used in conjunction with biological control efforts and may provide growers with a "safety net" should beneficials fail to keep the pests below economically damaging levels. The table below lists some well-known insect growth regulators.

Table 18.7 Selected insect growth regulators

Brand name	Supplier
Azatin	Hydro-Gardens, Olympic Horticultural Products
Neemazad	Thermo Trilogy
Neemix	Thermo Trilogy
Preclude	Whitmire Micro-Gen

18.10 Integrated Management

The fungus *Paecilomyces fumosoroseus* (PFR), the predatory mite *Neoseiulus cucumeris*, and an azadirachtin insecticide (neem) could be used alone or in combination for control of WFT in greenhouses (Grossman 1997).

Dan Gilrein (1999b) of Cornell Extension tested several biopesticides for their effect on western flower thrips, including Avid (contains abamectin), Conserve (spinosad), Sanmite (the synthetic chemical pyridaben), BotaniGard (*Beauveria bassiana*), a combination of BotaniGard + Azatin (neem) + M-Pede (insecticidal soap), and Alsa (a garlic product) (Gilrein 1999b). Although Conserve performed best, other treatments also provided control.

Researchers at Ohio State University are currently testing combinations of *B. bassiana* with the biorationals Azatin, Adept (an insect growth regulator), Conserve, Fulex SO-2000, Hot Pepper Wax, Garlic Barrier, yeast extract, and white sugar (Brownbridge et al. 2000). So far, they have shown that spray mixes combining *B. bassiana* with Adept, sugar, yeast extract, or Azatin produce higher levels of infection and mortality in western flower thrips.

Specifically, white sugar at a rate of 1 lb/100 gal and yeast extract at a rate of 0.25 % wt/volume increased mortality by about 20 %, presumably due to their ability to induce feeding. The opposite was true for plants treated with Hot Pepper Wax, an insect repellent and antifeedant. BotaniGard + Azatin showed excellent results: inclusion of Azatin at 1/10 or 1/4 the

Table 18.8 Control of western flower thrips on *Chrysanthemum* "Linda" after four weekly applications of biorational pesticides (Gilrein 1999b)

Treatment	Thrips/plant	Flower damage[a]
Avid	19.3	0.6
Conserve	3.6	0.6
Sanmite	17.3	0.8
BotaniGard ES	51.3	3.0
BotaniGard + Azatin + M-Pede	25.2	2.6
Alsa drench 4x rate	43.0	2.7
Alsa drench 1x rate	43.4	3.0
Water (control)	47.9	3.7

[a]Flower damage was based on a 0–5 scale, 0 = no damage and 5 = all flowers entirely brown

recommended rate led to a 50 % and 100 % increase, respectively, in infection and mortality (Brownbridge et al. 2000). Biorational pesticides for the management of greenhouse pests are summarized in Table 18.8. Please note that thrips are listed as target pests for each product in the Table 18.8.

18.11 Conclusions

Greenhouse thrips are tiny insects, but they demand serious attention on the part of the greenhouse grower. Integrated pest management offers a sustainable approach for dealing with greenhouse thrips, and safer pest control products facilitate the adoption of least-toxic control measures that dovetail very nicely with the IPM philosophy.

References

Allen WR (1991) Get the jump on thrips and TSWV. Greenhouse Grower, July, pp 78, 80, 82

Anon (1992) Informed sources. Greenhouse Manager, Aug, p 12

Anon (1993) Indicator plants give TSWV warning signs. Greenhouse Manager, Mar, p 70

Aylsworth J (1994) Taking care of number one. Greenhouse Grower, Feb, pp 52, 54, 56

Brownbridge M, Skinner M, Parker BL (2000) Enhancing the activity of insect-killing fungi for floral IPM. Ohio Florists' Association Bulletin, Jan, pp 14–16

Gill SA, Ross DS (1994) Insect microscreening: its time has come. Greenhouse Growerp, May, pp 77–82

Gill S et al (2000) Biological thrips control on bedding. GrowerTalks, Feb, pp 84–86

Gilrein D (1999a) Control thrips from the start. Greenhouse Management and Production, Jan, pp 72–74

Gilrein D (1999b) Tips on controlling western flower thrips populations. GMPro, Sept, pp 81–82

Grossman J (1996) Conference notes: screening out greenhouse pests. The IPM Practitioner, Sept, p 12

Grossman J (1997) Conference notes. The IPM Practitioner, Sept, p 13

Lindquist RK (1999) Evaluation of conserve for Western flower thrips control. Ohio Florists' Association Bulletin, Mar, p 8

Murphy BC, Morisawa TA, Newman JP, Tjosvold SA, Parella MP (1998) Fungal pathogen controls thrips in greenhouse flowers. California Agriculture, May–June, pp 32–36

Parella MP (1995) Thrips management guide – part 1: prevention and control. GrowerTalks, Apr, pp 30, 32, 34, 36, 38

Pundt LP, Sanderson J, Daughtrey M (1992) Petunias are your tip-off for TSWV. GrowerTalks, Nov, pp 69, 71–72

Sanderson J (1999) Thrips tips. GrowerTalks, July, p 128

Shipp JL, Gillespie TJ (1993) Influence of temperature and water vapor pressure deficit on survival of *Frankliniella occidentalis* (Thysanoptera: Thripidae). Environmental Entomology, Aug, pp 726–732

19 Whiteflies and Their Management

Abstract

This chapter focuses on integrated pest management for greenhouse whitefly on both vegetable and ornamental crops. The techniques for good greenhouse supervision, monitoring of whiteflies, biological controls, biopesticides, insect growth regulators, controlled atmosphere, and integrated management of whiteflies are discussed.

Keywords

Whiteflies • Scouting • Monitoring • Biocontrol • Biorational pesticides • Controlled atmosphere • Integrated management

19.1 Introduction

Whiteflies began showing resistance to synthetic insecticides early on, and by the 1980s they were very serious greenhouse pests. Not only do they feed on plants, but they also produce honeydew, which detracts from the plants' appearance and attracts other insects and sooty mold. Whiteflies can also transmit plant viruses. The mere presence of whiteflies in a greenhouse will stop customers from buying your product. Several species of whiteflies attack greenhouse plants, and they typically have a wide host range and resist insecticides. Greenhouse tomatoes and poinsettias are especially plagued by whiteflies. The most common whiteflies on greenhouse crops are the greenhouse whitefly (*Trialeurodes vaporariorum*), sweet potato whitefly (*Bemisia tabaci*), and the silverleaf whitefly (*Bemisia argentifolii*) (Fig. 19.1).

The various whitefly species and biotypes look very much alike, but they have subtle physiological differences. These differences can cause them to respond differently to control strategies. Because control measures must be selected according to the type of whitefly present, accurate identification is critical for successful control.

19.2 Crop Scouting and Trapping

Plants should be visually inspected for signs of a whitefly infestation, e.g., off-color or stunted plants. A hand lens is useful for systematically inspecting a number of individual plants for the presence of eggs, nymphs, or adults. Both the upper and lower leaf surfaces should be inspected (Table 19.1). It is important to check the greenhouse in the same pattern on each scouting trip. Locations where whiteflies are found should be

P.P. Reddy, *Sustainable Crop Protection under Protected Cultivation*,
DOI 10.1007/978-981-287-952-3_19

Fig. 19.1 Adult silverleaf whiteflies, *Bemisia argentifolii*. Note wings held closer to sides of the body, pupae without spines (*left*), and Greenhouse whitefly, *Trialeurodes vaporariorum*. Note wings held wider and more flat on body and pupae with long spines (*right*), feeding on the underside of leaves

Table 19.1 Monitoring and scouting techniques for aphids

	Whiteflies
Scouting:	
Inspect underside of leaf	x
Observe for:	
Chlorotic (yellow) spots on upper leaf surface	x

flagged so that population development and control efforts can be evaluated. Scouting frequency should be increased during warm weather as whitefly populations multiply faster then.

Trapping with yellow sticky cards, both inside and outside the greenhouse, is essential for a successful whitefly management program. The cards are used to detect and monitor population levels. As a general rule, one to four cards spaced evenly throughout 1000 square feet of greenhouse are sufficient (McHugh 1991). A generally acceptable threshold for whiteflies is 0.5 per card per day when the crop is young and two per card per day as the crop reaches maturity (Gill 2000). Traps should be hung level with the tops of the plants since whiteflies are most attracted to young foliage. Doors, vents, and other openings where whiteflies can enter the greenhouse are other good sites to hang yellow sticky cards.

Researchers in California have successfully used silver-painted pot spaces and silver polyethylene mulch to control whiteflies on greenhouse poinsettias (Price 1999). The reflective materials were used in conjunction with yellow sticky cards or tape and resulted in significantly enhanced trapping of whiteflies, relative to controls with sticky traps only.

Greenhouse plastics themselves may have significant influence on the initial attraction of insects into greenhouses. A study from the late 1990s showed that silverleaf whiteflies preferred to enter greenhouses covered with film that transmitted higher levels of ultraviolet light (Costa and Robb 1999).

19.3 Biology

At 70 °F, the greenhouse whitefly life cycle takes 6–10 days for eggs to hatch, 3–4 days as nymph I, 4–5 days as nymph II, 4–5 days as nymph III, and 6–10 days for the pupa. Adults can live for 30–40 days (Fig. 19.2).

19.4 Biological Control

Several types of beneficial organisms are available for biological control of whiteflies. The parasitic wasp *Encarsia formosa* preys on immature whiteflies and is commonly used for greenhouse whitefly. *Encarsia* wasps kill whitefly nymphs in one of two ways: they either lay an egg inside the nymph, providing food for their young, or they kill the nymph right away and feed on the fluids inside of it (Price 1999). Greenhouse whitefly pupae that have been parasitized by *E. formosa* turn black; silverleaf whitefly pupae turn amber brown (Price 1999). The cost of *E. formosa* is comparable to foliar pesticides (Price 1999).

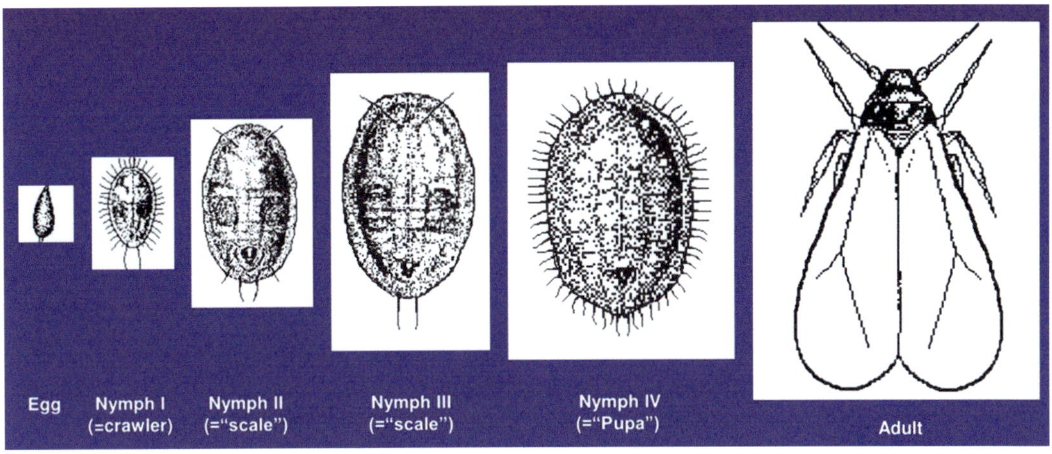

Fig. 19.2 Life cycle of greenhouse whitefly

Sweet potato and silverleaf whiteflies are not well controlled by *E. formosa*. Two other wasp parasites, *E. luteola* and *Eretmocerus californicus*, are commercially available for control of these species, but *E. formosa* remains the mainstay of most whitefly biological control programs due to the expense and intermittent availability of these other beneficials (Anon 1995a). Researchers are also examining different strains of *E. formosa* to determine their effectiveness against sweet potato and silverleaf whiteflies.

Biological control should be used only at low whitefly levels (less than one nymph per ten cuttings) and must be regularly evaluated (Daughtrey and Casey 1998).

Mark has also compared the effectiveness of *Encarsia formosa* and *Eretmocerus californicus* on silverleaf whitefly. In a 1996 study, both parasitoids controlled whiteflies at a 99 % control rate. However, fewer numbers of *E. californicus* had to be released, which would result in cost savings for growers. Another benefit, fewer parasitoids meant more food for each, so the reproduction rate was much higher (Grossman 1996).

Delphastus pusillus, sometimes called the whitefly destroyer, is a very small, black ladybird beetle that attacks all stages of whiteflies, but prefers eggs and nymphs. The females lay their eggs within clusters of whitefly eggs. Adults can consume 160 eggs or 12 large nymphs every day. A larva consumes 1000 whitefly eggs during its development. These beetles perform best at temperatures between 65 and 90 °F, with relative humidity above 70 %. These predators can be used in combination with *Encarsia* species (Cloyd 1999). A complete listing of biological controls for whiteflies has been presented in Table 19.2.

19.5 Biopesticides

Some microorganisms also control whiteflies. For instance, the fungus *Beauveria bassiana* (trade names Naturalis-O and BotaniGard) is effective against eggs and immature and adult whiteflies. Thorough coverage of leaf undersides and correct timing of applications result in best control.

Another fungus, *Paecilomyces fumosoroseus* (trade name PFR-97), is now commercially available. It controls whiteflies, aphids, and spider mites. Both *B. bassiana* and *P. fumosoroseus* need high humidity for best results.

A complete listing of biopesticides available for whitefly management is presented in Table 19.3.

Table 19.2 Beneficial organisms used in greenhouses

Organism	Supplier	Application/comments
Chrysopa carnea	Natural Pest Controls, Beneficial Insectary, Caltec, Arbico, A-1 Unique Insect Control, Praxis, Rincon-Vitova, Hydro-Gardens	1 lacewing/5–30 aphids; 1000 eggs/200 ft^2. Apply every 1–3 weeks as needed. May arrive as eggs, immatures, or adults
Chrysoperla rufilabris	Arbico, Beneficial Insectary, IPM Labs., A-1 Unique Insect Control, Nature's Control, Praxis, Rincon-Vitova	
Chrysoperla spp. (predator)	M&R Durango, Florikan, Green Spot	
Coleomegilla maculata	Arbico	1/ft^2; shipped as larvae and eggs
Deraeocoris brevis (predator)	Green Spot	
Harmonia axyridis (Asian lady beetle)	Green Spot	Temps. should be 70–85 °F; humidity around 70 %
Hippodamia convergens (lady beetle) (predator)	A-1 Unique Insect Control, Arbico, Caltec, IPM Laboratories, Natural Pest Controls, Nature's Control, Harmony Farm Supply, Hydro-Gardens, Praxis, Green Spot	Release at dusk near an immediate food source. Spray plants with water prior to release
Orius insidiosus (minute pirate bug) (predator)	Florikan, IPM Labs., Harmony Farm Supply, Arbico, Hydro-Gardens, Praxis, Koppert, Intl. Tech. Services, Green Spot	1/10 sq. ft. (preventive), 1 every 2 ft^2 when pests are present. Temperature should be 70–90 °F. *Orius* are dormant September–April. Reapply every 2–3 weeks. Very susceptible to pesticides. Works well in combination with *Neoseiulus cucumeris*

Table 19.3 Biopesticides effective against whiteflies

Brand name	Supplier	REI	Application/comments
Beauveria bassiana[a] Naturalis-O	SePro	4 h	Apply when insects first appear and repeat every 7–10 days. Good spray coverage is needed. Not compatible with other fungicides
Beauveria bassiana[a] BotaniGard	Mycotech	12 h	See above

[a]*Beauveria bassiana* – fungus that works through contact; exposure to nontarget insects should be avoided

19.6 Biorational Pesticides

Several least-toxic, or biorational, pesticides have been evaluated for their effectiveness against the different whitefly species. These include neem-based formulations (Neemazad and Azatin are two registered products), insecticidal soap (M-Pede), and horticultural oil. Enhanced whitefly control is achieved with thorough spray coverage. Wider plant spacing and removal of dead lower leaves improve pesticide coverage and pest control.

A 1995 study conducted in Florida compared the effectiveness of Sunspray Ultra-Fine Spray Oil, M-Pede insecticidal soap, a sucrose ester extract surfactant derived from tobacco, and Garlic Barrier for killing and repelling silverleaf whiteflies (*Bemisia argentifolii*) on tomatoes. Sunspray provided the best control, followed by M-Pede and the tobacco surfactant. The Garlic Barrier did not provide any control (Williams and Pat 1995).

A complete listing of biorational pesticides available for whitefly management is presented in Table 19.4.

Table 19.4 Biorational pesticides

Brand name	Supplier	REI	Application/comments
Azadirachtin – extract of neem seed; IGR that works through contact or ingestion			
Azatin	Green Spot	4 h	Apply when pests first appear
Neemazad	Thermo Trilogy	12 h	Cannot be applied through irrigation. Low rate can be used as a preventative
Horticultural oil – includes dormant and summer superior oils			
All Seasons	Green Spot	4 h	Use on sunny days to promote rapid drying and decrease chance of phytotoxicity. Not compatible
Hot Pepper Wax – contains capsaicin, paraffin, and mineral oil			
Hot Pepper Wax	Green Spot	4 h	Also contains herbal essential oils. Not compatible with beneficials
Hot Pepper Wax	Hot Pepper Wax, Inc.	0 h	
Insecticidal soap – contains potassium salts of fatty acids			
M-Pede	Mycogen	12 h	Phytotoxicity is often a concern, esp. after repeated applications
Safer	Green Spot	4 h	See above
Insecticidal Soap	Olympic		
Neem oil – multipurpose organic insecticide/fungicide/miticide; kills eggs and larval and adult stages of insects			
Trilogy 90EC	Thermo Trilogy	4 h	Apply at first signs of damage. Repeat every 7–10 days as needed
Triact 90EC	Thermo Trilogy	4 h	For ornamental crops only
Soybean oil			
Golden Natural Spray Oil	Stoller	12 h	

19.7 Insect Growth Regulators

Insect growth regulators (IGRs) are another least-toxic pesticide control option for pests. IGRs typically kill insects by disrupting their development. They have a complex mode of action that precludes insects from rapidly developing resistance. IGRs can work in one of several ways:

- They can mimic juvenile hormones, so that insects never enter the reproductive stage of development.
- They can interfere with the production of chitin, which makes up the shell of most insects.
- They can interfere with the molting process.

IGRs usually work through ingestion, so good spray coverage is essential. They generally do not affect nontarget species, such as humans, birds, fish, or other vertebrates. For most IGRs there are minimal reentry restrictions. IGRs typically take several days to have an effect on pest populations. Because IGRs do not affect mature insects, adult beneficials released into the greenhouse after an IGR application are not likely to be affected. The use of IGRs is generally prohibited by organic certification organizations because the products are synthesized. The table below lists some well-known insect growth regulators (Table 19.5).

IGRs can sometimes be used in conjunction with biological control efforts and may provide growers with a "safety net" should beneficial fail to keep the pests below economically damaging levels.

Table 19.5 Selected insect growth regulators effective against whiteflies

Brand name	Supplier
Adept, Dimlin	Uniroyal Chemicals
Azatin	Hydro-Gardens, Olympic Horticultural Products
Citation, Precision	Novartis
Distance	Valent
Enstar II	Wellmark Intl.
Neemazad, Neemix	Thermo Trilogy
Preclude, Pyrigro	Whitmire Micro-Gen

19.8 Controlled Atmosphere

Changing the composition of the atmosphere in the greenhouse by either reducing oxygen or increasing carbon dioxide appears to provide some control of greenhouse whiteflies, especially adults. Reduced-oxygen experiments by Dr. Susan Han at the University of Massachusetts resulted in 100 % adult mortality after less than 2 h of exposure, though 8-h treatments were needed to control most (about 80 %) of the eggs and pupae (Anon 1995b).

Horticulturists at North Carolina State University reported that whitefly population levels were lower in greenhouses where carbon dioxide enrichment occurred daily for about 8 h. The likely reason for the population reductions is that plants grown in atmospheres with high carbon dioxide levels tend to have higher concentrations of carbohydrates in the plant tissue relative to nitrogen, resulting in a nitrogen-dilute diet for the whiteflies. Lower dietary nitrogen would slow the growth and reproduction of the pests, without adversely affecting crop yields or quality (Tripp and Peet 1993).

19.9 Integrated Management

Researchers at Ohio State University showed that mortality rates of silverleaf whitefly were higher when BotaniGard was used together with Adeptô (an insect growth regulator), insecticidal soaps, horticultural oil, Fulex SO-2000, glycerol, and yeast extract (Brownbridge et al. 2000). In most cases, the rate of infection and kill was faster than with BotaniGard alone. This is important because it would allow more biopesticides and biorationals to be used retroactively. When several substances are used together, they are applied at lower-than-recommended rates (one-tenth to one-half). The researchers are trying to develop management guidelines on spray-tank mixes that would enhance fungal efficacy (Brownbridge et al. 2000).

Eretmocerus eremicus was effective for silverleaf whitefly control and is best used in combination with the insect growth regulators Precisionô or Applaudô. Enstarô was not compatible with natural enemies tested.

References

Anon (1995a) Biological pest control. Greenhouse Product News, July, p 17

Anon (1995b) Controlled atmosphere to manage whitefly. Cut Flower Q, July, pp 14–16

Brownbridge M, Skinner M, Parker BL (2000) Enhancing the activity of insect-killing fungi for floral IPM. Ohio Florists' Association Bulletin, Jan, pp 14–16

Cloyd RA (1999) Know your friends: *Delphastus pusillus*: whitefly predator. Midwest Biological Control News, Oct, p 3

Costa HS, Robb KL (1999) Effects of ultraviolet-absorbing greenhouse plastic films on flight behavior of *Bemisia argentifolii* (Homoptera: Aleyrodidae) and *Frankliniella occidentalis* (Thysanoptera: Thripidae). J Econ Entomol 92:557–562

Daughtrey M, Casey C (1998) Highlights from SAF's pest conference. Grower Talks, Apr, pp 44, 46

Gill S (2000) Pest control: whitefly control for cut flower growers. Cut Flower Q 12(1):26–30

Grossman J (1996) Conference notes. The IPM Practitioner, Mar, p 14

McHugh J (1991) Monitoring the first line of defense. Greenhouse Grower, Feb, p 66

Price RP Jr (1999) Reflective mulches and yellow sticky tape control whiteflies in greenhouse poinsettia (*Euphorbia pulcherrima*). As reported in Williams, Greg and Pat. From the 1999 ASHS conference HortIdeas, Aug, p 85

Tripp K, Peet M (1993) New use for CO_2: slowing whiteflies. American Vegetable Grower, Nov, pp 43–44

Williams G, Pat M (1995) Oil, soap, surfactant, and garlic vs. whiteflies on tomatoes. HortIdeas, May, pp 55–56

20 Pest and Predatory Mites

Abstract

Mites feature prominently in greenhouses, both as pests causing economic injury to crops and as predatory mites used in biological control of mite and insect pests.

Spider mites, false spider mites, tarsonemid mites, and eriophyid mites injure both vegetable crops and ornamental plants. Acarid mites damage the bulbs of flowers and stored roots of many crops. Their damage symptoms, biology, and management are discussed.

The predatory phytoseiid and laelapid mites are used in the biological control of mite pests and thrips on many crops in greenhouses. Their biology and use in biocontrol of insect and mite pests of greenhouse crops are discussed.

Keywords

Pest mites • Predatory mites • Damage • Biology • Management • Biocontrol

20.1 Introduction

Greenhouses are maintained under relatively stable environmental conditions for optimal growth of plants. Unfortunately, these conditions also favor the rapid growth of pests, especially small pests such as mites, which develop rapidly and can increase their population quickly over a short period before notice by growers. Cultural methods commonly adopted in greenhouses also encourage pest development. For example, crops are often cultivated in monoculture in greenhouses, which facilitate the dispersal of pests among plants and rapid development of pest populations. The lack of natural enemies in the enclosed artificial environment and the rapid development of pesticide resistance in greenhouses are also important factors for the pest status of plant mites.

Mites, although much smaller in size than many insects, feature prominently in greenhouses, both as pests causing economic injury to crops and as predatory mites used in biological control of mite and insect pests.

P.P. Reddy, *Sustainable Crop Protection under Protected Cultivation*,
DOI 10.1007/978-981-287-952-3_20

20.2 Pest Mites

Spider mites (Tetranychidae) are among the most important pests on many greenhouse plants. False spider mites (Tenuipalpidae), which are related to spider mites, sometimes also become pests on greenhouse plants. Several species of tarsonemid mites and eriophyid mites injure both vegetable crops and ornamental plants. Acarid mites of the genus *Rhizoglyphus* damage the bulbs of flowers and stored roots of many crops (Table 20.1).

20.2.1 Spider Mites

Spider mites belong to the family Tetranychidae of the order Prostigmata. They are so named because many members of this family produce silk webbing on host plants. Spider mites are without doubt the most important mites attacking plants. Most spider mite species are polyphagous. They occur on virtually every major food crop and ornamental plant. They include some of the most important pests on economic plants.

Most spider mites attack outdoor plants and a few of them regularly occur in greenhouses. Only *Tetranychus urticae* and *T. cinnabarinus* are widespread and frequently reach economic injury levels on many plants in greenhouses. A few species of the genus *Bryobia* are occasionally found in greenhouses and sometimes cause injury to plants. A couple of species of *Eutetranychus* also occasionally attack greenhouse plants, as does one species of *Panonychus* and another of *Oligonychus*.

20.2.1.1 Two-Spotted Spider Mite, *Tetranychus urticae*

The two-spotted spider mite is the most common name and is also known informally by many other names (e.g., the glasshouse spider mite, the yellow spider mite). Not very appropriately, it is often called "red spider mite" or "red spider." It is the most polyphagous species of spider mites and has been reported from over 150 host plant species of some economic value. It attacks over 300 plant species in greenhouses.

20.2.1.1.1 Damage Symptoms

Two-spotted spider mites often feed on cell chloroplasts on the undersurface of the leaf. The upper surface of the leaf develops characteristic whitish or yellowish stippling, which may join and become brownish as mite feeding continues. As mites move around, their webbing can span leaves and stems. Heavy damage may cause leaves to dry and drop, and the plant may be covered with webbing and may die prematurely.

20.2.1.1.2 Life Cycle

Development occurs between 12 °C and 40 °C. Developmental time from egg to adult decreases with increasing temperature (Fig. 20.1)

Table 20.1 Common pest mites of importance under greenhouse conditions

Pest mites	Mite species	Affected crops
Spider mites	*Tetranychus urticae* Two-spotted spider mite	Many vegetables (e.g., tomato, cucurbits, capsicum, French bean, strawberry, etc.) and ornamentals (rose, carnation, cyclamen, gerbera, etc.)
	Tetranychus cinnabarinus Carmine spider mite	Mainly tomato and carnation
Tarsonemid mites	*Polyphagotarsonemus latus* Broad mite	Many vegetables and ornamentals
	Phytonemus pallidus Cyclamen mite	Mainly cyclamen and strawberry
Eriophyid mites	*Aculops lycopersici* Tomato russet mite	Tomato
Acarid mites	*Rhizoglyphus echinopus* Bulb mite	Mainly lily
	Rhizoglyphus robini Robine bulb mite	Attack bulbs of *Narcissus*, *Eucharis*, lilies, orchids, gladioli, hyacinth, and tulip

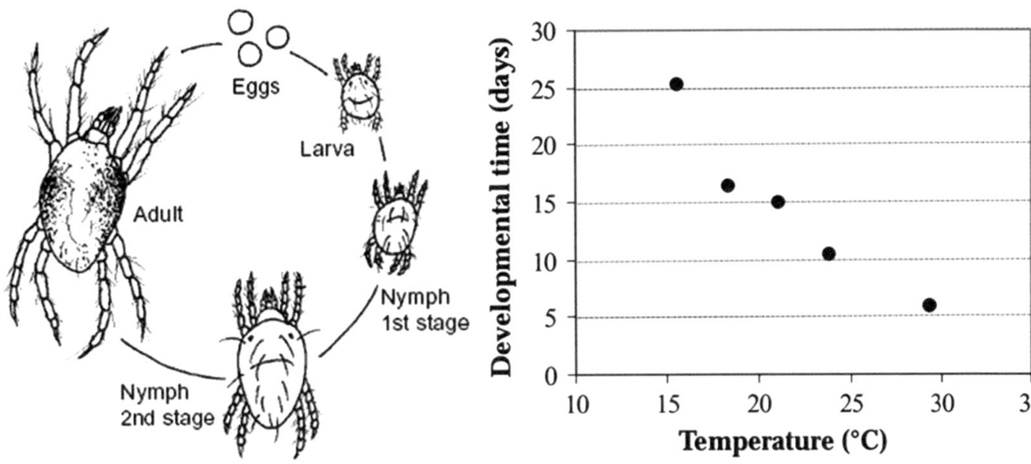

Fig. 20.1 Life cycle of (left), and developmental time for Tetranychus urticae from egg to adult in relation to temperature (Drawn from data in Carey and Bradley 1982)

and is less than a week at optimal temperatures for development (30–32 °C). Under a diurnal temperature cycle of 15–28 °C, developmental time is about 16 days. Males develop slightly faster than females.

20.2.1.1.3 Management

(a) *Biological control*: biological control of *T. urticae* by phytoseiid mites is now widely used by the greenhouse industry in Europe, some parts of Asia and Africa, Australasia, and North America. The most commonly used species are *P. persimilis*, *P. macropilis*, *Neoseiulus californicus*, *N. fallacis*, *N. longispinosus*, and *Galendromus occidentalis*. They are often released repeatedly for biological control but also inoculatively (such as in "pest in first" methods, in which predators and spider mites were introduced together early in the season). The rates at which predators are released vary with the density of the spider mites, crop species, the temperature, and other environmental conditions.

Predatory midges are also useful biocontrol agents against spider mites. A common species, *Feltiella acarisuga*, is commercially available for use in many countries. Release of cocoons is carried out in the early morning or late evening, when it is cool and humid. Adults of *F. acarisuga* are mosquito-like but smaller in size. They mate within 24 h of emergence and locate spider mite colonies, where they lay about 30 eggs over a 5-day life span. Development occurs over a range of 15–25 °C and is slowed down at 27 °C. The eggs and larvae are killed at 30 °C and above. The larvae hatch from eggs in a couple of days and feed on all stages of spider mites. They feed for 4–6 days and consume a total of about 150 spider mite eggs. They develop best when food is abundant but can also pupate at a reduced size in times of food shortage. This predator can control spider mites on tomato and cucumber, and the effects are enhanced by high humidity and a source of sugar droplets or honeydew. Repeated releases may be needed, and rates will depend on prey density and crop and environmental conditions. This predator can also be used year round due to the absence of diapauses.

Lady beetles of the genus *Stethorus* are specialist predators of spider mites and are also useful for the control of *T. urticae* in greenhouses. A common species, *S. punctillum*, is a voracious predator and is commercially available. This species can consume over 1000 spider mite eggs over a developmental span of 2–3 weeks in greenhouses. It is able to find small colonies of spider mites and has very good dispersal ability. It is active and performs well within 33–90 % RH and 20–30 °C. In greenhouses, it can establish well on pepper and cucumber, but not on tomato.

Because of their voracious appetite, these lady beetles are best released in "hot spots" of spider mite infestation.

Some other generalist predators also feed on spider mites and may be of some use. Predatory Hemiptera used in biocontrol of whiteflies (e.g., *Macrolophus caliginosus*) and other insects are also useful biocontrol agents against spider mites to some degree. Lacewings are generalist predators and also used to control spider mites by periodic releases. A predatory ant, *Tapinoma melanocephalum*, is known to attack *T. urticae* on *Salvia splendens* in central Florida greenhouses and has been shown to be a significant predator of *T. urticae*.

Thrips are generally considered to be harmful, but *Scolothrips sexmaculatus* attack spider mites and are used in biological control of spider mites. This species is adapted to hot and dry conditions. It is commercially available for spider mite control. Other thrips are also known as facultative predators of spider mites.

Entomophagous fungi may also be the biological control of two-spotted spider mites. Two species, *Entomophthora thaxteriana* and *E. adjarica*, can cause a heavy epizootic in populations of *T. urticae* and may be useful in greenhouses when humidity can be maintained near saturation for a period of time.

(b) *Chemical control*: chemical control of spider mites is becoming more and more difficult due to the rapid development of resistance in mites and decrease in the number of registered acaricides for use. For example, clofentezine resistance in *T. urticae* was recorded in Australia in 1987 after mites in Queensland greenhouse roses had been exposed to 40 applications of clofentezine over a 10-month period; clofentezine resistance in this strain was extremely high (>2500×) and conferred high level of cross-resistance to the chemically unrelated compound hexythiazox. However, some chemicals appear to be less susceptible to resistance development in mites. *T. urticae* was subjected to 4, 6, or 15 times of artificial selection of resistance to abamectin, and no increased resistance to the pesticide was detected; abamectin is one of the most widely used chemicals in greenhouses. However, development of resistance of *T. urticae* to this pesticide has now been reported in some populations.

(c) *Integrated management*: more and more greenhouse pests are now under integrated pest management (IPM) programs, and *T. urticae* is one of these. The key to the success of IPM programs is careful monitoring of pest populations and the application of control measures only when necessary. That is, control measures are applied only when spider mite densities exceed the action threshold. In IPM programs, the use of pesticides is minimized, and the use of biological control and other environmentally sound methods is encouraged. It is important that chemical control, when it is used, should be integrated with other control methods such as biological control.

Chemical control of *T. urticae* can be integrated with biological control by using selective chemicals that are less or not toxic to natural enemies or by using chemicals in selected areas of the crop. In greenhouses in Belgium, for example, a stable equilibrium between *T. urticae* and the predator *P. persimilis* is achieved by selective use of chemicals and creating an asynchronous development in the predator population. The predators are introduced into one end of the greenhouse, while acaricides (Torque 50 % [fenbutatin oxide] and hexythiazox [Nissorun 10 %]) are sprayed into the other end. Only 3300 predators per 100 m^2 combined with three acaricide treatments applied to half the plants are required to control the pest for 30 weeks. Once the system is established, it is self-regulating and so the use of acaricides is needed only initially. Likewise, it may also be possible to apply chemical control to part of the plants while giving biological control a chance in other parts. For example, integrated control might be possible for spider mites on roses if different injury levels are assigned to the upper and lower canopies and sprays are confined to upper canopies. The upper portion of a rose canopy has an extremely low injury level because it bears the flowers and foliage which are cut for sale. Although low densities of spider mites may not affect the quantity of the product, they may damage its aesthetic appearance. However, the lower canopy can have a much higher injury level and aesthetic damage is of no concern on the lower canopy. Confining pesticide applications to only a portion of the canopy may create refugia for pesticide-susceptible individuals of *T. urticae* that may breed with resistant ones and thereby retard the development of pesticide resistance. This also allows *P. persimilis* to be used for biological control of spider mites on lower canopies.

Pesticide-resistant predatory mites can also be used in IPM, and some strains of *P. persimilis* and *G. occidentalis* have been developed for controlling *T. urticae* in greenhouses.

Some plant cultivars are naturally less susceptible to spider mites and plant resistance may be used as a component in spider mite IPM. This will reduce the need for control measures. Sometimes, resistant plants have lower yield and other undesirable features, but plant breeding is helping to overcome some of these.

Cultural and physical methods may also be used. For example, spider mites thrive in hot, dry conditions. Increasing the humidity level by misting plants can reduce the growth of spider mites. However, care should be taken because increasing humidity may increase the possibility of fungal diseases.

20.2.1.2 Carmine Spider Mite, *Tetranychus cinnabarinus*

This species is commonly known as the carmine spider mite, the red spider mite, the cotton spider mite, and the carnation mite. This species attacks both vegetables (especially tomatoes, cucumbers, eggplants) and ornamentals (e.g., carnation, gerbera) in greenhouses.

20.2.1.2.1 Symptoms

Symptoms caused by *T. cinnabarinus* vary slightly on different plants but in general are very similar to symptoms caused by *T. urticae*. The carmine spider mite is often more phytotoxic than the two-spotted spider mite, especially on tomato plants. The necrotic symptoms consist of premature chlorosis of infested leaflets which subsequently wither and die.

20.2.1.2.2 Life Cycle

Development is faster at higher temperatures and can be completed in less than a week at 35 °C (Fig. 20.2).

20.2.1.2.3 Management

Control methods for *T. cinnabarinus* are similar to those for *T. urticae*, although on some plants (e.g., tomato), *T. cinnabarinus* is sometimes more difficult to control because they induce damage at lower population densities than *T. urticae*.

Phytoseiulus persimilis is the most effective predator of *T. cinnabarinus* and *T. urticae*, although some preference for *T. cinnabarinus* has been reported. *N. californicus* and *N. longispinosus* have also been used effectively against *T. cinnabarinus*.

Predatory insects are also effective natural enemies, including predatory midges (e.g., *Feltiella*), lady beetles (e.g., *Stethorus*), predatory Hemiptera (e.g., *Orius*), and predatory thrips (e.g., *Scolothrips*). Spiders are also considered significant predators. Entomophagous fungi (e.g., *Hirsutella thompsonii*) can also be effective if humidity can be maintained very high for a period of time and temperature is lower than 37 °C.

Chemical and other control methods are also similar to those used against *T. urticae*.

20.2.2 Tarsonemid Mites

Tarsonemid mites belong to the family Tarsonemidae of the order Prostigmata. Some phytophagous tarsonemids are pests of agricultural crops and are important in greenhouses. Most plant-feeding species belong to a few genera in the Tarsonemidae, except *Polyphagotarsonemus*, which is a member of the Pseudotarsonemoidinae.

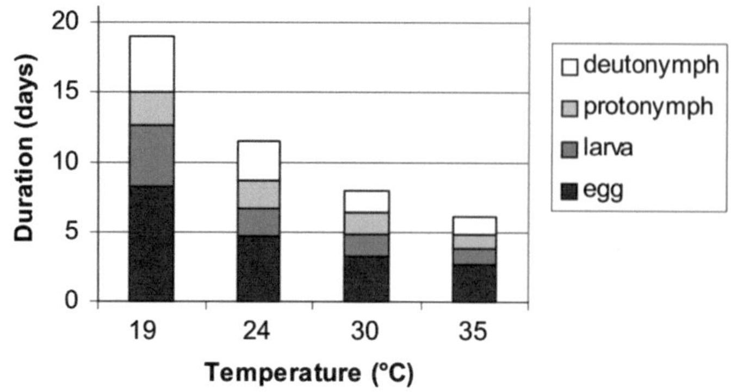

Fig. 20.2 Developmental time for *Tetranychus cinnabarinus* from egg to adult in relation to temperature at RH of 80 % (Drawn from data in Hazan et al. 1973)

20.2.2.1 Broad Mite, *Polyphagotarsonemus latus*

This species is widely known as the broad mite. It is also known as the yellow (tea) mite, white mite, citrus silver mite, or tropical mite. It is a major mite pest throughout the tropics and also in greenhouses in temperate regions. As its name suggests, *P. latus* is polyphagous; it attacks plants belonging to 60 families, among which are economic crops and ornamental plants such as pepper, tomato and cucumber, African violet, begonia, chrysanthemum, cyclamen, dahlia, *Gloxinia*, *Fuchsia*, *Gerbera*, *Hibiscus*, *Impatiens*, and ivy (*Hedera*).

20.2.2.1.1 Symptoms

Broad mites are often found on young leaves and feed mostly on the undersurface of the leaves. Damaged leaves of Gerbera may split or crack open and have a rugged appearance. Injured flowers have distorted and discolored rays. There is often sudden curling and wrinkling of leaves followed by discoloration or blistering. Plant growth may stop and survival of the plant may be threatened when severely injured. Damage of cucumber, eggplants, and *Solanum aviculare* includes crinkling, cracking, discoloration, malformations, swelling, and necrosis. The symptoms can persist for many weeks after the removal of mites.

20.2.2.1.2 Life Cycle

Developmental time decreases with temperature and is often less than a week in greenhouses (Fig. 20.3). On pepper, the developmental period from egg to adult averages 4.1 and 4.1 days at 25 °C for males and females, respectively. Adult female and male longevity is 11 and 15 days, respectively. Adult females lay 25 eggs.

20.2.2.1.3 Management

(a) *Biological control*: several phytoseiid species are known to attack the broad mite, and some of them have been demonstrated to be effective in greenhouses.

Neoseiulus barkeri, which is commercially available in many countries, can effectively reduce populations of *P. latus* from more than 100 mites per leaf to zero in a week on *Capsicum*, when released at the rate of ten or more predatory mites per plant. Three weekly releases of five predatory mites per main stem provide adequate protection of the plant from mite injury for over 7 weeks. Another commercially available species, *N. cucumeris*, is also effective in a similar way.

Neoseiulus californicus and *Euseius ovalis* feed and reproduce very well on *P. latus*. The latter, when released at the predator/prey ratio of 1:20, can cause a rapid decline in adult population of *P. latus*. They are promising candidates for use in greenhouses for *P. latus* control.

Neoseiulus agrestis can consume *P. latus* and reproduce on it in the laboratory, but it cannot reduce the abundance of *P. latus* on azalea plants in the greenhouse. Augmentative releases may work for this species. *Neoseiulus longispinosus* and *Typhlodromus peregrinus* are able to consume adults and immature stages of *P. latus*, and their potential as biocontrol agents should be investigated.

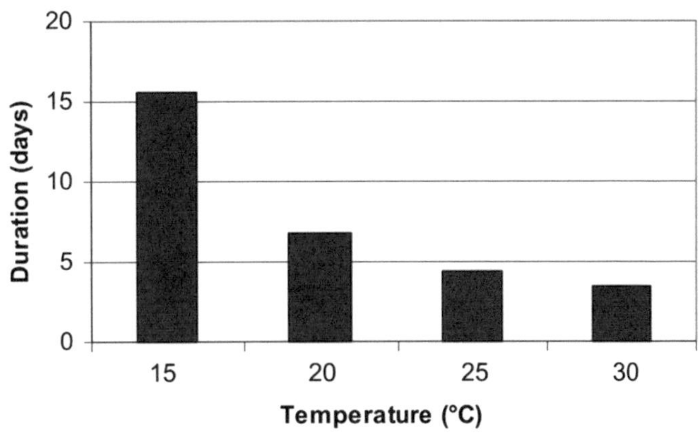

Fig. 20.3 Developmental time for *Polyphagotarsonemus latus* from egg to adult at different temperatures on *Phaseolus vulgaris* (Drawn from data in Lee et al. 1992)

Conidia of *Beauveria bassiana* can cause 88 % mortality of *P. latus* in controlled laboratory conditions and may be tried in greenhouses.

(b) *Chemical control*: because biological control works well for this species, the use of chemicals should be minimized. Some chemicals may sometimes induce mite problems instead of suppressing them. An example is the use of dichlorvos, cypermethrin, fluvalinate, monocrotophos, methyl-O-demeton [demeton-O-methyl], formothion, thiometon, and ethion on *Capsicum* in India. This mite may be difficult to control using nonsystemic pesticides on certain plants because mites may be able to hide in curly leaves, which make full coverage of sprays difficult.

In India, dicofol, bromopropylate, azocyclotin, and avermectin (abamectin) can eliminate the mite population in 2 weeks, and abamectin (at 18 g/l, applied at 25 ml/100 l water) remains effective for up to 3 weeks. Wettable sulfur is also known to be effective.

In Thailand, 0.07 % prothiofos, 0.075 % formetanate, 0.072 % triazophos, and 0.071 % methiocarb provide effective control of *P. latus*.

In China, liuyangmycin (an antibiotic preparation from *Streptomyces griseolus*) applied at 25 ppm gave good control of this mite on *Capsicum*, with an efficiency equivalent to control by dicofol applied at 200 ppm.

In Korea, dicofol EC, chinomethionat WP, pyridaben WP, and pyraclofos WP gave effective control of this mite on pepper. Elsewhere, bromopropylate 50 % at 0.5 %, hexythiazox 10 % at 0.4–0.5 %, dicofol 21 % + tetradifon 7.5 % at 1.5–2 %, and endosulfan 35 % at 2 % are recommended for preventive control of this mite on *Capsicum*.

In Brazil, two applications (spaced 7 days) of abamectin (5.4 g a.i./ha) and triazophos (400 g a.i./ha) gave significant control for up to 28 days.

(c) *Integrated management*: selective acaricides such as abamectin can be used together with biological control. Organic material such as manipueira, a liquid extract from cassava roots, can provide effective control of *P. latus* when diluted in water (1:3) and sprayed three times at weekly intervals. Plant resistance is not yet employed as a component in broad mite control, but there is good potential. In Cuba, for example, a double haploid of sweet pepper that has higher mean fruit weight and yield is known to be tolerant of *P. latus*.

20.2.2.2 Cyclamen Mite, *Phytonemus pallidus*

This species is most commonly known as the cyclamen mite. It is also known as the strawberry mite because it is a major pest of strawberries. It is a pest of many ornamental plants such as cyclamen, African violet, azalea, begonia, carnation, chrysanthemums, gerbera, and ivy.

20.2.2.2.1 Symptoms

This species prefers young leaves or flower buds. Eggs are often laid within unopened buds and between adpressed halves of the dorsal surfaces of folded leaves. On leaves they feed on the upper side. Infested leaves are twisted, curled, distorted, brittle, and reduced in size. Damaged gerbera produces bronzed patches along the midribs and slight curling of the leaf. Rays of flower are deformed when flowers are attacked in the bud stage. Damaged strawberry leaves have wrinkled upper surfaces, irregular folding, and fluting of the leaf margin; veins bulge upward like blisters. Mildly damaged plants have an unnatural dense appearance because petioles fail to elongate. Severely damaged leaves become brittle, turn brown or silvery, and eventually die, and flowers and young fruits become brown near the base and may turn black and die.

20.2.2.2.2 Life Cycle

The life cycle is completed in 1–3 weeks depending on temperature and is usually completed in 10–14 days in greenhouses.

20.2.2.2.3 Management

(a) *Biological control*: *Neoseiulus cucumeris* is used in many parts of Europe to control of *P. pallidus* on strawberries and can provide effective control at lower cost than chemical control. It should be released at 10–20 mites per plant or the predator/prey ratio of 1:10 as soon as the symptoms of mite damage are noticed. Another two phytoseiid species, *Neoseiulus reductus* in Russia and *N. reticulatus* in the USA, are also effective for *P. pallidus* control on strawberries. Like *N. cucumeris*, *N. californicus* can also significantly reduce *P. pallidus* densities on strawberries as well as provide effective control of *T. urticae* on the same crop in greenhouses. In Northwest USA, *N. fallacis* and *N. cucumeris* are both effective predators of *P. pallidus* on strawberries, but the former performs slightly better than the latter regardless of whether *T. urticae* is present on the crop. Although *N. fallacis* can provide more rapid control of both pests, *N. cucumeris* seems to give longer-term regulation at lower densities.

(b) *Chemical control*: on strawberries, effective control can be achieved using endrin at 0.025 % or endosulfan at 0.05 % at 4-week intervals before the opening of the first flower; triazophos, amitraz, and pyridaben with two applications at 7-day intervals after fruit harvest and leaf mowing; pirimiphos-methyl applied at 1.6 kg/ha; and abamectin at 1.12 and 2.24 kg a.i./ha.

On cyclamen, control can be achieved using endrin at 0.025 % or endosulfan at 0.05 % or dicofol at 0.03 % at 4-week intervals at the first sign of damage.

On *Saintpaulia*, dienochlor, triazophos, fenbutatin oxide, and oxamyl are effective when applied at first sign of damage.

On *Peperomia* and *Saintpaulia*, mites can be controlled by drenching plants in 0.05 % schradan at the rate of 142 ml per pot (10.2 cm) at 3- or 4-week intervals.

(c) *Integrated management*: it helps greatly to use healthy plants, with good hygiene in plant production. Strawberry planting material can be kept free of mites or completely disinfected on acquisition by immersing them in warm water at 45 °C for 13–15 min and then in cold water. Immersion of plants in hot water at 43.5 °C for 30 min kills all the mites. A better alternative is to treat loosely stacked plants with saturated air at 44.5 °C for 1 h. After the treatment, plants should be dried before packing and planted as soon as possible.

Immersion of cyclamen, *Crassula*, and *Saintpaulia* in water at 15 °C for 7 min is effective against this mite but may also injure plants.

It is important to monitor the level of infestation and introduce predatory mites at the right time and rates. Biological control should be integrated with chemical control for other pests by careful and selective use of pesticides to prevent harmful effects on predatory mites.

Plant resistance has not yet been used in cyclamen mite control, but there is evidence that the potential exists. Some varieties of strawberry are known to have higher levels of resistance to cyclamen mite.

20.2.3 Eriophyid Mites

Eriophyid mites belong to the superfamily Eriophyidae of the order Prostigmata. Some rust mites and gall mites are important pests on economic plants.

20.2.3.1 Tomato Russet Mite, *Aculops lycopersici*

This species is most commonly known as the tomato russet mite. It is also known as the tomato rust mite. This species is a common pest on tomatoes worldwide. Its occurrence and damage are increasing on greenhouse tomatoes in many European countries, e.g., Israel, Italy, France, Spain, Germany, Moldova, Bulgaria, and Hungary. This species is exceptionally polyphagous for an eriophyid. It is also found on eggplant, tobacco, potato, *Datura*, and many other Solanaceae (with the exception of *Convolvulus* species), but only occasionally on chilies and petunia.

20.2.3.1.1 Symptoms and Damage

The infestation of tomato plants starts from the lower leaves, which curl up and turn silvery on the lower surfaces. Later, infested leaves turn brown, withered, and paper like, but they do not wilt as they do when they are diseased by the late blight. As the mite population increases, they spread upward over the stem and upper leaves. The lower parts of the stem lose their surface hairs before damage appears on the upper leaves. The infested stem changes from green to brown, and its surface cracks longitudinally, displaying a characteristic rusty-brown russeted aspect. As infestation continues, the browning of leaves and stems increases and some injured leaves start to drop. With the defoliation of the plant, fruit production is reduced and fruits sunburned, although they do not usually russet. In very heavily infested plants, young leaves may be distorted and fruits bronzed when attacked.

Mite feeding reduces the rate of photosynthesis by host plants. This is caused by the destruction of the guard cells and subsequent reduction of leaf gas exchange.

20.2.3.1.2 Life History and Biology

The life cycle of this mite passes through egg, larva, nymph, and adult stages. The generation time is about 1 week at 21–25 °C. Males develop slightly faster than females. Females live for several weeks and lay ten to 53 eggs. Fertilized eggs produce both males and females, whereas unfertilized eggs give rise to males only.

The optimal conditions for mite development are 21–27 °C and 30 % RH, conditions that greenhouses often provide. At higher temperatures, lower humidity is required; high humidity can cause increased mite mortality.

20.2.3.1.3 Management

(a) *Biological control*: natural enemies of the tomato russet mites are many and include both predatory insects and mites.

Two species of predatory thrips, *Leptothrips mali* and *Scolothrips sexmaculatus*, are known to be associated with this mite, and the former actively feeds on *A. lycopersici* but is hindered by glandular hairs on tomato. Their predation effects on *A. lycopersici* are unknown.

A tydeid mite, *Pronematus ubiquitus*, is known to be associated with *A. lycopersici* but nothing is known about its effectiveness as a predator of *A. lycopersici*. Another tydeid mite, *Homeopronematus anconai*, however, is known to feed on all stages of the tomato russet mite, but it cannot develop to the adult stage when fed solely on this prey; it needs to feed on alternative foods such as fungal spores to complete development. The ability to feed on alternative foods may be an advantage because its densities may be sustained by alternative foods when eriophyid mites are low in density.

A stigmaeid mite, *Agistemus exsertus*, has been evaluated in the laboratory with promising results and is recommended for releases onto mite-infested plants for biological control. However, this species is not commercially available.

Euseius concordis develops and reproduces well on the tomato russet mite and is naturally associated with the mite on tomato plants. However, it is hindered by the webbing produced by spider mites which occur with the eriophyid on tomato plants. Only *Neoseiulus fallacis* seems to be a good biocontrol agent of this mite. It attacks all stages of the pest, survives well (92 %), develops rapidly (6.3 days at 22 °C), and reproduces well on *A. lycopersici*. This commercially available predator should be effective when released at a rate of 1000 per acre when the tomato seedlings are transplanted into the greenhouse.

Pathogenic bacteria such as Bt (*Bacillus thuringiensis*) may be a new biological weapon for mite control. For example, an exotoxin from *B. thuringiensis* shows promising experimental results in the biocontrol of *A. lycopersici* on tomato plants in Uzbekistan.

(b) *Chemical control*: the chemicals that provide good control include binapacryl, bromopropylate, cyhexatin, fenvalerate, flubenzimine, fluvalinate, ethion, dichlorvos, dicofol, formothion, karathane, lambda-cyhalothrin, methamidophos, milbex, profenofos, propargite, pyridaphenthion, and zineb. Abamectin and sulfur are very effective and widely used and may facilitate biological control of other pests on tomato, especially *T. cinnabarinus* by predatory mites.

Various kinds of oils are also more environmentally friendly than synthetic pesticides and can be used effectively against the tomato russet mite.

(c) *Integrated control*: in integrated pest control, it is essential to monitor the pests. It is useful to regularly search for bronzing on lower leaves and stems and then check damaged leaves and other leaves immediately above them for mites. Damage is typically first observed when green fruits reach 5 cm; rarely is it first observed when more than 25 % of the fruit are ripe. For effective control, treatment is necessary immediately when damage symptoms begin to spread. A thorough cleanup of the greenhouse and good sanitation will help to prevent the carry-over of this mite.

Some varieties of tomato are known to be less susceptible to tomato russet mite damage and may be used in IPM programs.

20.2.4 Acarid Mites

Acarid mites belong to the family Acaridae of the order Astigmata. The genus *Rhizoglyphus* contains several species of pests attacking root crops. Some *Tyrophagus* species can cause economic damage to plants, including both ornamental flowers and vegetables grown in greenhouses.

20.2.4.1 Robine Bulb Mite, *Rhizoglyphus robini*

No widely accepted common names have been proposed for this species, although some people call it the robine bulb mite. This is a cosmopolitan species. It is often associated with bulbs, roots, and seeds of many plant species and is known to attack bulbs of *Narcissus*, *Eucharis*, lilies, orchids, gladioli, hyacinth and tulip, tubers of dahlia, and corms of *Freesia* in storage, in greenhouses, and in the field.

20.2.4.1.1 Damage Symptoms

Mite infestation of bulbs and corms starts with penetration through the basal plate or outer skin layers. Bruised bulbs and fungus-infected bulbs allow more rapid mite establishment than healthy ones. On *Fusarium*-infected Rakkyo bulbs, *R. robini* achieves exponential population growth after 14 days, whereas on healthy bulbs it takes 60–90 days to achieve exponential population growth. Infested plants have dark-brown streaks on roots and develop distorted growing tips and leaves. On an injured bulb, *R. robini* can increase rapidly in number and quickly reduce the whole bulb to a rotten pulp.

20.2.4.1.2 Life History

The life cycle takes 7–27 days at 13–26 °C excluding the hypopial stage and is 9 days at 25 °C.

20.2.4.1.3 Management

Several species of predatory mites have been tested for biological control of *R. robini*. *Hypoaspis aculeifer*, which is commercially available, is effective against this mite in small-scale lily bulb propagation at the predator/prey ratio of 1:20 but in large greenhouse plots at the ratios of 1:2 to 1:5. However, as many as three predators per bulb mite are needed for the elimination of the bulb mite population. In small-scale lily bulb propagation, the ascid mite *Lasioseius bispinosus* is also able to control the bulb mite, so is the parasitoid mite *Parasitus fimetorum*, but for the latter only when the growing medium is peat. Another acarid mite *Protogamasellus minutus* and a digamasellid mite *Dendrolaelaps sayedi* also feed on *R. robini*, but their effectiveness against this mite is unknown.

This species can be controlled using pirimiphos-methyl combined with cultural measures such as hot water treatment (at 39 °C or 41 °C for 2 h) and storage of the bulbs at −2 °C. In peat growing media, a combination of hot water treatment and the release of *H. aculeifer* are also effective against the bulb mite in lilies during the propagation phase. Soaking bulbs of lilies in 548 ppm dicofol for 30 min can significantly reduce mite numbers. Flooding soil with water for 5 days kills 96.1 % of the mite infesting gladiolus and all the mites are killed after 14 days. *R. robini* is known to be sensitive to sulfur, azocyclotin, cyhexatin, methidathion, phosmet, profenofos, and prothiofos, as well as to hot and dry conditions.

20.2.4.2 Bulb Mite, *Rhizoglyphus echinopus*

This species is known generally as the bulb mite. This is a cosmopolitan species. It attacks bulbs and roots of many species, including *Freesia*, *Gladiolus*, hyacinth, lily, iris, *Narcissus*, and tulips.

(a) *Symptoms*: damage symptoms are also similar to those caused by *R. robini*. Damaged roots of *Freesia* and *Gladiolus* develop dark-brown streaks and are often mined internally. Healthy corms grown in heavily infested soil will have distorted growing tips and leaves.
(b) *Management*: control methods used against *R. robini* should be generally applicable to *R. echinopus*.

When released at relatively high predator/prey ratios, *Hypoaspis aculeifer* should provide effective control of *R. echinopus*. A female deutonymph of *Hypoaspis aculeifer* can consume 60 eggs, 132 larvae, 20 protonymphs, eight deutonymphs, or four adults of *R. echinopus*. When feeding on the larvae of *R. echinopus*, an adult female of *H. aculeifer* can lay 2.8 eggs per day and a total of 114 eggs during her life.

Another laelapid mite, *Hypoaspis vacua*, can also develop successfully when feeding on nymphs of *R. echinopus*. Immature stages of *H. vacua* consume an average total of 33 nymphs of *R. echinopus*, whereas adult females consume 13 nymphs daily.

Rhizoglyphus echinopus is insensitive to many pesticides (at least 9 pyrethroids, 6 organochlorines, 4 formamidines, 14 specific acaricides, diflubenzuron, nicotine, and abamectin) but is known to be susceptible to dieldrin, endrin, aldrin, deltamethrin, chlorpyrifos, Diazinon, azinphos-ethyl, and carbofuran.

20.2.5 General Management of Pest Mites

In general, commonsense (Table 20.2) and chemical control (Table 20.3) measures should be taken whenever appropriate. These are applicable to the control of many mite pests.

Table 20.2 Commonsense pest control methods in greenhouse crop production (Modified from Dole and Wilkins 1999)

Production stage	Commonsense pest control methods in greenhouse crop production
Before production	Use pest-free plants and containers Clean/sterilize benches and irrigation system Remove weeds in and outside (within 3–9 m) of greenhouses Remove fallen plant material, media, debris, and extra plants Use exclusion screens
During production	Use pest-resistant species or cultivars Inspect incoming plant material carefully; isolate new ones, if possible. Monitor pest populations regularly Produce plants at the optimal growing conditions Change clothes and boots before entering greenhouses
If a problem occurs or is likely to occur	Decide on the threshold of tolerance for each pest Remove infested plants or plant parts promptly Use beneficials when and where appropriate Modify crop environment to discourage pest growth Use effective chemicals when and where appropriate

Table 20.3 Chemical control measures for greenhouse mites

Chemical	Trade name	Effective on
Abamectin	Avid	Broad mites, cyclamen mites, eriophyid mites, false spider mites, two-spotted spider mites, mites (general)
Bifenthrin	Talstar	Broad mites, eriophyid mites, false spider mites, two-spotted spider mites
Bifenazate	Floramite	Red spider mites, two-spotted mites
Chlorfenapyr	Pylon	Broad mites, cyclamen mites, eriophyid mites, false spider mites, two-spotted spider mites
Chlorofentazine	Ovation	Two-spotted spider mites
Deltamethrin	DeltaGard	Mites (general)
Dicofol	Kelthane	Broad mites, cyclamen mites, red and two-spotted spider mites
Etoxazole	TetraSan	Red and two-spotted spider mites
Fenbutatin oxide	Vendex	Two-spotted spider mites
Fenpyroximate	Akari	Red and two-spotted spider mites
Fluvalinate	Mavrik	Two-spotted spider mites
Hexythiazox	Hexagon	Broad and two-spotted spider mites
Pyridaben	Sanmite	Two-spotted spider mites
Propargite	Omite	Red and two-spotted spider mites
Spinosad	Conserve	Red and two-spotted spider mites
Spiromesifen	Forbid	Broad mites, cyclamen mites, eriophyid mites, false spider mites, two-spotted spider mites

20.3 Predatory Mites

20.3.1 Introduction

The predatory mites of the family Phytoseiidae and Laelapidae are used in the biological control of mite pests and thrips on many crops in greenhouses (Table 20.4).

20.3.2 Phytoseiid Mites

Phytoseiid mites belong to the family Phytoseiidae of the order Mesostigmata. They are predators of spider mites and other small mites and insects on plants. Several members of this family are of great importance in the biological control of spider mites and thrips in greenhouse crop production. Effective biocontrol agents (most commercially available species) that are commonly used in greenhouses belong to the genera *Neoseiulus* and *Phytoseiulus* in the Amblyseiinae.

Some 20 species of phytoseiids have been made commercially available for pest control and many of these have been applied on greenhouse plants. Several genera have been recorded from, tried, or used in greenhouses: *Phytoseiulus*, *Neoseiulus*, *Galendromus*, *Typhlodromus*, *Typhlodromalus*, *Amblyseius*, and *Euseius*. Seven species have been proven to be of importance in greenhouses and have been commonly used: *Phytoseiulus persimilis*, *Neoseiulus cucumeris*, *N. barkeri*, *N. californicus*, *N. fallacis*, *Iphiseius degenerans*, and *Galendromus occidentalis*.

20.3.2.1 *Phytoseiulus persimilis*

This mite is a specific predator of *Tetranychus* spider mites and shows reduced reproduction and survival on other spider mites and phytophagous mites. It has been collected in association with tetranychid spider mites of the genera *Tetranychus*, *Eutetranychus*, and *Panonychus* and other predatory mites such as *Neoseiulus californicus* and *Iphiseius degenerans*.

Table 20.4 Commercially available predatory mites of importance in biological control of pest mites under greenhouse conditions

Beneficial mites	Mite species	Family	Pests attacked
Phytoseiid mites	*Phytoseiulus persimilis*	Phytoseiidae	Spider mites (*Tetranychus*)
	Neoseiulus californicus	Phytoseiidae	Spider mites (*Tetranychus*)
	N. cucumeris	Phytoseiidae	Thrips, broad mite (*Polyphagotarsonemus latus*), cyclamen mite (*Phytonemus pallidus*)
	Iphiseius degenerans	Phytoseiidae	Thrips, spider mites
Laelapid mites	*Hypoaspis aculeifer*	Laelapidae	Bulb mites (*Rhizoglyphus*), sciarid flies, fungus gnats, thrips
	H. miles	Laelapidae	Bulb mites (*Rhizoglyphus*), sciarid flies, fungus gnats, thrips

20.3.2.1.1 Life History and Biology

Development from the egg to adult takes 3.6 days for males and 4.1 days for females at 26 °C. An adult female can consume 10–20 *Tetranychus* spider mite eggs per day and lay as many as 5 eggs per day and up to 80 eggs during her life. The sex ratios of offspring are often highly female biased (>80 % daughters).

It has been introduced to many countries and is now used throughout the world for the control of *Tetranychus* spider mites on crops such as cucumber, pepper, tomato, eggplant, strawberry, and cut flowers in greenhouses. This mite is a specific predator of *Tetranychus* spider mites and shows reduced reproduction and survival on other spider mites and phytophagous mites (type I).

20.3.2.1.2 Use in Biocontrol

Phytoseiulus persimilis can provide effective control of *T. urticae* in greenhouses but control is unsatisfactory at very high temperatures and low humidities. In greenhouses in many countries, *Tetranychus* mites can be controlled by *P. persimilis* using several release systems, and the use of pesticides against them is greatly reduced or completely discontinued.

(a) *Cucumber*: *Phytoseiulus persimilis* has been used successfully for greenhouse spider mite control on cucumbers in many countries and for many years. Earlier studies in the 1960s in the UK on spider mite control using *P. persimilis* on this crop led to great success of this species in the greenhouse industries. The control program uses a leaf damage index system (a score of 4 for a leaf covered with mites). When the mean leaf damage index reaches 0.4, two predators are released to every other plant near the infestation site, and predators will overrun the spider mites in 30 days, eliminate them in 7 weeks, and continue to survive for at least 3 weeks. In the second year during late April and early May, 10–20 spider mites per plant should be introduced first and two predators introduced to every five plants 10 days later. This "pest in first" program will ensure successful control of spider mites during the growing season.

This "pest in first" method was later optimized in Europe during the 1980s with the development of a mathematical modeling of the population dynamics of *T. urticae* and *P. persimilis* on cucumber under greenhouse conditions. As a result, the optimization of the "pest in first" method enables the number of *T. urticae* introduced to be reduced 20-fold and that of *P. persimilis* three- to fourfold.

In Austria, *T. urticae* are controlled by *P. persimilis*. In Bulgaria, spider mite populations are monitored, and *P. persimilis* is introduced onto cucumbers at a predator/prey ratio of 1:20 when there are five to six spider mites per leaf and 20 % leaf damage. This method has been successfully used against tetranychid mites. In curative control, timing and rate of predator release are important. Releasing too many predators may result in the lack of food for predators and releasing too few may result in poor control.

Extreme temperature and humidity in greenhouses can affect predator performance. In greenhouses in Egypt, for example, relative humidity can be within 22–68 %, and the temperature fluctuates and can get as high as 43 °C, which can kill *P. persimilis*. The greenhouse conditions can be modified to favor *P. persimilis*. For example, regular overhead misting can reduce populations of *T. urticae* (which do well under dry and hot conditions) and promote the growth of *P. persimilis* on cucumber greenhouses. A study in Ohio, USA, has showed that the western flower thrips *Frankliniella occidentalis* can also be significantly reduced by a regular misting program. The role of *P. persimilis* in IPM has been expanded through the development of resistant strains. In the former USSR, strains resistant to malathion, pirimiphos-methyl, high-temperature, and high-temperature plus organophosphorus compounds, respectively, have been developed for use in the control of *T. urticae* on cucumbers in greenhouses. In Japan, a resistant strain of *P. persimilis* can control *T. urticae* on cucumbers under seven sprays of fungicides and four sprays of fenitrothion. The control is successful when they are introduced at the rate of one *P. persimilis* female per ten *T. urticae* females.

In Turkey, where *Tetranychus cinnabarinus* also attacks cucumbers in greenhouses, *P. persimilis* provides effective control when released at the 1:5 predator/prey ratio.

(b) *Tomato*: biological control of spider mites on tomato is generally less successful than on cucumber. One reason is the interference of predator searching behavior by the exudating hairs on tomato leaves and stems.

A practical program similar to that used on cucumber has been used in the UK. When the leaf damage index reaches 0.2, five predators should be introduced to every fifth plant. For a new crop, one-fifth of the plants should be inoculated with 30 spider mites 3 weeks before planting. Four predators should then be introduced on every infested plant 10 days later.

In Bulgaria, *P. persimilis* has been successfully used against spider mites, when it is introduced onto tomatoes at the predator/prey ratio of 1:10 when there are two to three spider mites per leaf.

In greenhouses in Belgium, temperatures frequently increase much above 30 °C with mean RH below 57 % in summer. There is a higher density of glandular trichomes on leaflets at higher temperatures and a larger percentage of the predators are stuck to leaflet trichomes, especially on some cultivars. Increasing the humidity by using a humidifying system and reducing the temperature by whitewashing the roof can reduce mite damage. However, it is better to use predators that are reared on tomato rather than on beans. Strains adapted to tomato have been selected and made available commercially for spider mite control on tomatoes.

Pesticide-resistant strains of *P. persimilis* have also been used on greenhouse tomatoes. In Bulgaria, a strain resistant to organophosphorus pesticides is used in integrated control of tomatoes in the greenhouses. Predators can be used for spider control, while chemical control is employed against other pests.

A relatively new approach has been the use of bean (*Phaseolus vulgaris*) as indicator plants for spider mites in greenhouse tomato production. Although spider mite population growth rates are the same on both host plants, spider mites can become established 5 weeks earlier on beans because of the lower-temperature threshold on beans (7 °C lower) than on tomatoes. *T. cinnabarinus* can be controlled when *P. persimilis* is released at the threshold of 12 spider mites per tomato leaf. For some reason, *P. persimilis* is more efficient on the tomatoes than on the beans, which helps to keep the two predator/prey systems on the two hosts out of synchrony, giving better persistence of the system.

(c) *Bell pepper*: when *P. persimilis* is introduced to sweet peppers at the predator/prey ratio of 1:10 when the initial spider mite density is two to three per leaf, successful control can be achieved. In Bulgaria, *P. persimilis* provides effective control of *T. urticae* and *T. turkestani* on pepper when released at a rate of 70,000–80,000 individuals per hectare, reducing the need for chemical control. In Poland, *P. persimilis* is effective against both *T. urticae* and *T. cinnabarinus* developing separately or together on plants of two cultivars of greenhouse sweet pepper. In unheated greenhouses in Sicily, releases of *P. persimilis* from February onward provide good control of *T. urticae* on chili, but control can be disrupted by the application of chinomethionat against powdery mildew from March.

(d) *Ornamental plants*: ornamental plants are grown for cosmetic purposes and can tolerate lower levels of mites than other plants. *P. persimilis* has been used to control spider mites on a variety of ornamental plant species with success and is sometimes used with spot treatment of mite infestations using acaricides.

On greenhouse roses, *T. urticae* has been effectively controlled by *P. persimilis* in many countries. When *P. persimilis* reaches the level of one predator for every ten spider mites, it can eliminate spider mites in a few weeks' time and then disperse them. In very large greenhouses, *P. persimilis* can persist in the lower canopies and can tolerate spot treatment of upper foliage using selective chemicals such as abamectin.

Gerbera, being bushier with a more humid microenvironment, favors the development of *P. persimilis*. In greenhouses in Sicily, Italy, natural populations of *P. persimilis* move in and provide natural control of *T. urticae* on *Gerbera*, provided pesticides toxic to the predator are not used.

In the UK, *T. urticae* on chrysanthemums is successfully controlled by releasing ten *P. persimilis* females per plant 3–4 weeks after treatment of the plants with aldicarb.

In China, *T. urticae* on *Salvia splendens*, *Ageratum conyzoides*, *Zantedeschia aethiopica*, and *Pelargonium lateripes* in greenhouses are successfully controlled by releases of *P. persimilis* either at the seedling stage or when mites just begin to increase. Such timing of releases is both effective and economic. The release rate ranges from one to 50 mites/plant depending on the size, pest density, and plant species. A second release is sometimes necessary when the spider mite populations begin to increase on some plants.

In Iowa, USA, *P. persimilis*, along with *Neoseiulus californicus*, is released to control *T. urticae* on greenhouse-grown poplar (*Populus* spp.) with spot treatment using pesticides when needed. Spider mites are suppressed to acceptable levels and pest management costs are reduced by 81 % compared to chemical control.

In Poland, *P. persimilis* provides good control of *T. urticae* on orchids in greenhouses.

In New Zealand, *P. persimilis* provides good control of *T. urticae* on *Cymbidium* but is not effective against *T. urticae* on greenhouse carnations due to its inability to maintain traction on the waxy surface of the leaves and stems.

20.3.2.2 *Neoseiulus cucumeris*

20.3.2.2.1 Life History and Biology

At 25 °C, the egg-to-adult development is completed in 8–9 days when feeding on thrips larvae and about 7 days when feeding on acarid mites. Females produce on an average of 53 eggs during the oviposition period at the rate of 1.9 eggs/day. The intrinsic rate of increase is 0.203 females/female per day and the population is able to double in 3.41 days.

This species can feed on pollen, and the availability of pollen on greenhouses enhances the development and reproduction of *N. cucumeris*, although pollen feeding reduces the predation rate on prey.

In temperate areas, the effectiveness of *N. cucumeris* in autumn and winter may be limited by the reproductive diapauses induced under short-day conditions. The critical day length for inducing diapauses at 22 °C is 12.45 h, and most *N. cucumeris* undergo diapauses only when exposed to diapause-inducing conditions throughout their juvenile development. A New Zealand strain with low incidence of diapauses has been selected and improved and is now widely used in greenhouses around the world.

20.3.2.2.2 Use in Biocontrol

This mite is an effective predator of some tarsonemid mites and some spider mites that do not produce webbing. It has also been tested against *Bemisia tabaci* in the laboratory. It can complete its development to the adult-stage feeding on a combination of eggs and first- and second-instar larvae of *B. tabaci* with a 72 % survival rate and can also reproduce on eggs and larvae of *B. tabaci*. An adult *N. cucumeris* can consume an average of 6.6 eggs, 1.9 first-instar larvae, or 0.9 s-instar larvae of *B. tabaci* per day. This species should be tried for whitefly control in greenhouses.

N. cucumeris has been used most successfully against thrips on vegetables and ornamentals in greenhouses. Often repeated releases and high predator/prey ratios are required for adequate control. Preventive releases are also very effective.

(a) *Cucumber*: in the former USSR, larvae of *Thrips tabaci* are controlled on cucumbers in greenhouses when *N. cucumeris* is released at predator/prey ratios of 1:2. In Turkey, effective control is achieved by releasing predators at the rate of four to five individuals per plant as soon as thrips are detected on host plants. In the UK, good establishment is possible by either a single release of 250 predators/plant or three consecutive releases at 50 predators/plant at two weekly intervals, starting at the first sign of infestation. *N. cucumeris* can reduce populations of *T. tabaci* but cannot eliminate the population of the thrips. In Canada, *N. cucumeris* provides effective control of *T. tabaci* and *Frankliniella occidentalis* on seedless cucumber in greenhouses in British Columbia. Adult predators can persist on plants for 7 weeks in the virtual absence of thrips prey and can increase numerically in response to increases in thrips populations.

Neoseiulus cucumeris can keep *Frankliniella occidentalis* populations at a low level but sometimes do not provide effective control, despite repeated releases. This is overcome by the development of controlled release system (CRS), which provides better distribution of the predator than the traditional loose bran system. The CRS consists of a specially formulated bran-based population of *N. cucumeris* in a waxed paper pack of specific porosity and gives more rapid establishment. The predators continue to breed and emerge from the pack for at least 6 weeks. The technique provides better control of *F. occidentalis* and is also less expensive than the traditional method. The best control is prevention. When *N. cucumeris* is introduced in good-quality culture sachets immediately after planting, thrips populations do not develop.

(b) *Bell pepper*: the effectiveness of *N. cucumeris* in greenhouses is inconsistent. In the Netherlands, control of thrips on peppers is less successful than on cucumber. In the USA, 2 introductions of 10– 25 predators per plant during the growing season are not effective against *F. occidentalis* in greenhouses. In Spain, inundative releases of 50–400 predators/plant on different occasions and controlled releases are unable to control *F. occidentalis* on sweet pepper in unheated greenhouses. However, the combined use of *N. cucumeris* and predatory bugs (*Orius*) can provide effective control of *F. occidentalis* on sweet peppers in greenhouses.

The effectiveness *of N. cucumeris* against thrips of the genus *Thrips* is more consistent. In greenhouse sweet peppers in Japan, *N. cucumeris* released 3 times at 1-week intervals starting 3 days after planting reduces *T. palmi* population to one-third to one-fifth of its original size for 6 weeks after planting. In New Zealand, three releases of 10–140*N. cucumeris* per plant against *Thrips tabaci* and *T. obscuratus* result in low populations of thrips and high predator numbers for 12 weeks, with no apparent thrips damage to plants.

(c) *Ornamental plants*: control *of F. occidentalis* on chrysanthemums using *N. cucumeris* has been effective in both Europe and North America. In the USA, three predatory mites per leaf are released when the density of adult thrips is high, and pollen is released with the mites to encourage predator survival when thrips are low in number. In the UK, releasing 100 *N. cucumeris* in bran per m^2 every other week gives effective biological control of *F. occidentalis* on chrysanthemums. In bed-grown year-round chrysanthemums in greenhouses, effective control can be achieved by introducing three predators per plant in the third, fourth, and fifth weeks after planting.

In the UK, *N. cucumeris* is also effective against *F. occidentalis* on cyclamen in greenhouses. In the USA, it provides control of *Frankliniella tritici* and *F. occidentalis* on ornamental bedding plants in greenhouses when breeding sachets containing 50 mites are introduced at the rate of 125 sachets per 200 m^2 of growing area.

20.3.2.3 *Neoseiulus californicus*

Also known as *Amblyseius californicus, N. californicus* is a type II phytoseiid species, mainly feeding on *Tetranychus* spider mites and also other mites and pollen. Like type III phytoseiids, it can also feed on small insects.

20.3.2.3.1 Life Cycle

The egg-to-egg generation time is 9.5 days at 25 ± 1 °C when reared on broad mites, 2 days longer than when reared on *Tetranychus urticae* under the same conditions. The proportion of females in the offspring is 51.2 %, and each female lays on an average of over two eggs per day. The larvae are active and feed on prey. The nymphs of *N. californicus* can attack 13 eggs of *T. urticae* and 11 eggs of *T. cinnabarinus*, consuming over 86 % of the eggs killed. Adult females can resist starvation for over 10 days and can resume oviposition when food is available. Female fecundity is 65 eggs when feeding on nymphs of *T. urticae*.

20.3.2.3.2 Use in Biocontrol

This species is from arid and semiarid climates and more tolerant to low humidity and low prey density. In perennial greenhouse-grown crops, *P. persimilis* and *N. californicus* have complementary effects, and a combination of the two can enhance long-term biological control of spider mites.

This species is also an effective biocontrol agent of broad mites and cyclamen mites.

20.3.2.4 *Iphiseius degenerans*

Also known as *Amblyseius degenerans*, this widespread species is a generalist predator (type III). It feeds on small mites, insects, and pollen.

20.3.2.4.1 Life Cycle

Development from egg to adult takes 5.7 days at 25 °C. Females lay 2.2 eggs per day or a total of 68 eggs during 30 days of the oviposition period. The eggs are relatively tolerant to desiccation. Deutonymphs of *I. degenerans* consume 21.1 *Tetranychus pacificus* per day, whereas those of *P. persimilis* consume 7.5 preys per day. It is a fast-moving species.

20.3.2.4.2 Use in Biocontrol

This species has recently been employed for the biological control of thrips in greenhouses. Releases of ten predators per plant eliminate citrus thrips *Scirtothrips citri* from potted citrus trees in greenhouses in the USA. In the Netherlands, *I. degenerans* is introduced to greenhouse sweet pepper in potted flowering castor bean bushes, which serves as a reservoir for predatory mites to colonize sweet peppers.

20.3.3 Laelapid Mites

The family Laelapidae is a member of the superfamily Dermanyssidae in the order Mesostigmata. Species of *Hypoaspis* are free-living predators common in soil and have been employed in the biological control of soil-inhabiting mites and insects in greenhouses. Only two species, *Hypoaspis aculeifer* and *H. miles*, have been used in greenhouses to control a number of soil-inhabiting pest insects and mites.

20.3.3.1 *Hypoaspis aculeifer*

This mite is a polyphagous predator, feeding on other mites, small insects, and nematodes.

20.3.3.1.1 Life History and Biology

The developmental time varies with types of food and temperatures. The egg-to-adult development takes ten to 13 days at 26 °C with *Tyrophagus putrescentiae* as food. With larvae of *Rhizoglyphus robini* as prey, this species reaches adulthood in 11–12 days, and females lay an average of 114 eggs during an oviposition period of 41 days.

20.3.3.1.2 Use in Biocontrol

This species is an effective biological control agent against *Rhizoglyphus* bulb mites. It can also be released to control sciarid flies (*Bradysia* spp.) in growing media of crops such as poinsettia in greenhouses. It can also be released to reduce densities of thrips pupae and other acarid mites (*Tyrophagus*) in the soil or growing media.

20.3.3.2 *Hypoaspis miles*

This mite is a polyphagous species and feeds on small insects, mites, and nematodes.

20.3.3.2.1 Life History and Biology

Developmental time from the egg to adult varies with temperature and types of food and takes 34 days at 15 °C but 9 days at 28 °C when feeding on *Bradysia* larvae. The developmental threshold temperature is between 10 °C and 12 °C. Females lay on an average of 44 eggs in 53 days when feeding on larvae of *Lycoriella* and 22 eggs in 69 days when feeding by *Tyrophagus*. Males live twice as long as females.

20.3.3.2.2 Use in Biocontrol

Hypoaspis miles is an effective predator of sciarid larvae. When released at 55 mites per pot, it provides satisfactory control of *Bradysia* in pot-grown *Cyclamen* and poinsettias in greenhouses.

This species also feeds on thrips pupae, shore fly larvae (*Scatella*), and acarid mites (*Rhizoglyphus* and *Tyrophagus*) and can be released to reduce densities of these pests in the soil or growing media.

References

Carey JR, Bradley JW (1982) Developmental rates, vital schedules, sex ratios and life tables for *Tetranychus urticae, T. turkestani* and *T. pacificus* (Acarina: tetranychidae) on cotton. Acarologia 23:333–345

Dole JM, Wilkins HF (1999) Floriculture: principles and species. Prentice-Hall, Simon & Schuster, Upper Saddle River

Hazan A, Gerson U, Tahori AS (1973) Life history and life tables of the carmine spider mite. Acarologia 15:414–440

21 Strawberry

Abstract

In greenhouses, strawberry plants are susceptible to threat from various pests and diseases. A number of precautionary and protective measures are required in order to achieve maximum production. The major pests (spider mites, armyworm, aphids, whiteflies, Japanese beetle, loopers, thrips, weevils, slugs) and diseases (gray mold, anthracnose, powdery mildew, red stele, *Verticillium* wilt, leaf scorch, angular leaf spot) of strawberry and their symptoms, biology, survival, spread, and management are discussed.

Keywords

Strawberry insects • Mites • Diseases • Biology • Spread • Management

21.1 Introduction

Strawberries are grown in greenhouses in many parts of the world (Fig. 21.1). In Europe they have been grown successfully and farmers can get more than 120 t/ha/per year. In China, there are over 24,000 ha of strawberries grown over the winter in energy-saving greenhouses which rely mostly on solar energy to heat the greenhouses.

The advantages of growing strawberries under greenhouse conditions are:

- The extension of the fruiting season to approximately 26 weeks (May to mid-autumn).
- Improvement in class 1 grade fruits and reduced wastage. Class 1 grade (supermarket saleable fruits) for outdoor production is generally 50–70 %, while under protection it is nearer to 80–90 %.
- As the fruit is produced in controlled conditions, there can be reduced use of pesticides by up to 50 % against *Botrytis*, downy mildew, and black spot.
- Provision of an improved environment for the use of biological control.
- Increased opportunities for organic production due to the reduced incidence of disease and the enhanced use of predators.
- Guaranteed conditions to carry out production and harvesting operations and no need to stop the pickers in poor conditions or to have long waits for early dew to burn off.

P.P. Reddy, *Sustainable Crop Protection under Protected Cultivation*,
DOI 10.1007/978-981-287-952-3_21

Fig. 21.1 Strawberry under protected cultivation

21.2 Varieties

21.2.1 Early and Midsummer Varieties

- Elvira
- Honeoye – susceptible to mildew when under stress
- Pantagruella

21.2.2 Midsummer Varieties

- Pegasus – resistant to wilt, crown rot, and mildew
- Elsanta – generally susceptible to disease
- Hapil – susceptible to mildew and *Verticillium* wilt
- Cambridge favorite – susceptible to *Botrytis* in wet condition
- Alice – resistant to wilt, crown rot, and mildew

21.2.3 Late Summer Varieties

- Symphony – good disease resistance
- Florence – good disease resistance especially mildew and some resistance to vine weevil
- Sophie
- Rhapsody

21.3 Diseases

Disease management begins before field preparation. Proper site selection, fumigation, drainage, and use of resistant cultivars and certified and vigorous planting stock will help to establish a healthy stand. Root and crown diseases are caused by pathogens present in the soil or water and/or organisms present on infected transplants. Soil fumigation and use of certified transplants are the major ways to avoid disease problems.

21.3.1 Gray Mold, *Botrytis cinerea*

21.3.1.1 Symptoms

Botrytis fruit rot can be a serious preharvest and postharvest disease. Gray mold is a very common disease, causing a growth of fuzzy gray mold. Symptoms include blossoms turning brown and dying; misshapen fruit; patches of rot on fruit which enlarge and often affect the entire fruit; masses of gray mycelium on the surface of rotting tissue; no leakage of fluid from fruit; and fruit becomes dried and mummified (Fig. 21.2). The pathogen infects flowers during wet or foggy weather, but the disease does not become apparent until the fruit ripens or after harvest in storage.

21.3.1.2 Spread

Disease emergence is favored by extended periods of high humidity or leaf wetness during flowering and moderate temperatures. The conidia of this disease are wind and splash dispersed.

21.3.1.3 Management

Remove dead and dying leaves, buds, flowers, and fruits promptly. Do not leave dead plant material lying around. Reduce humidity by improving ventilation and do not overcrowd plants. Use plastic mulch to reduce fruit contact with soil; apply appropriate fungicides; plant in areas where wind will rapidly dry wet plants and fruit; plow crop debris into soil after harvest.

Fig. 21.2 Gray mold symptoms on strawberry

Excess nitrogen has been shown to increase fruit rot when weather conditions are favorable. To avoid overfertilization, schedule fertilizer programs according to leaf tissue nutrient analysis reports. Research has demonstrated that increasing nitrogen levels beyond an optimum level does not increase yield but does increase fruit rot problems.

Allow adequate spacing between plants to improve airflow in the canopy. However, manage plant spacing for optimum yields rather than to manage disease. Planting in raised beds improves drainage and also increases airflow, resulting in lower disease levels. Plastic mulch helps protect against plant and soil-surface contact and weeds that may harbor *Botrytis* inoculum and reduces moisture within the canopy. Drip irrigation provides a direct source of water and eliminates excess moisture on fruit and leaves.

Rapid removal of field heat and keeping fruit at around 1–2 °C and increasing carbon dioxide levels during shipping (12–15 % concentration in gastight storage bags) when harvested will help keep *B. cinerea* problems down.

Antagonistic fungi such as *Trichoderma harzianum* and *Gliocladium roseum* have been used in Europe and Brazil as alternatives to fungicides. In the northeastern matted row production region, bees have been used efficiently to deliver beneficial fungi to flowers resulting in reduced gray mold incidence.

Effective control requires applications of fungicides such as captan, thiram, Switch, and Elevate during flowering.

21.3.2 Anthracnose, *Colletotrichum fragariae*, *C. gloeosporioides*, and *C. acutatum*

21.3.2.1 Symptoms

21.3.2.1.1 Leaf Spot

Round black or light-gray lesions on leaves; numerous spots may develop but leaves do not die.

21.3.2.1.2 Runners and Petioles

Dark-brown or black sunken, circular lesions on stems, petioles, and runners; plants may be stunted and yellow; plants may wilt and collapse and internal tissues discolored red (Fig. 21.3).

21.3.2.1.3 Crown Rot

Youngest plant leaves wilt during water stress in the early afternoon and recovers in the evening; wilting progresses to the entire plant; plant death; reddish-brown rot or streak visible when crown is cut lengthways (Fig. 21.3).

21.3.2.1.4 Bud Rot

Damp, firm dark-brown to black rot on buds; plants with single buds may die; plants with multiple crown may wilt as disease progresses.

21.3.2.1.5 Flower Blight

Dark lesion extending down the pedicel which girdles the stem and kills the flower; flowers dry out and die; infection after pollination may result in small, hard, deformed fruit (Fig. 21.3).

Fig. 21.3 *Top left,* strawberry crown revealing discoloration; *top right* ,anthracnose lesion on strawberry runner; *bottom left,* blight symptoms on flowers; *bottom right,* infected fruit displaying darkening sunken lesions

21.3.2.1.6 Fruit Rot

Light-brown water-soaked spots on ripening fruit which develop into firm dark-brown or black round lesions (Fig. 21.3).

21.3.2.2 Survival and Spread

Seedlings that are planted in infected soil become infected by splashing water and soil; fungus survives in soil for up to 9 months.

21.3.2.3 Management

Use anthracnose-free plants. Use of drip irrigation and between row straw mulch will also help to lessen the spread of disease within fields. Early season fruit with infections should be culled and removed from fields. Anthracnose fruit rot may be partly controlled with protective fungicide applications from flower bud emergence to harvest.

Washing roots with running water prior to planting can remove conidia and appressoria of the pathogen. Dipping plants in Quadris may reduce the severity of the disease. Quadris, Cabrio, Switch, and captan provide fair to good control, although resistance seems to be developing to the strobilurin (Quadris and Cabrio) fungicides. In severe cases, replanting with clean stock may be necessary.

Fumigating soil may help to reduce soil inoculum; solarizing soil may destroy soil inoculum; rotate to nonhost crops if fumigation or solarization is not possible; wash all soil from plant crowns prior to planting; weed around plants regularly; plant only disease-free transplants; do not use excessive amounts of nitrogen fertilizer.

21.3.3 Powdery Mildew, *Sphaerotheca macularis*

21.3.3.1 Symptoms

White patches of fungus occur on the undersides of leaves. These patches can gradually enlarge until entire lower surfaces are covered. The disease causes the leaf edges to roll upward, exposing the whitened, fungus-covered lower surfaces. Purple to reddish blotches can also appear on the undersides. Occasionally, leaf petioles, flowers, and fruits may also be attacked by the powdery mildew fungus (Fig. 21.4).

Fig. 21.4 Powdery mildew symptoms on lower surface of strawberry leaf and fruit

21.3.3.2 Survival and Spread

Fungus overwinters on leaves; spores are spread by wind.

21.3.3.3 Management

Because the powdery mildew fungus apparently survives the winter on living leaves, removing leaves that appear infected before flowering begins may help reduce disease severity. Many widely available commercial cultivars are highly resistant to powdery mildew. Applying fungicides (e.g., sulfur) at regular intervals from the early flowering period through the growing season also can control the disease and limits its carryover from one growing season to the next. However, use of fungicides to control powdery mildew on the foliage has not significantly increased yields, even for highly susceptible cultivars.

Avoid overhead irrigation; remove dead leaves at harvest to decrease overwintering of the mycelium.

A biological control agent for powdery mildew is commercially available: *Ampelomyces quisqualis* (AQ10), a biofungicide which is a selective fungal hyperparasite, works well early in the season but is less effective in inland growing regions. It works well only under low-to-moderate disease pressures and normally requires multiple applications. Other non-fungicidal materials such as potash soap and chitosan (Elexa) are available but have only limited efficacy. Abound, applied early, provides good disease control but is very expensive and may stunt plants. Captan provides only poor to fair control of powdery mildew. Rally works fairly well on powdery mildew, although there is resistance in the field. Potassium bicarbonate works well but requires multiple applications and cannot be mixed with other pesticides or fertilizers. Mono potassium phosphate (MKP) works fairly well against powdery mildew disease.

21.3.4 Red Stele, *Phytophthora fragariae*

21.3.4.1 Symptoms

Symptoms typically appear a year after planting. However, when plants are severely infected before being planted or where soil conditions strongly favor the disease, symptoms can appear during the first growing season.

Plants with severely infected root systems are often stunted in aboveground growth (Fig. 21.5). Young leaves may sometimes be bluish green and older leaves may be red, orange, or yellow. Less severely infected plants may show no aboveground symptoms except reduction in growth rate, production of runners, and size of fruit. Symptomatic plants tend to be distributed in irregular patches and the disease is often most severe where the soil is wettest.

The soilborne fungus that causes red stele attacks the roots. Young roots rot first at the tip. Above the tip, the stele turns red (Fig. 21.5). As the disease progresses the lateral roots are killed, giving the main roots a "rattail" appearance (Fig. 21.5). Eventually the crowns may also take on a red discoloration. The fungus reduces future productivity of plants most when soils are wet following the fruiting period, because new adventitious roots are normally produced at this time.

Fig. 21.5 *Top left,* stunted infected strawberry plants; *top right,* strawberry roots infected with red stele; *bottom,* strawberry roots showing red stele symptoms

21.3.4.2 Survival and Spread

The pathogen can survive in cuttings and spread to new plants, disease emergence favored by wet or moist soils and cool, wet weather conditions; younger plants generally show more damage to roots. The fungus thrives well in damp conditions; hence, plants should be grown in well-drained soil or raised beds.

21.3.4.3 Management

Improving soil drainage and reducing soil compaction will help to control red stele. Planting material that has been tested and certified free of the red stele fungus should be used. Applying appropriate soil fungicides after the fruiting period ends can protect the development of new roots during the post-fruiting period.

Cultural control of the fungus includes locating strawberry fields on well-drained soil, using raised beds to provide optimum drainage, and using less susceptible cultivars. Using drip irrigation and managing irrigation schedules to minimize soil saturation near plant crowns are key methods to reduce losses from this pathogen. Aliette and Ridomil Gold also provide good control of these diseases, although there is resistance to these materials.

In warmer areas, solarization has been shown to be effective for the control of soilborne pathogens and weeds. Solarization is carried out after the beds are formed and can be effective if weather conditions are ideal (30–45 days of hot weather that promotes soil temperatures of at least 122 °F). The effectiveness of solarization can be increased by incorporating the cruciferous crop residue (in particular broccoli or mustards) into the soil or application of Metam sodium (40 gal/acre) before solarization.

Only resistant cvs should be planted in a field where red stele is known to have caused losses within the last 5–10 years. Resistant cvs include Allstar, Earliglow, Guardian, Midway, Redchief, Scott, Sparkle, and Surecrop and the day-neutral cvs Tribute and Tristar.

21.3.5 *Verticillium* Wilt, *Verticillium dahliae*

21.3.5.1 Symptoms

Plants are most severely infected during the first year after planting. The outer leaves show interveinal browning and ultimately collapse (Fig. 21.6). Inner leaves are stunted but remain

Fig. 21.6 Symptoms of *Verticillium* wilt on strawberry

green until the plant dies. Initial symptoms appear rapidly in late spring, especially in conjunction with periods of environmental stress, such as drought and/or onset of high temperatures. Symptoms may continue through summer and fall, although disease spread from plant to plant during a growing season is probably minimal. The disease may affect plants in clusters of varying size, with healthy plants and diseased ones often interspersed.

21.3.5.2 Favorable Conditions

Cool, overcast weather interspersed with warm, bright days is most favorable for the development of *Verticillium* wilt. Infection and disease development may occur when soil temperature is from 21 °C to 24 °C.

21.3.5.3 Management

Remove the plant and its surrounding soil immediately when the symptoms are spotted. Crop rotation helps to avoid this problem, as does never planting strawberries in the same soil twice nor growing them on sites on which potatoes, tomatoes, or chrysanthemums have been cultivated.

A number of cultivars are moderately to highly resistant to *Verticillium* wilt. The following June bearing varieties are reported to be resistant to *Verticillium* wilt: Allstar, Catskill, Delite, Earliglow, Guardian, Lester, Redchief, Scott, Sunrise, Surecrop, and Tennessee Beauty. The everbearing varieties, Tribute and Tristar, are also reported to be resistant.

Biofumigation is another soil treatment that can reduce *Verticillium* numbers in the soil. Broccoli crop residues release chemicals that both directly reduce *Verticillium* propagules and affect the soil microbial diversity, which can suppress the pathogen. While mustards and other cruciferous plants show similar effects, broccoli appears to be one of the best choices for this soil biofumigation treatment. A crop rotation that includes broccoli will have the same suppressive effect, since the harvested broccoli florets are not needed for biofumigation to take place.

Rotations with broccoli and/or Brussels sprouts followed by incorporation of the crop residue for 2 years prior to planting strawberry resulted in reduced disease incidence.

Satisfactory chemical control has been obtained by preplant soil fumigation or by preplant soil drenches with several types of compounds.

21.3.6 Leaf Scorch, *Diplocarpon earlianum*

21.3.6.1 Symptoms

The earlier symptom is the appearance of numerous small, 1–5-mm diameter, purplish, irregularly shaped blotches on leaves (Fig. 21.7). The centers of these blotches become brownish, not white or gray as in common leaf spot. When blotches are numerous they will coalesce, and the entire leaf blade turns purplish or reddish. These discolored leaves later dry up and their margins

Fig. 21.7 Advanced scorch symptoms on strawberry leaves

curl up, giving a scorched or burned appearance. When conditions are particularly favorable for disease development, plants of susceptible cultivars can become seriously weakened and defoliated by late summer. In such situations, disease symptoms can appear on petioles, fruit peduncles, pedicels, calyxes, flowers, and fruits as well as leaves. Scorched calyxes ("dead cap" or "dead burr") can cause fruit to receive a lower market grade. Furthermore, leaf scorch can become progressively more damaging to perennial plantings of susceptible cultivars each year; yields may be severely reduced by the second year after the disease initially appeared. Disease emergence is favored by wet foliage for extended periods of time.

21.3.6.2 Management

- Choose sites with full sun and good soil drainage and air circulation. This fungus requires long periods of continuous wetness to infect plants; thus, any practice that promotes quick drying of leaves and fruit will reduce disease.
- Remove weeds to improve air circulation around plants.
- Plant in rows or narrow beds, no wider than 12–18 in., to promote good air movement in and around plants. Patches grow with time as new runners are produced. Use renovation to maintain narrow beds.
- Renovate strawberry beds every year after harvest. Following renovation, rake and remove old leaves.
- Irrigate through drip irrigation or a soaker hose. If overhead sprinkling is your only option, water early in the morning on a sunny day so leaves dry quickly after irrigation.
- Apply nitrogen fertilizers after renovation. Avoid early spring applications of nitrogen which encourage overly lush growth which reduces air flow promoting a moist microclimate in the canopy that is optimal for the growth of leaf scorch.
- Choose disease-resistant cultivars suitable for your location.
- Plant only disease-free plants purchased from reliable nurseries.
- Remove older or infected leaves before setting runners in new plantings.
- Removing and burning all debris at renovation (after harvest) helps to reduce overwintering inoculum of leaf pathogens.
- Follow a fungicide spray schedule recommended for control of leaf diseases and fruit rots to aid in control. Thoroughly cover all aboveground plant parts with spray, especially undersides of leaves.

21.3.7 Angular Leaf Spot, *Xanthomonas fragariae*

21.3.7.1 Symptoms

Very small water-soaked lesions on lower surfaces of leaves which enlarge to form dark-green or translucent angular spots which ooze bacteria; lesions may coalesce to form reddish spots with a chlorotic halo (Fig. 21.8).

21.3.7.2 Survival and Spread

The bacterium survives in crop debris and overwintering plants and can survive for long periods on plant debris but cannot live free in soil; bacteria can be spread by splashing water.

Fig. 21.8 Angular leaf spot lesions on strawberry leaf

21.3.7.3 Management

Use only certified planting stock; rotate crops and avoid overhead irrigation, chemical controls generally ineffective. Preventive, weekly applications of copper fungicides at 0.3 lb of metallic copper per acre (such as copper hydroxide, copper oxychloride, basic copper sulfate, cuprous oxide) were effective in reducing disease symptoms without phytotoxicity to the plants.

Some of the common diseases of strawberry and their symptoms and management are presented in Table 21.1.

Commonly grown disease-resistant strawberry cultivars are presented in Table 21.2.

Table 21.1 Strawberry fruit rot diseases and their management

Disease	Symptoms on berry	Cultural control	Chemical control
Gray mold fruit rot	Turns light brown; no sharp border with healthy tissue. Becomes covered by velvety gray growth	Space plants properly. Avoid fertilizing in spring. Mulch. Minimize cultivation until harvest	Spray with mix of systemic and protectant fungicides from bloom until harvest
Anthracnose fruit rots (black spots)	Tan to light-brown lesions; become circular, tan to dark brown, compact, shrunken. Pink to buff spore masses in humid weather. Attack ripe or green fruits	Mulch. Remove plant debris. Plant resistant cultivars	Spray protective fungicides from bloom until harvest
Rhizopus rot (leak)	Gradually turns light brown. White, fluffy fungal growth with black spore bodies (pinhead size) in humid weather. Rapid softening, collapse, leakage	Cool fruit rapidly after harvest. Store fruit below 43 °F	Spray protectant broad-spectrum fungicides or mixtures during ripening
Rhizoctonia fruit rot (hard rot)	Soil adheres to infected area. Infected area brown. Only ripe berries are attacked. Rot occurs on the side of fruit touching soil	Mulch	Protectant fungicides when disease appears on fruits grown through plastic mulch
Powdery mildew	Immature berries won't ripen. Ripe berries – white fungal growth on the surface. Attacks blossoms and leaves also	Resistant cultivars. Plant in areas with full sun and good air circulation	Protectant and systemic fungicides during growing season in places where powdery mildew is a problem
Stem-end rot	Discoloration at calyx end; distinct line between healthy and diseased tissue. Small green fruits – don't ripen, browning. Large green fruits – pale red or brownish. Ripe fruits – little color change. Soft, watery fruits	Resistant cultivars. Field cleanup at renovation time (weeds favor this disease)	Spray as for gray mold fruit rot
Leather rot	Dull, lifeless green fruits – infected areas brown to dark brown. Maturing fruits – infected areas bleached to light purple. Infected areas tough; edges slightly softened. Bitter taste	Mulch	Mixture of systemic and protectant fungicides

Table 21.2 Disease-resistant strawberry cultivars commonly grown in Illinois

Cultivar	Reaction of strawberry cultivars to different diseases[a]				
	Red stele	*Verticillium* wilt	Leaf spot	Leaf scorch	Powdery mildew
Allstar	VR[a]	R	R	R	R
Canoga	I	I	R	R	–
Cardinal	S	S	R	R	R
Catskill	S	VR	S	R	R
Delite	R[b]	R	R	S-R	S
Earliglow	R[b]	T-R	S-R	R	I
Guardian	R[b]	T-R	S-R	R	S-R
Honeoye	S	S	R	R	–
Jewel	S	S	R	R	–
Lester	R	R	R	R	R
Midway	R[b]	S-I	S	S	I
Pocahontas	S	S	S-R	S-I	R
Raritan	S	S	S	S	S
Redchief	R[b]	R	S-R	R	S-R
Scott	R	S-R	S-R	R	R
Sparkle	S-R	S	S	S-I	R
Sunrise	R[b]	R	VS	R	R
Surecrop	R[b]	VR	S-R	S-R	–
Tennessee Beauty	S	R	R	S-R	S
Everbearing					
Tribute	VR	T-R	T	T	R
Tristar	R	R	T	T	R

[a]*VS* very susceptible, *S* susceptible, *I* intermediate, *T* tolerant, *R* resistant, *VR* very resistant, – unknown. Resistant characteristics of the cultivar usually preclude the need for other controls
[b]Resistant to several races of the red stele fungus

21.4 Insect and Mite Pests

Several insects injure strawberries either directly by attacking the fruit or indirectly by reducing plant vigor. Unless control measures are followed, damage may be severe enough to drastically reduce yields and fruit quality. Insect and mite pressure varies according to location and season. While the single most serious pest of strawberries is the two-spotted spider mite, armyworms, aphids, and whiteflies can also cause concern during this stage of strawberry development. Some general predators, such as big-eyed bugs, minute pirate bugs, damsel bugs, spiders, and ladybird beetles, provide limited biological control of these pests.

The most common insect and mite pests and the damage symptoms are as follows.

21.4.1 Spider Mites, *Tetranychus urticae*, and *T. turkestani*

21.4.1.1 Symptoms

The two-spotted spider mite (Fig. 21.9), a key pest of strawberries in all California growing areas, expresses its damage as stippling, scarring, and bronzing of the leaves and calyx. Foliage develops yellow flecking and mottling and fine webbing between leaves. The other symptoms include webbing covering leaves; mites may be visible as tiny moving dots on the webs or

Fig. 21.9 Spider mite on strawberry leaf

underside of leaves, best viewed using a hand lens, usually not spotted until there are visible symptoms on the plant; leaves turn yellow and may drop from plant.

21.4.1.2 Survival

Spider mites thrive in dusty conditions; water-stressed plants are more susceptible to attack.

21.4.1.3 Management

Spraying plants with a strong jet of water can help reduce buildup of spider mite populations; if mites become problematic, apply insecticidal soap to plants; chemical insecticides may actually increase mite populations by killing off natural enemies and promoting mite reproduction.

Conserve or release natural enemies like predatory mites. Biological control agent *Phytoseiulus persimilis* (Fig. 21.10) gives effective control of spider mites if introduced early. It is a specialist predator and is effective early in the season. *Neoseiulus californicus* (Fig. 21.10), which tolerates warmer temperatures and a wide range of humidities, is a generalist predator and a predominant species later in the season. There are several other natural enemies like minute pirate bug (*Orius tristicolor*); a small, black lady beetle (*Stethorus* spp.); a small, black rove beetle (*Oligota oviformis*); big-eyed bugs (*Geocoris* spp.); brown lacewings (*Hemerobius* spp.), green

Fig. 21.10 Spider mite predators

lacewings (Chrysopa spp.); six-spotted thrips (*Scolothrips sexmaculatus*); damsel bugs (*Nabis* spp.); a cecidomyiid fly maggot (*Feltiella acarisuga*); and a predaceous midge.

Horticultural oil kills by suffocation and is effective against all stages of mites, including eggs. Some miticides are only active against the mobile stages of mites. These would include Dicofol (Kelthane), abamectin (Agri-Mek), Bifenazate (Acramite), and Vendex (Fenbutatin-oxide). Hexythiazox (Savey) and similar miticides not yet registered for use in strawberries kill the mite eggs and some of the hexapod nymphs and sterilize adult females.

The following chemicals also give effective control of spider mites:

- Agri-Mek is a fairly effective miticide.
- Savey, an ovicide, provides very good control of eggs and is a good tool for IPM programs.
- Acramite is also a good material for use in IPM.
- Brigade and Danitol provide good mite control.
- Bifenthrin, plant oils, and fatty acids make a good alternative.

21.4.2 Armyworm, *Spodoptera exigua*

21.4.2.1 Symptoms

The greatest damage from beet armyworm occurs to summer- and fall-planted strawberries. Egg clusters of 50–150 eggs may be present on the

Fig. 21.11 Beet armyworm larva

leaves; egg clusters are covered in a whitish scale which gives the cluster a cottony or fuzzy appearance. Newly hatched beet armyworms are foliage feeders, skeletonizing the upper or lower leaf surfaces adjacent to their egg mass. Young larvae feed on foliage before attacking fruit. Larger larvae can attack the crowns of young plants and kill them. Young larvae are pale green to yellow in color, while older larvae are generally darker green with a dark and light line running along the side of their body and a pink or yellow underside (Fig. 21.11).

21.4.2.2 Biology

An insect can go through 3–5 generations a year.

21.4.2.3 Management

Weed control in and near the field minimizes armyworm populations because the adult moths are attracted to weeds for egg laying. Organic methods of controlling armyworms include biological control by natural enemies which parasitize the larvae and the application of *Bacillus thuringiensis.* Biological control is provided by the ichneumonid parasite, *Hyposoter exiguae*, and armyworms often become diseased with a virus that can cause high mortality.

Lorsban provides fair to good control of worm pests in strawberries and has only a moderate effect on beneficial insect and mite species. Lannate provides excellent control of cutworms and beet armyworms, and Brigade and Danitol also provide good worm control.

21.4.3 Aphids (Peach Aphid, Potato Aphid, Melon Aphid, and Strawberry Aphid), *Myzus persicae, Macrosiphon euphorbiae, Aphis gossypii,* and *Chaetosiphon fragaefolii*

21.4.3.1 Symptoms

Ahids are small soft-bodied insects on the underside of leaves and/or stems of plant, usually green or yellow in color (Fig. 21.12) but may be pink, brown, red, or black depending on species and host plant. Heavy aphid infestation cause leaves to turn yellow and/or distorted and have necrotic spots on leaves and/or stunted shoots. Aphids secrete a sticky, sugary substance called honeydew which encourages the growth of sooty mold on the plants.

Aphids transmit several viruses which can be very damaging to strawberry; the insects rarely reach a high enough population to cause severe damage directly, but the spread of viruses is a major concern in strawberry production.

21.4.3.2 Management

If aphid population is limited to just a few leaves or shoots, then the infestation can be pruned out to provide control; check transplants for aphids before planting; use tolerant varieties if available; reflective mulches such as silver-colored plastic can deter aphids from feeding on plants; sturdy plants can be sprayed with a strong jet of water to knock aphids from leaves; insecticides are generally only required to treat aphids if the infestation is very high – plants generally tolerate low- and medium-level infestation; insecticidal soaps or oils such as neem or canola oil are usually the best method of control.

Cultural controls for aphids include row covers and dust reduction in and around fields. Removing weeds in strawberry and nearby fields that harbor aphids also will help to manage populations.

Biological and natural control of aphids is accomplished using parasitic wasps, lacewings (especially nymphs), big-eyed bugs, minute pirate bugs, damsel bugs, and ladybugs; ladybugs are often released into fields to control aphids. In

Fig. 21.12 Strawberry aphid

Fig. 21.13 High numbers of greenhouse whitefly on strawberry leaf

some circumstances biological control is sufficiently effective to provide an economically viable alternative. Thus, growers should monitor aphid populations and avoid applications of insecticides/miticides that will destroy natural enemies.

When aphid pest pressure reaches a trigger level (sometimes when 30 % of young trifoliate leaves are infested), chemical treatments are made. Diazinon, malathion, and Lorsban, all organophosphate insecticides, provide only fair control of aphids. Azadirachtin works only fairly well, requires multiple applications, and is very expensive.

21.4.4 Whiteflies, *Trialeurodes vaporariorum*

21.4.4.1 Symptoms

Whiteflies are always present in strawberry fields (their occurrence has become more widespread in the past few years for unknown reasons) (Fig. 21.13). Whiteflies suck plant juices and, at high population levels, can excrete large amounts of honeydew on which a sooty mold fungus grows. Occasionally populations build up to damaging numbers causing economic loss.

21.4.4.2 Management

An area-wide approach to whitefly management is essential for control because of the broad host range and overlapping of crop host cycles. This approach involves breaking the continuous generation cycles of this pest by cultural or chemical means. For example, multiple-year plantings should be avoided in areas where whiteflies are especially problematic, and hosts which are no longer being harvested should be destroyed before new strawberry plantings are established. Trap crops have proven only marginally effective and can even serve as bridge hosts; the use of sticky tapes as a control measure is not very effective. Released parasitic wasps are costly and few survive in strawberry fields.

Admire provides moderately good control of whiteflies. Lannate provides fair control of adult whiteflies, but coverage is an issue since this is a contact insecticide. Danitol in combination with malathion provides only fair control of adult whiteflies, and only a limited number of applications may be made; also, these materials are harsh on beneficials.

21.4.5 Japanese Beetle, *Popillia japonica*

21.4.5.1 Symptoms

Leaves skeletonized (only veins remaining); flowers and buds damaged; plant damage may be extensive; adult insect is a metallic green-bronze beetle with tufts of white hair protruding from underwing covers on each side of the body; adult beetles are approximately 13 mm in length; larvae are cream-white grubs which develop in the soil (Fig. 21.14).

Fig. 21.14 Japanese beetle damage on strawberry leaf

Fig. 21.15 Cabbage looper damage

21.4.5.2 Biology

One beetle generation every 1–2 years; pheromone traps may actually attract more beetles to home gardens and should generally be avoided; a beetle overwinters as larvae in soil; a beetle has an extensive range of over 300 host plants.

21.4.5.3 Management

If beetles were a problem in the previous year, use floating row covers to protect plants or spray kaolin clay; adult beetles can be handpicked from plants and destroyed by placing in soapy water; parasitic nematodes can be applied to soil to reduce the number of overwintering grubs; insecticidal soaps or neem oil can help reduce beetle populations.

21.4.6 Loopers (Cabbage Looper), *Trichoplusia ni*

21.4.6.1 Symptoms

Young larvae feed primarily on the undersides of leaves, skeletonizing them. High populations can damage fruit but this is very uncommon. Large or small holes in leaves; damage often extensive; caterpillars are pale green with white lines running down on either side of their body; caterpillars are easily distinguished by the way they arch their body when moving (Fig. 21.15); eggs are laid singly, usually on the lower leaf surface close to the leaf margin, and are white or pale green in color. Insects overwinter as pupae in crop debris in soil; an adult insect is a dark-colored moth; caterpillars have a wide host range.

21.4.6.2 Management

Looper populations are usually held in check by natural enemies; if they do become problematic, larvae can be handpicked from the plants.

Loopers are commonly controlled by parasitic wasps *Hyposoter exiguae, Copidosoma truncatellum,* and *Trichogramma* spp. and by outbreaks of nuclear polyhedrosis virus.

An organically acceptable control method is the application of *Bacillus thuringiensis* which effectively kills younger larvae. Chemical sprays may damage populations of natural enemies and should be selected carefully.

21.4.7 Thrips (Western Flower Thrips), *Frankliniella occidentalis*

21.4.7.1 Symptoms

Thrips feeding on strawberry blossoms causes the stigmas and anthers to turn brown and wither prematurely, but not before fertilization has occurred. As fruit develops, thrips feeding may cause a russeting (type I bronzing) of the fruit around the cap, but this injury is seldom economic. Damage to strawberry flowers, including browning of anthers and stigmas; insects may be numerous on fruit and may cause bronzing around the cap; the insect is small (1.5 mm) and slender and best viewed using a hand lens; adult thrips are pale yellow to light brown and the nymphs are smaller and lighter in color (Fig. 21.16).

Fig. 21.16 Western flower thrips

21.4.7.2 Survival and Spread

Thrips populations build up in spring on weeds and a number of crops and migrate to strawberries when crops are harvested or when weeds dry up.

21.4.7.3 Management

Avoid planting next to onions, garlic, or cereals where very large numbers of thrips can build up; use reflective mulches early in growing season to deter thrips; apply appropriate insecticide if thrips become problematic.

Naturally occurring minute pirate bugs *(Orius* spp.) feed on thrips. *Orius* spp. are also available commercially.

21.4.8 Weevils (Strawberry Root Weevil, Rough Strawberry Root Weevil, and Black Vine Weevil), *Otiorhynchus* spp.

21.4.8.1 Symptoms

Adult vine weevils eat notches in the edges of leaves, while plump, creamy white larvae with brown heads cause more damage to the roots, on which they feed. This can kill the plants. Plants are stunted and dark in color with closely bunched leaves. An adult insect is a snot-nosed beetle which can vary in color from dark to light brown depending on species; larvae are cream-white-colored grubs which feed on roots below ground (Fig. 21.17). Damage within a field usually affects a circle of plants; weevil attack is sporadic and can be very damaging.

21.4.8.2 Management

Keep strawberry beds free of weeds and grass; avoid planting close to woodland or blackberry or elderberry which may harbor weevil populations; pesticide sprays or dusts which contain pyrethroids are effective at controlling strawberry weevils. Apply chemical control Acetamiprid or Thiacloprid as a liquid drench applied to the compost or the biological control, nematode *Steinernema kraussei*.

21.4.9 Slugs, *Arion hortensis*, and *Deroceras reticulatum*

21.4.9.1 Symptoms

Slugs feed on ripe fruit and produce rough holes that render the fruit unmarketable. Slugs also feed on the leaves of strawberries, and the effects of the rasping feeding are ragged holes in the leaves. Irregularly shaped holes in leaves and stems; rough holes in ripe fruit (Fig. 21.18); if infestation is severe, leaves may be shredded; slime trails present on rocks, walkways, soil, and plant foliage; several slug and snail species are common garden pests; slugs are dark gray to black in color and can range in sizes from 2.5 cm to 10 cm. Slugs prefer moist, shaded habitats and will shelter in weeds or organic trash; adults may deposit eggs in the soil throughout the season; damage to plants can be extensive.

21.4.9.2 Management

Practice good garden sanitation by removing garden trash, weeds, and plant debris to promote good air circulation and reduce moist habitat for slugs; handpick slugs at night to decrease population; spread wood ashes or eggshells around plants; attract mollusks by leaving out organic matter such as lettuce or grapefruit skins, destroy any found feeding on lure; sink shallow dishes filled with beer into the soil to attract and drown the mollusks; chemical controls include ferrous phosphate for organic gardens and metaldehyde (e.g., Bug-Geta) and carbaryl (e.g., Sevin bait) for nonorganic growers.

Fig. 21.17 Rough strawberry grub and weevil

Fig. 21.18 A common garden slug

22 Tomato

Abstract

Improved flavor and shelf life are the driving force for the increased demand for greenhouse-grown tomatoes. Greenhouse tomatoes are harvested and marketed over a 30-week period. The major insect and mite pests (whitefly, serpentine leaf miner, red spider mite, tobacco caterpillar, thrips, fruit borer) and diseases (early blight, late blight, gray mold, leaf mold, powdery mildew, leaf spot, root rots, bacterial canker, soft rots, pith necrosis, bacterial speck, bacterial spot, tobacco mosaic virus, tomato mosaic virus, cucumber mosaic virus, tomato yellow leaf curl virus, tomato spotted wilt virus) and root-knot nematodes and their symptoms, biology, spread, survival, and management are discussed.

Keywords

Tomato • Pests • Diseases • Nematodes • Symptoms • Spread • Biology • Management

22.1 Introduction

Productivity of tomato in greenhouse (Fig. 22.1) is increased manifold in comparison to open field (Tables 22.1 and 22.2). Improved flavor and shelf life are the driving force for the increased demand for greenhouse-grown tomatoes. Consumers love the taste of a fresh tomato, and they are willing to pay a premium price, especially in the winter. Unlike field-grown tomatoes which require harvesting and marketing over a short period of time (3–4 weeks), greenhouse tomatoes are harvested and marketed over a 30-week period.

22.2 Varieties

Some of indeterminate elite varieties/hybrids which have performed exceedingly under low-cost greenhouses are DARL-303, HT-6 Sun-7611, NS-1237, Naveen, NS-4130, Abiman, COTH-1, NDT-5, NDT-120, Pusa Divya, Meenakshi, Himsona, Badshah, Vaishali and Lakshmi. Two varieties are suitable for greenhouse conditions developed from Pantnagar, viz., Pant polyhouse bred tomato-2 and Pant polyhouse hybrid tomato-1.

P.P. Reddy, *Sustainable Crop Protection under Protected Cultivation*,
DOI 10.1007/978-981-287-952-3_22

22.3 Insect and Mite Pests

22.3.1 Whitefly, *Bemisia tabaci*

22.3.1.1 Symptoms

Whitefly infesting vegetables under greenhouse conditions is a vector of tomato leaf curl virus (Fig. 22.2). Adults and nymphs suck the sap from leaves resulting in curling of leaves (Fig. 22.2). If the infestation is severe, sooty mold develops on the leaves due to excretion of honeydew by the whitefly, which reduces the photosynthetic activity of the plant and yield gets reduced drastically.

22.3.1.2 Management

- Raise the seedlings in polyhouse using plastic seedling trays.
- Spray imidacloprid 200 SL (0.3 ml/L) or thiamethoxam 25 WP (0.3 g/L) in nursery at 15 days after sowing.
- Drench the base of the seedlings with imidacloprid 200 SL (0.3 ml/L) or thiamethoxam 25 WP (0.3 g/L) before transplanting.
- Remove the leaf curl-infested plants as soon as disease symptoms are expressed. This helps in reducing the source of inoculum of the disease.
- After transplanting give need-based sprays of imidacloprid 200 SL (0.3 ml/L) or thiamethoxam 25 WP (0.3 g/L) at 15 days after planting, and do not repeat after fruiting stage as this may leave harmful residues in fruits.
- Install yellow sticky traps coated with adhesive or sticky glue at crop canopy level for monitoring adult whitefly population.
- If the traps indicate the whitefly activity, spray dimethoate 30 EC at 2 ml/L or neem seed kernel extract 4 % (NSKE) or pongamia or neem oil (8–10 ml/L) or neem soap (10 g/L).
- Inoculative releases of parasitic wasp *Encarsia formosa* and *Eretmocerus eremicus* have been used successfully to control greenhouse whitefly. *Delphastus pusillus* is a small beetle that feeds on whitefly eggs and is ideal for complementing *Encarsia* and *Eretmocerus* (Portree 1996).

Fig. 22.1 Tomato under protected cultivation

22.3.2 Serpentine Leaf Miner, *Liriomyza trifolii*

22.3.2.1 Symptoms

Serpentine leaf miner is a tiny metallic fly with a characteristic yellow patch behind the compound eye. Damage starts in nursery itself. Initially the adult female punctures the developing leaf and feeds on the oozing sap. Larvae mine leaves in a serpentine manner (Fig. 22.3) and on pupation they drop to soil. Severe damage leads to drying and dropping of leaves.

Table 22.1 Performance of tomato varieties under polyhouse and open-field conditions in NEH region (Barapani)

Varieties	Polyhouse yield (t/ha)	Open-field yield (t/ha)	Varieties	Polyhouse yield (t/ha)	Open-field yield (t/ha)
BT-117-5-3-1	34.200	11.500	Selection-2	23.300	7.383
KT-10	28.360	11.740	Selection-1	20.098	8.403
BT-10	29.400	11.165	KT-15	21.160	5.165
Arka Alok	26.000	5.790	H-24	14.317	5.875
BT-12	30.240	10.100	Arka Abha	19.350	7.033

Table 22.2 Yield of tomato under greenhouse at different locations in India

Location	Yield (t/ha)
Pune	124
Coimbatore	186
Bangalore	152
Solan	95
Hisar	211

22.3.2.2 Management

- Apply neem cake at 250 kg/ha to beds while planting and repeat after 25 days to prevent pupal emergence from soil.
- Spray neem seed powder extract (4 %) or neem soap 1 %.
- Frequent spraying of synthetic pesticides should be avoided as it may cause resurgence of the pest. At the most, one spray of deltamethrin 2.8 EC at 1 ml/L or cypermethrin 25 EC at 0.5 ml/L or triazophos 40 EC at 2 ml/L may be given if required.

22.3.3 Fruit Borer, *Helicoverpa armigera*

22.3.3.1 Symptoms

This is a major problem in improperly maintained polyhouses. Eggs are laid on young leaves. Initially larvae feed on tender leaves and scrape the leaf surface; later they bore into fruits (Fig. 22.4). Generally the posterior end of the body remains outside the hole. The damaged fruits become unmarketable.

22.3.3.2 Management

- Spray *Ha*NPV at 250 LE/ha+1 % jaggery along with sticker (0.5 ml/L) during evenings when the larvae are young.
- For grown-up larvae spray indoxacarb 14.5 SC at 0.5 ml/L or thiodicarb 75 WP at 1 g/L.

22.3.4 Tobacco Caterpillar, *Spodoptera litura*

22.3.4.1 Symptoms

This pest is also important in ill-managed polyhouses. Eggs are laid in clusters on foliage. Young larvae feed gregariously on leaves. Mature larvae migrate and cause extensive damage to leaves and fruits. They hide in soil and crop debris during daytime.

22.3.4.2 Management

- Collection and destruction of egg masses and gregarious larvae.
- Spray *Spodoptera* NPV 250 LE/ha+1 % jaggery along with sticker (0.5 ml/L) during evenings.
- Use poison baiting. Mix 10 kg of rice bran or wheat bran with 2-kg jaggery by adding a little water in the morning. In the evening add 250 g of methomyl or thiodicarb formulation and sprinkle over the bed. Caterpillars get attracted to fermenting jaggery, feed, and get killed (Fig. 22.5).

22.3.5 Thrips, *Frankliniella occidentalis*, and *Thrips tabaci*

There are two species of thrips that are common pests in greenhouse vegetable crops, the western flower thrips (*Frankliniella occidentalis*) and the onion thrips (*Thrips tabaci*) (Howard et al. 1994; Portree 1996).

22.3.5.1 Symptoms

Thrips (Fig. 22.6) feed by opening wounds on the plant surface and sucking out the contents of the plant cells; the feeding results in small whitish streaks on the leaves and fruit and can cause distortions in the young developing fruit (Howard et al. 1994; Portree 1996).

The adult thrips congregate in the flowers and regular monitoring of the flowers will allow for the early detection of thrips. Yellow and/or blue sticky traps placed throughout the crop, as with

Fig. 22.2 *Left* – whiteflies on tomato leaf. *Right* – tomato leaf curl disease vectored by whiteflies

Fig. 22.3 Serpentine leaf miner damage on tomato leaves

Fig. 22.4 Tomato fruit borer damage

the other insect pests, will help in the early detection of thrips infestations. Avoid using yellow traps if *Aphidius* spp. are being used for the control of aphids in the crop.

In addition to causing direct feeding damage and resultant yield loss, both thrips species are vectors of tomato spotted wilt virus (TSWV) which can be a serious disease problem in peppers and tomatoes (Howard et al. 1994; Portree 1996). One of the main control measures for minimizing the spread and infection of TSWV within the crop is to control the thrips vectors.

22.3.5.2 Management

There are a number of predators available for biological control of thrips: predatory mites *Amblyseius degenerans, A. cucumeris, Hypoaspis miles*, and *H. aculeifer* and predatory bugs, *Orius insidiosus*, and other *Orius* species.

22.3.6 Red Spider Mite, *Tetranychus urticae*

22.3.6.1 Symptoms

Red spider mites are highly polyphagous. They can cause serious damage by their fast multiplication rate in a very short time. They thrive under high temperature and dry weather. They are generally found on the lower surface of older leaves. However, when the infestation is very high, they attack all parts of the plant and are observed in colonies covered by white-silky webs (Fig. 22.7). Adults and nymphs lacerate the leaves causing yellowing and discoloration. Drying and dropping of attacked leaves is also observed when the infestation is severe.

Fig. 22.5 Dead tobacco caterpillar after baiting

Fig. 22.6 The western flower thrips

Fig. 22.7 Red spider mite damage on tomato leaves

22.3.6.2 Management

- Remove the mite-infested leaves and destroy before spraying.
- Spray need-based application of acaricides like abamectin 1.9 EC at 0.5 ml/L or dicofol 18.5 EC at 2.5 ml/L or Fenazaquin 10 EC at 1 ml/L in rotation with plant products like *Pongamia* oil or neem oil (8–10 ml/L) or neem soap (10 g/L).

Note Botanicals have very short residual toxicity, particularly when the temperature is high. In addition, these may cause phytotoxicity. Hence, sprays with botanicals should be done with caution and sprays to be given in the evenings when the temperature is low.

22.3.7 Flow Chart for Insect Pest Management

The calender of operations for integrated pest management of tomato is presented in the flow chart (Fig. 22.8).

22.4 Diseases

High humidity and warm temperatures are congenial for disease development in greenhouse. Common diseases of greenhouse tomato are as follows:

22.4.1 Early Blight, *Alternaria solani*

22.4.1.1 Symptoms

Early blight of tomato is a common foliar disease of tomato. This disease causes direct losses by the infection of fruits and indirect losses by reducing plant vigor.

A collar rot of the young plants before or after transplanting may be the first symptom. In mature plants small irregular brown spots, with or without a yellow halo and concentric rings, appear mainly on leaves. Severely infected leaves are ragged and senescent. Similar spots without a yellow ring appear along the stem, leafstalks, peduncles, and calyx. On fruits, brown to black spots with a leathery surface appear at the stem end (Fig. 22.9). Severely infected plants may be defoliated (Sherf and McNab 1986).

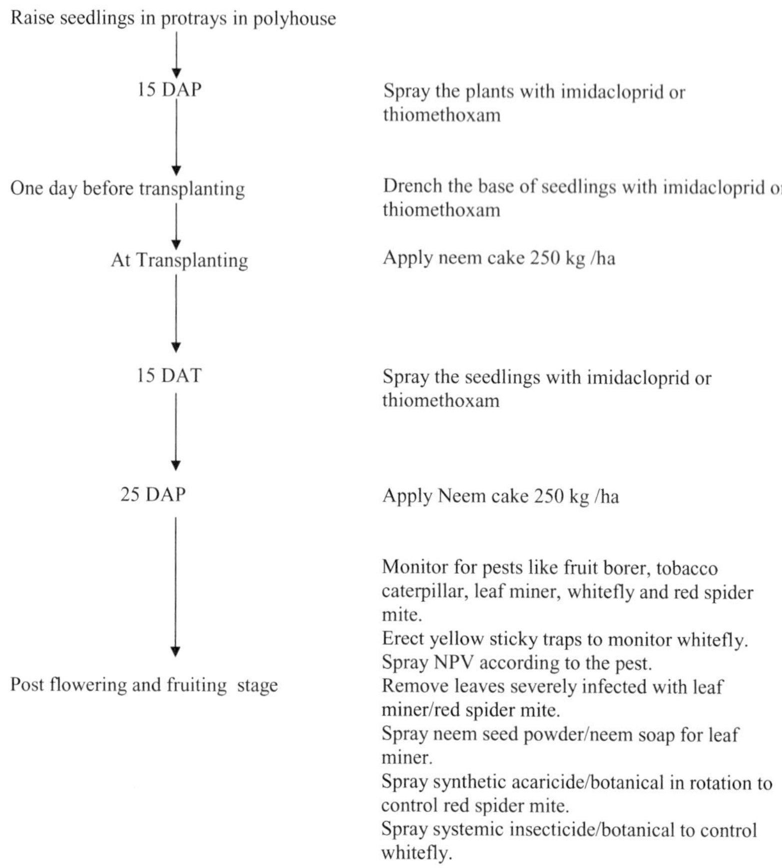

Fig. 22.8 Flowchart for insect pest management in tomato

22.4.1.2 Management

The disease can be managed by spray of Bordeaux mixture (1:1:100).

22.4.2 Late Blight, *Phytophthora infestans*

22.4.2.1 Symptoms

Late blight of tomato is the most devastating disease. It infects and destroys the leaves, stems, and fruits (Fig. 22.10). Its infection is most rapid during conditions of high moisture and moderate temperatures (20–25°C).

22.4.2.2 Management

- Nursery soil treatment with *Trichoderma* culture at 2 %.
- Soil drenching with *Trichoderma* culture at 2 % before planting.
- Seedling treatment with *Trichoderma* culture at 10 % for 15 min.
- Soil drenching with Bordeaux mixture (1:1:100) after 15 days of planting.
- Spray of Bordeaux mixture (1:1:100) at fortnightly interval after 1 month of planting.
- In severe infection, application of Ridomil/Matko at 2 g/L manages the disease effectively.

22.4.3 Gray Mold, *Botrytis cinerea*

22.4.3.1 Symptoms

Gray mold is an important disease of tomato and its infection occurs after flowering and fruit set. In case of severe infection, flowers and

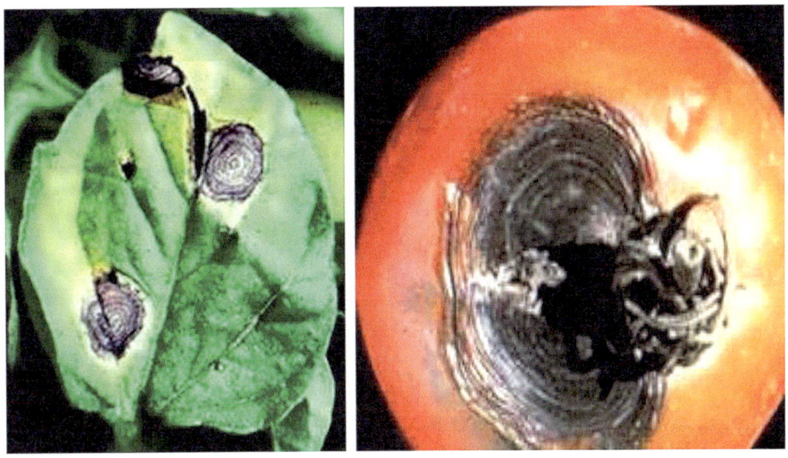

Fig. 22.9 Symptoms of early blight on tomato leaf and fruit

Fig. 22.10 Symptoms of late blight on tomato leaves and fruits

pedicel start dying and in later stage fruits also start rotting. The infection starts from the peduncle end and infected fruits are not suitable for consumption. A light-gray fuzzy growth appears on stems and leaves. Soft rot of the stem end of the fruit can also occur (Fig. 22.11). Gray mold is most severe in greenhouses with moderate temperatures, high humidity, and stagnant air. Gray mold development is favored by cool and humid conditions. These conditions also stress the plants, making them susceptible to disease.

22.4.3.2 Survival and Spread

The gray mold fungus survives on plant debris and in the soil. The source of contamination to a new crop is from spores carried by wind coming from host plants outside the greenhouse or spores spread on air currents in the greenhouse.

22.4.3.3 Management

Increasing ventilation and air circulation to reduce humidity levels can be helpful, as well as timely fungicide applications. Spray Bordeaux mixture (1:1:100) after 1 month of planting. In case of severe infestation, a spray of Kavach (chlorothalonil) at 1.5 g/L minimizes the infection.

Plants should be supplied with adequate calcium by liming acidic soils and maintaining uniform soil moisture. A calcium-to-phosphorus ratio of 2:1 or higher in leaf petiole tissue decreased plant susceptibility.

Fig. 22.11 Gray mold symptoms on tomato

22.4.4 Leaf Mold, *Fulvia fulva (Cladosporium fulvum)*

22.4.4.1 Symptoms

This causes light-green to yellow spots on the upper surface of mature leaves in humid greenhouses with poor air circulation. Soon the sporulating fungus growth appears as an olive-green velvety growth on the underside of the yellow spots (Fig. 22.12). These leaves drop prematurely as the disease progresses upward on the plant. The pathogen survives for several months on the greenhouse frame, on the materials used for cropping, and in plant debris. It is disseminated by wind or splashed by waterdrops. The optimal temperature for infection is 20–25 °C. If weather conditions are favorable, leaf mold has several cycles in a season and can destroy the crop completely. There are several races of the pathogen (Jones et al. 1993).

22.4.4.2 Spread and Survival

Wind, splashing irrigation water and rain, workers, tools, and insects readily disseminate spores. Contaminated seed can also initiate epidemics. The pathogen survives between tomato crops in and on crop residues and in the soil as spores (conidia) or dormant resting structures (sclerotia).

22.4.4.3 Management

Lowering greenhouse humidity, planting resistant varieties, and applying fungicide promptly can be helpful in leaf mold management.

Eliminate crop debris and culls by deep tillage soon after harvest to reduce pathogen survival. Promote rapid drying of leaf foliage by avoiding overhead irrigation and dense plantings. Planting parallel to the prevailing wind direction can hasten leaf drying and reduce leaf mold severity. In greenhouse production, thoroughly sterilize production areas by steam to eliminate *F. fulva*. Minimize wetting the foliage when irrigating tomatoes. Maintain night temperatures higher than outside temperatures.

Fungicides (Captan, copper fungicides, Famoxadone/cymoxanil) control leaf mold but should be used in combination with as many cultural control strategies as possible to be most effective.

22.4.5 Powdery Mildew, *Oidium neolycopersici*

22.4.5.1 Symptoms

It causes powdery white lesions on the adaxial tomato leaf surface (Fig. 22.13). The fungus can also infect abaxial surfaces, petioles, and the calyx but the fruit remains uninfected. Severe infections lead to leaf chlorosis, premature senescence, and a marked reduction in fruit size and quality. In severe outbreaks the lesions coalesce and the disease is debilitating.

22.4.5.2 Management

- Increase air circulation and spacing between plants.
- Fungicide sprays also can be effective if used when symptoms are first noticed. Effective active ingredients include benomyl, Bitertanol,

Fig. 22.12 Leaf mold symptoms on tomato

Fig. 22.13 Powdery mildew on tomato leaf

Bupirimate, carbendazim, Fenarimol, pyrazophos, thiabendazole, Triforine, and various sulfur preparations.

- Foliar sprays of *Sporothrix flocculosa* reduce the development of *O. neolycopersici* on tomato in the glasshouse.

22.4.6 *Cercospora* Leaf Spot, *Cercospora* sp.

22.4.6.1 Symptoms

Infected leaf has brown spots with curled leaf margin and later on, it withers. As the disease progresses, the leaflets turn yellow and also with curled margins. Spots on the leaf petioles, stems, and flower parts become elongated and have dark-brown color. In heavily infested plant, defoliation may occur.

22.4.6.2 Management

The disease can be managed by the application of Bordeaux mixture (1:1:100).

22.4.7 Corky Root Rot, *Pyrenochaeta lycopersici*

22.4.7.1 Symptoms

The pathogen damages mostly tomato but also eggplant, melon, etc. Initially tomato leaves turn dull green and growth is stunted. Later, leaves take on a bronze color and curl downward. Necrosis of the leaflets follows. Young roots are brown and poorly developed. Scattered lesions appear on the surface of the larger roots which become corky with cracks of different sizes (Fig. 22.14). Yield may be severely reduced.

22.4.7.2 Survival and Spread

The pathogen survives on the infected root debris due to the presence of minute sclerotia. It is a cool weather disease. In subtropical countries it progresses during the winter and plants start to recover by early spring (Malathrakis et al. 1983).

22.4.7.3 Management

A compost with a low concentration of ammonium nitrogen and a high concentration of calcium reduced corky root disease severity.

Fungivorous nematodes (*Aphelenchus avenae*) reduced corky root disease severity when added to *P. lycopersici*-infested soil.

In greenhouse trials, bacterial antagonists such as *Bacillus subtilis* and *Streptomyces graminofaciens* have been found to effectively suppress corky root disease of tomatoes and enhance plant growth, resulting in higher yields (Bochow 1989).

The commercial biofungicide Binab TF, based on the antagonists *Trichoderma polysporum* and *T. harzianum*, is primarily used in the greenhouse to control soilborne fungal diseases in tomato. Another product, Mycostop, a commercial formulation of the antagonist *Streptomyces griseo-*

Fig. 22.14 Symptoms of corky root disease on tomato

viridis strain K61, has proven effective against *P. lycopersici* when applied with irrigation water (Minuto et al. 2006).

22.4.8 Crown and Root Rot, *Fusarium oxysporum* f. sp. *radicis-lycopersici*

22.4.8.1 Symptoms

In plastic greenhouses, a yellowing of the lower leaves appears in infected plants during late winter, when many fruits have already set. In severe infections the whole plant becomes chlorotic and wilts (Fig. 22.15). A dry lesion up to 10 cm long appears on part of or all around the collar. There is a brown discoloration on the root system, predominantly at the end of the main root, the base of the stem, and the vascular region of the central root. A large number of microconidia, which disseminate the pathogen, appear on the infected stem. The fungus survives by chlamydospores which develop in the soil. The disease is favored by cool weather (Jarvis et al. 1983).

22.4.8.2 Management

Plant growth promotion and antifungal activity of *Paenibacillus* strains showed its suitability as potential bio-inoculant agent.

Under greenhouse conditions, the fungicide hymexazol reduces the percentage of dead plants at 6.4 %.

Fig. 22.15 Symptoms of crown and root rot on tomato

22.4.9 Bacterial Canker, *Clavibacter michiganensis* ssp. *michiganensis (Corynebacterium michiganense* ssp. *michiganense)*

22.4.9.1 Symptoms

Initially, infected plants show a sudden unilateral wilting of leaflets, entire leaves, or shoots. Young plants are more susceptible to wilting. Stem vessels at the side of the wilted leaves develop a yellow-brown discoloration. In the more severely infected places, the cortex splits and several centimeter-long cankers may develop. Such plants usually die prematurely. Systemic infection leads to yellow or brown discoloration of vascular strands and infected seeds are often shriveled and black. Bird's-eye spots, up to 6 mm in diameter, often appear on fruits (Fig. 22.16).

22.4.9.2 Survival and Spread

The pathogen is a typical seedborne organism. It can also survive for several months on cultivation equipment, on plant debris, and in the soil. It can also maintain large populations on leaves of tomato and other plant species. It may infect at 16–36 °C, with optimum at about 24–28 °C. It is disseminated by seed or transplants, which

Fig. 22.16 Symptoms of bacterial canker on tomato

remain symptomless until transplanted. In greenhouses, it spreads mostly during cultural practices (Gleason et al. 1993).

22.4.9.3 Management

In protected crops, strict hygiene measures such as early detection, isolation, and eradication of infected plants, destruction of crop residues, rinsing hands/gloves and pruning tools with a disinfectant after working each row, and disinfection of structures and equipment are essential to manage canker.

Improved management of bacterial canker in tomato has been shown to be possible by the use of healthy seeds, seed treatment, appropriate cultural practices, chemical sprays where needed, and hygiene and sanitation (Gleason et al. 1993).

A substantial reduction of infection can be achieved by acid extraction of seeds or treatment of seeds with acid or other disinfectants or hot water (Dhanvantari 1994).

Cultural practices such as deep plowing to bury infected crop residue after harvest to accelerate decomposition and crop rotation away from solanaceous crops for at least 2 years are recommended to reduce the incidence of canker (Gleason et al. 1991). Production of tomato transplants in greenhouses planted in soilless medium in plastic trays has been found to be feasible and more reliable.

Chemical sprays with copper-containing compounds have been found useful in reducing foliar blight and fruit spotting (Shoemaker 1992).

22.4.10 Soft Rots, *Erwinia carotovora* ssp. *carotovora*, *E. carotovora* ssp. *atroseptica*

22.4.10.1 Symptoms

Infected plants are stunted; their lower leaves show yellowing at the edges and on the veins and become flaccid. Initially the pith turns yellow to light brown, but later it disintegrates. The stem becomes hollow, splits, and may exude bacterial slime. Brown to black blotches may also appear along the stem and the leafstalks. A yellow to light-brown discoloration usually appears along the vascular system (Fig. 22.17). Plants with severe stem rot may wilt and die, but very often even plants with split stems survive and yield normally. Plants with lush growth, grown under conditions of high RH, are more susceptible. Infection starts from leaf scars on the lower part of the stem but may also appear in plants which have never been pruned (Malathrakis and Goumas 1987).

22.4.10.2 Host Range

The pathogen causes various rotting, necrotic, and systemic diseases of several floral crops, such as chrysanthemum, cyclamen, and *Saintpaulia*, in greenhouses.

22.4.10.3 Spread

The pathogen comes from affected stock plants and is disseminated by cultural practices.

Fig. 22.17 Symptoms of bacterial soft rot on tomato stem

22.4.10.4 Management

Infected plants should be discarded and knives disinfected (Fletcher 1984). Avoid injuring plants. Disinfect hands and tools when pruning tomato plants and wash contaminated clothing. Prevent the occurrence of insect wounds by controlling the pests. Planting should be taken up in well-drained soil.

22.4.11 Pith Necrosis, *Pseudomonas viridiflava*, *P. corrugata*, and *P. cichorii*

22.4.11.1 Symptoms

Tomato pith necrosis affects mature tomato plants. Symptoms include a brown discoloration and necrosis of the pith, which eventually leads to hollow chambers in the stem. The pith browning usually extends far up the plant (Fig. 22.18). Profuse adventitious roots are associated with the stem where the pith is affected. Gray or dark-brown lesions may appear on the surface of the stem. Affected plants may turn chlorotic and wilt.

22.4.11.2 Management

Tomato pith necrosis may be controlled by avoiding excessive nitrogen rates. Copper hydroxide was effective in disease reduction (66–72 %). Plant activators resulted in disease reductions ranging from 58 % for acibenzolar-S-methyl to 20 % for harpin.

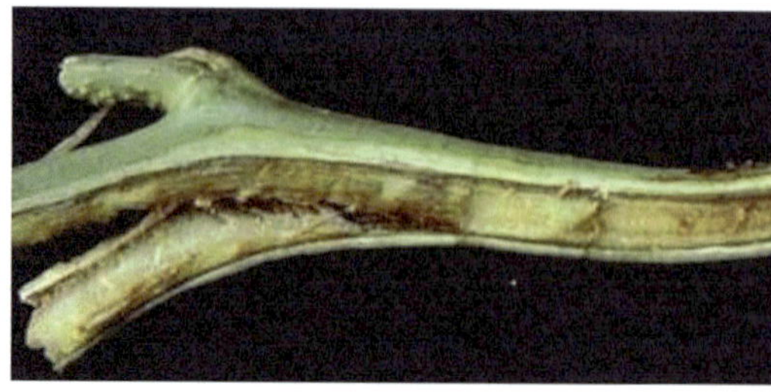

Fig. 22.18 Symptoms of pith necrosis in tomato

22.4.12 Bacterial Speck, *Pseudomonas syringae* pv. *tomato*

22.4.12.1 Symptoms

Bacterial speck causes small dark-brown spots with bright yellow halo on tomato leaves. Necrotic tissues tear off and leaves appear ragged. Small dark-brown spots develop on stem and petioles. Spots may coalesce to cause dark-brown–black blotches on the surface of the infected plant parts. Small (up to 1 mm) black spots also appear on the fruits (Fig. 22.19). Severely infected leaves turn yellow and finally dry out. The pathogen survives on plant debris in the greenhouse or outdoors, as well as on seeds.

22.4.12.2 Survival and Spread

The bacterium is seedborne and can survive in crop residue for up to 30 weeks. It spreads with heavy rainfall, by mechanical means and through seeds, especially under conditions of high humidity and low temperature (18–24 °C). They are splashed from plant to plant by waterdrops from condensation and infect plants through stomata and injuries. Infection requires free water on plant surfaces (Gitaitis et al. 1992).

22.4.12.3 Management

Seed inoculation with *Azospirillum brasilense* combined with a single streptomycin foliar treatment and two foliar bactericide applications at five-day intervals (a third or less of the recommended commercial dose) reduced disease severity in tomato seedlings by over 90 % after 4 weeks and significantly slowed disease develop-

Fig. 22.19 Symptoms of bacterial speck on tomato fruit

ment under mist conditions. Treatment of tomato seeds that were artificially inoculated with *P. syringae* pv. *tomato*, with a combination of mild chemothermal treatment [seeds were incubated in a circulating water bath at 35 ± 2 °C for 1 h. One gram of seeds was incubated in 1 liter distilled water containing: cupric acetate, 2.0 g; glacial acetic acid, 1 ml; pentachloronitrobenzene, 23 % (w/v); 6 % 5-ethoxy-3 (trichloromethyl)-1,2,4-thiadiazole, 4.5 ml; and Triton x-100, 0.2 ml]. This treatment has been reported to eradicate *P. syringae* pv. *tomato* from tomato seeds (Kritzman 1993) and *A. brasilense* seed inoculation, and later, a single foliar application of a copper bactericide nearly eliminated bacterial leaf speck even when the plants were grown under mist for six weeks (Bashan and de-Bashan 2002).

22.4.13 Bacterial Spot, *Xanthomonas vesicatoria*

22.4.13.1 Symptoms

Bacteria attack the foliage, stems, and fruits of tomatoes (Fig. 22.20). Lesions may form on fruit, including the peduncle, but the major crop loss results from shedding of blossoms and young, developing fruit. Fruits that remain are usually nonmarketable because of poor quality. They may have lesions and are often misshapen and damaged from excessive exposure to the sun as a result of defoliation. This can result in sunscald on the fruit. Diseased leaves drop prematurely resulting in extensive defoliation. As newly emerging leaves become infected and defoliation of older, diseased leaves continues, plants possess leaves mostly on their upper stems. Because the diseased foliage often remains on tomato plants, affected plants may have a scorched appearance.

22.4.13.2 Management

Crop rotation should be used in an attempt to avoid carryover on volunteers and crop residue (tomato and pepper). Use disease- and pathogen-free transplants.

Biological control of the bacterial spot can be achieved with antagonistic bacteria such as *Pseudomonas putida* B56 and *P. syringae* Cit7 (Wilson et al. 1997).

Spraying of copper–mancozeb mixtures two or more times per week controlled bacterial spot.

Actigard (acibenzolar-S-methyl) and copper-mancozeb combination gives effective control. Actigard needs to be applied every 14 days. The first application should be started as early as possible after transplanting. Reduce copper-mancozeb applications if you are using Actigard.

22.4.14 Tobacco Mosaic Virus (TMV)

22.4.14.1 Symptoms

Typical mosaic marks appear on the foliage and yield is significantly reduced. The most common symptoms are light and dark-green mottling and dark-brown discoloration on tomatoes (Fig. 22.21), making them unfit for consumption.

22.4.14.2 Management

- **Remove and destroy infected plants:** pull plants with mosaic symptoms immediately. Remove the debris from the garden area and bury them if possible.

Fig. 22.20 Bacterial spot on tomato leaves and fruits

Fig. 22.21 Tobacco mosaic virus on tomato leaf

- **Keep your garden weed-free:** some weed may be harboring the virus. These represent sources of inoculum.
- **Always wash your hands thoroughly and disinfect tools:** before handling plants, wash hands with soap and water, especially if you are a smoker.
- **Plant resistant varieties of tomato:** suggested varieties include "Big Pick," "Celebrity," "President," and "Park's Extra Early Hybrid." Other resistant varieties will be labeled TMV resistant.

22.4.15 Tomato Mosaic Virus (ToMV)

Tomato mosaic virus is distributed throughout the world and is found wherever tomato is grown (Howard et al. 1994; Portree 1996). Tomato mosaic virus has a very wide host range that includes a number of greenhouse crops including: petunia, snapdragon, and pepper (Howard et al. 1994). In tomato crops the virus is very easily spread from plant to plant during routine handling of the plants (Portree 1996).

22.4.15.1 Symptoms

Typical symptoms include alternating light-green to dark-green mottling or mosaic of the leaves and the fern leaf symptom, where the blades of the leaflets undergo a distorted development into a very narrow shape (Fig. 22.22) (Howard et al. 1994; Portree 1996).

ToMV can also reduce fruit quality when fruit symptoms occur. Infected fruits develop a blotchy appearance which can increase in severity to form sunken areas that are dark brown to black in color (Howard et al. 1994; Portree 1996). Severe infections can cause significant yield losses (Howard et al. 1994; Portree 1996).

ToMV is both seed- and soilborne and can survive in the crop residue for up to two years (Portree 1996). There are a number of strains of the ToMV and the symptoms expressed by the plants can vary with the strain (Howard et al. 1994; Portree 1996).

22.4.15.2 Management

Most modern commercial cultivars contain resistance to ToMV and infected plants rarely develop severe symptoms. The symptoms of ToMV are often more severe under low light conditions, and the fern leaf symptom can develop on some of the resistant cultivars under low light conditions. The infected plants improve and grow out of the symptom as the light levels increase. There is no appreciable yield loss in most cases.

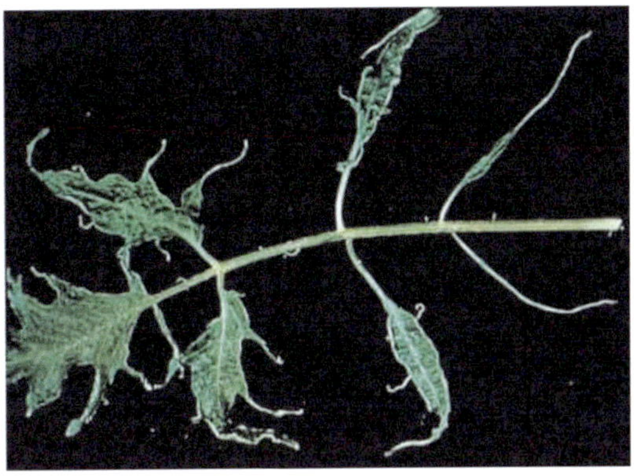

Fig. 22.22 Shoestring leaf symptom of ToMV

Other control measures include the routine washing of work cloths and disinfection of pruning tools by dipping in a 10 % trisodium phosphate solution. A thorough greenhouse cleanup should be undertaken at the end of the crop year; growing in soilless substrates and ensuring that the soil floors of the greenhouse are completely covered and sealed off with plastic serve to limit the year-to-year spread of the virus.

22.4.16 Cucumber Mosaic Virus (CMV)

22.4.16.1 Symptoms

The leaves are affected more than the fruit; plants are often stunted, have short internodes, and may have extremely distorted and malformed leaves (Fig. 22.23), known as fern leaf. The most characteristic symptom of CMV is extreme filiformity, or shoestringing, of leaf blades. CMV has an extensive host range and is transmitted by aphids in a nonpersistent manner.

22.4.16.2 Management

Control aphids early in the season to reduce initial infection and spread. Remove and destroy diseased plants early in the season. Control aphids.

Fig. 22.23 Symptoms of cucumber mosaic virus on tomato leaves

22.4.17 Tomato Yellow Leaf Curl Virus (TYLCV)

22.4.17.1 Symptoms

Infected tomato plants initially show stunted and erect or upright plant growth; plants infected at an early stage of growth will show severe stunting. The critical symptom for diagnosis is when the leaves become small and curl upward; they show strong crumpling and interveinal and marginal yellowing. The internodes of infected plants become shortened and growth is stunted, resulting in plants having a bushy appearance (Fig. 22.24). Flowers formed on infected plants commonly do not develop and fall off (abscise). Fruit production is dramatically reduced, particularly when plants are infected at an early age.

22.4.17.2 Spread

TYLCV is primarily transmitted by the sweet potato whitefly (*Bemisia. tabaci*) and the biotype B (or silverleaf) whitefly (*Bemisia argentifolii*).

22.4.17.3 Management

Symptomatic plants should immediately be carefully removed, bagged, and discarded to prevent

Fig. 22.24 *Left:* typical foliar symptoms of tomato yellow leaf curl. *Upper right:* new leaflets with symptoms of yellowing between veins and an upward curling of their margins. *Lower right:* adult and nymph of *Bemisia* whitefly, a vector of TYLCV

the spread of whiteflies on them that may be carrying the virus. Remove weeds and avoid growing other solanaceous crops near tomato plants whenever possible, as these plants may serve as a reservoir of the virus. UV-reflective mulch can also be effective at deterring whiteflies from landing on crops. Insect exclusion screens rated for whitefly can be used for individual plants and are now being used for screenhouse construction in regions where TYLCV is prevalent.

Imidacloprid-based systemic insecticides (e.g., Admire Pro, Provado) are effective for both adult and nymph stages of whitefly. Pyrethrins and pyrethroid-based insecticides effectively control adult whiteflies. Spirotetramat (Movento), with methylated seed oil plus silicone surfactant, and insecticidal oils, soaps, and other extracts are effective at suppressing all stages.

Many TYLCV-resistant tomato varieties are available for both field and greenhouse production systems. These varieties come in many fruit shapes (globe, Roma, cherry, etc.) and as both determinate and indeterminate plant types.

22.4.18 Tomato Spotted Wilt Virus (TSWV)

22.4.18.1 Symptoms

The symptoms appearing on leaves, petioles, stems, and fruit vary, depending on the stage at which plants become infected. Young leaves may show small, dark-brown spots and eventually die. Dark-brown streaks appear on stems and leaf petioles. Growing tips are usually severely affected with systemic necrosis and greatly stunted growth. The plant may exhibit one-sided growth (Fig. 22.25). Tomato fruit set on severely infected plants displays very characteristic symptoms: immature fruits have mottled, light-green rings with raised centers; the unique orange and red discoloration patterns on mature fruits make them unmarketable. The virus is spread primarily by thrips, particularly the western flower thrips.

22.4.18.2 Management

- Control the virus by controlling thrips vector through insecticides.
- Do weed control to minimize the population of thrips and virus.

22.5 Nematodes

22.5.1 Root-Knot Nematodes, *Meloidogyne* spp.

22.5.1.1 Economic Importance

M. incognita was responsible for 30.57–46.92 % loss in fruit yield of tomato, while *M. javanica* caused 77.5 % loss in yield.

22.5.1.2 Symptoms

Aboveground symptoms are stunting, yellowing, wilting, reduced yield, and premature death of plants. Belowground symptoms are swollen or knotted roots (root galls) (Fig. 22.26) or a stubby root system. Root galls vary in size and shape depending on the nematode population levels and species of root-knot nematode present in the soil.

Fig. 22.25 Wilting caused by TSWV

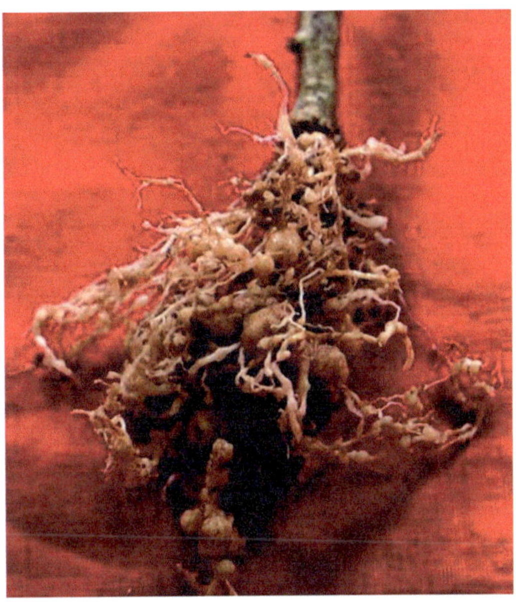

Fig. 22.26 Heavy galling of tomato roots infected with *Meloidogyne incognita*

22.5.1.3 Management

Soil solarization to control root-knot nematode is simple and economic in its application and leaves no toxic residues for humans, animals, and plants. These attributes make solarization suitable for application in a wide variety of production system. Soil solarization should be done during the hottest period of the year. It offers a satisfactory and environmentally friendly solution for the control of root-knot nematode. This method is easy to be used into organic, conventional, and integrated control growing system, and it is more economic than chemical treatments.

Addition of dazomet at 1 kg a.i./ha to the nursery beds gives effective control of root-knot nematodes. Further, addition of enriched FYM/compost/vermicompost (with *Trichoderma harzianum/T. viride* + *Paecilomyces lilacinus* + *Pseudomonas fluorescens* each at 2 kg/t of organic amendment) at 5 t/ha helps in the management of root-knot nematodes.

After 15–30 days of planting, addition of enriched neem cake (with *Trichoderma harzianum/T. viride* + *Paecilomyces lilacinus* + *Pseudomonas fluorescens* each at 2 kg/t) at 50 g/m^2 gives effective management of nematodes under polyhouse conditions. Further, drip application of neem cake suspension enriched with bioagents (20 kg neem cake + *Trichoderma harzianum/T. viride* + *Paecilomyces lilacinus* + *Pseudomonas fluorescens* each at 2 kg) helps in keeping the polyhouse beds free from root-knot nematodes.

22.6 IPM Package for Management of Insect Pests, Diseases, and Nematodes

The IPM package gives the possibility to manage the key pest and diseases in tomato crops in greenhouses that causes serious damage reducing the quantity and the quality of these products. The IPM package developed for protected crops gives sustainable management of the most important pests and diseases and helps growers to use methods that minimize environmental, health, and economic risks. The IPM package consists of an innovative combination of three different methods for controlling the main pests in greenhouses:

- Implementing the best monitoring system for the detection of key pest and diseases
- Selecting the soil solarization as an eco-friendly method to control root-knot nematodes
- Using the integrated fungicidal control programs to maximize economic return on tomato production in greenhouses

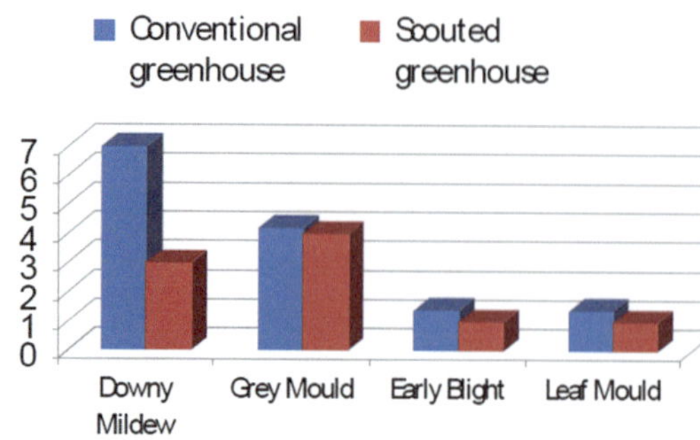

Fig. 22.27 Number of treatments (X - axix) in conventional and scouted greenhouse for control of diseases

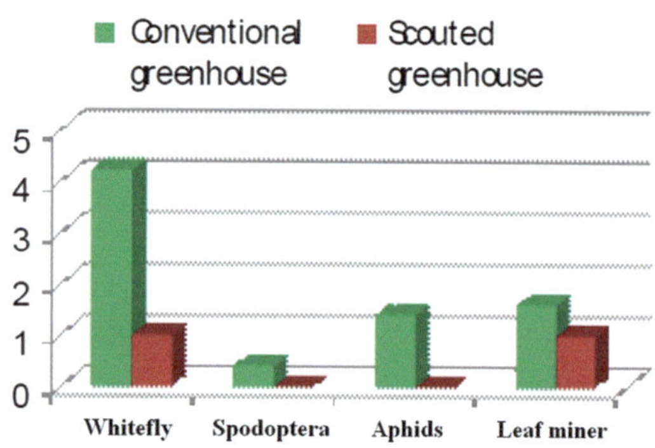

Fig. 22.28 Number of treatments (X - axix) in conventional and scouted greenhouse for control of tomato pests (Y - axis)

22.6.1 Implementation of IPM Scouting Programs

Regular monitoring is the framework of good integrated pest management (IPM) programs and must be in place before others control alternatives can be implemented. The backbone of a good IPM program is the scout who looks regularly for pests and plant abnormalities, records what is found, summarizes this information, and reports it to the grower for the final pest management decision. Since so many vegetable crop growers are still mismanaging pesticides because of the lack of sound monitoring practices, the primary objective of this study was to demonstrate the benefits and economic feasibility of using scouts in IPM programs. Observation done during the scouting program indicated that the monitoring of pests and diseases is an effective tool in timing pesticide applications and minimizing pesticide input.

The chemical treatments done for the control of different diseases and insect pests are shown in Figs. 22.27 and 22.28. In the conventional greenhouses the number of treatments used per crop cycle is quite variable depending on the type of greenhouses, farmer practice, and period of the year. In the case of tomatoes grown under protected cultivation, surveys showed that farmer's approaches to chemical control were quite variable.

The number of pesticide sprays per crop cycle varied between 7.6 for insecticides and 14 for fungicides. In scouted greenhouses these figures are quite different and the number of insecticides and fungicide application is 2 and 9, respectively. During the period of tomato cultivation in conventional and scouted greenhouses, 22 and 11 treatments were compared. The information about the economic analysis of the benefits of IPM scouting program as compared to the growers' conventional pests and diseases management is presented in Fig. 22.29. The economic effect of scouting program is very clear.

The economic analyses indicate that scouting saved 46.6 % of the applied pesticide volume as compared to the non-scouted area and resulted in overall economic savings of 29.37 % (or 6967 lek/ha = about US$ 650).

22.6.2 Soil Solarization to Control Root-Knot Nematodes

Tomato is susceptible to infection by soilborne diseases and severe yield losses can result especially by the root-knot nematode. The effective-

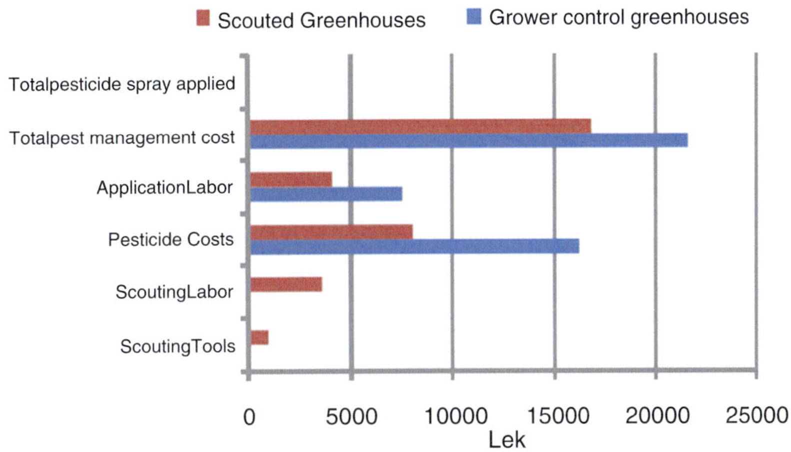

Fig. 22.29 Cost of scouting in conventional and scouted greenhouse

ness of solarization, a soil disinfection technique that uses passive solar heating, to control the incidence of root-knot nematode under greenhouse condition was studied. Solarization was accomplished by the application of 0.05–0.06-mm clear polyethylene sheets to moist soil for four weeks during the hot season.

In closed plastic greenhouses the mulching of moistened soil with transparent polyethylene sheets for four weeks during the hot season induces an increase of temperature. Maximum soil temperatures achieved under solarization were 51 °C in the alluvial soil and 52.4 °C in the heavy clay soil type at depth 20 cm. In the non-solarization unheated greenhouse soil temperature achieved was 39 °C (Fig. 22.30).

The effect of soil solarization treatments on the population densities of the root-knot nematode *M. incognita* was very clear. The larval density per 100 ml of soil was markedly reduced compared with the untreated control (Fig. 22.31). The soil solarization method had a significant influence to reduce the severity of root damage. Root galling index is the most important symptom for the root-knot nematode. This index was dramatically reduced in both solarized greenhouses compared with untreated control where the highest root galling index was found (Fig. 22.32).

The growth of tomato plants, as indicated by visual assessment and plant height measurements, was significantly improved by the soil solarization. Tomato fruit yields in solarized greenhouses were fourfold higher than those obtained in untreated control (Fig. 22.33). Such significant increases were consistent with effective control of nematode provided by solarization.

Solarization is also known to improve plant growth and yield through the release of nutrient induced by high soil temperatures. This technique at the same time showed good results to reduce the infection of a severe tomato disease like corky root disease (*Pyrenochaeta lycopersici*).

22.6.3 Integrated Fungicidal Control Programs

The percentage of severity of gray mold, downy mildew, leaf mold, and early blight diseases was significantly higher on the low cost and control treatments compared to the standard and intensive fungicide treatment (Table 22.3).

22.6.3.1 Economic Costs of Spray Programs

At the end of the season, the economic costs were significantly lower in the low cost and control treatments, moderate in standard treatments, and higher in intensive fungicide treatments (Table 22.4).

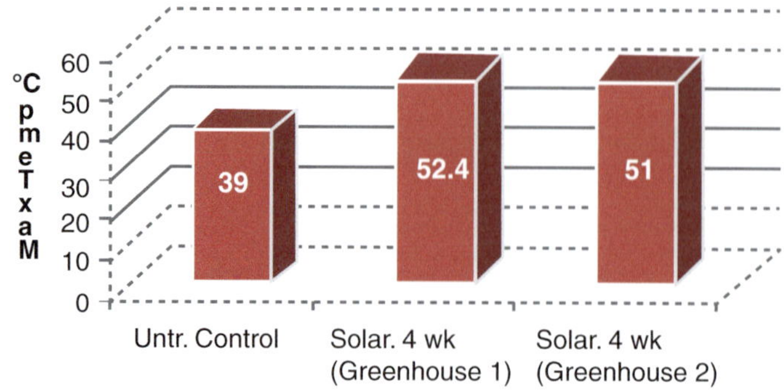

Fig. 22.30 Maximum temperatures attained during soil solarization

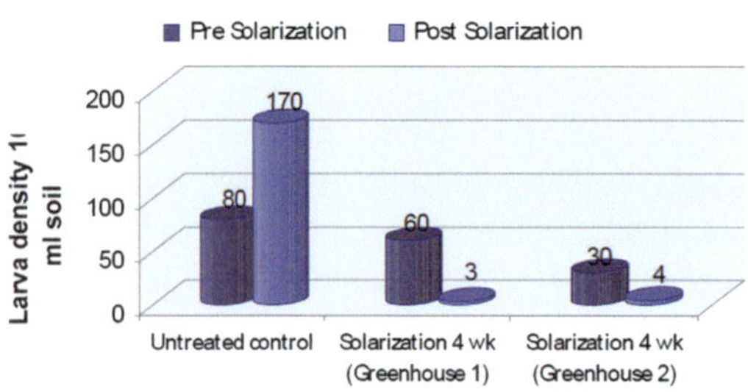

Fig. 22.31 Effect of soil solarization on the population densities of root-knot nematode larvae in soil

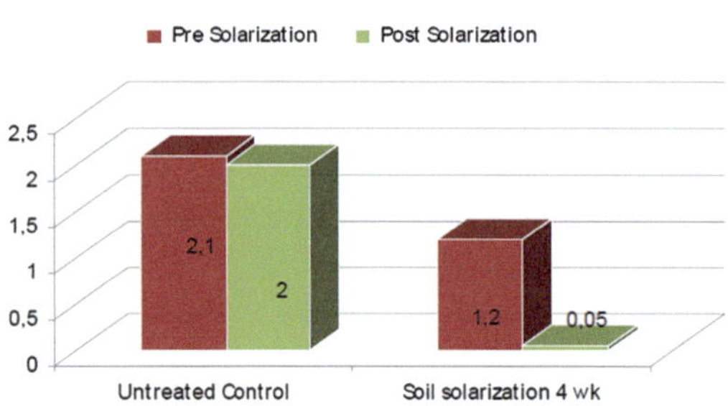

Fig. 22.32 Effect of soil solarization on root gall index (X - axis) by root-knot nematode

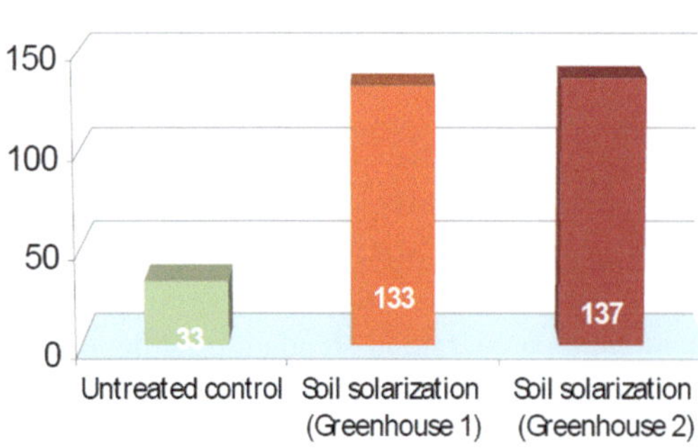

Fig. 22.33 Effect of soil solarization on tomato yield in tons/ha (X – axix)

22.6.3.2 Environmental Impact of Spray Programs

Aside from fungicide costs, growers are concerned with environmental impacts associated with their production practices, specifically pesticide applications. Environmental Impact Quotients (EIQ) are a scientific means of assessing pesticide fate in the environment based on toxicological and environmental data; impacts against arthropod natural enemies, honeybees, and birds; leaching potential; and related human health issues. EIQ values for fungicides used in this study have already been calculated in the paper by Kovach et al. (1992). In general, lower EIQ scores indicate products with overall lower environmental impact. Perhaps what is more important than the raw EIQ score is the Field Use Rating (FUR), which takes into account the EIQ score x % active ingredient x rate per ha to get a more accurate picture of environmental impact. When comparing spray programs it is important to realize that the EIQ FUR score needs to be multiplied by the number of applications made over the season to get an accurate overall understanding of environmental impacts.

The program EIQ values were lower in control, moderate in low cost and standard treatments, and higher in intensive treatments (Table 22.5).

The standard fungicide program remains the best recommendation. It controls the main tomato diseases better than the others treatments and had no significant differences with intensive program. However, it was more expensive than low cost control but cheaper and had the lower environment impact than intensive program.

Table 22.3 The effect of fungicide program on yield and percent infection of main diseases

Fungicide program	Fruit wt. (g)	Downy mildew rating	Gray mold rating	Leaf mold rating	Early blight rating
Low cost	120.6	8.0	10.70	24.6	4.83
Standard	151.3	1.7	1.50	3.17	1.67
Intensive	163.4	1.0	1.33	2.67	1.0
Control	114.8	9.2	12.30	27.3	5.33

Table 22.4 Composition, cost, and frequency of four fungicide programs

Fungicide combinations	Segment cost (Leks)	No. of applications	Segment cost/ season	Program cost/season
Low cost (Dithane + Kocide)	162	8	1296	1296
Standard (Daconil + Acroblu)	294	4	1176	2448
Standard (Equation pro + Teldor)	318	4	1272	
Intensive (Daconil + Acroblu)	294	8	2352	4578
Intensive (Equation pro + Teldor)	318	7	2226	
Control (Acrobat)	270	3	810	1301
Control (Bavistin)	75	2	150	
Control (Acroblu)	114	1	114	
Control (Daconil)	180	1	180	
Control (Stoper)	47	1	47	

Table 22.5 Rank of fungicide programs by environmental impact and economic and disease factors

Fungicide combinations	EIQ FUR[a]	No of applications	Partial program EIQ	Program EIQ
Low cost (Dithane + Kocide)	55.2	8	441.60	441.60
Standard (Daconil + Acroblu)	103.44	4	413.76	445.92
Standard (Equation pro + Teldor)	8.04	4	32.16	
Intensive (Daconil + Acroblu)	103.44	8	827.52	883.80
Intensive (Equation pro + Teldor)	8.04	7	56.28	
Control (Acrobat)	32.77	3	98.31	233.13
Control (Bavistin)	14.04	2	28.08	
Control (Acroblu)	45.70	1	45.70	
Control (Daconil)	57.74	1	57.74	
Control (Stoper)	3.30	1	3.30	

[a]Environmental Impact Quotients for Field Use Rating (EIQ FUR)

References

Bashan Y, de-Bashan LE (2002) Reduction of bacterial speck (*Pseudomonas syringae* pv *tomato*) of tomato by combined treatments of plant growth-promoting bacterium, *Azospirillum brasilense*, streptomycin sulfate, and chemo-thermal seed treatment. Eur J Plant Pathol 108:821–829

Bochow H (1989) Use of microbial antagonists to control soil-borne pathogens in greenhouse crops. Acta Horticult 255:271–279

Dhanvantari BN (1994) Further studies on seed treatment for tomato bacterial canker. In: Proceedings of the 10th annual tomato disease workshop, pp 49–51

Fletcher JT (1984) Diseases of greenhouse plants. Longman Inc, New York

Gitaitis R, McCarter S, Jones J (1992) Tomato transplants produced in Georgia and Florida: their diseases and control. Plant Dis 76:651–656

Gleason ML, Braun EJ, Carlton WM, Peterson RH (1991) Survival and dissemination of *Clavibacter michiganensis* sub sp. *michiganensis* in tomatoes. Phytopathology 81:1519–1523

Gleason ML, Gitaitis RD, Ricker MD (1993) Recent progress in understanding and controlling bacterial canker of tomato in eastern North America. Plant Dis 77(11):1069–1076

Howard RJ, Garland JA, Seaman WL (1994) Diseases and pests of vegetable crops in Canada. The Canadian Phytopathological Society and the Entomological Society of Canada, Ottawa, Canada

Jarvis WN, Thorpe HJ, Meloche RB (1983) Survey of greenhouse management practices in Essex County, Ontario, in relation to Fusarium foot and root rot of tomato. Plant Dis 67:38–40

Jones JP, Stall RE, Zitter TA (eds) (1993) Compendium of tomato diseases. APS Press, St Paul

Kovach J, Petzoldt C, Degni J, Tette J (1992) A Method to measure the environmental impact of pesticides. IPM Program, Cornell University, New York State Agricultural Experiment Station, Geneva

Kritzman G (1993) A chemi-thermal treatment for control of seed-borne bacterial pathogens of tomato. Phytoparasitica 21:101–109

Malathrakis NE, Goumas DE (1987) Bacterial soft rot of tomato in plastic greenhouses in Crete. Ann Appl Biol 111:115–123

Malathrakis NE, Kapetanakis GE, Linardakis DC (1983) Brown root rot of tomato, and its control in Crete. Ann Appl Biol 102:251–256

Minuto A, Spadaro D, Garibaldi A, Gullino L (2006) Control of soil-borne pathogens of tomato using a commercial formulation of *Streptomyces griseoviridis* and solarisation. Crop Prot 25:468–475

Portree J (1996) Greenhouse vegetable production guide for commercial growers. Province of British Columbia, Ministry of Agriculture, Food and Fisheries, Victoria, BC, 117 pp

Sherf AF, McNab AA (1986) Vegetable diseases and their control. Wiley, New York

Shoemaker PB (1992) Evaluation of different materials for tomato bacterial canker. Fungic Nematic Tests 47:155

Wilson M, Campbell HL, Jones JB, Suslow TV, Cuppels DA (1997) Biological control of bacterial speck of tomato. Phytopathology 86:S49

Bell Pepper

23

Abstract

In greenhouses, capsicum fruits are harvested with full color maturation, and fruit yields are greater and of high quality and usually produced at a time of the year when production in the field is not possible and market prices for capsicum are highest. Capsicum being a relatively long duration (9–10 months) crop in polyhouse, the plant parts (vegetative, floral, and fruit) are more exposed to adverse effect of pests and diseases on the yield, quality, and market value of the produce. Hence their identification and management at right stage of the crop should be given importance. The major pests and diseases, their symptoms, and their management in capsicum are discussed. The major focus has been given on adoption of integrated approach in managing pests and diseases that help to reduce the pesticide load and cost of chemicals and avoid the resurgence of pests and diseases.

Keywords

Capsicum • Pests • Diseases • Nematodes • Symptoms • Biology • Spread • Management

23.1 Introduction

Greenhouses can be a means to economically maintain a warm environment during cool season, to protect bell pepper plants from rain, wind, and high solar radiation, and to retain pollinators and beneficial insects while excluding unwanted insect pests (Fig. 23.1). In greenhouses, capsicum fruits are harvested with full maturation color and fruit yields are greater and of high quality and usually produced at a time of the year when production in the field is not possible and market prices for capsicum are highest (Tables 23.1 and 23.2).

Colored capsicums are in great demand in urban markets. The demand is mostly driven by hotel and catering industry. The traditionally grown green capsicum, depending upon variety and season, usually yields 20–40 t/ha in about 4–5 months. In greenhouse, the crop duration of green and colored capsicums is about 7–10 months and yields about 80–100 t/ha. The advantages of protected cultivation are:

P.P. Reddy, *Sustainable Crop Protection under Protected Cultivation*,
DOI 10.1007/978-981-287-952-3_23

Fig. 23.1 Colored capsicum under greenhouse

Table 23.1 Yield of capsicum under greenhouse at different locations in India

Location	Yield (t/ha)
Pune	203
Coimbatore	148
Bangalore	147
Solan	79
Hisar	302

Table 23.2 Performance of capsicum in various protected structures

Treatments	Days to 50 % flowering	No. of fruits/plant	Average fruit weight (g)	Fruit yield/plant (kg)
Greenhouse	63.00	10.40	100.05	93.733
Double wall polyench	82.33	7.30	73.33	60.453
L Polycarbonate	71.33	8.50	83.5	71.927
Polyench	74.33	10.40	67.69	64.213
Open	79.00	8.91	42.19	47.193

- Higher productivity resulting in increased yield
- Provides better growing environment to plants
- Protects from rain, wind, and high temperatures and minimizes the damage of insect pests and diseases, thereby improving the quality and yield
- Facilitates year-round production coupled with yield enhancement by 2–3 times compared to open cultivation

23.2 Varieties Suitable for Greenhouse Production

- Indra
- Bharat
- Tanvi
- Manhattan
- Swarna
- Natasha
- Angel
- California Wonder
- Arka Mohini
- Pusa Deepti
- Solan Hybrid-1

23.3 Diseases

23.3.1 Damping-Off, *Pythium* spp, and *Rhizoctonia solani*

23.3.1.1 Symptoms

Infection takes place at the base of the young seedlings just above the ground level which leads to wilting and later death of seedlings (Fig. 23.2). Any damage caused to seedlings while transplanting can also lead to damping-off or seedling wilt besides fresh infection in main field or infection that is carried from nursery.

23.3.1.2 Management

Drench the base of the plant with carbendazim (1 g/L) or Metalaxyl MZ (2 g/L) or copper oxychloride (3 g/L) or Captan (3 g/L) at about 25–50 ml/plant.

Fig. 23.2 Damping-off of capsicum seedlings

Fig. 23.3 External and internal *dark brown* discoloration of pepper stems due to *Fusarium solani*

Maintain proper sanitation practices in greenhouse. Provide optimal soil temperature, watering, etc. to the plants. Avoid the conditions where young plants are stressed.

23.3.2 Fusarium Stem and Fruit Rot, *Fusarium solani*

23.3.2.1 Symptoms

Soft, dark brown or black cankers are formed on the stem, usually at nodes or wound sites. These may girdle the stem in later stages of disease development. There is a dark brown discoloration of the internal portion of the stem that may extend a considerable distance (Fig. 23.3). The lesions may eventually develop cinnamon or light orange-colored, very small (<1 mm diameter), flask-shaped fruiting structures known as perithecia, which are the fruiting bodies of the fungus.

Fig. 23.4 *Cercospora* leaf spots on capsicum leaf

23.3.2.2 Management

- Clean greenhouse and follow good sanitation practices.
- Maintain good air circulation and avoid high relative humidity conditions (above 85 %).
- Avoid wounding fruit and excessive wounding to the stems.

23.3.3 Cercospora Leaf Spot, *Cercospora capsici*

23.3.3.1 Symptoms

Cercospora appears initially as tiny yellow spot on leaf surface leading to increased dark grey spots (Fig. 23.4) which spreads on entire leaf resulting in dropping of leaf.

23.3.3.2 Management

Spray chloroth alonil (2.5 g/L) or Mancozeb (2.5 g/L) or carbendazim (1 g/L).

23.3.4 Gray Mold, *Botrytis* sp

23.3.4.1 Symptoms

- Water-soaked spots that rapidly expand into large yellowish green or grayish brown.
- Irregular lesions on fruits which are soft and spongy in texture (Fig. 23.5).
- Velvet-like fungus mycelium and spores are produced on the lesion surface under cool and humid conditions.

- The fungus produces overwintering structures called sclerotia in addition to other types of spores.

23.3.4.2 Management

- Pre-storage dry heat treatment.
- Hot air treatment at 38 °C for 48–72 h.
- Hot water treatment at 50–53 °C for 2–3 min.

23.3.5 Anthracnose, *Colletotrichum capsici*

23.3.5.1 Symptoms

Small or large lesions typical of *Colletotrichum* infections (anthracnoses) are formed on leaves and fruits, but in other cases the disease may develop as purplish or brown patches without the formation of definite lesions (Fig. 23.6). Stems and petioles may be girdled, and necrosis of inflorescences causes dieback and shriveling. Other symptoms include water-soaked lesions on fruit that becomes soft and slightly sunken and concentric rings within the fruit spots.

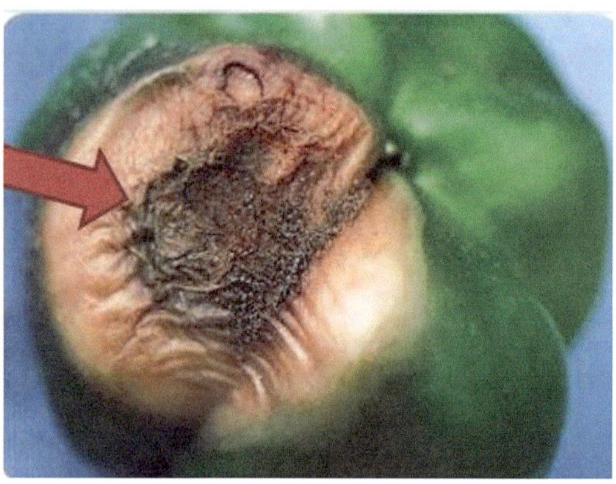

Fig. 23.5 Grey mold on sweet pepper fruit (Note peeling away of epidermis)

23.3.5.2 Management

Avoid alternate host and infected plant debris and optimize the humidity in polyhouse.

Various fungicides have been found to be effective, including 0.2 % Mancozeb, 0.1 % Ziram, Blitox 50 (copper oxychloride), 0.1 % Bavistin (carbendazim), and 0.5 % or 1 % Bordeaux mixture. Benlate (benomyl) and Delsene M (carbendazim + maneb) (Alabi and Emechebe 1990) are used as seed dressings.

23.3.6 Powdery Mildew, *Leveillula taurica*

23.3.6.1 Symptoms

The disease initially appears as tiny yellow spots on surface of leaf and powder like material on the lower surface leading to a powdery growth covering the entire lower surface of leaf which leads to drying and dropping of leaves at later stages (Fig. 23.7). The disease reduces growth of leaves and fruits leading to low quality and quantity of the produce.

23.3.6.2 Management

Spray pongamia/neem oil (7 ml/L) + Sulfur WDG-80 (2 g/L) or wettable sulfur (2 g/L) or hexaconazole (0.5 ml/L) or Myclobutanil (1 g/L) or Dinocap (1 ml/L) or azoxystrobin (0.5 ml/L) or Penconazole (0.5 ml/L) or Flusilazole (0.5 ml/L).

Fig. 23.6 Anthracnose symptoms on leaf and fruit of sweet pepper

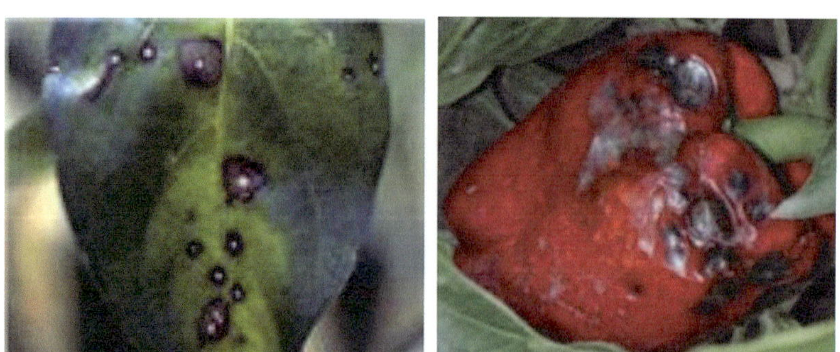

Fig. 23.7 Powdery mildew on bell pepper leaves. *Left* – Profuse sporulation on under surface of leaves. *Right* – Necrotic lesions develop at later stages of disease

Fig. 23.8 Symptoms of *Phythopthora* blight on capsicum

23.3.7 *Phythopthora* Blight, *Phytophthora capsici*

23.3.7.1 Symptoms

This disease appears during fruiting and flowering stage resulting in tiny oil like spot on leaf surface resulting in rotting and blackening of plants (Fig. 23.8). Later plant weakens and dies in 2–3 days. Heavy and continuous rainfall coupled with high humidity favor disease appearance and its quick spread. *Phythopthora* disease is relatively more severe in net houses which may lead to 40–80 % crop damage.

23.3.7.2 Management

Spray copper hydroxychloride (3 g/L) or Bordeaux mixture (1 %) or Metalaxyl MZ (2 g/L) or Dimethomorph + Mancozeb (1 g + 2.5 g/L) or Fosetyl aluminum (2 g/L) or azoxystrobin (0.5 ml/L).

Severely infected plant parts should be destroyed. It is better to avoid capsicum cultivation in severely affected net houses.

23.3.8 Bacterial Wilt, *Ralstonia solanacearum*

23.3.8.1 Symptoms

Wilting of leaves, sometimes only few branches of the plant, occurs during hot days followed by recovery throughout the evening and early hours of the morning. The wilted leaves maintain their green color and do not fall as disease progresses. Under favorable conditions complete wilt will occur (Fig. 23.9). The vascular tissues in the lower stem of wilted plants show a dark brown discoloration (Fig. 23.9). A cross section of the stem of a plant with bacterial wilt produces a white, milky strand of bacterial cells in clear water.

Fig. 23.9 Bacterial wilt on bell pepper. *Left* – Leaves droop without yellowing. *Right* – Vascular bundle rotting and white substance formed

23.3.8.2 Management

- Plant varieties which are tolerant/resistant, if available.
- Seedlings must be free from infection by *R. solanacearum*.
- Do not grow crops in soil where bacterial wilt has occurred.
- Remove wilted plants to reduce spread of the disease from plant to plant.
- Control root-knot nematodes since they could facilitate infection and spread of bacterial wilt.
- Soil amendments (organic manures) can suppress bacterial wilt pathogen in the soil.
- Rotation is of limited value since the disease has a wide host range, but still it is recommended to avoid continuous planting of solanaceous crops.

23.3.9 Bacterial Spot, *Xanthomonas vesicatoria*

23.3.9.1 Symptoms

The symptoms of bacterial spot are similar to those of bacterial speck. The spots on the fruits are initially raised and at the end look scabby (Fig. 23.10). The pathogen survives on plant debris in the greenhouse or outdoors, as well as on seeds. They are splashed from plant to plant by water drops from condensation and infect plants through stomata and injuries. Infection requires free water on plant surfaces (Gitaitis et al. 1992).

23.3.9.2 Management

- Buy certified seed from a reputed source or use seed treatments to reduce transmission.
- Produce disease-free plants by raising transplants in an area where production does not occur. Inspect all purchased transplants carefully, and if transplants originate in southern states, they should be certified.
- Rotate fields to avoid carryover on volunteer plants or crop residue.
- Keep fields free from volunteer plants, weeds, and cull piles.
- Crop rotation with nonhost crops.
- Avoid working in fields when bacterial diseases are present and the fields are wet.
- Apply appropriate bactericides or combination of pesticides.

Fig. 23.10 Bacterial spot on capsicum leaf and fruit

23.3.9.2.1 Chemical Recommendations

- **Copper Hydroxide**: Begin applications when disease first threatens and repeat at 7–10-day intervals as needed. Do not apply spray solution with pH less than 6.0 to prevent phytotoxicity.
- **Mancozeb plus Copper Hydroxide**: Begin applications when disease threatens and repeat at 7–10-day intervals as needed. Use higher rates and 3–7-day interval when disease pressure is severe.
- **Streptomycin Sulfate (Agrimycin)**: Use 200 ppm only up to transplant (REI 12 h).

23.3.10 Bacterial Soft Rot, *Erwinia carotovora*

23.3.10.1 Symptoms

- The fleshy fruit peduncle is highly susceptible and is frequently the initial point of infection.
- Initially, the lesions on the fruit are light to dark-colored, water-soaked, and somewhat sunken (Fig. 23.11).
- The affected fruit hangs from the plant like a water-filled bag.

23.3.10.2 Management

- Use chlorinated water to reduce populations of soft rot bacteria and to reduce the risk of infection during washing.
- Allow fruit to dry thoroughly.

Fig. 23.11 Bacterial soft rot on capsicum fruit

- During packing and storage, the fruit should be kept clean and maintained in a cool, dry place.

23.3.11 Pepper Mild Mottle Virus (PMMV)

23.3.11.1 Symptoms

- Symptoms include the development of obvious bumps on the fruit as well as color streaking and green spotting as the fruit matures.
- Fruit tend to have pointed ends and may also develop sunken brown areas on the surface (Fig. 23.12).

23.3.11.2 Management

- The virus is spread by the routine handling of the young plants, especially at transplanting.

Fig. 23.12 Pepper mild mottle virus symptoms

- Do minimum mechanical activities during cultivation.
- Grow resistant varieties.

23.3.12 Tobacco Mosaic Virus

23.3.12.1 Symptoms

- The symptoms of infection first appear on the leaf as a necrosis along the main veins accompanied by wilting and leaf drop.
- New growth on the plants may exhibit mosaic symptoms as well as distorted growth.

23.3.12.2 Management

- Use disease-free seed.
- Grow resistant cultivars.
- Destroy any infected plants.

23.3.13 Tomato Spotted Wilt Virus (TSWV)

23.3.13.1 Symptoms

The virus may cause sudden yellowing and browning of young leaves which later become necrotic. Long necrotic streaks appear on stems extending to the growing tips. Fruits formed after infection display large necrotic streaks and spots, while younger fruit may be completely necrotic. The virus is spread primarily by thrips, particularly the western flower thrips.

23.3.13.2 Management

- Control the virus by controlling the thrips vector through insecticides.
- Do weed control to minimize the population of thrips and virus.

23.3.14 Cucumber Mosaic Virus (CMV)

23.3.14.1 Symptoms

Severe mosaic forms on the foliage; older leaves may have large necrotic rings, fruit may be malformed, and conspicuous yellow concentric rings or spots often appear on the fruit from infected plants.

23.3.14.2 Transmission

The virus is very readily transmitted by mechanical means. In nature, it is transmitted by several species of aphid vectors like *Aphis craccivora*, *A. gossypii*, and *Myzus persicae*.

23.3.14.3 Management

- Removal and destruction of infected plants from the field.
- Removal of all the weeds in and around the chili field.
- Spraying 0.05 % Nuvacron or 0.02 % Metasystox at 10-day interval to control aphid vectors.
- Soil application of carbofuran at 1.5 kg a.i./ha at the time of sowing seeds in the seed bed followed by 1–2 foliar sprays of 0.05 % Nuvacron.
- Spraying of 1.5–1.8 % mineral oil at 5–6-day interval has been found effective in reducing the disease incidence.

23.4 Insect Pests

23.4.1 Thrips, *Frankliniella occidentalis*

23.4.1.1 Symptoms

Thrips cause upward curling of leaves, suck sap, and reduce leaf growth, plant growth, yield, and market value of produce. It also reduces leaf area

Fig. 23.13 Thrips damage on leaves and fruit of bell pepper

and hinders absorption of nutrients and water by the plants. Other symptoms include white or brown spots on leaves, leaf scars, and distorted buds, leaves, or flowers. Increased infestation leads to blackening and drying of leaves and irregular fruit bearing (Fig. 23.13).

23.4.1.2 Management

Remove affected plant parts including leaves, flowers, and fruits. Keep the plots clean by removing all the dropped plant parts. Spray pongamia oil (5–8 ml/L) or neem seed kernel extract (NSKE 4 %) or pongamia/neem soap (7 g/L) or fipronil (1 ml/L) or chlorpyrifos (2 ml/L) or acephate (1.5 g/L) or imidacloprid (0.5 ml/L). Drench soil with chlorpyrifos (4 ml/L) or imidacloprid (0.5 ml/L).

23.4.2 Spider Mite, *Tetranychus urticae*

23.4.2.1 Symptoms

Young larvae and adults feed on leaves, bud, and fruits and suck sap from plant parts which in turn causes downward curling of leaves (Fig. 23.14). The size of leaf, fruit, and plants gets reduced; fruit and flower drop affecting the market value of the produce. This pest infestation increases with increased temperature coupled with high humidity.

23.4.2.2 Management

Remove the pest damaged plant parts including leaves, flowers, and fruits and spray pongamia oil (5–8 ml/L) or pongamia/neem soap (8–10 g/L) or dicofol (2 ml/L) or wettable sulfur (2 ml/L) or abamectin (0.5 ml/L) or Ecomite or Propargite or chlorfenapyr (1 ml/L) or Fenazaquin (1 ml/L).

23.4.3 Aphids, *Myzus persicae*

23.4.3.1 Symptoms

Nymphs and adult aphids suck sap from leaf veins and younger leaves resulting in reduced plant growth and decrease in yield. Its infestation not only causes curling of leaves (Fig. 23.15) but also spreads viral diseases.

23.4.3.2 Management

Keep a close watch on the plants at regular intervals for aphids' infestation. Spray pongamia/neem soap (8–10 g/L) or imidacloprid (0.5 ml/L) or Thiomethoxam (0.5 g/L) or Dimethoate (2 ml/L).

Fig. 23.14 *Left* – Spider mite damage on sweet pepper. *Right* – Adults and egg of spider mite

Fig. 23.15 *Left* – Leaf distortion after aphid attack. *Right* – Aphids on leaf

Fig. 23.16 Tobacco caterpillar damage on capsicum leaves

23.4.4 Tobacco Caterpillar, *Spodoptera* sp

23.4.4.1 Symptoms

Fruit borers are very active during night. The adults lay eggs on fruits, flowers, and leaves in large numbers, and the nymphs that come out of eggs feed on fruits and leaves (Fig. 23.16) causing heavy destruction of crops and severely affecting the quality of the produce. Whenever night temperature is low, coupled with cool and high humidity, the infestation is increased. Since eggs are laid in group, the larva also feeds gregariously on leaves at one place, which can be easily identified and destroyed.

23.4.4.2 Management

Pick and destroy nymphs and adult insects. Generally eggs are laid and hatch in groups, which is easy to identify from a distance. Hence they should be identified and destroyed immediately. Spray thiodicarb (1 ml/L) or carbaryl (3 g/L) or Indoxacarb (1 ml/L) or Rynaxypyr (0.5 ml/L) or Chlorofenfur (1.5 ml/L) or fipronil (1

Fig. 23.17 Whitefly infestation on bell pepper leaves

ml/L). In addition to the sprays, grownup adults should be subjected to Methomyl baiting,[1] which is a safe, healthy, and effective practice.

23.4.5 Whitefly, *Trialeurodes vaporariorum*

23.4.5.1 Symptoms

Whiteflies damage the plant by sucking sap from the leaves (Fig. 23.17). Large infestations can cause leaf yellowing and a general decline in the plant. Sooty mold is commonly found in association with whitefly. As with aphids, whitefly feeding also results in honeydew formation which can reduce fruit quality. The presence of the honeydew and sooty mold can necessitate that the fruit be washed prior to going to market. The presence of sooty mold on the leaves can reduce the productivity of the leaf by reducing the amount of light reaching the leaf surface (Howard et al. 1994).

23.4.5.2 Management

The parasitic wasps, *Encarsia formosa* and *Eretmocerus eremicus*, are effective against whitefly with parasitized whitefly scale becoming yellow or black in color, depending on the parasite. Scale parasitized by *Encarsia formosa* is black in color. *Delphastus pusillus* is a small beetle that feeds on whitefly eggs and is ideal for complementing *Encarsia* and *Eretmocerus* (Portree 1996).

23.4.6 Leaf Miners, *Liriomyza* spp

23.4.6.1 Symptoms

Thin, white, winding trails on leaves; heavy mining (Fig. 23.18) can result in white blotches on leaves and leaves drop from the plant prematurely; early infestation can cause fruit yield to be reduced; adult leaf miner is a small black and yellow fly which lays its eggs in the leaf; larvae hatch and feed on leaf interior.

23.4.6.2 Management

Check transplants for signs of leaf miner damage prior to planting; remove plants from soil immediately after harvest; only use insecticides when leaf miner damage has been identified as unnecessary spraying will also reduce populations of their natural enemies.

23.4.7 Flow Chart for Insect Pest Management in Capsicum (Fig. 23.19)

The calender of operations for insect pest management in capsicum is presented in flow chart (Fig. 23.19).

[1] Methomyl Baiting Procedure: Prepare a mixture of 10 kg paddy husk and 1 kg jaggery solution and store for 6–8 h. Add ½ kg Methomyl to the mixture. Small sized balls of mixture are made which are spread near the root zone of the plants and also around polyhouse/net house to avoid infestation of fruit borer. It should be applied during night hours, and the domestic or pet animals should not be allowed to move in and around net-/polyhouse overnight.

Fig. 23.18 Damage caused by leaf miners

Fig. 23.19 Flow chart for insect pest management in capsicum

Raise seedlings in protrays in polyhouse

↓

At Transplanting — Apply neem cake 250 kg /ha

↓

15 DAP — Spray Acephate for thrips

↓

30 DAP — Apply neem cake 250 kg /ha

↓

Post flowering and fruiting stage — Monitor for pests like fruit borer, tobacco caterpillar, thrips and yellow mite.
Spray NPV according to the pest.
Spray Acephate/Fipronil/Ethofenprox/Acephate + Pongamia oil in rotation for thrips control.
Spray synthetic acaricide/botanical in rotation to control yellow mite.

Fig. 23.20 Root-knot nematode infestation on bell pepper

23.5 Nematodes

23.5.1 Root-Knot Nematodes, *Meloidogyne* spp

23.5.1.1 Symptoms

Root galls caused by root-knot nematodes on sweet pepper are frequently small (Fig. 23.20).

23.5.1.2 Management

23.5.1.2.1 Physical Methods

Burning of paddy husk or saw dust on infested nursery proved to be reasonably effective.

23.5.1.2.2 Cultural Methods

Several cultural methods like selection of nematode-free nursery sites; destruction of infested roots after crop harvest; crop rotation with antagonistic crop (marigold) or intercropping with marigold, onion, and garlic; flooding; fallowing; and deep summer plowing either alone or in combination proved to be reasonably effective and economical to check multiplication of root-knot nematodes on capsicum. Plowing not only leads to disturbance and instability in nematode community but also causes their mortality by exposing them to solar heat and desiccation. Use of plastic sheets for covering soil in greenhouse further enhances nematode reduction. Such an approach also helps in reducing the intensity of weeds, fungi, and bacteria in the soil.

23.5.1.2.3 Chemical Methods

Bare-root dip treatment of capsicum seedlings in 0.1 % Carbosulfan/Monocrotophos for 6 h eliminates root-knot infection.

23.5.1.2.4 Biological Methods

(a) **Antagonistic Bacteria**: Application of a nematicidal *B. thuringiensis* strain to capsicum in Puerto Rico reduced galling in roots due to *M. incognita* and increased yield significantly (Zuckermann et al. 1993).

(b) **Antagonistic Fungi**: Application of "Royal 350" (*Arthrobotrys irregularis* cultured on oat seed medium) at 140 g/m^2 a month before transplantation of capsicum resulted in good protection against root-knot nematodes. Among the egg parasites, efficacy of *Paecilomyces lilacinus* (commercially formulated as "Biocon" in the Philippines) has been found to be comparatively higher in suppressing the population of *Meloidogyne* spp on capsicum.

Treatment of the greenhouse bed with the formulation of *Pochonia chlamydosporia* at 50 g/m^2 was significantly effective in reducing the galling index and number of nematodes in roots and soil and increasing the percent parasitization of eggs by bioagent and also yields. Seed treatment with *P. fluorescens* alone and the nursery bed treatment with *P. chlamydosporia* alone were effective (Naik 2004).

23.5.1.2.5 Host Resistance

Capsicum cvs Mississippi-68, Santanka, Anaheim Chile, and Italian Pickling are reported to be resistant to root-knot nematodes. Black Indica, Naharia, and Pant C1 were reported to be resistant to *M. incognita*, while All Big was moderately resistant. California Wonder and Naharia were resistant to *M. javanica*, while Early California Wonder as moderately resistant. Naharia was reported to be resistant to *M. arenaria*.

The southern root-knot nematode (*M. incognita*)-resistant 'Carolina Wonder' and 'Charleston Belle' are recommended for use by greenhouse growers.

23.5.1.2.6 Integrated Methods

Integration of *P. fluorescens* and *P. chlamydosporia* in the greenhouse bed has proved significantly effective in reducing the root-galling index (*M. incognita*) and number of nematodes in the roots and soil and increasing the yield of the capsicum crop (Naik 2004).

In capsicum, the seedling stand was good where the combination of neem-based *P. fluorescens* and *T. harzianum* was used (4.4), followed by seed treatment with *P. fluorescens* + soil application of *T. harzianum* (Naik 2004).

References

Alabi O, Emechebe AM (1990) Field evaluation of seed treatment and foliar spray fungicides for the control of cowpea brown blotch induced by *Colletotrichum capsici*. Samaru J Agric Res 7:151–158

Gitaitis R, McCarter S, Jones J (1992) Disease control in tomato transplants produced in Georgia and Florida. Plant Dis 76:651–656

Howard RJ, Garland JA, Seaman WL (1994) Diseases and pests of vegetable crops in Canada. The Canadian Phytopathological Society and the Entomological Society of Canada, Ottawa

Naik D (2004) Biotechnological approaches for the management of wilt disease complex in capsicum (*Capsicum annum* L) and egg plant (*Solanum melongena*) with special emphasis on biological control. PhD thesis, Kuvempu University, Shimoga

Portree J (1996) Greenhouse vegetable production guide for commercial growers. Province of British Columbia Ministry of Agriculture, Fisheries and Food, victoria

Zuckermann BE, Dicklow MB, Ascosta N (1993) A strain of *Bacillus thuringiensis* for the control of plant parasitic nematodes. Biocontrol Sci Technol 3:41–46

24 Cucumber

Abstract

Cucumber is one of the most important greenhouse crops particularly because it can be grown round the year. As many as three crops under subtropical conditions do make it economically very profitable. However, due to such intensive production systems, the crop is often severely infested with several insect and mite pests and diseases including the fungal, bacterial, and viral diseases. Their symptoms and damage, biology, spread, and management are discussed. Integrated pest management (IPM) in greenhouses is one of the most important approaches for successful pest control.

Keywords

Cucumber • Gherkins • Pests • Diseases • Symptoms • Biology • Spread • Management

24.1 Introduction

Cucumber and gherkins are being extensively grown in greenhouses in many countries. There are two types of cucumbers grown under protected cultivation. The continental or burpless cucumber is the main type of cucumber and has long fruit (Fig. 24.1), whereas the Lebanese (mini) cucumber has shorter fruit (parthenocarpic). Compared with field cucumbers, these two types are seedless and do not require pollination to produce fruits. To ensure they are not pollinated, they must be kept apart from field crops of conventional cucumbers. Compared with field crops, protected cultivated cucumbers have flesh that has less total acids (0.11 % versus 0.17 %) and more sugars (3 % versus 2.8 %). They are smoother and greener and have a finer texture. The skin is more tender, not peeled, and is easier to digest. The crops produced in greenhouses fetch higher income due to good-quality produce and higher productivity per unit area and also due to being available in off season (Table 24.1). The cropping period gets extended, so regular supply could be maintained for a long time.

24.2 Greenhouse Cucumber Varieties

Commonly grown greenhouse cucumber varieties are presented in Table 24.2.

P.P. Reddy, *Sustainable Crop Protection under Protected Cultivation*,
DOI 10.1007/978-981-287-952-3_24

Fig. 24.1 Cucumber under protected cultivation

Table 24.1 Greenhouse production of cucumber and income

Variety/crop	Area under greenhouse	Duration of crop	Total yield (t/1000 m^2)	Av. price of green fruit Rs/kg	Gross income (Rs/1000 m^2)
First crop of parthenocarpic cucumber (var. Kian)	1000 m^2	08.09.2011–12.01.2012	50	30	1,50,000
Second crop of parthenocarpic cucumber (var. Hilton)	1000 m^2	28.01.2012–25.5.2012	75	35	2,62,500

Between crops, disinfect empty greenhouses with 1 % chlorine (dilute sodium hypochlorite, as the 12.5 % chlorine product, by 12.5 times). Do not plant too closely. Practice good hygiene, such as removal of prunings, old leaves, and rejected fruits.

24.3 Diseases

24.3.1 Anthracnose, *Colletotrichum orbiculare*

24.3.1.1 Symptoms

On cucumber leaves the spots start as water-soaked areas and expand into brown spots which are roughly circular, reaching about ¼ to ½ inch in diameter. Small, growing leaves may be distorted and severe spotting may cause entire leaves to blight (Fig. 24.2).

Leaf petiole and stem lesions are shallow, elongate, and tan. Lesions on fruit are roughly circular and sunken and contain pinkish spore masses in moist weather. It is favored by high relative humidity and temperatures.

24.3.1.2 Survival and Spread

The fungus may live for 2 years in the absence of a suitable host. The fungus can be seedborne, and this is often the source of primary inoculum. The spores depend upon water for spread and infection; warm and humid rainy weather at frequent intervals is necessary for disease development. The spotted or striped cucumber beetle can carry the spores from plant to plant within a greenhouse.

Table 24.2 Greenhouse cucumber varieties and their characteristics

Greenhouse cucumber varieties	Varietal characteristics
Carmen	Excellent yields, bitter-free fruit, excellent disease resistance, ideal for organic gardeners
Swing	Very good yields, long dark green fruit, very good slicing cucumber, excellent disease resistance
Baribal	Early maturing, heavy yields, short dark green fruit
Prima Top	Fairly compact, good disease resistance, long bitter-free fruits
Picolino	Mini cucumbers, juicy and crisp, good disease resistance

24.3.1.3 Management

The registered active ingredient, mancozeb, can be sprayed to control anthracnose. The resistant pickling cucumber cvs include Calypso, Chipper, Galaxy, Carolina, and Explorer, and the most popular resistant slicing cucumbers are Poinsett and Highmark II.

24.3.2 Gray Mold, *Botrytis cinerea*

24.3.2.1 Symptoms

Botrytis or gray mold is a disease which is one of the first problems encountered by the new greenhouse grower, mainly because of poor ventilation.

The disease symptoms include soft watery rot on flowers, stems, leaves, and fruits (Fig. 24.3) and may also cause damping-off in seedlings. It will grow on debris on the greenhouse floor.

The spores are borne on stalks and result in a gray fuzzy appearance. The disease often first affects the plant through pruning wounds and old flowers and then spreads to the rest of the plant.

24.3.2.2 Management

Gray mold is most prevalent in cool moist climates and optimum temperatures are 18–23 °C. High relative humidity will increase the disease. Check the relative humidity with a hydrometer.

A most important means of control is to adopt good hygiene. Do not space plants too closely. Remove old leaves, prunings, and debris from the greenhouse. Do not overwater.

The plants must be well ventilated especially in mid afternoon, so that moist air is removed. This will reduce humidity from 4.00 to 9.00 am, as high relative humidity at this time will result in high spore production. Botrytis will also be less prevalent in a tall greenhouse which has low relative humidity and where the polythene cover has good anti-condensate properties.

If necessary, use registered chemicals that contain the active ingredients chlorothalonil or mancozeb. If possible, spray immediately after removal of leaves and shoots and alternate with different chemicals.

24.3.3 Gummy Stem Blight, *Didymella bryoniae* or *Mycosphaerella cucumis*

24.3.3.1 Symptoms

Symptoms on leaves begin as water-soaked areas and become light brown and irregular in outline. Stem symptoms may include cankers which can girdle the stem and cause total plant loss. These lesions may occur where leaves or fruit have been removed. Symptoms also begin as water-soaked areas, often at the blossom end, becoming brownish and exuding gum (Fig. 24.4). Lesions caused by this fungus are characterized usually by the abundant production of small black fruiting bodies.

Infection may be by water splash, but in the greenhouse it is more likely by aerial dispersion of spores. High relative humidity and condensation from the polymer roof plus low night temperatures will increase this disease.

24.3.3.2 Management

The registered active ingredients, mancozeb or chlorothalonil, can be sprayed to control gummy stem blight.

24.3.4 Powdery Mildew, *Sphaerotheca fuliginea*

24.3.4.1 Symptoms

Powdery mildew is seen as a white growth under the leaves and later attacks the whole leaf and the upper leaves (Fig. 24.5). It is worse in autumn and winter.

Fig. 24.2 Cucumber anthracnose symptoms on leaf fruits

Fig. 24.3 Gray mold symptoms on cucumber

24.3.4.2 Management

Many varieties have some tolerance to powdery mildew but may produce lower yields. Registered pesticides that contain the active ingredient Triadimefon may be needed to control this disease. The relative humidity should also be not too high, and good ventilation is needed. Infection begins when surface moisture is present on the leaves, but then the fungus grows best under dry conditions.

24.3.5 Root Rot, *Pythium* spp

24.3.5.1 Symptoms

Pythium root rot destroys fine feeder roots, so that the plant has difficulty obtaining water and nutrients. Plants wilt and appear nutrient deficient. Roots are yellow to brown and the base of the stem may rot (Fig. 24.6). Root rots are induced by overwatering and poor aeration.

24.3.5.2 Management

Pythium may be controlled in hydroponic crops by using calcium hypochlorite at 1.5 g/1000 L (1 ppm) regularly in the water supply and nutrient solutions. *Pythium* can also be in the bore water and can be treated in a holding tank with a high rate of calcium hypochlorite (20 ppm chlorine) for 1 day before the water is applied to the cucumbers. Application of nutrients in warm water in winter may help to reduce the disease. Do not replant into slabs in which plants were infected with the disease.

24.3.6 Black Root Rot, *Phomopsis sclerotioides*

24.3.6.1 Symptoms

The disease has been recorded in several countries of northwestern Europe and elsewhere. It infects cucumber and melon, causing a brown rot in the cortical tissue of the root system. Soon a large number of sclerotia develop and the infected tissues turn black (Fig. 24.7). Severely infected plants wither and die. Infection is favored by cool weather. The pathogen survives in the soil for several years by means of sclerotia (Blancard et al. 1991).

24.3.6.2 Management

Reduce plant density and increase ventilation. Soil treatment with thiram at 5–7.5 g/m^2 or carbendazim at 2–4 g/m^2 before "macking up" the beds of fresh substrate and spraying the fungicides over beds at intervals prevented the substrates from quick reinfestation by the pathogen.

Fig. 24.4 Gummy stem blight on cucumber

Fig. 24.5 Powdery mildew symptoms on cucumber leaf

Application of bark pellets containing *Trichoderma* spp. to greenhouse soil infested with *Phomopsis sclerotioides* reduced black root rot of cucumber for 2 months.

24.3.7 Scab, *Cladosporium cucumerinum*

24.3.7.1 Symptoms

Scab mostly attacks cucumber, but also squash, melon, etc. It causes nearly circular or angular leaf spots on the leaves, which look water soaked. Fruit infection is more serious. Initially, water-soaked lesions about 1 cm long, with gummy exudations, develop. A corky tissue usually develops around the lesions, which finally develop a scabby appearance (Fig. 24.8). The pathogen survives on plant debris, and spores are air disseminated. The disease development is favored by temperatures of about 15–25 °C and RH over 86 % (Blancard et al. 1991).

24.3.7.2 Management

For disease control, greenhouse ventilation is the best control measure. The disinfestation of greenhouse soil and frames is also very important. Regular application of dithiocarbamates, iprodione, benzimidazoles, etc. is recommended as well. There are several resistant cvs against some races of the pathogens.

24.3.8 Downy Mildew, *Pseudoperonospora cubensis*

24.3.8.1 Symptoms

Symptoms are usually seen first on the lower, older leaves. Initial symptoms of downy mildew typically consist of angular, yellow spots on the upper leaf surfaces (Fig. 24.9). On the undersides of such spots, a purplish-gray fungal growth may be visible when there is high relative humidity or moist conditions prevail (Fig. 24.9). As the disease progresses, the yellow spots enlarge and become necrotic or brownish in the center, with the browning spreading to the margins of the spots. Such spots may merge to form large brown areas on the leaves. Diseased leaves are killed in this manner, if the disease is allowed to develop unchecked. Lack of photosynthetic tissue results in stunting of plants, reduced fruit size, and poor fruit set.

Fig. 24.6 Symptoms of *Pythium* root rot on cucumber

Fig. 24.7 Black root rot symptoms on cucumber

Fig. 24.8 Scab symptoms on cucumber fruit

24.3.8.2 Management

The following control measures are recommended:

- Avoid dew formation by providing adequate heating and ventilation. This is critical to reducing the conditions for infection and disease development.
- Remove all sources of infection, such as infected leaves and plants, and discard them, in garbage bags, away from the site.
- Ensure that plants are sufficiently spaced and that the canopy is well pruned and thinned to provide for adequate air circulation.
- Overwatering not only leads to overly soft, more vulnerable plants, but also to production

Fig. 24.9 *Left* – Yellow angular spots defined on upper surface of cucumber leaf. *Right* – Fungal growth of *Pseudoperonospora cubensis* on lower surface of cucumber leaf

of droplets of moisture at the margins of leaves (guttation) early in the morning that provide perfect infection sites for the downy mildew pathogen.

- Any cultural practice (e.g., misting) that increases surface leaf moisture will increase disease development when downy mildew spores are present in the air.
- If only a few spots are evident on a few leaves of one plant, the disease is in the early stages of development. Implement appropriate control procedures immediately. Ensure good coverage, particularly with protectant fungicides. Wherever possible, rotate fungicides to reduce the chances of developing resistance in the fungus to the material applied. Generally, use systemic fungicides in combination with protectant fungicides.

24.3.9 Angular Leaf Spot, *Pseudomonas syringae* pv *lachrymans*

24.3.9.1 Symptoms

Symptoms on leaves begin as small dark water spots which enlarge and become light brown and finally a pale bleached color. The larger lesions are restricted by the leaf veins giving an angular appearance to them (Fig. 24.10). Old lesions frequently become tattered and holes develop on the leaf. Yellow haloes may be seen around the lesions.

Fruit lesions are often slightly raised and corky and may produce a white crusty deposit. Under high humidity, there is often a bacterial gum exudate from the scabs. Lesions caused by this disease do not contain the black fruiting bodies characteristic of gummy stem blight.

24.3.9.2 Survival and Spread

Angular leaf spot on cucumber is seedborne and survives beneath the seed coat and on the infected plant debris, and upon germination the cotyledons become infected. It can also survive in crop residues for 10 months and in the soil for about 140 days at 5–15 °C. It is spread by rainfall and insects, and through cultural practices. Disease development is favored by warm temperatures (24–27 °C) and high relative humidity.

The bacterium is splashed from the soil by water and infects plants. It spreads from plant to plant during the cultural practices (Fletcher 1984). The bacteria are spread in water and are more common under overhead irrigation. It is worse in areas with high relative humidity.

24.3.9.3 Management

Strict hygiene, soil disinfestation, use of healthy seeds, and reduction of the wetness period are recommended measures against bacterial diseases in greenhouses. Reduction of nitrogen fertilizers is also important for tomato soft rot. Copper fungicides are the most effective chemicals. Some resistant cultivars have also been released for tomato speck and bacterial spot of tomato, but none of them is suitable for greenhouses (Fletcher 1984).

The registered active ingredients, Propineb and copper oxychloride, can be sprayed to control angular leaf spot.

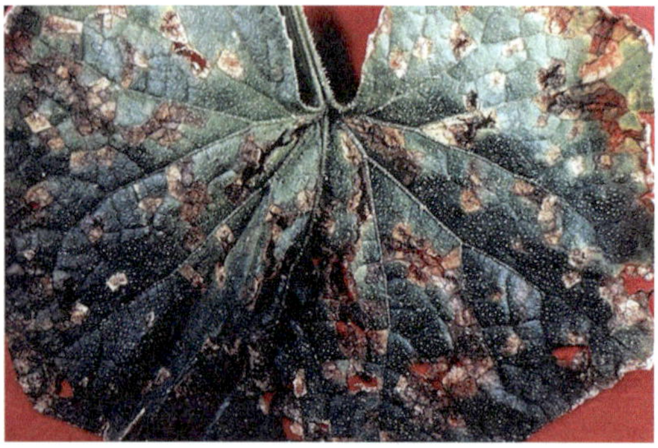

Fig. 24.10 Angular leaf spots of cucumber

Fig. 24.11 Symptoms of cucumber mosaic virus on leaves and fruits

24.3.10 Cucumber Mosaic Virus (CMV)

24.3.10.1 Symptoms

Fruits and leaves are both affected, as the leaves become mottled, distorted, and wrinkled, and their edges begin to curl downward (Fig. 24.11); the internal part of the fruit turns brown, making it unfit for consumption or processing.

24.3.10.2 Transmission

Cucumber beetles and over 60 species of aphids can pick up CMV from infected plants while feeding. Some of the most efficient aphid vectors are the cotton aphid, *Aphis gossypii*, and the green peach aphid, *Myzus persicae*.

24.3.10.3 Management

Apart from the aphid vectors, CMV is easily transmitted on garden tools and gardeners' fingers. Avoid handling healthy plants after working with suspected infected ones until tools or hands have been washed with soapy water. Destroy suspect plants promptly to reduce the risk of transmission. Keep the garden weed-free. Groundsel and chickweed are particularly likely to harbor CMV. Choose resistant cultivars such as 'Bush Champion', 'Crispy Salad', 'Jazzer F', 'Paskia Fi', 'Petita', and 'Country Fair'.

Fig. 24.12 Symptoms of cucumber green mottle mosaic virus

24.3.11 Cucumber Green Mottle Mosaic Virus (CGMMV)

24.3.11.1 Symptoms

In the early stage of infection, pale yellow spots develop on the leaves at the top of the plant. Stunting is common in the infected plants. In the late stage of infection, leaf mosaic and fruit mottling can be seen (Fig. 24.12).

In cucumber (*Cucumis sativus*) the type strain causes leaf mottling, blistering, and distortion (Fig. 24.12), with stunted growth. Yield losses may be 15 % (Fletcher et al. 1969); fruits are usually unmarked (Smith 1972), but some strains cause severe fruit mottling and distortion (Inouye et al. 1967). Some Asian cvs show no leaf symptoms but suffer yield losses (Kooistra 1968).

24.3.11.2 Management

Cucumber seeds heat treated at 70 °C for up to 3 days were free from active virus particles and were still able to germinate (Fletcher et al. 1969). Transplants and grafted cucurbits should be tested for the virus before being sold to growers. If the virus is found, all infected plant and plant debris must be removed and destroyed and the facilities cleaned.

24.3.12 Cucumber Vein Yellowing Virus (CVYV)

24.3.12.1 Symptoms

The virus causes pronounced vein clearing, chlorosis (Fig. 24.13), and finally general necrosis. A light to dark green mosaic appears on the fruit. The virus is transmitted mechanically and by the whitefly *Bemisia tabaci*, in a semi-persistent manner (Mansour and Al-Musa 1993).

24.3.12.2 Transmission

Cucumber vein yellowing virus (CVYV) is transmitted by *Bemisia tabaci*. Although a single whitefly can transmit the disease, the efficiency of transmission was low. At least 15–20 insects per plant were required to cause an infection of 55 % of inoculated plants. The minimum acquisition and inoculation feeding periods were 30 and 15 min, respectively.

24.3.12.3 Management

In protected crops, control relies on preventive and cultural practices (use of pest-free seedlings, adequate glasshouse window screens, double doors, treatment of infected vegetable residues and the introduction of a rest period of at least 1 month between two cucurbit crops, monitoring of *B. tabaci* populations). Control of CVYV depends on the control of its whitefly vector, *B. tabaci*. Regarding chemical control, *B. tabaci* appears to develop resistance to all groups of pesticides that have been developed for its control. A rotation of insecticides that offer no cross resistance should therefore be used to control *B. tabaci* infestations. The parasite *Encarsia formosa* is used as a biological control agent to control *T. vaporariorum*, but it is less efficient against *B. tabaci*. Repeated introductions of larger numbers of *E. formosa* are necessary if eradication is required. The predatory beetle *Delphastus pusillus* is very effective against *B. tabaci* (Anonymous 2000).

Fig. 24.13 Symptoms of cucumber vein yellowing virus

24.3.13 Cucurbit Yellow Stunting Disorder Virus (CYSDV)

24.3.13.1 Symptoms

Interveinal chlorotic spots appear on mature leaves. The yellow spots enlarge and may eventually coalesce, resulting in the yellowing of the entire leaf except for the veins which remain green. The leaves may roll up and turn brittle (Fig. 24.14).

24.3.13.2 Transmission

Cucurbit yellow stunting disorder virus is spread from plant-to-plant exclusively by the whitefly vector, *Bemisia tabaci*. All biotypes of *B. tabaci* known to exist in North America can transmit the

Fig. 24.14 Symptoms of cucurbit yellow stunting disorder virus

virus efficiently, including biotypes A, B, and Q. Cucurbit yellow stunting disorder virus is not transmitted by the greenhouse whitefly (*Trialeurodes vaporariorum*).

24.3.13.3 Management

Currently, a number of management strategies are recommended to minimize the chance of this virus becoming established and/or causing significant losses to cucurbit production.

i) Before Growing Season

- Although the virus is not seed transmitted, it is important to use pathogen-free, high quality seed.
- Use virus- and whitefly-free transplants.
- Do not import cucurbit (or any potential whitefly host) transplants from areas known to have the virus.
- Manage whiteflies on transplants.

ii) During Growing Season

- Plant immediately after any cucurbit-free period (either an arranged regional host-free period or true winter season).
- Soil application of a neonicotinoid insecticide (imidacloprid, dinotefuran, or thiamethoxam) at transplanting. Recent research results indicate that dinotefuran may slow the spread and development of the disease, whereas imidacloprid and thiamethoxam may be less effective.
- Monitor whitefly populations throughout the growing season and implement insecticide application as needed. Rotate insecticides with different modes of action group members to minimize development of insecticide resistance.
- Practice good weed management.

iii) After Growing Season

- Sanitation is very important; remove and destroy old crops/volunteers on a regional basis.
- Implement a voluntary or enforced regional cucurbit-free period to eliminate the virus from the cropping system.

24.4 Insect and Mite Pests

24.4.1 Sciarid Fly, *Bradysia paupera*

24.4.1.1 Symptoms

It is similar to a sandfly in appearance and has long legs (Fig. 24.15). It may eat the roots of cucumbers planted in media such as coco peat. It has also been found on cucumber grown in rock wool. Mushroom flies are worse in spring and autumn. The larvae normally feed on organic matter and fungi. They will also hollow out the stems of young seedlings at the base of the plant and cause plants to fall over. They also attack older plants. The pest has caused severe damage in some greenhouses and may also spread *Pythium* fungus disease. The fly lays its eggs in algae. If the trickle outlet is placed further down into the pot, the surface of the media may be drier and result in fewer algae.

24.4.1.2 Management

Prevent entry to the growing area by using an insect-proof glasshouse or tunnel. Placement of insect-proof screens can increase the humidity in the structure, causing ventilation problems. It is recommended to use a protected cropping consultant/designer before retrofitting or building an insect-proof tunnel or glasshouse. This method will only be successful if growing media and incoming plants are free of fungus gnats.

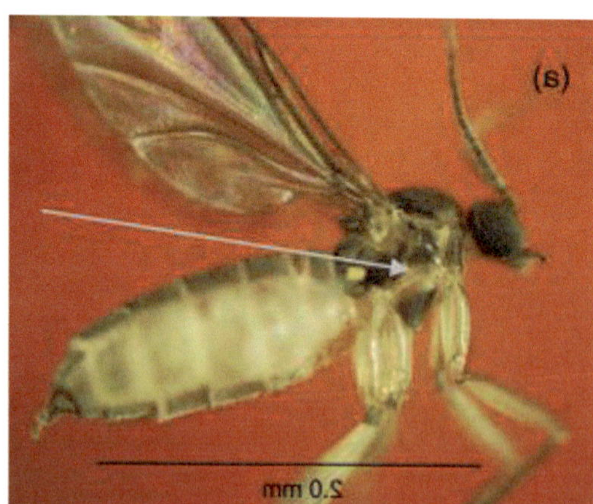

Fig. 24.15 Adult sciarid fly has a yellow band running below its wing (see arrow in photo) and is relatively large

Avoid excess watering. Fungus gnat numbers are lower when moisture levels are relatively low.

The active ingredient *Bacillus thuringiensis israelensis* is an organic product (under permit) that can be used to control fungus gnats. Other biocontrol agents include *Geolaelaps aculeifer*, *Stratiolaelaps scimitus*, *Dalotia coriaria*, and entomopathogenic nematodes (Table 24.3).

24.4.2 Whitefly, *Trialeurodes vaporariorum*

24.4.2.1 Symptoms

The main pest in greenhouse cucumbers is the greenhouse whitefly. It is a small sucking pest which may build up to large numbers (Fig. 24.16). Whiteflies damage greenhouse cucumbers by covering fruit and leaves with the sticky honeydew as they feed. High whitefly populations also reduce the vigor of plants. It is favored by warm temperatures and high relative humidity.

24.4.2.2 Management

- It is essential to start with a clean crop at the beginning of the season; therefore destroy all crop residues and dispose of them at a site remote from the greenhouse.
- If the previous crop was infested with whiteflies, leave the greenhouse entirely empty of plants for 5 days, with heat, to starve whiteflies; or, kill them with cold by allowing the greenhouse to freeze below −1 °C for a week.
- If a plant-free period is not possible between crops, use a short-residual fumigant such as naled (Dibrom). Apply at the end of the crop, before removing plant debris, and again to the empty greenhouse. Add insecticidal soap or dormant oil to the spray used to wash down the greenhouse structure and floor between crops to kill whitefly and other pests that often move to the walls or ceiling or hide in protected areas.
- Keep the greenhouse weed-free and maintain a 3-m-wide, weed-free border around the greenhouse.

Table 24.3 Biological management of sciarid fly

Biological management option	Low populations < 20 per week	Moderate populations 20–50 per week	High populations 50+ per week
Biological management option 1	Release *Geolaelaps aculeifer, Stratiolaelaps scimitus*, or *Dalotia coriaria* on a regular basis at a preventative rate[a]. Preferably, release when potting up and once 2 weeks later	Release *G. aculeifer, S. scimitus*, or *D. coriaria* at moderate rate[a] weekly for three consecutive weeks	Release *S. scimitus or G. aculeifer* and nematodes at a relatively high rate[a, b] for three consecutive weeks
Biological management option 2	Apply nematodes (*Steinernema feltiae*)[b] or *Bti* (*Bacillus thuringiensis israelensis*) on a regular basis at a preventative rate[b], preferably starting when first potting up. Alternate between *Bti* and nematodes at fortnightly intervals	Release nematodes three consecutive weeks at moderate rates[b]	If the population is particularly high, apply *Bti* once per week in addition to predatory mites and nematodes

[a]Spread 1 L of *G. aculeifer*, *S. scimitus*, or *D. coriaria* over 100–150 m^2 for preventative rates, 50–75 m^2 for moderate rates, and 30 m^2 for high rates

[b]Low pressure nematodes spray to drench. Use one 55 million pack of nematodes in 100–200 l of water covering 200, 150, and 100 m^2, at low, medium, and high rates, respectively. Nematodes may also be applied through irrigation lines, but filters must be removed

- Do not keep ornamental plants in a cucumber greenhouse as these are also whitefly hosts.
- If greenhouse whitefly numbers are high, hang yellow sticky tapes (up to 1 tape per plant) at the top of the plant canopy to trap adult whiteflies. If sweet potato whitefly is present, hang traps about 1 m below the top of the plants.
- If whiteflies are present on outdoor plants, screen all entry points.

It may be controlled by spraying with endosulfan. This has a 1-day withholding period. Whiteflies can be controlled by the parasitic wasp, *Encarsia formosa*, and the predatory beetle, *Delphastus*.

Fig. 24.16 Whiteflies on cucumber leaf

24.4.3 Red Spider Mite, *Tetranychus urticae*

24.4.3.1 Symptoms

Red spider mite is a major problem in warm weather, especially where the relative humidity is low. Once two-spotted mites are well established in a crop, major yield loss is inevitable. Two-spotted mite (*Tetranychus urticae*), also known as "red spider" (Fig. 24.17), is a destructive pest and is usually found on the underside of the leaves. In severe infestations, large numbers of mites and eggs occur on the leaves and are protected by webbing. The development from egg to adult varies from a fortnight in cool conditions, to 5 days during a heat wave. Two-spotted mites cause the leaves to become mottled, yellow, and shriveled. They prefer hot and dry conditions and the use of overhead sprinklers during high temperatures may reduce populations and also cool the plants.

Fig. 24.17 Adult male (smaller individual) and adult female of *T. urticae*

24.4.3.2 Management

A predatory mite (*Phytoseiulus persimilis*) can help to reduce numbers of two-spotted mite. These are fost, bright orange and the adults are a little larger than two-spotted mite. They are difficult to manage as they do not like cold or hot conditions, and it will take 6–8 weeks to control a well-established population of two-spotted mites. They must be introduced when the first two-spotted mites are seen. It is best to introduce them in late winter/early spring and keep plants cool by misting in summer. Introduce 100 predators per 1,000 plants each week throughout the growing season. Optimum temperatures are 25–30 °C, with high relative humidity. The predatory mite is tolerant to most fungicides and some miticides and insecticides. Propargite (active ingredient) may be used in conjunction with predatory mite to control two-spotted mite, especially in temperatures above 30 °C, which is tolerated more by two spider mites than the predatory mite. Other pesticides may have a severe effect on predatory mites and should not be used.

Predatory mites are killed by some classes of pesticides such as synthetic pyrethroids. Avoid using these chemicals.

If biological control is not used, spraying is necessary as soon as the mites appear on the leaves (1–2 per leaf). If the infestation is not noticed at an early stage, huge populations soon build up and are difficult to control. Applications of spray material must be thorough with both sides of the leaves wetted. The active ingredient abamectin (permit) and botanical oils (registered) can be used to control two-spotted mites, but do not overuse the botanical oils.

TSSM can be successfully managed on greenhouse cucumbers by a combination of biological and chemical control. When TSSM were treated with Bifenthrin and *Phytoseiulus persimilis*, populations remained at a low level (Saeid and Sultan 2010).

24.4.4 Thrips, *Frankliniella occidentalis*

24.4.4.1 Symptoms

These are small insects that may cause bent fruits and scarring. Whiteflies are small delicate insects (Fig. 24.18) that feed on the leaves and may cause curled fruit and black, sooty mold deposits on the leaves.

24.4.4.2 Management

Releasing the beneficial *Amblyseius cucumeris* was able to effectively control thrips in greenhouse cucumbers (Fig. 24.19).

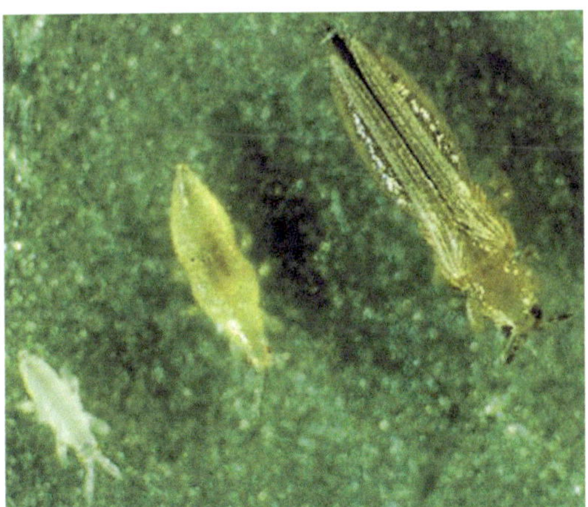

Fig. 24.18 First and second larval instars and adult of western flower thrips

The biological agent, *Encarsia formosa* wasps are about 0.6 mm in size, and they lay their eggs on the small whitefly nymphs. They can be controlled with the biological agent, *Typhlodromips montdorensis*, which is a mite.

They may be controlled with the pesticide Buprofezin under permit. The active ingredients imidacloprid (permit) and Spinosad (registered) can be used under permit to control thrips.

24.4.5 Aphids, *Aphis gossypii*

24.4.5.1 Symptoms

The melon aphid, *Aphis gossypii* (Fig. 24.20), can be a serious problem in greenhouse cucumbers. New infestations are usually detected first on cucumber leaves and stems. The melon aphid reproduces very rapidly on cucumber.

24.4.5.2 Management

All three biological control agents listed below should be released as soon as possible.

- Release parasitic wasp (*Aphidius matricariae*) to parasitize the melon aphid.
- Release aphid midges (*Aphidoletes aphidimyza*) between March 15 and August 15, at a high rate, weekly, until aphids are controlled.
- Release parasitic wasp (*Aphidius matricariae*) to parasitize the melon aphid.
- Release the lady beetle, *Harmonia axyridis*, to control all species of aphids. If the aphid infestation is widespread, also release field-collected *Hippodamia convergens* as well.

The active ingredients imidacloprid (under permit), botanical oils (registered), and Spinosad (registered) can be used to control aphids.

24.4.6 Slugs and Snails

24.4.6.1 Symptoms

Slugs and snails are a danger to young cucumber plants. They feed on cucumber leaves and fruits (Fig. 24.21) and destroy them in one night, and there is nothing more disheartening than seeing

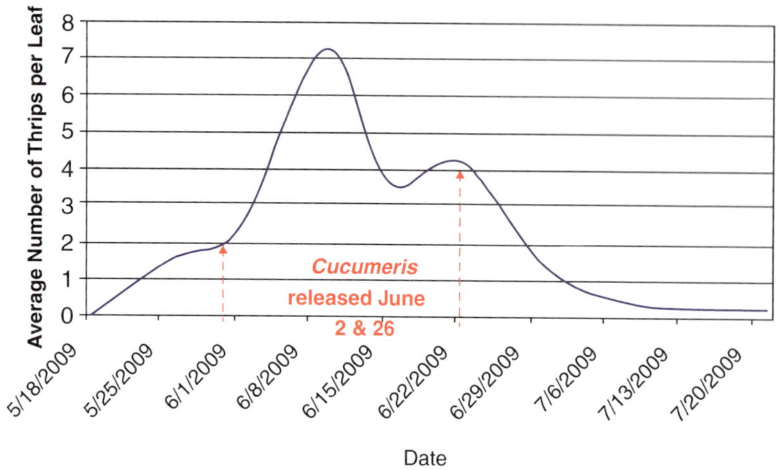

Fig. 24.19 Biological control of thrips on cucumber using *Amblyseius cucumeris*

Fig. 24.20 Aphid infestation on cucumber leaf

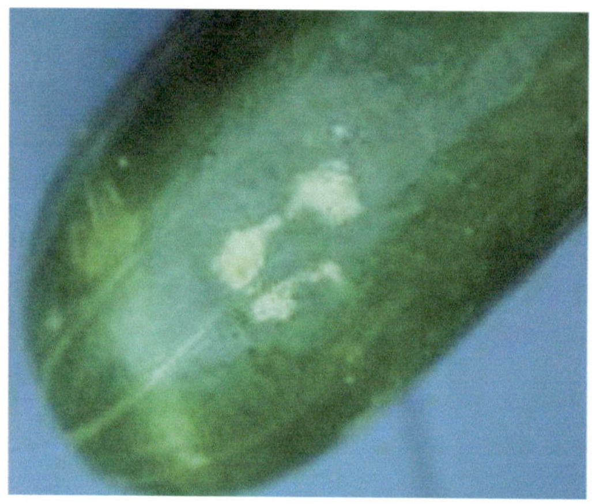

Fig. 24.21 Slug feeding damage on cucumber fruit

the heads of tray seedlings fall victim to the nocturnal slippery slimes.

24.4.6.2 Management

Slugs and snails love decaying leaves – indeed that is their main diet, not your plants. So remove pulled weeds, fallen leaves, and the like. Planting out small plants rather than seedlings and a tray or two of beer for any passing slugs is the solution. Snails and slugs love the malty smell of beer, yet alcohol is a poison for them.

24.5 Pest and Disease Management

Diseases commonly encountered in greenhouse cucumber production include Botrytis, or gray mold, powdery mildew, and mosaic viruses. Inadequate ventilation, particularly during periods of high humidity, increases damage from Botrytis. Powdery mildew is encouraged by high humidity and the excessive use of nitrogen fertilizers. Removal of a plant's lower leaves to increase air flow around the plant, controlling greenhouse humidity, and avoiding over application of nitrogen fertilizers can help reduce the damage caused by these plant diseases.

Many viruses are transmitted to crops from other hosts or plants by insect pests. Aphids, commonly encountered in greenhouse growing situations, are able to transmit a number of viruses that affect cucumbers.

Developing and following IPM practices is the most effective method for dealing with pests. Examples of IPM strategies for greenhouse cucumber production include using sticky traps to monitor for pest presence, controlling unwanted vegetation in and around the greenhouse to help exclude aphids from the area, and positively identifying insects as pests to determine the most appropriate control methods.

The best "cure" is prevention and sticking to the following points will help.

- Keep your plants as free of aphids as you can – they spread diseases.
- Avoid cold or overly wet growing conditions.
- Avoid splashing the fruit when you water.
- Avoid cucumbers resting on the ground. Provide a tile or flat stone for them to rest on.
- Keep the greenhouse or polytunnel well ventilated.
- Water regularly, but avoid overwatering – the soil should never be soggy.

References

Anonymous (2000) Current recommendations for eradication and containment, PHSI handbook of instructions. MAFF, York

Blancard D, Lecoq H, Pitrat M (1991) Maladies des cucurbitacées. Revue Horticole/INRA, Paris

Fletcher JT (1984) Diseases of greenhouse plants. Longman Inc, New York

Fletcher JT, George AJ, Green DS (1969) Cucumber green mottle mosaic virus, its effect on yield and its control in the Lea Valley, England. Pl Path 18:16

Inouye T, Inouye N, Asatani M, Mitsuhata K (1967) Studies on cucumber green mottle mosaic virus in Japan. Nogaku Kenkyu 51:175–186

Kooistra E (1968) Significance of the non-appearance of visible disease symptoms in cucumber (*Cucumis sativus* L) after infection with Cucumis virus 2. Euphytica 11:136–140

Mansour A, Al-Musa A (1993) Cucumber vein yellowing virus; host range and virus vector relationships. J Phytopathol 137:73–78

Saeid A, Sultan C (2010) Integrated control possibilities for two-spotted spider mite *Tetranychus urticae* Koch (Acarina: Tetranychidae) on greenhouse cucumber. Int J Acarol 36:259–266

Smith KM (1972) A textbook of plant virus diseases. Longman, London, 684 pp

25 Cole Crops

Abstract

Cultivation of cole crops under controlled environments or greenhouse is one of the most promising measures. Temperature plays a vital role with regard to growth and development of cole crops. The major insect and mite pests (flea beetle, cabbage worm, diamondback moth, thrips, cabbage maggot, cutworms, aphids, tarnished plant bug, slugs), diseases (black rot, black leg, wire stem/damping-off, leaf spot, downy mildew, club root, stem rot, white rust, powdery mildew, broccoli head rot), and root-knot nematodes (their symptoms, biology, spread, and management) are discussed.

Keywords

Cabbage • Cauliflower knol-khol • Pests • Diseases • Nematodes • Symptoms • Biology • Spread • Management

25.1 Introduction

The performance of cole crops (cauliflower, knol-khol, and cabbage) under protected environment and in open-field conditions was studied. The seedlings transplanted outside the greenhouse germinated late due to low temperature and chilling injury in winters. Greenhouse culture leads to 10–15 times higher yield than that of outdoor cultivation, depending upon the greenhouse design, availability of environment control facilities, cropping systems, greenhouse management, and crop type. The cauliflower (*cv* Aghani), knol-khol (*cv* Earliest white), and cabbage (*cv* Pride of India) gave higher yield inside the greenhouse, and the same was lowest in outside of greenhouse. No. of leaves and spread in both directions was higher inside the greenhouse as compared to outside of the greenhouse. Early flowering was also obtained under protected environment. The growth and yield of all these vegetable crops were found encouraging due to utilization of high rate of carbon dioxide inside the greenhouse (Chandra et al. 2000). Circumference of curd, gross weight of plant, and curd weight for cauliflower, knol-khol, and cabbage were found higher inside the greenhouse, and the same was found lower in open field (Table 25.1) (Agrawal et al. 2003).

P.P. Reddy, *Sustainable Crop Protection under Protected Cultivation*,
DOI 10.1007/978-981-287-952-3_25

Table 25.1 Performance of cauliflower, knol-khol, and cabbage inside the low-cost greenhouse and in open-field condition

	Cauliflower		Knol-khol		Cabbage	
Observation	Inside	Outside	Inside	Outside	Inside	Outside
No. of leaves	18.20	15.42	19.18	16.65	11.20	10.25
Spread EW (cm)	85.16	80.92	40.35	38.12	42.90	48.18
Spread NS (cm)	92.00	89.20	45.70	42.23	39.32	42.70
Length of leaves (cm)	49.68	38.10	30.12	22.80	40.22	31.69
Width of leaves (cm)	28.75	24.62	20.15	19.50	32.17	29.26
Circumference of curd (cm)	49.65	44.60	27.28	23.95	48.63	42.35
Av. gross wt. of plant (kg)	2.240	1.860	0.506	0.415	2.123	1.885
Av. curd wt. (kg)	0.985	0.796	0.291	0.220	1.540	1.206
Yield (kg/m^2)	5.47	4.42	3.61	2.22	8.55	6.70

25.2 Diseases

25.2.1 Black Rot, *Xanthomonas campestris*

25.2.1.1 Symptoms

Black rot lesions first appear at margins of leaves. The tissue turns yellow and the lesion progresses toward the center of the leaf, usually in a V-shaped area with the base of the V toward the midrib. Lesions may coalesce along the leaf margin to give plant a scorched appearance. The veins become dark and discoloration frequently extends to the main stem and proceeds upward and downward (Fig. 25.1).

25.2.1.2 Survival and Spread

Black rot bacterium is capable of surviving in the soil for 3 years without another cole crop present. Pathogen is spread via infected seed or by splashing rain or irrigation water and insect movement; disease emergence favored by warm and humid conditions.

25.2.1.3 Management

The following practices are recommended:

- Hot water treatment of seeds at 50 °C for 30 min (Shekhawat et al. 1982).
- Seed treatment with 0.1 % mercuric chloride for 30 min, 100 ppm Plantomycin (Kishun 1984), 100 ppm Agrimycin-100 (Rao and Kishun 1986), 100 ppm streptocycline, or 0.01 % Aureomycin (Trivedi and Patel 1972).
- Drenching of seed beds with 5 % commercial formalin or 1 % Bordeaux mixture.
- Spraying cauliflower with 100–200 ppm Agrimycin-100 (Bera 1986) and 500 ppm carbendazim (Saini and Parashar 1981) and cabbage with erythromycin (Mondel and Mukherjee 1978) was found effective in reducing the disease incidence.
- Strict field sanitation will help in checking the disease.
- Rouging of diseased plants helps in reducing the disease incidence.
- Destroy *Brassica* weeds and thoroughly incorporate plant debris and plant-resistant varieties.
- Good air circulation and water drainage is critical in controlling this disease, along with avoiding water on the crop in the afternoon and evenings.

25.2.1.3.1 Biomanagement

Bacillus subtilis was found effective in managing black rot of cabbage when applied as seed treatment, seedling root dip treatment, and soil drenching (Bhattacharya 1996).

Jalali and Parashar (1995) reported that Hsb-19 strain of *Bacillus* was an effective and safe means to suppress the progression of black rot disease. They recorded higher protection than

Fig. 25.1 Symptoms of *black rot*on cabbage leaves

that obtained with the application of streptocycline when the antagonist was applied 24–74 h before inoculation with *X. campestris* pv *campestris*. This provided 83.15 % disease control over the pathogen inoculated check.

25.2.2 Black Leg, *Phoma lingam*

25.2.2.1 Symptoms

Early signs of blackleg appear as small spots on leaves of young plants. On stems the spots are more linear and often surrounded by purplish borders. Stem lesions at the soil line usually extend to the root system causing dark cankers (Fig. 25.2). The fibrous root system may be destroyed although new roots sent out above the lesion may keep the plant alive. Many plants wilt abruptly and die. Damping-off of seedlings; round or irregularly shaped gray necrotic lesions on leaves with dark margins; lesions may be covered in pink masses in favorable weather conditions.

25.2.2.2 Favorable Conditions

Favored by warm, wet conditions; higher temperatures result in the development of more visible symptoms.

25.2.2.3 Survival and Spread

This organism is capable of surviving in the soil for 3 years without another cole crop present. This disease can be seedborne. The pathogen infects seedlings, forms pycnidia, and produces abundant amounts of spores which exude from the pycnidia in long coils and are splashed to nearby plants to initiate new infections. The disease is favored by wet, rainy weather.

Fig. 25.2 Symptoms of *black leg* on crucifers

25.2.2.4 Management

Use disease-free seed or treat with hot water to remove fungus prior to planting; remove and destroy crop debris after harvest or plow deeply into soil.

Use clean, certified seed or seed which has been hot water treated. Practice a 4-year crop rotation, destroy *Brassica* weeds, and thoroughly incorporate plant debris. Good air and water drainage is critical in controlling this disease, along with avoiding water on the crop in the afternoon and evenings.

25.2.3 Wire Stem/Damping-Off, *Rhizoctonia solani*

25.2.3.1 Symptoms

The soilborne fungi *Rhizoctonia* and *Pythium* cause two diseases of broccoli including damping-off and wire stem.

Pre-emergence damping-off occurs when seeds are attacked and decay, as well as when plants germinate, but fail to emerge. Post-emergence damping-off occurs when the stem of 2–5-cm tall plants is attacked. A water-soaked area completely encircles the stem near the soil line and the seedling wilts and topples over. Wire stem results from an extension of the damping-off process, but new infections may occur on plants 10–15 cm tall. The stem above and below the soil line darkens, and the outer cortex tissue decays and sloughs off in sharply defined area encircling the stem (Fig. 25.3). The stem is thin and wiry at the lesion but remains erect. The plant may survive, but will perform poorly.

25.2.3.2 Favorable Conditions

Disease emergence in seedlings is favored by cool temperatures.

25.2.3.3 Management

Management strategies include planting pathogen-freeseed or transplants that have been produced in sterilized soil, applying fungicide to seed to kill fungi, shallow planting of seeds, or delaying planting until soil warms.

For damping-off and wire stem in seedbeds, only sterilized soil or soil that has not previously had *Brassicas* for several years should be used. Seeds should be hot water treated and also treated with a suitable fungicide. Plant density should permit adequate light and air penetration. Factors such as deep planting, reduced seed vigor, and excessively cold, hot, moist, or saline soils that delay seed emergence should be avoided. Deficiencies of calcium, potassium, and nitrogen or excessive nitrogen may promote disease. A field rotation with non-*Brassica* crops should be practiced for at least 3 years. Avoid mounding of soil onto lower leaves when cultivating.

25.2.3.3.1 Biomanagement

Good control of damping-off was obtained through soil incorporation of wheat bran preparation of *Trichoderma harzianum* in commercial nursery. Seed treatment with 0.2 % carbendazim or benomyl or *Trichoderma* spp. was found to be as effective as soiltreatment for managing the disease. Seed treatment with *Trichoderma* spp. was found to be as effective as soil application for managing pre- and post-emergence damping-off in cabbage (Dar et al. 1997).

Seed treatment or soil application of *Aspergillus niger*, which multiplied in the rhizosphere of cauliflower, gave effective control of damping-off (Majumder 2000).

25.2.4 Leaf Spot, *Alternaria brassicae*, and *Alternaria brassicola (Mycosphaerella brassicicola)*

25.2.4.1 Symptoms

A. brassicae (gray leaf spot) causes small and light brown or gray lesions and *A. brassicola* (black leaf spot) causes larger and darker lesions. These diseases are seed- and soilborne. Small black spots (1–2 mm in diameter) appear on leaves, later turning into a tan color with target-like concentric rings (Fig. 25.4). When the spots dry out, the tissue falls from leaves, resulting in a "shot-hole" appearance. Cool temperatures, rain, and high humidity favor the development of this disease. Spots usually are most conspicuous on the outer, older leaves. The disease causes small brown sunken lesions and decay of broccoli heads, under very wet conditions. The spots enlarge in storage to sunken, black areas. *Alternaria* is a secondary fungus; it usually invades the plant after it has been injured by other pests or management practices.

25.2.4.2 Favorable Conditions

Disease emergence is favored by warm, rainy weather. The pathogen requires cool, moist

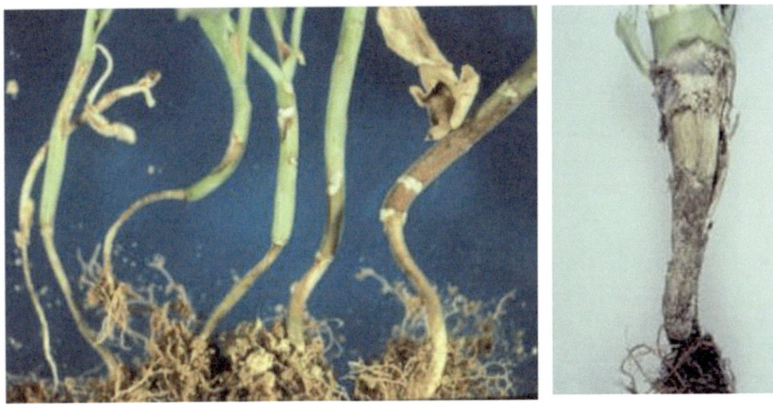

Fig. 25.3 Constricted stem of broccoli seedlings caused by damping-off/wire stem

Fig. 25.4 Symptoms of *black spot* on broccoli leaves

conditions to survive; disease symptoms typically develop in the fall and the peak of the infection occurs in winter.

25.2.4.3 Management

Seed treatment with hot water maintained at 50 °C for 30 min is effective. Giving wider spacing (60 × 50 cm) and removal of basal leaves from time to time help to reduce the disease incidence. Seed treatment with *Trichoderma viride* and *Pseudomonas fluorescens* and soil treatment of nursery beds with neem cake enhance seed germination and seedling stand. Spraying with Captan or copper oxychloride both at 0.3 % is effective in controlling the disease.

Practice long rotations between cole crops, avoid overhead irrigation, and make sure to incorporate plant debris. Good air circulation is needed in the field, as well as in storage. Keep storage temperature at 0 °C and relative humidity at 92–95 %.

25.2.5 Downy Mildew, *Hyaloperonospora parasitica*

25.2.5.1 Symptoms

Small angular lesions develop on upper surface of leaves which enlarge into orange or yellow necrotic patches. White fluffy growth develops on undersides of leaves (Fig. 25.5).

Once infected, the plant shows white, fuzzy masses in patches on the underside of leaves, stems, and heads. The tops of leaves turn purple and then later turn yellow or brown. Internally, the broccoli heads may show brown and black streaks on the main stalk and branches leading to the florets.

25.2.5.2 Survival

This fungus overwinters in seed and can survive for at least 2 years. It also overwinters on infected plants and can survive in the soil for at least 1 year.

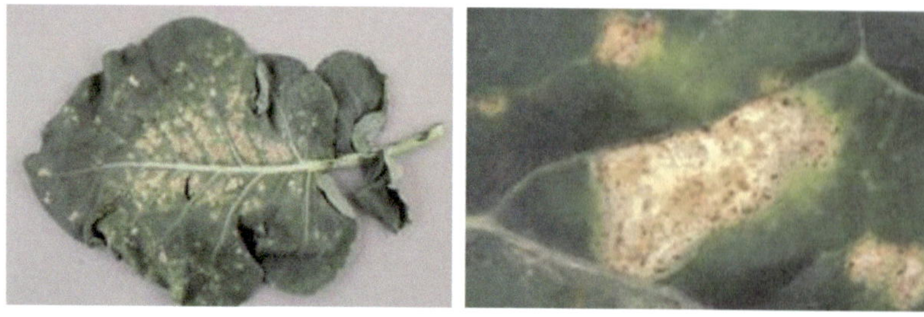

Fig. 25.5 Downy mildew symptoms on broccoli leaves

25.2.5.3 Favorable Conditions

Infection is favored by wet, cool weather, especially during prolonged periods of leaf wetness, such as during dew or fog.

25.2.5.4 Management

Good air and water drainage is critical in controlling this disease, along with avoiding water on the crop in the afternoon and evenings. Crop rotation with non-*Brassica* plants and incorporating plant debris will also aid in controlling this disease. Spraying with Ridomil MZ or Aliette both at 0.2 % or copper oxychloride at 0.3 % gives effective control of the disease.

25.2.6 Club Root, *Plasmodiophora brassicae*

25.2.6.1 Symptoms

Club root is a soilborne disease which affects broccoli. Early infections are difficult to detect as symptoms begin underground. Symptoms include small to large swellings and other malformations of the roots (Fig. 25.6). As a result of these swellings, water and nutrient flow are restricted within the plant, which causes the aboveground parts to wilt, turn color, and look stunted. Wilting is most common on warm sunny days; plants may show little wilting early in the morning or late at night.

The club root fungus enters the plant through the many fine hairs on the roots. The extent of the disease is affected by many factors. Moist, cool soils usually produce more diseased plants than dry, warm soil. The disease also thrives best in acid soils; that is when the pH is below 7. Once land becomes infested with this disease, it will remain so for several years.

25.2.6.2 Survival and Spread

When club root-affected plants rot and breakdown in the fall, the fungus spores are released into the soil, where they may live for 10–20 years, ready to infect any cole crop subsequently planted. Since the fungus spores are in the soil, movement of the soil by any means (boots, tools, wheels, wind and water, etc.) also spread the disease. The disease can also be spread by movement of contaminated soil and irrigation water to uninfected areas.

25.2.6.3 Management

Once the pathogen is present in the soil, it can survive for many years, and elimination of the pathogen is economically unfeasible. Plant only certified seed and avoid field-grown transplants unless produced in a fumigated bed. Applying lime to the soil can reduce fungus sporulation. Apply lime to raise the pH of the soil to at least 7.2.

25.2.7 Stem Rot (White Mold), *Sclerotinia sclerotiorum*

25.2.7.1 Symptoms

Leaves when infected turn light green and later yellow and are shed prematurely. Black soft rot often appears on petioles and midrib touching the soil and is covered with fluffy growth of the fungus. Stem at the ground level shows girdling. Fluffy mycelium of the fungus grows on these infected plant parts with numerous sclerotia embedded.

25.2.7.2 Favorable Conditions

Cool (20–25 °C) and humid (95–100 % RH) conditions are ideal for disease initiation and spread (Sharma and Sharma 1985).

Fig. 25.6 Club root-affected cabbage roots

Fig. 25.7 *White rust* on crucifer leaf

25.2.7.3 Management

Avoiding excess moisture with less frequent irrigation and reducing the plant populations by following wider spacing will create microclimate less congenial for disease development. Application of FYM or saw dust reduces the disease incidence. Soil amendment with sunflower inflorescence, rapeseeds, and pine needles reduce the disease incidence (Gupta et al. 1986). Remove diseased leaves and destroy them.

Application of benomyl and MBC both at 0.1 % was found effective in reducing the disease incidence. Soil application of *T. harzianum* (TH 4) and*A. niger* (Sanjeevini and Kalisena) was highly effective (Sharma et al. 2001).

Fig. 25.8 Powdery mildew on cabbage leaf

25.2.8 White Rust, *Albugo candida*

25.2.8.1 Symptoms

White pustules on cotyledons, leaves (Fig. 25.7), stems, and/or flowers which coalesce to form large areas of infection; leaves may roll and thicken.

25.2.8.2 Survival and Spread

Fungus can survive for long periods of time in dry conditions; disease spread by wind. Oospore-contaminated seeds and zoospore-infected flower buds perhaps are the most important sources of primary infection of white rust.

25.2.8.3 Management

Rotate crops; plant only disease-free seed; apply appropriate fungicide if disease becomes a problem. Spray fungicide metalaxyl (Ridomil), Blitox, wettable sulfur, Maneb, and Captan.

25.2.9 Powdery Mildew, *Erysiphe cruciferarum*

25.2.9.1 Symptoms

Small white patches appear on upper and lower leaf surfaces (Fig. 25.8) which may also show purple blotching; patches coalesce to form a dense powdery layer which coats the leaves; leaves become chlorotic and drop from plant.

25.2.9.2 Favorable Conditions

Disease emergence is favored by dry season, moderate temperatures, low humidity, and low levels of rainfall.

25.2.9.3 Management

Plant-resistant varieties; rotate crops; remove all crop debris after harvest; remove weeds; avoid excessive application of nitrogen fertilizer which encourages powdery mildew growth; powdery mildew can be controlled by application of sulfur sprays, dusts, or vapors.

25.2.10 Broccoli Head Rot, *Pseudomonas marginalis*

25.2.10.1 Symptoms

Symptoms appear after periods of rain when heads remain wet for several days. The bacteria are splashed up from the soil to the head. When heads are colonized by the bacteria, some areas appear water soaked (because a bio-surfactant is released by the bacteria) in contrast to unaffected areas where the waxy surface of the florets causes water to form in beads. Small black lesions may develop in these water-soaked florets. During long periods of wetness, decay spreads rapidly, resulting in a sunken area on the head (Fig. 25.9).

Fig. 25.9 Head rot on broccoli

Head rot develops most rapidly at high temperatures (28 °C). Frost injury and infection by downy mildew may also bring rise to this disease.

25.2.10.2 Management

Avoid high levels of nitrogen and avoid applying pesticides during head formation. Use resistant cultivars whenever possible and large plant spacings to increase air movement through the crop. Cultivars Green Defender and Shotgun consistently appeared highly resistant, while cv. Green Valiant was rated moderately resistant (Canady et al. 1991).

25.3 Insect Pests

25.3.1 Flea Beetle, *Phyllotreta crucifera*

25.3.1.1 Symptoms

Flea beetles are small shiny black beetles, about 2 mm in length. They are very active early in the growing season, especially during periods of dry sunny weather. Flea beetles can seriously damage seedlings and transplants and to a lesser extent larger plants, by chewing small pinholes through the leaves (Fig. 25.10). There is one generation per year. The larvae live in the soil and feed on roots.

Small holes or pits in leaves that give the foliage a characteristic "shot-hole" appearance; young plants and seedlings are particularly susceptible; plant growth may be reduced; if damage is severe, the plant may be killed; the pest responsible for the damage is a small (1.5–3.0 mm) dark-colored beetle which jumps when disturbed; the beetles are often shiny in appearance. Younger plants are more susceptible to flea beetle damage than older ones; older plants can tolerate infestation.

25.3.1.2 Survival

Flea beetles may overwinter on nearby weed species, in plant debris, or in the soil; insects may go through a second or third generation in 1 year.

25.3.1.3 Management

Plant seeds early to allow establishment before the beetles become a problem – mature plants are less susceptible to damage; application of a thick

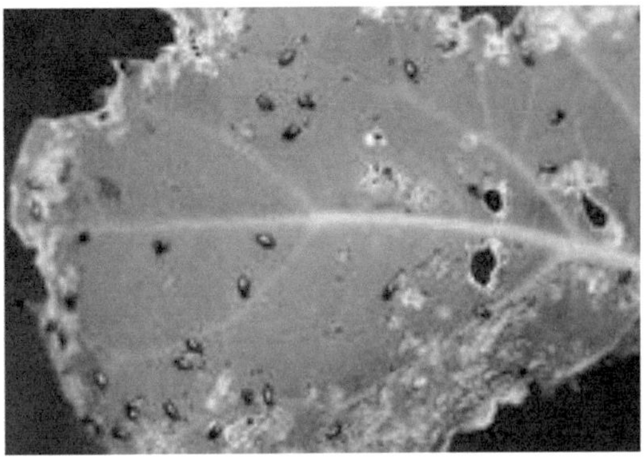

Fig. 25.10 Damage to broccoli leaf caused by crucifer flea beetle

layer of mulch may help prevent beetles reaching surface; application on diatomaceous earth or oils such as neem oil are effective control methods for organic growers. Using white or yellow sticky traps every 4.5–9.0 m and making sure to destroy plant debris are also good cultural control practices.

Biological control options for flea beetle include using a braconid wasp that will parasitize and kill adult flea beetles and using nematodes that attack the larvae.

If using chemical controls, scout plants frequently and treat when the threshold has been reached. One flea beetle per plant (up to the sixth leaf stage) is the threshold number. After the sixth leaf stage, feeding will not interfere with plant growth. Application of insecticides containing carbaryl, spinosad, bifenthrin, and permethrin can provide adequate control of beetles for up to a week but will need to be reapplied.

25.3.2 Cabbage Worm, *Pieres rapae*

25.3.2.1 Symptoms

The adult of the cabbage worm is a white butterfly, easily seen going from plant to plant laying eggs during the summer. The eggs hatch into velvety-green larvae with one thin yellow stripe down the centre of its back (Fig. 25.11). The cabbage worm larvae do not loop when they walk. They are generally the most prevalent of the caterpillars found on cole crops.

Large ragged holes in leaves (Fig. 25.11); green-brown frass (insect feces) on leaves; caterpillar is green in color and hairy, with a velvet-like appearance; may have faint yellow to orange stripes down back; slow-moving compared with other caterpillars.

Butterfly larvae cause damage by feeding on plants; can be distinguished from other caterpillars by its sluggish movement; in large numbers larvae can cause extensive damage very quickly.

25.3.2.2 Biology

The cabbage worm overwinters as a chrysalis in or near cruciferous crops. In Virginia, the adults emerge in March. Mating and egg laying occur within 24 h of emergence. Eggs are laid singly on the underside of the outer leaves of the host plant. Hatching occurs in 4–8 days. The larvae pass through 5 instars in 12–33 days and then form a chrysalis. Adults emerge in 8–20 days and have a life span of approximately 3 weeks. Female cabbage worm butterflies generally lay 200–300 eggs. In southwestern Virginia, the cabbage worm has 2–3 generations per year on cabbage. In the piedmont of North Carolina, it has 5–7 generations per year.

25.3.2.3 Management

Handpick caterpillars from plants and destroy; scrape eggs from leaves prior to hatching; apply appropriate insecticide if infestation is very heavy.

25.3.2.3.1 Biological Control

(a) ***Cotesia glomerata* (Hymenoptera: Braconidae):** *C. glomerata* (formerly *Apanteles glomeratus*) was the most common larval parasite of imported cabbage worm in southwestern Virginia. It attacks the first three instars of cabbage worm, ovipositing 20–50 eggs per host. Parasite larvae emerge *en masse* from the late fifth instar and spin characteristic yellow cocoons. Masses of yellow cocoons of parasite are seen on *Brassica* leaves.

(b) ***Pteromalus puparum* (Hymenoptera: Pteromalidae):***P. puparum* is the most com-

Fig. 25.11 *Left*, cabbage worm larva on broccoli leaf; *right*, broccoli plant damaged by cabbage worm while feeding

mon parasite of cabbage worm in southwestern Virginia. In surveys conducted from 1981 to 1984, parasitism of cabbage worm by *P. puparum* in southwestern Virginia was greater than 50 %. *P. puparum* is a gregarious internal parasite of the cabbage worm, attacking the newly formed pupa. Larvae develop within the host and emerge as adults through a small hole cut in the pupal case.

(c) ***Bacillus thuringiensis***:*B. thuringiensis* is a Gram-positive, spore-forming bacterium, which is very pathogenic to a large number of lepidopterous larvae. Different caterpillar species exhibit varying responses when fed these crystals and/or *B. thuringiensis* spores. The imported cabbage worm is highly susceptible to *B. thuringiensis*.

25.3.3 Diamondback Moth, *Plutella xylostella*

25.3.3.1 Symptoms

The diamondback moth is much smaller than the previous insects. Three to six generations of 1.1 cm yellow-green larvae may develop each year. The larvae squirm actively when disturbed and produce many small holes on the host plant. This pest can bore into the heads of cabbage. Diamondback moths do not survive the winter in this region. Adult moths migrate in throughout the growing season. Hence, there is often an overlap in generations, and all stages may be present at one time.

Young larvae feed between upper and lower leaf surface and may be visible when they emerge from small holes on the underside of the leaf; older larvae leave large, irregularly shaped shot holes on leaf undersides, may leave the upper surface intact; larvae may drop from the plant on silk threads if the leaf is disturbed; larvae are small (1 cm) and tapered at both ends; larvae have two prolegs at the rear end that are arranged in a distinctive V shape (Fig. 25.12).

Larvae take between 10 and 14 days to mature and spin a loose, gauze-like cocoon on leaves or stems to pupate.

25.3.3.2 Biology

It overwinters in the pupal stage, with adults emerging in early to mid-May. Mating occurs at dusk on the day of emergence and lasts about 1 h. Egg laying begins shortly after dusk and reaches its peak about 2 h later. Eggs are oviposited singly or in small groups (two to eight), mainly on the upper surface of the host plant leaves. Hatching occurs in 4–8 days. The larval period varies from 9 to 30 days, during which the larva passes through four instars. The mature larva constructs its cocoon typically on the lower leaves. Adults emerge in 5–15 days. The life span of the adult averages 2 weeks and during that time the female oviposits an average of 159 eggs. In Virginia, three generations of diamondback moth are known to occur on cabbage; however, the potential for more generations exists in areas of southwestern Virginia or the Carolinas where cabbage is in continuous cultivation or wild Cruciferae are present.

25.3.3.3 Management

Cultural controls include pheromone emitters to disrupt mating and evening overhead sprinkler irrigation.

If using chemical controls, scout plants frequently and treat when the threshold level has been reached. To determine the threshold level,

Fig. 25.12 Diamondback moth larva, pupa, and adult and damage to cabbage head

count the number of plants out of 25 randomly selected plants that have one or more caterpillars and then multiply by 4. This will give the percentage of plants infested. For broccoli, the threshold guidelines are 20–30 % before heading and 5–10 % after heading.

Larvae can be controlled organically by applications of *B. thuringiensis* or Entrust; application of appropriate chemical insecticide is only necessary if larvae are damaging the growing tips of the plants.

25.3.3.3.1 Biological Control

Diadegma insulare (=*Diadegma insularis*) (Hymenoptera: Ichneumonidae) is the major parasitoid of diamondback moth in Virginia. The female attacks diamondback moth larvae, especially the later instar larvae. A solitary parasite emerges from the prepupa shortly after the host has spun its cocoon. This parasitoid can control diamondback moth populations.

25.3.4 Thrips (Western Flower Thrips, Onion Thrips), *Frankliniella occidentalis*, and *Thrips tabaci*

25.3.4.1 Symptoms

Thrips (*T. tabaci*) are slender, yellow-brown insects about 1 mm long (Fig. 25.13). They feed by puncturing the leaves and sucking up the exuding sap. This causes the appearance of dark warts or blisters on the leaves. They also feed on broccoli heads, damaging them and making them unmarketable. They overwinter on refuse and weeds. Populations increase quickly when the air temperature is over 21 °C.

If population is high, leaves may be distorted; leaves are covered in coarse stippling and may appear silvery; leaves speckled with black feces; insect is small (1.5 mm) and slender and best viewed using a hand lens; adult thrips are pale yellow to light brown and the nymphs are smaller and lighter in color.

Thrips transmit viruses such as tomatospotted wilt virus; once acquired, the insect retains the ability to transmit the virus for the remainder of its life.

25.3.4.2 Management

Destroy refuse and control weeds. It is critical to control them at early head formation (7.5 cm leaf ball).

If using chemical controls, check plants frequently and treat when damage is first observed.

Use reflective mulches early in growing season to deter thrips; apply appropriate insecticide if thrips become problematic.

25.3.5 Cabbage Maggot, *Delia radicum*

25.3.5.1 Symptoms

The cabbage maggot or cabbage fly adults fly close to the ground near *Brassica* plants and lay elliptical white eggs on the stems of crops or in nearby crevices in the soil. The adult is a two-winged, ash grey fly, with black stripes on the midsection. It is half the size of a housefly, but has longer legs. Eggs hatch in 3–7 days. Larvae

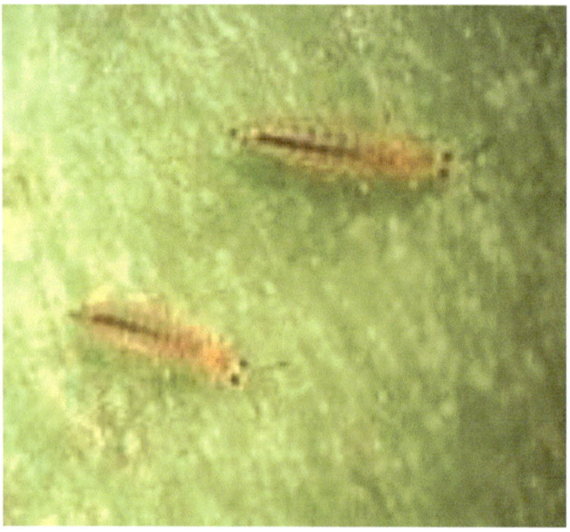

Fig. 25.13 Western flower thrips

are white, legless maggots (Fig. 25.14) that enter the roots and feed by rasping the plant tissue with a pair of hooklike mouth parts and tunneling into the roots. Feeding damage by the cabbage maggot causes roots to be misshapen and allows the entry of decay organisms and other species of maggots, resulting in stunted or killed young plants. Maggots mature in 3–4 weeks and pupate. The pupae are 6 mm long, oval, hard shelled, and dark brown. Pupae overwinter in the soil near the roots of the host plant. Adult flies emerge in 2–3 weeks; the first generation usually emerges in late May to early June. The presence of adult flies can be determined by looking for eggs which are laid at the base of plants. Generally, there are two to three generations a year.

25.3.5.2 Management

Natural enemies for the cabbage maggot include ground beetle, rove beetle, spiders, harvestmen or daddy longlegs, and ants.

If using chemical controls, scout plants frequently and treat when damage is first observed.

25.3.6 Cutworms, *Agrotis ipsilon*

25.3.6.1 Symptoms

Cutworms are grayish, fleshy caterpillars up to 5 cm long, which curl up when disturbed. Plants may be chewed off above or below ground level

Fig. 25.14 Cabbage maggot larvae on cauliflower stem

and may be damaged higher up by climbing cutworms. Most of the cutworm damage is to newly set plants in the field, but they are often found attacking seedlings in plant bed and greenhouses. Late infestation of variegated cutworm occasionally occurs.

25.3.6.2 Management

Prepare the soil 2 weeks before planting to cultivate in cover crops and destroy weeds. Check plants frequently and treat when damage is first observed.

25.3.7 Aphids, *Brevicoryne brassicae*

25.3.7.1 Symptoms

The cabbage aphid is a major pest of cole crops worldwide. Aphids are small, soft bodied, slow-moving insects. A colony consists of winged and wingless adults and various sizes of nymphs. Aphids may be black, yellow, or pink, but mostly are various shades of green. They are often found in large colonies on the undersurface of leaves (Fig. 25.15); however, aphids will feed on heads, flower stalks, as well as leaves, resulting in

Fig. 25.15 Cabbage aphids

unmarketable produce. Aphids feed by piercing plants and sucking out plant sap, resulting in distorted plant parts and a slowing of plant growth. The plants may be covered by a sticky substance, called honey dew, which is excreted by the aphids.

25.3.7.2 Management

There are many natural enemies that will feed on aphids, thus helping to reduce the populations of this pest in the field. Natural enemies that produce larvae which will feed on aphids include syrphid flies, lacewings, and the predaceous midge. Adults and larvae of minute pirate bugs, big-eyed bugs, lady beetles, soldier beetles, and parasitic wasps like *Diaeretiella rapae* will also consume aphids.

Cultural controls include using high-pressure sprinkler irrigation to knock the insects off of plants.

If using chemical controls, check plants frequently and treat when damage is first observed.

25.3.8 Tarnished Plant Bug, *Lygus lineolaris*

25.3.8.1 Symptoms

Adult tarnished plant bugs are light brown to reddish brown in color and about 5–6 mm in length (Fig. 25.16). They occur throughout the season and are very active and quick moving. They feed on broccoli heads causing dry, shriveled, and grayish to brown florets, which reduces the marketability of the head. Bacterial and fungal rots may invade these damaged tissues.

Fig. 25.16 Tarnished plant bug on leaves

Fig. 25.17 Slug damage on cabbage head

25.3.8.2 Management

Keep plantings and adjacent areas weed-free. Check plants frequently and treat when damage is first observed. For vegetable crops, dimethoate, endosulfan, and several of the synthetic pyrethroids are labeled for tarnished plant bug control.

25.3.9 Slugs

25.3.9.1 Symptoms

Slugs exist in various sizes up to 10 cm. They eat holes in the leaves and leave a trail of mucus, which makes plants unsightly (Fig. 25.17). The control of slug populations has been a continuing problem in the cole crop industry.

25.3.9.2 Management

Slugs prefer areas which are cool, moist, and high in organic matter. Since slugs can overwinter fairly easily, cultural practices aimed at controlling them should begin at least 1 year before the susceptible crop is put in. If urea (4 kg/ha) is sprayed on this cultivated strip, slug movement may be further impeded. The salt irritates the slugs as they move over it. Repeated applications are necessary. Slugs are usually more numerous in heavy, moist soils; sandy soils usually have fewer slug problems.

25.4 Nematodes

25.4.1 Root-Knot Nematodes, *Meloidogyne* spp.

25.4.1.1 Symptoms

Galls on roots which can be up to 3.3 cm in diameter but are usually smaller; reduction in plant vigor; yellowing of plants which wilt in hot weather.

25.4.1.2 Favorable Conditions

Galls can appear as quickly as a month after planting; nematodes prefer sandy soils and damage in areas of field or garden with this type of soil is most likely.

25.4.1.3 Management

Plant-resistant varieties if nematodes are known to be present in the soil; check roots of plants mid-season or sooner if symptoms indicate nematodes; solarizing soil can reduce nematode populations in the soil and levels of inoculum of many other pathogens.

References

Agrawal N, Mehta N, Sharma HG, Dixit A, Dubey P (2003) Cultivation of cole crops under protected environment. Karnataka J Agril Sci 16(2):332–334

Bera SC (1986) Control of black rot disease of cauliflower by some chemicals. Pesticides 20:51–52

Bhattacharya AK (1996) Management of black rot of cabbage caused by *Xanthomonas campestris* pv *campestris* (Pammel) Dowson. M.Sc. thesis, Assam Agricultural University, Jorhat

Canady CH, Wyatt JE, Mullins JA (1991) Resistance in broccoli to bacterial soft rot caused by *Pseudomonas marginalis* and fluorescent *Pseudomonas* species. Plant Dis 75:715–720

Chandra P, Sirohi PS, Behera TK, Singh AK (2000) Cultivating vegetables in polyhouse. Indian Hortic 45(3):17

Dar GH, Zargar MY, Beigh GM (1997) Biocontrol of *Fusarium* root rot in the common bean (*Phaseolus vulgaris* L) by using symbiotic *Glomus mosseae* and *Rhizobium leguminosarum*. Microbial Ecol 34:74–80

Gupta PK, Agarwal RK, Sharma SL (1986) Efficacy of soil amendments in managing 'stalk rot' (*Sclerotinia sclerotiorum*) of cauliflower. Indian J Plant Path 4:129–131

Jalali I, Parashar RD (1995) Biocontrol of *Xanthomonas campestris* pv *campestris* in *Brassica juncea* with phylloplane antagonist. Plant Dis Res 10:145–147

Kishun R (1984) Seed treatment for control of cabbage black rot. J Turkish Phytopath 13:81–88

Majumder S (2000) Biological control of *Rhizoctonia solani* and *Pythium aphanidermatum* on vegetables by *Aspergillus niger* v Tieg Strain AN27. Ph.D. thesis, Bidan Chandra Agriculture University, West Bengal, p 110

Mondel R, Mukherjee N (1978) Sensitivity of plant bacteria to some antibiotics, antibactericides and other drugs. Z Pfenkh Pflaschutz 85:607–616

Rao ANS, Kishun R (1986) A note on the control of bacterial black rot [*Xanthomonas campestris* (Pammel) Dowson] of cabbage (*Brassica oleracea* var *capitata* Linn) by seed treatment. Pesticides 20:47

Saini LC, Parashar RD (1981) Comparative efficacy of stable bleaching powder with other antibacterial formulations in controlling black and soft rot of cauliflower. Indian Phytopath 34:465–469

Sharma RC, Sharma SL (1985) Correlation of meteorological factors with stalk rot of cauliflower caused by *Sclerotinia sclerotiorum*. Indian J Ecol 12:327–330

Sharma P, Singh L, Adlakha D (2001) Antagonistic potential of *Trichoderma* and *Aspergillus* species on *Sclerotinia sclerotiorum* (Lib) de Bary causing rots in cabbage and cauliflower. Pestic Inform 2:41–44

Shekhawat PS, Jain ML, Chakravarti BP (1982) Detection and seed transmission of *Xanthomonas campestris* pv *campestris* causing black rot of cabbage and cauliflower and its control by seed treatment. Indian Phytopath 35:442–447

Trivedi BM, Patel PN (1972) In: Patel PN (ed) Plant bacteriology: vol. I. Bacterial diseases of plants in India. Indian Council of Agricultural Research, New Delhi, p 260

26 Lettuce

Abstract

The productivity and quality of lettuce under greenhouse is much higher compared to open field, and it is available from November to May. Various insect pests, such as cutworms, aphids, slugs and snails, American bollworm, *Plusia* looper, thrips, and whitefly, may attack lettuce. Lettuce is also affected by several diseases, such as Septoria leaf spot, Sclerotinia rot, bacterial rot complex, downy mildew, powdery mildew, and various virus diseases being the most important. Lettuce is also susceptible to attacks by nematodes, of which the root-knot nematodes, *Meloidogyne* spp, are the most important. Their symptoms, biology, spread, and management are discussed.

Keywords

Lettuce • Pests • Diseases • Nematodes • Symptoms • Biology • Spread • Management

26.1 Introduction

Greenhouse-grown lettuce (Fig. 26.1) is of better quality than outdoor crops, and it is available from November to May when the outdoor crops are not available. The variety 'Kwiek' is suitable for sowing in late August for winter lettuce and 'Emerald' and 'Kloek' in September/October for late winter and spring supplies. The productivity of lettuce under greenhouse is much higher compared to open field (Table 26.1).

26.2 Varieties

26.2.1 Butterhead Vatieties

Floribibb, Florida butter crisp, Ermosa, Margarita, Red Rouge, Carmona, Ostinata, Rex

26.2.2 Loose Leaf Varieties

Red (New Redfire, Vulcan), Green (Two Star), Oakleaf green (Saladbowl), Oakleaf red (Cerize)

P.P. Reddy, *Sustainable Crop Protection under Protected Cultivation*,
DOI 10.1007/978-981-287-952-3_26

Fig. 26.1 Rows of lettuce grown under protected cultivation

26.3 Diseases

26.3.1 Downy Mildew, *Bremia lactucae*

26.3.1.1 Symptoms

Bremia lactucae causes the most serious disease of lettuce under protected cultivation. It persists as oospores in soil and is air dispersed as sporangia. Lettuce plants can be infected at the seedling stage and all the way through the growing period. Lesions on leaves are first discolored (Fig. 26.2) and finally rot after sporulation of the pathogen.

26.3.1.2 Management

The treatments applied to seedling compost or soil against fungal pathogens is not very effective against *B. lactucae* oospores. If possible, soils in which significant downy mildew was seen on a previous lettuce crop should be avoided. Treatment of seedlings is essential, usually with sprays of dithiocarbamates (zineb or mancozeb), and continues after planting out. Because the dithiocarbamates are the fungicides most likely to present residue problems, it is most usual to use them on seedlings, or in the first 2 weeks after planting, and then use a systemic fungicide on the crop. Lettuce cultivars with resistance to *B. lactucae* are available. This is often through the combination of several vertical resistance genes, but numerous corresponding pathotypes of *B. lactucae* exist. It is not advised to use a resistant cultivar, without chemical treatment, relying on the absence of the matching pathotype, as the pathotypes which occur on the crop may not correspond to the resistance of the plant. However, under such circumstances, the number of treatments could be reduced, if no downy mildew is seen.

26.3.1.2.1 Problems with Fungicide Resistance

Resistance to phenylamides has been found in *B. lactucae*. It is recommended in general not to use phenylamides more than twice on a crop, preferably only once. They should not be used curatively, i.e., when disease is visibly present. The number of treatments is more important than the timing in reducing resistance risk.

26.3.1.2.2 Main Fungicides

Sprays Cymoxanil, folpet, fosetyl-aluminum, mancozeb, maneb, metalaxyl, oxadixyl, propamocarb, thiram, zineb.

26.3.2 Bottom Rot, *Thanatephorus cucumeris, Sclerotinia* spp, and *Botryotinia fuckeliana*

26.3.2.1 Symptoms

Several fungi [*Thanatephoruscucumeris* (anamorph *Rhizoctonia solani*), *Sclerotinia sclerotiorum*, and *S. minor*] cause a collar rot of lettuce, the inoculum coming from sclerotia in the soil. These rots then spread to the basal leaves and may eventually destroy the heart of the lettuce (Fig. 26.3). *Botryotinia fuckeliana* (anamorph *Botrytis cinerea*) can also persist as soilborne sclerotia, but, unlike the others, produces abundant air-borne conidia which can infect leaves directly, especially if these are damaged. Because of the close relationship between *Botryotinia* and *Sclerotinia*, the same fungicides are generally

Table 26.1 Yields of lettuce under greenhouses and open fields

Crop	Greenhouses			Open fields
	Yield/crop (t/ha)	No. of crops/year	Total yield t/ha/year	Total yield t/ha/year
Lettuce	31.3	10	313.0	52

Fig. 26.2 Downy mildew symptoms on lettuce leaf

effective against them (in particular, those of the dicarboximide group). *T. cucumeris*, a basidiomycete, requires quite distinct fungicides. The fungicides used against *B. lactucae* have little useful action against any of the fungi considered here.

26.3.2.2 Management

With soilborne fungi, it is essential to ensure that both the compost used for seedling production and the soil are free from the inoculum. Seedling compost should be based on new peat. If there is any indication that the soil may be infested, it should be treated by steam sterilization (to 10 cm depth for 15 min) if possible. Such treatment will also destroy *Olpidium brassicae*, the fungal vector of several lettuce viruses (lettuce big-vein varicosavirus, lettuce ring necrosis virus) and *Pythium* spp which may cause damping-off.

If steam sterilization is not an option, mepronil, pencycuron, or tolclofos-methyl can be used as a specific preplanting treatment against *T. cucumeris*. Alternatively, a single treatment may be used, immediately after planting, with mepronil or pencycuron or tolclofos-methyl.

For partial control of the sclerotia of *Sclerotinia* spp, the antagonist *Coniothyrium minitans* may be applied before planting. Any plants affected by *Sclerotinia* spp, and the surrounding soil, should be carefully removed to reduce any soil inoculum for future crops.

Foliar sprays can be applied to minimize airborne infection by *B. fuckeliana* and the spread of *Sclerotinia* spp between young plants.

26.3.2.2.1 Problems with Resistance

Resistance to benzimidazoles (as Benomyl) and dicarboximides (e.g., iprodione, procymidone, vinclozolin) is fairly common in *B. fuckeliana*, in view of their use on the many different hosts it attacks. It is possible then to substitute a fungicide from a different group, e.g., Tolylfluanid. If as many as three or four treatments are made, it is advisable to alternate fungicides from different groups.

26.3.2.2.2 Main Fungicides

Seed Treatments Mancozeb, maneb, mepronil, thiram.

Soil Treatments *Coniothyrium minitans*, dazomet, mepronil.

Sprays Benomyl, dichlofluanid, dicloran (or fog), iprodione, mepronil, procymidone, thiram, tolclofos-methyl, tolylfluanid, vinclozolin.

26.3.3 *Lettuce Big-Vein Varicosavirus*

26.3.3.1 Symptoms

The most common symptom in big-vein affected lettuce is a clearing along the veins (Fig. 26.4). Infected head of lettuce plants grow like leaf lettuce. Many lettuce types can be severely stunted. The virus is associated with the soilborne fungus *Olpidium brassicae*. The fungus persists in soils for very long periods of time. It is spread by soil and irrigation water. It also transmits the uncharacterized Lettuce ring necrosis virus.

Fig. 26.3 *Left* – Bottom rot, *Thanatephorus cucumeris*. *Right* – Lettuce drop, *Sclerotinia sclerotiorum*

Fig. 26.4 Lettuce big-vein symptoms

Lettuce big-vein varicosavirus can be especially severe when lettuce is grown under cool conditions (<16 °C) and the soil moisture is favorable for spread of *O. brassicae*.

26.3.3.2 Management

There are attempts to control the fungal vector by steaming of the soil. Use of uncontaminated water is recommended. It is possible to use Carbendazim against *O. brassicae*.

26.3.4 *Cucumber Mosaic Cucumovirus* (CMV)

26.3.4.1 Symptoms

Cucumber mosaic cucumovirus is usually a minor pathogen of lettuce, but it can be a significant problem in glasshouses. Symptoms can be very difficult to differentiate from those induced by *Lettuce mosaic potyvirus*. They include stunted plant growth and yellow mottling, distortion, and necrotic spots on leaves (Fig. 26.5). CMV is transmitted by aphids in the nonpersistent manner.

26.3.4.2 Management

Because CMV is acquired and transmitted by aphids in a matter of seconds and because of its wide host range, it is a difficult virus to control. Thus, spread may occur rapidly and regular control of aphids is necessary. Lettuce should not be grown near autumn crops of cucumber or near infected plants. CMV cannot be spread by mechanical means. The most effective method of control is the use of resistant cultivars.

26.3.5 *Lettuce Mosaic Potyvirus* (LMV)

26.3.5.1 Symptoms

Lettuce mosaic potyvirus is one of the commonest and potentially most damaging pathogens of lettuce. Symptoms vary depending on the type of lettuce and cultivar, the growth stage when infected, and environmental conditions. Seedling leaves originating from infected seeds appear irregular in shape and develop a light green mottle or mosaic (Fig. 26.6).

26.3.5.2 Management

All types of lettuce are susceptible. The virus is seed transmitted and sap transmissible. Effective control is provided by the use of virus-free seed and/or resistant cultivars. An annual lettuce-free period of 2 weeks is recommended. The virus is also transmitted in a nonpersistent manner by many species of aphids; therefore their control is needed. Alternate hosts should be removed, in particular the weed host *Senecio vulgaris*.

Fig. 26.5 Lettuce CMV symptoms

Fig. 26.6 LMV symptoms

26.4 Insect Pests

26.4.1 Aphids, *Myzus persicae*

26.4.1.1 Symptoms

Lettuce in protected cultivation is attacked by aphids, mainly *Myzus persicae*, *Macrosiphum euphorbiae*, *Aulacorthum solani*, *Nasonovia ribisnigri*, and *Hyperomyzus lactucae*. These aphids feed on the lower side of the leaves and even in the heart of the plant. They cause poor growth, malformation, and discoloration (Fig. 26.7). Their physical presence is not acceptable on the marketed product, even after treatment when dead insects and shed skins remain. *Myzus persicae* is also the vector of lettuce mosaic virus and cucumber mosaic virus, and one of the objectives of aphid control is to treat early so as to prevent any virus transmission within the crop.

26.4.1.2 Management

The need to treat against aphids is determined by their appearance in the crop. Yellow sticky traps can be used to monitor aphid flight, but this is not as reliable as monitoring the plants. Treatments should not be carried out routinely. Nevertheless, because of the virus-vector risk, it is important to be vigilant in aphid detection during the period immediately after planting. This is particularly important at times when aphids are most likely to be viruliferous. A single early treatment may be sufficient, but if more aphids appear, up to three treatments may be needed. Exclusion of aphids by insect-proofing is technically feasible but costly.

The range of insecticides which can be used is very wide. Pirimicarb is favored as a specific product against aphids, but some strains of some species are known to be resistant. *Myzus persicae* and *Nasonovia ribisnigri* strains found on protected crops in Europe are likely to be resistant to commercially available insecticides. Therefore, insecticides with different modes of action should be used in alternation. Biological control agents should be used wherever possible.

26.4.1.2.1 Main Insecticides

Sprays Cypermethrin, deltamethrin, dimethoate, dichlorvos, heptenophos, malathion, mevinphos, pirimiphos-methyl, pymetrozine.

Fogs Diazinon, dichlorvos, pirimicarb (aerosol dispenser), pirimiphos-methyl.

Smoke Diazinon, pirimicarb (smoke generator), sulfotep (fumigant/vapor-releasing product).

Fig. 26.7 Lettuce leaves infected with aphids

26.4.2 Noctuids, *Agrotis ipsilon*

26.4.2.1 Symptoms

The most important insect pests of lettuce in protected cultivation, apart from aphids, are the larvae of noctuid moths, especially *Agrotis ipsilon*, *A. segetum* (cutworms, i.e., soil-inhabiting noctuid larvae feeding on the roots), and *Autographa gamma*, *Mamestra brassicae*, and *Mythimna unipuncta* (leaf-feeding noctuid larvae, whose eggs are laid directly on the foliage) (Fig. 26.8). However, these pests are only sometimes important, occurring especially in crops grown in soil rather than in soil-less culture.

26.4.2.2 Management

Trapping adult moths can be used to determine the time of application of irrigation and/or sprays. Where there is regular irrigation, there are few problems with cutworm damage. For the soil noctuids, a single treatment (bait, granule, or spray) should be done if a pest problem is detected, or if pheromone-trap catches indicate a risk, just before or after planting. When pyrethroids are used, it is recommended to avoid the products to be used later against aphids or leaf-feeding noctuids.

For leaf-feeding noctuids, a single treatment is recommended, when damage is seen. This should not normally be applied later than the 15–17-leaf stage, but emergency treatments are available for later use.

26.4.2.2.1 Main Insecticides

Against Soil Noctuids Acephate, *Bacillus thuringiensis* s sp *aizawai*, bifenthrin, chlorpyrifos, deltamethrin, diazinon, fenitrothion, malathion, tralomethrin.

Sprays (Against Leaf-Feeding Noctuids) Acephate, alpha-cypermethrin, *Bacillus thuringiensis* s sp *aizawai*, cypermethrin, deltamethrin, lambda-cyhalothrin.

26.4.3 Leaf Miners, *Liriomyza* spp

26.4.3.1 Symptoms

Several *Liriomyza* sppinfest lettuce: *L. bryoniae*, *L. trifolii*, and *L. huidobrensis*. The last was introduced into the EPPO region in 1987 and is still absent, or under eradication, in some countries. Where the pest is considered established, however, chemical control is needed. *Liriomyza trifolii* has limited potential to survive outdoors in northern Europe.

Phytomyza horticola may also attack lettuce. This leaf-mining species (Fig. 26.9) is polyphagous and very common in Europe.

26.4.3.2 Management

Since leaf miners in general are mainly pests of protected crops, at least in northern Europe, the first point in the strategy is to exclude them by general hygienic precautions and to use seedlings free from infestation. Weekly observations with, for example, sticky traps should be used to detect imminent outbreaks. Seedlings can be covered with insect nets (0.8-mm mesh) during the hardening period before planting. A single spray when the pest is detected may be sufficient, but treatment may be repeated at 3–5-day intervals if necessary. Leaf miner resistance can sometimes make control difficult. Biological control with *Diglyphus isaea*, *Dacnusa sibirica*, and *Opius pallipes* is also possible and recommended as GPP. The mines in the leaves disfigure the crop for marketing but are

Fig. 26.8 Cut worm damage on lettuce

Fig. 26.9 Leaf miner damage on lettuce leaf

usually confined to the lower leaves, so that they can be trimmed off at harvest.

26.4.3.2.1 Main Insecticides

Sprays Abamectin, cypermethrin, cyromazine, dichlorvos, mevinphos, oxamyl, propoxur, pirimiphos-methyl.

26.4.4 Slugs

26.4.4.1 Symptoms

Slugs (*Agriolimax* spp, *Arion* spp, *Deroceras* spp) may inflict considerable damage to lettuce seedlings and established plants by feeding and by forming slimy tracks. They can be particularly troublesome in autumn.

26.4.4.2 Management

General hygiene is important. All plant debris, leftover plastic, growing medium, etc. that may serve as hiding places or oviposition sites should be removed. The soil surface should be clod-free and free from weeds, also along the walls. Moist conditions favor the development of slugs. Biological control is possible with nematode, *Phasmarhabditis hermaphrodita.* If problems with slugs persist, molluscicides formulated as baits can be used until 1 week before sowing or planting, but not during the growing period.

26.4.4.2.1 Main Molluscicides

Metaldehyde, Methiocarb.

26.5 Nematodes

26.5.1 Root-Knot Nematodes, *Meloidogyne* spp

26.5.1.1 Symptoms

Root-knot nematodes (*Meloidogyne incognita, M. hapla, M. arenaria, M. javanica*) are polyphagous nematodes causing characteristic knots (galls), swellings, and other malformations on the roots of lettuce (Fig. 26.10), which results in poor growth and occasional wilting leading to poor yields. The smaller discrete galls with adventitious root proliferation formed by *M. hapla* are distinctive. Although *M. hapla* prefers coarse-textured soils, it also occurs in organic

Fig. 26.10 Root-knot nematodes, *Meloidogyne incognita* and *M. hapla* on lettuce roots

soils, and it is a major pest of lettuce which is commonly grown in this kind of soil.

26.5.1.2 Management

Nematode-free planting material and non-infested soil are normally sufficient to keep lettuce free from these nematodes. Weeds should be thoroughly controlled, throughout the areas where lettuces may be grown. The broad host range of *Meloidogyne* spp adds to the problem of selecting suitable rotation crops. Steam sterilization of the soil is an effective curative measure. It is not GPP to treat soil systematically with nematicides. Such treatments should be limited to what is strictly necessary and may be subjected to official limitations.

26.5.1.2.1 Main Nematicides

Preplanting soil treatment, allowing for the above considerations (that nematicide use should be restricted rather than recommended): dazomet, oxamyl.

26.6 Good Plant Protection Practices (GPP)

26.6.1 Difficulties in Protected Cultivation of Lettuce Crop

- As a monoculture under protected conditions, lettuce is liable to rapidly spreading pest attacks.
- The product is the leafy vegetable as such, and market standards tolerate no spotting, discoloration, visible insect damage, etc. (in contrast to other important vegetables in protected cultivation like tomatoes or cucurbits, where fruits are the marketed product and a certain amount of damage to the foliage can be tolerated).
- There is no practicable biological control of pests in lettuce at present.
- Residue tolerances are very low, as the part of the plant treated is the part consumed, and it is consumed fresh (without a significant storage period).

26.6.2 Guidelines for Good Plant Protection Practices

GPP in lettuce implies good management of the protected environment, for the incidence of pests can be much affected by general cultural conditions. In particular, crops should be well spaced, conditions favoring very high RH should be avoided, plants should be well watered during hot dry periods to prevent marginal leaf necrosis (which favors *Botrytis* rot) but should not be watered late in the day when the crop might stay wet for a long time (possibly overnight), infected plants should be carefully removed (avoiding spore dispersal as far as possible), and general hygiene should be high. Special attention should be given to the control of weeds (using herbicides or other means of control) around the glasshouses, as they can be reservoirs for many pests and also sources of weed seeds.

Soil sterilization is a common practice for pest control (including weed control) in protected crops. It is GPP to use steam sterilization for this purpose.

Viruses are also important pests of lettuce, and control of their vectors (soil fungi, aphids) is essential. Some viruses can be partially controlled by using tolerant cultivars. Use of virus-free planting material is essential. Plants should be regularly inspected for virus symptoms and infected plants immediately removed.

Lettuces for protected cultivation are mostly grown from seedlings raised in small peat pots (or similar containers) or peat blocks and transplanted as a whole into the final growing medium. Seedlings should be produced in high-quality compost. They should be raised in an isolated location, away from lettuce crops or other crops (to prevent infection especially by downy mildew and viruses). The seeds should meet normal certification standards for *lettuce mosaic potyvirus* or tolerant cultivars should be used. Normal precautions should be taken against damping-off diseases, including the use of seed treatments.

Cultural conditions can markedly affect disease risk in lettuce. Planting large or damaged plants or planting into dry soil should be avoided to reduce *Botryotinia fuckeliana*. Watering is carried out mostly early in crop growth, creating a reserve in the soil, so that there is less need to water later on, when watering is likely to encourage downy mildew. Balanced nutrition with adequate potassium is also believed to reduce *Botrytis* risk.

Finally, it may be noted that most glasshouse lettuces are grown in soil, so that appropriate precautions have to be taken to avoid carry-over of pests in the soil, including pathogens and weed seeds. However, some lettuces are grown in soilless media or by liquid-film methods. In many countries in the EPPO region, lettuce is grown under the protection of plastic tunnels. Under these conditions, general hygiene standards need to be maintained at a very high level, but the basic strategies to control pests remain broadly the same.

27 Rose

Abstract

With the advent of state-of-the-art greenhouse cultivation in early 1990s, large-scale cultivation of export quality cut flower of roses in protected condition started, thereby totally altering production dynamics. Cut flower trade is worldwide dominated by Hybrid Tea roses. Roses are attacked by many pests and diseases that can lower quality and yields. Among the insect pests, thrips, white fly, red spider mites, tobacco budworm, and aphids are important. The major diseases affecting roses are dieback, black spot, rust, and powdery mildew. Their symptoms, biology, spread, and management are discussed.

Keywords

Rose • Insects • Mites • Diseases • Symptoms • Biology • Spread • Management

27.1 Introduction

Previously commercial rose cultivation in India was mainly in open-field conditions. However, with the advent of state-of-the-art greenhouse cultivation in early 1990s, large-scale cultivation of export quality cut flower in protected condition started, thereby totally altering production dynamics (Fig. 27.1). Cut flower trade is worldwide dominated by Hybrid Tea roses.

27.2 Varieties

The rose varieties suitable for growing under greenhouse cultivation are presented in Table 27.1.

27.3 Diseases

27.3.1 Dieback, *Diplodia rosarum*

27.3.1.1 Symptoms

This is the most serious of the rose diseases in India. As the term dieback implies, the disease causes the death of the plant from top downward. The disease starts from the pruned surface of the twigs (Fig. 27.2). Initially it may be observed to the extent to a few centimeters below the pruned end, but in the severe cases, the disease spreads further and kills the entire plant.

27.3.1.2 Management

Soil drenching with 2 g/L bavistin, benomyl, or demosan, has been reported to control the dis-

P.P. Reddy, *Sustainable Crop Protection under Protected Cultivation*,
DOI 10.1007/978-981-287-952-3_27

Fig. 27.1 Rose under protected cultivation

Table 27.1 Rose varieties used for cut flower purpose

Flower color	Variety
Red	Jaguar, Gabriella, Sasha, Grand Gala, First Red, Dallas, EG Hill, Happiness
Pink	Kiss, Europe, Prophyta, Royal, Nobles, Pink, Aristocrat, Better Times
Bicolor	Amour, Rodeo, Confetti, Ambience, Lionides, Yellow Gloria
Orange	Indian Puma, Candid, Mercedes, Jazz, Orange Delight, President Herbert
Yellow	Golden Time, Golden Gate, Frisco, Golden Rapture, Golden Scepter
Cream	Prestige, Vivaldi, Verselia, Florence
Purple	Jakaranka, Souvenier
White	Eskimo, Double White Killarney, White Pearl

ease. Spraying with 0.2 % captan or 0.2 % mancozeb or 0.2 % copper oxychloride immediately after pruning and then twice at 10-day interval is also effective for controlling dieback of rose.

27.3.2 Black Spot, *Diplocarpon rosae*

27.3.2.1 Symptoms

This disease is also called as leaf blotch, leaf spot, blotch, and star sooty mold. Characteristic black spot 2–12 mm in diameter develops on upper leaf surfaces. These leaf spots are circular or irregularly coalescent with characteristic feathery, radiate, fibrillose margins of subcuticular mycelial strands (Fig. 27.3).

27.3.2.2 Management

The infected leaves as they are observed should be clipped off and burnt. Preventive sprays of Ferbam (0.1 %) at fortnightly intervals and Benlate or Bayleton (0.1 %) applied just before the appearance of the spots helps in managing the disease. Bavistin (0.1 %) was found to be the most effective fungicide against leaf spot caused by *Diplocarpon rosae*.

27.3.3 Powdery Mildew, *Sphaerotheca pannosa* var *rosae*

27.3.3.1 Symptoms

It is a major disease of the rose all over the world. The disease affects all the aerial parts of the plant, though the leaves are found to be affected more. The younger leaves get curled, exposing the lower surface, and such leaves are likely to be purplish than the normal leaves with raised blister-like areas developing in these leaves, which becomes coated with the white powdery growth of the fungus (Fig. 27.4).

27.3.3.2 Management

Apply chlorothalonil, propiconazole, piperalin, fenarimol, sulfur, copper hydroxide, neem oil, azoxystrobin, kersoxim methyl, ziram, or triforine.

27.3.4 Rust, *Phragmidium mucronatum*

27.3.4.1 Symptoms

The disease is easily identified by the yellow orange rust pustules which develop profusely on the lower surface of older leaves (Fig. 27.5). In greenhouses it is not very destructive. Several species of *Phragmidium* have been reported to infect

Fig. 27.2 Symptoms of dieback on rose stem and leaves

Fig. 27.3 Symptoms of black spot on rose leaf

rose, but *P. mucronatum* is the most common. It is an autoecious, macrocyclic fungus producing telia by the end of the crop season in the same place as uredospores. They serve as overwintering structures and initiate infection during spring. Free water and temperature 9–27 °C are necessary for the uredospores to germinate (Horst 1989).

27.3.4.2 Management

- Frequently remove all rust-infected leaves and badly infected plants and destroy by burning, rapid composting, or burying. Severely infected roses can be dramatically pruned.
- At the end of the growing season, carefully clean up and destroy all crop debris. Sterilize benches and propagation rooms with an appropriate greenhouse disinfectant.
- Keep the humidity within the greenhouse at less than 80 %. Increasing air movement by adding fans will prevent moisture from condensing on the foliage.
- Practice only surface watering and avoid splashing water onto foliage. If overhead irrigation is necessary, early morning irrigation is recommended when plants will dry quickly.
- Space plants to allow for good air circulation.
- The fungicides Myclobutanil, Tebuconazole, and Triticonazole are approved for rust control.

Rose diseases under greenhouse conditions, their symptoms, and management are presented in Table 27.2.

27.4 Insect and Mite Pests

27.4.1 Thrips, *Scirtothrips dorsalis*

27.4.1.1 Symptoms

Thrips are a serious problem on roses grown in polyhouses almost round the year except during November–December and July–August. Both nymphs and adults suck cell sap from tender leaves, growing buds, and flowers. Curled leaves with brown streaks appear on the young leaves. When they feed on the young buds, scorchy appearance of sepals and petals is seen. Later the deformed flowers with burnt margins are very common.

When thrips populations are high on roses, flower buds may become deformed and fail to open. Petals may be covered with brown streaks

Fig. 27.4 Powdery mildew symptoms on rose

and spots (Fig. 27.6). Western flower thrips can also vector certain tospoviruses including impatiens necrotic spot virus and several strains of tomato spotted wilt virus.

27.4.1.2 Management

- Spray insecticides like imidacloprid 200 SL at 0.4 ml/L or fipronil 5 SC at 1.5 ml/L or acephate 75 SP at 1.5 g/L or dimethoate (0.15 %) at 10–15-day interval.
- Drench the beds after raking up with chlorpyrifos 20 EC at 4 ml/L for killing resting pupae.
- Spray jatropha or pongamia oils at 10 ml/L.

27.4.2 Two Spotted Spider Mite, *Tetranychus urticae*

27.4.2.1 Symptoms

The mite is found initially in colonies on lower surface of mature leaves. In case of severe infestation, population spreads to entire foliage including flowers and buds (Fig. 27.7) covering them with webs. Hot and dry conditions (mainly summer) favor mite multiplication. Cream-colored nymphs and red colored adults with two lateral black spots feed on leaves resulting in discoloration, bronzing, and drying. Damage to buds and flowers leads to discoloration, retardation, and

Fig. 27.5 Rose leaves infected with rust pustules

drying. Due to their feeding, white specks appear on the leaves, and these specks coalesce and appear as white patches. Ultimately, affected leaves become mottled.

27.4.2.2 Management

- Cutting and burning of infested plant parts.
- Proper ventilation and watering to reduce temperatures.
- Spray abamectin 1.9 EC at 0.5 ml/L followed by flufenoxuron 10 DC at 1 ml/L or fenazaquine 10 EC at 1 ml/L or difenthiuron 50 SC at 0.6 ml/L during export season.
- Spray dicofol 18.5 EC at 2.5 ml/L or wettable sulfur 80 WP at 3 g/L or profenofos 50 EC at 1 ml/L or Ethion 50 EC at 1 ml/L followed by

Table 27.2 Rose diseases and their management in greenhouse

Disease/pathogen	Symptoms	Management
Anthracnose, *Sphaceloma rosarum*	Dark purple to black spots are bordered by a narrow, dull brown band. Centers of spots turn gray and fall out. Spots form on canes, similar to leaf spots	Maintain good sanitation. Black spot control procedures (below) also control anthracnose
Black spot, *Diplocarpon rosae*	Brown to black round spots with feathery edges form on leaves. Leaves yellow and fall. Small, purplish spots form on canes	Remove infected canes. Remove and destroy fallen leaves. Water in a manner that keeps foliage surfaces dry. Apply one of the following to protect new foliage: chlorothalonil, propiconazole, neem oil, mancozeb, thiophanate methyl, thiophanate methyl + mancozeb, chlorothalonil + thiophanate methyl, copper hydroxide, ziram, captan, trifloxystrobin, or triforine. Triadimefon can stunt some cultivars
Botrytis blight, *Botrytis cinerea*	Small water-soaked lesions form on petals. Gray fungal growth covers infected petals. Stubs left after harvest become infected. The fungus then moves down to girdle the cane	Space plants and provide ventilation to avoid excessively high humidity. Remove fading flowers and yellowing leaves. Apply chlorothalonil, dichloran, trifloxystrobin, iprodione, fenhexamid, or azoxystrobin to protect healthy tissue. Heat and ventilate to maintain low humidity
Cankers, *Coniothyrium fuckelii*, *Cryptosporella umbrina*, *Coniothyrium wernsdorffiae*, *Cylindrocladium scoparium*	Reddish-brown spots on canes turn light to dark brown and become covered with tiny black dots. Cankers girdle and kill the cane	Do not plant stock with cankers. Remove infected canes making the cut immediately above a bud. Apply a fungicide (see black spot) after pruning. Sterilize the shears with bromine disinfectant between cuts. Maintain even soil moisture and fertilization
Crown Gall, *Agrobacterium tumefaciens*	Small white to cream-colored galls form on stems. Galls may enlarge to 15 cm in diameter. Galls can form on roots or stems	Do not plant infected material. Steam sterilized beds where infected plants were grown. Remove and destroy infected plants. Apply *Agrobacterium radiobacter* to protect healthy plants
Damping-off, *Pythium* spp.	Cuttings fail to root, defoliate, and die. Roots are killed	Pot and propagate in pasteurized media. Use clean, disinfested tools. Discard infected plants. Do not leave cuttings in mist beds for excessive periods. Pot as soon as rooted
Downy Mildew, *Peronospora sparsa*	Purplish-brown spots form on leaves. Leaves yellow and fall. Small spots or long purplish areas may form on canes and may kill twigs	Heat and ventilate to maintain low humidity. Water in a manner that keeps leaf surfaces dry. Apply potassium phosphate, potassium salts of phosphorus acid, trifloxystrobin, mancozeb, azoxystrobin, dimethomorph, or mancozeb + thiophanate methyl

(continued)

Table 27.2 (continued)

Disease/pathogen	Symptoms	Management
Powdery Mildew, *Sphaerotheca pannosa*	Spots on leaves, stems, and flower parts expand and become covered with white fungal growth. Small dead spots form on some cultivars	Apply chlorothalonil, propiconazole, piperalin, fenarimol, sulfur, copper hydroxide, neem oil, azoxystrobin, kersoxim methyl, ziram, or triforine. Triadimefon can stunt many cultivars used in greenhouse production
Viruses – rose mosaic, mottle, yellow mosaic, ring pattern, tobacco streak, rose rosette, rose wilt, spring dwarf, color break or strawberry latent ringspot virus	Leaves may exhibit mosaic, mottling, yellow line, or ring patterns. Veins may turn yellow	Destroy infected plants. Plant only healthy, virus-free plants. Maintain good insect and mite control

Fig. 27.6 Thrips damage on rose flowers

Fig. 27.7 Red spider mite damage on rose

pongamia or neem or jatropha oil at 5 ml/L during off-season.

- Release predatory mite *Amblyseius tetranychivorus* at 20 mites/plant when *T. urticae* populations are at moderate levels.
- Spray *Verticillium lecanii* formulation at 5 g/L during evening hours.

27.4.3 Whitefly, *Bemisia tabaci*

27.4.3.1 Symptoms

The pest is active during January–June. Small and fragile white adults lay eggs on lower surface of young leaves. Nymphs and adults suck cell sap from lower side of leaves resulting in yellowing and dropping of leaves (Fig. 27.8). In case of severe infestation sooty mold develops on honeydew secreted by nymphs and significantly affects the growth of the plants resulting in production of small flowers.

27.4.3.2 Management

- Removal and burning of heavily infested shoots.
- Clean cultivation and use of insect proof nets.
- Spray triazophos 40 EC at 1.5 ml/L or methomyl 40 SP at 2 ml/L or acephate 75 SP at 1.5 g/L at 15-day interval alternating with pongamia oil at 10 ml/L for nymphs.
- A management package of spraying Dichlorvos 76 EC at 1 ml/L followed by lambda cyhalothrin 5 EC at 1 ml/L or deltamethrin 2.8 EC at 1 ml/L at 5–7-day interval and installing yellow sticky traps effectively controls whitefly adults.

Fig. 27.8 Whiteflies on rose leaves and flower bud

Fig. 27.9 Tobacco bud worm damage on rose flower

Fig. 27.10 Aphids on rose flower bud

27.4.4 Tobacco Budworm, *Spodoptera litura*

27.4.4.1 Symptoms

Adult moths lay eggs in groups on lower surface of leaves, and damage is seen throughout monsoon period in polyhouses. Early instars are gregarious in nature and feed on leaves by scraping chlorophyll. Black colored mature larvae damage growing buds and flowers during nights resulting in qualitative loss to flowers (Fig. 27.9).

27.4.4.2 Management

- Collection and destruction of egg masses and gregarious early instar larvae.
- Spray quinalphos 25 EC at 2 ml/L or endosulfan 35 EC at 2 ml/L at 15-day interval.
- Spray indoxacarb 14.5 SC at 1 ml/L or thiodicarb 75 WP at 1 g/L, if the incidence is high.
- Spray *Sl* NPV at 250 LE/ha followed by neem formulations at 1–2 ml/L.
- Spread poison bait made of wheat/rice bran, jaggery, and chlorpyrifos (10:1:0.5) for killing grownup larvae.

27.4.5 Aphids, *Macrosiphum rosae*

27.4.5.1 Symptoms

Large, dark green or pink brown aphids feed on buds (Fig. 27.10), shoots, and leaves. Colonies may persist throughout the year but are most numerous and troublesome in late spring and early summer. Foliage of infested plants is coated with sticky honeydew on which sooty mold grows. Aphid populations tend to increase most rapidly on soft and sappy growths.

27.4.5.2 Management

Use of nitrogenous fertilizers should be restricted. Systemic insecticides like dimethoate (0.15 %) should be sprayed to control aphids.

Reference

Horst RK (ed) (1989) Compendium of rose diseases. APS Press, St Paul

28 Carnation

Abstract

Carnation is one of the most popular cut flowers of commercial importance in the world market ranking next only to rose. The major insect and mite pests (thrips, spider mites, bud borer, aphids), diseases (powdery mildew, black spot, gray mold, foot rot, root rot, stem rot and wilt, rust, bacterial wilt, slow wilt/bacterial stunt), nematodes (root-knot, spiral), and their symptoms, biology, spread, and management are discussed. Maintenance of adequate aeration inside the polyhouse and proper hygiene of the beds, implements, and workers would help prevent most of the pests and diseases.

Keywords

Carnation • Pests • Diseases • Nematodes • Symptoms • Biology • Spread • Management

28.1 Introduction

Carnation is a half-hardy perennial with branching stems and tumid joints. Leaf blades are simple, entire, and usually narrow. Each stem forms a terminal flower and hence inflorescence is generally a terminal cyme. The flowering shoots can be marketed in one or two forms like standard and spray or miniature. Perpetual flowering and greenhouse carnations are of modern origin (Fig. 28.1). The present-day flower shop carnations are the result of a series of complicated crosses over a period of many years.

28.2 Types and Varieties

At present enormous varieties are available with many attractive colors, remarkable ranges of shapes and sizes, and delicate fragrance.

28.2.1 Standard Type

Flowering buds formed on short lateral shoots arising from the axis of the upper leaves are removed to leave one large, terminal flower on a long leafy stem. It has one large flower on an individual stem (Table 28.1).

P.P. Reddy, *Sustainable Crop Protection under Protected Cultivation*,
DOI 10.1007/978-981-287-952-3_28

Fig. 28.1 Protected cultivation of carnations

28.2.2 Spray Type

Terminal flower bud is removed at an early stage to encourage more even development of the lateral flowers, which then produce a multiple flowered stem (Table 28.1).

28.2.3 Dianthine Type

They are more or less similar to the spray varieties, but in this type flowers are smaller and produce more flowers on long stems. Market of these types of carnations is small.

28.2.4 Mignon and Microtype

These types have much in common with the normal spray varieties. The stems, however, are smaller and yield is higher than spray varieties. Demand for these types in the market is less.

Carnation cultivars grown in different parts of India are presented in Table 28.2.

Table 28.1 Type and varieties of rose suitable for greenhouse cultivation

Color	Varieties of standard-type carnation
Red	Aicardi, Barbarot, Balance, Corba, Domingo, Gaudina, Hilwarda, Kooij, Santamaria, Selecta, Turbo
Pink	Charmant Pink Dona, Famosa Pink Dover, Pink Shiva, Pink America, Pink Monterama
White	Angelica, Baltico, Hydra White Liberty, Madame Coletta, Viking
Yellow	Diana, Hermes, Salamanca, Tobago, Yellow Solar
Cream	Kamar, Sablia, Sagres, Shanteng
Cerine	Cherry Solar, Dona, Maxi, Rubesco, Shiva
Orange	Malaga, Nabita, Nova Scollia, Olymbia, Solar
Green	Green Elegance, Lady Green, Lenny, Nokia, Pradorova
Lidae	Farida, Remo
Fancy	Alibaba, Garuda, Rindez Sonya
Color	Varieties of spray-type carnations
Red	Etna, Karma, Rony
Pink	Anne Lies, Barbara, Silvery Pink
Yellow	Alicetta, Lior, Odeon
White	Excel, Royalette, Tibet
Others	Exquisite, Kissi, Scarlet Elegance
Micro	Eolo, Pink Eolo, Wiko
Mini	Lima, Onia, Roland

28.3 Insect and Mite Pests

28.3.1 Thrips, *Frankliniella schultzei*

28.3.1.1 Symptoms

Both the nymphs and adults suck the sap from leaves, buds, and flower (Fig. 28.2). Damaged buds show brown streaks and get distorted in severe cases with burnt margins leading to quality reduction. They excrete brown droplets, which afterward turn black. Leaves may fade and shrivel in case of heavy infestation, and foliage becomes silvery.

28.3.1.2 Favorable Conditions

Mainly temperature affects the thrips population.

28.3.1.3 Management

- Spray dimethoate 30 EC at 2 ml/L or oxy-demeton methyl 25 EC at 2 ml/L at 15-day interval.
- Drench the soil with chlorpyrifos 20 EC at 4 ml/L to reduce pupae.
- Spray imidacloprid 200 SL at 0.4 ml/L or fipronil 5 SC at 1 ml/L or acephate 75 SP at 1.5 g/L.
- Spray Asatap (acephate) at 1.0 gm/L or Decis (deltamethrin) at 0.5 ml/L or Exodust at 1.0 g/L.

Table 28.2 Carnation cultivars grown in different parts of India

State/region	Cultivars
Maharashtra – Pune	Aristo, Flair, Fambo, Stylo, Scania, William Sim
Himachal Pradesh – Solan	Arthur Sim, Candy, Espana, Pamir, White Candy
Punjab – Ludhiana	Arthur Sim, Sam's Pride, Scarlet Elegance
Tamil Nadu – Kodaikanal	Arthur Sim, Scania, Scarlet Elegance, Star Light
West Bengal – Kalimpong	Arthur Sim, Dusty, Scania

28.3.2 Two-Spotted Spider Mite, *Tetranychus urticae*

28.3.2.1 Symptoms

It is a devastating pest on carnation grown under protected conditions especially during September–November and February–May. Mite colonies are found on lower surface of leaves, covered by dirty webs. Affected leaves get discolored and dry up. Infestation spreads to buds, flowers (Fig. 28.3), and entire plant in severe cases. They suck sap from the leaves which results in tiny yellow or white speckles. Once the foliage of a plant becomes bronze, it often drops prematurely. Attacked plants lose vigor and get stunted in growth and dry in severe cases.

28.3.2.2 Favorable Conditions

- High temperature and low relative humidity
- More populated crop (dense crop)

28.3.2.3 Management

- Cutting and burning of infested plant parts.
- Thorough spraying of water on the plant dislodges the webs.
- Spray wettable sulfur 80 WP at 3 g/L or profenofos 50 EC at 1 ml/L or ethion 50 EC at 1

Fig. 28.2 Thrips damage on carnation. *Left*, healthy; *right*, infested

Fig. 28.3 Mites damage on carnation flower bud

Fig. 28.4 Bud borer infestation on carnation

ml/L at 15-day interval followed by 10 ml/L pongamia or neem oil.
- Spray abamectin 1.9 EC at 0.5 ml/L followed by flufenoxuron 10 DC at 1 ml/L or fenazaquin 10 EC at 1 ml/L during summer when infestations are high.
- Release predatory mite, *Amblyseius tetranychivorus* at 20 mites/plant.

28.3.3 Bud Borer, *Helicoverpa armigera*

28.3.3.1 Symptoms

Eggs are laid singly on growing buds during September–October, and damage continues up to February–March. Larvae bore into young buds and feed on developing petals making them hallow (Fig. 28.4). Larvae also attack open flowers resulting in dropping of petals and significant yield loss. Plant growth is affected adversely.

28.3.3.2 Favorable Conditions

- Warmer climate in polyhouse
- Fluctuation of relative humidity in the polyhouse
- Fluctuation of temperature in the polyhouse

28.3.3.3 Management

- Collection and destruction of larvae.
- Spraying of quinalphos 25 EC at 2 ml/L or methyl parathion 50 EC at 1 ml/L or endosulfan 35 EC at 2 ml/L at 15-day interval.
- Spray indoxacarb 14.5 SC at 1 ml/L or thiodicarb 75 WP at 1 g/L if the incidence is severe.
- Application of *Ha*NPV at 250 LE/ha followed by neem formulations at 1–2 ml/l or neem seed kernel extract (40 g/L).
- Spray *Bacillus thuringiensis* at 2 g/L.
- Release 25 million *Trichogramma* egg parasitoids/ha.
- Set up Helilure sex pheromone traps at 10 No. /ha.
- Spray spinosad at 0.75 ml/l or thiodicarb at 0.5 ml/L.

28.3.4 Aphids, *Myzus persicae*

28.3.4.1 Symptoms

These are the sucking insects. Feeding usually occurs on buds and undersides of leaves. Feeding on young leaves resulted in distorted leaves as they continue to grow. Older leaves may display patches of chlorotic spots. Nymphs and adults suck the sap from the leaves, stems (Fig. 28.5), and flower buds in colonies. Aphids excrete honeydew, which results in black fungus development on plants.

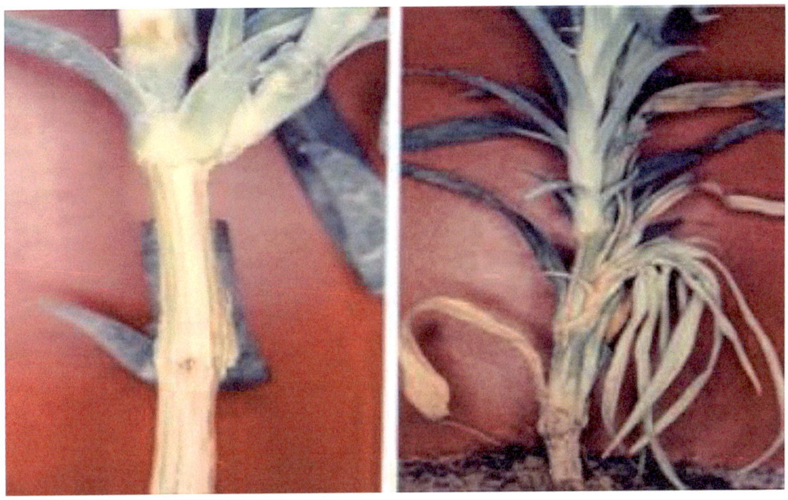

Fig. 28.5 Aphid infestation on carnation

28.3.4.2 Favorable Conditions

- Cloudy weather favors rapid buildup of aphid population.
- Low temperature favors development of young aphids.

28.3.4.3 Management

Spray the plants with thiamethoxam 1 ml/L or imidacloprid @ 2 m l/L or Asatap (acephate) at 0.5–1.0 gm/L of water.

28.4 Diseases

28.4.1 Powdery Mildew, *Oidium dianthi*

28.4.1.1 Symptoms

On the top of the leaves, white, powdery fungal growth develops consisting of hyphal threads and spore carriers. The fungus is also observed on the underside of leaves. Malformation of young leaves takes place due to infection of powdery mildew. Brown spots develop on full-grown leaves.

28.4.1.2 Management

Spray thiophanate methyl (Roko) at 2.0 g/L or bitertanol at 2.0 g/L of water.

28.4.2 Black Spot, *Alternaria dianthi*

28.4.2.1 Symptoms

This mostly infects carnation cuttings during mist propagation and in wet parts of greenhouses. Small purple spots on the leaves are the first symptoms. Soon they enlarge, and their center turns brown and then black due to the masses of spores which develop (Fig. 28.6). Stem infection usually appears on the knots (Strider 1978). The leaf tissues surrounding this spot turn yellow; severe infection leads to premature death of the leaves.

28.4.2.2 Management

Spray chlorothalonil at 1.5 g/L of water.

28.4.3 Gray Mold, *Botrytis cinerea*

28.4.3.1 Symptoms

The fungus is usually identified by the development of fuzzy, grayish spore over the surface of the rotted tissues. The fungus causes a brown rotting and blighting of affected tissues. Blooms are worst affected resulting in poor quality (Fig. 28.7).

28.4.3.2 Management

Spray carbendazim/capton at 2 g/L of water.

Fig. 28.6 Black spot-infected carnation flower buds

Fig. 28.7 Blossom blight caused by *Botrytis cinerea*

28.4.4 Foot Rot, *Phytophthora* sp

28.4.4.1 Symptoms

Withering and yellowing of foliage, leaf death, external browning of stems, and internal browning at nods appear. Stem and root rot may take place. Lower leaves become purple and dry, shortening and blackening of stems take place, and rotting of stems can be observed (Fig. 28.8).

28.4.4.2 Management

Apply Alleite at 1.5 g/L or chlorothalonil at 1.5 g/L of water.

28.4.5 Root Rot, *Pythium* sp

28.4.5.1 Symptoms

Its infection results in stunted growth and ultimately drying of the entire plant (Fig. 28.9).

28.4.5.2 Management

Spray carbendazim (Bavistin) at 2.0 g/L or Alleite at 1.5 g/L or chlorothalonil (Kavach) at 1.5 g/L of water.

28.4.6 Stem Rot and Wilt, *Fusarium* sp

28.4.6.1 Symptoms

Infection takes place through wound. The lower leaves start becoming yellow followed by withering of leaf bases and yellowing of midribs and eventually the branch wilts. The pathogen also causes browning, discoloration, and shredding of vascular bundles (Fig. 28.10).

28.4.6.2 Management

Spray Alleite at 1.5 g/L or Kavach (chlorothalonil) at 1.5 g/ L of water. Soil drenching with carbendazim at 0.1 % or difenoconazole at 0.05 %

Fig. 28.8 Phytophthora blight of carnation

Fig. 28.9 Root rot of carnation. *Left*, healthy plant; *right*, infected plants

or *Pseudomonas fluorescens* as soil application at 25 g/m^2 and foliar application at 0.5 % at monthly intervals or soil drenching with *Bacillus amyloliquefaciens* at 0.5 % at monthly intervals is also effective.

28.4.7 Rust, *Uromyces dianthi* (= *Uromyces caryophyllinus*)

28.4.7.1 Symptoms

The disease is more severe on leaves, but other green plant parts are infected as well. Initially, small light green spots appear. They gradually turn to powdery brown blisters due to the development of uredospores (Fig. 28.11). Severely infected plant parts are twisted.

Healthy crops are infected by uredospores transferred from neighboring crops. They are wind or watersplash disseminated and germinates readily on free water. The cycle of the pathogen lasts about 2 weeks. In greenhouses, where leaves may remain wet for several hours, there may be many disease cycles per crop season (Strider 1985).

28.4.7.2 Management

Good hygiene should be adhered to, including:

- Ensure that cuttings taken are from uninfected plants.
- Sterilize border soil to destroy any overwintering spores.
- Ensure that glasshouses are well ventilated to prevent damp/humid patches occurring which can encourage infection. Space plants out to encourage good airflow.
- Do not wet the leaves when watering.
- Remove and destroy infected leaves (do not put on domestic compost heaps).
- Ensure that hands and implements are cleaned before using on uninfected plants.
- The use of the fungicide mancozeb can provide protection from infection and, used fortnightly from the first signs, can prevent spread to some extent.
- Use of a high potash fertilizer can help plants recover from an attack.
- A myclobutanil-based fungicide can be effective on infestations. Alternatives to myclobutanil are penconazole, mancozeb (also used for prevention), or triticonazole.

28.4.8 Bacterial Wilt, *Burkholderia caryophylli*

28.4.8.1 Symptoms

The upper parts of established plants turn pale and wilt (Fig. 28.12). The stem develops elongated, discolored stripes, and split. The roots are rotted partially and the cortical tissue becomes sticky and shows discoloration. The leaves are twisted. The roots remain discolored.

Fig. 28.10 Stem rot of carnation

Fig. 28.11 Carnation plant showing rust pustules

Fig. 28.12 Symptoms of bacterial wilt of carnation

28.4.8.2 Management

Disease-free mother plants should be used and rooting beds and soil should be fumigated. Sanitary procedures which can prevent further infestation include picking up infected host debris and careful handling of tools. Checking for signs of infected plant before plant cutting and checking the soil for the presence of bacteria will ensure that further dissemination is significantly decreased, if not prevented.

28.4.9 Slow Wilt/Bacterial Stunt, *Dickeya dianthicola*

28.4.9.1 Symptoms

Infected plants become gray green and may be stunted without any obvious wilting. Plants eventually wilt and in a period of 6–8 months may die. Vascular tissues, and pith mainly at the base of the stem, may show a yellow discoloration. Occasionally, stem cracks and root rot may occur (Fig. 28.13) (Fletcher 1984).

28.4.9.2 Management

Soil disinfestation, use of resistant cvs, grafting on resistant rootstocks, use of clean propagating material, and application of strict hygienic conditions are recommended against wilts (Sherf and Macnab 1986).

28.5 Nematodes

28.5.1 Root-Knot Nematode, *Meloidogyne incognita*

28.5.1.1 Symptoms

Root-knot nematode is a serious pest. Infected plants usually appear stunted and tend to wilt on warmer days. When such plants are dug, the root galls are generally conspicuous and easily identified (Fig. 28.14).

28.5.1.2 Management

Preplant treatment of beds with dazomet followed by the application of neem cake (1 kg/m^2, 15 days later) along with *Paecilomyces lilacinus* or *Pochonia chlamydosporia* significantly

Fig. 28.13 *Dickeya dianthicola* causing disease on carnation. (**a**) Stem necrosis on carnation. (**b**) Brown discoloration of vascular bundles. (**c**) Brown discoloration of stems. (**d**) Field symptoms. (**e**) Brown discoloration of vascular bundles

Fig. 28.14 Carnation roots infected by root-knot nematodes

reduced populations of *M. incognita* and the mortality of plants and suppressed the nematode infection for nearly 2 years. The antagonistic fungi established themselves better in the beds treated with Dazomet than in untreated beds. Chlorpyrifos and carbofuran (each applied twice in 6 months) significantly reduced nematode populations in roots and soil. However, there was a buildup of nematode populations in beds treated with these two chemicals after 1 year. On a long-term basis, soil management with preplant treatment of Dazomet, followed by the application of oil cakes plus antagonistic fungi, was more effective against *M. incognita* than post-plant treatment with carbofuran, carbosulfan, and chlorpyrifos on carnation grown in polyhouses.

28.5.2 Spiral Nematode, *Helicotylenchus dihystera*

28.5.2.1 Symptoms

Helicotylenchus dihystera was found to be the most predominant nematode as revealed by the community analysis of carnation rhizosphere samples. An initial inoculum level of 1000 nematodes of *H. dihystera* per plant was found to be pathogenic to carnation. *H. dihystera* was responsible for the reduction in growth parameters as well as for delay in flowering. Staining of the affected carnation root system showed the presence of this nematode pathogen in the cortical tissues.

28.5.2.2 Management

Pseudomonas fluorescens and *Paecilomyces lilacinus* (at 2.5 g and 10 g/m^2, respectively) effectively reduced the population of *Rotylenchulus reniformis* and *H. dihystera*. These treatments were followed by farm yard manure, neem powder, and carbofuran.

References

Fletcher JT (1984) Diseases of greenhouse plants. Longman Scientific and Technical, London, 351 pp

Sherf AF, MacNab AA (1986) Vegetable diseases and their control. John Wiley and Sons, New York, 728 pp

Strider DL (1978) Alternaria blight of carnation in the greenhouse and its control. Plant Dis Rep 62:24–28

Strider DL (ed) (1985) Diseases of floral crops. Vols 1 and 2, Praeger Special Studies, New York

29 Gerbera

Abstract

Gerberas are especially known for their wide spectrum of colors and shapes, and the popularity of this flower has also increased worldwide. Protection of the crop is of major importance when growing gerbera plants. First, focus on prevention of diseases and insect pests so that they do not have the chance to develop. The major pests (whitefly, aphid, leaf miner, thrips, mites), diseases (powdery mildew, collar rot, root rot, stem rot, leaf spot), nematodes (root-knot, spiral), and their symptoms, biology, spread, and management are discussed. Hygiene should be on top priority as well as a requirement. The use of chemicals should be the last resort. Integrated pest management in gerbera is feasible. The combination of biological products, natural predators, and insect growth regulators has had a lot of success in the last few years.

Keywords

Gerbera • Pests • Diseases • Nematodes • Symptoms • Biology • Spread and management

29.1 Introduction

Gerbera (*Gerbera jamesonii*), being a shade-loving plant, needs about 50 % shade for quality bloom production. Moreover, the covered structure protects the plant from storms, winds, frosts, dews, gales, and other adverse effects of the environment (Fig. 29.1). It also helps to maintain a constant and optimum temperature and humidity as required by the plant during the period of growth. It provides an active protection against disease and pests and ensures round the year production of cut blooms.

A day temperature of 22–25 °C and night temperature of 12–16 °C are ideal for cultivation. Poor light during winter adversely affects the flower production. The average yield under greenhouse is around 200 cut flowers/m^2/year, with 85 % of flowers being first grade quality, while it is 180 cut flowers/m^2/year under open condition with only 15–20 % of flowers being of first grade.

P.P. Reddy, *Sustainable Crop Protection under Protected Cultivation*,
DOI 10.1007/978-981-287-952-3_29

Fig. 29.1 Gerbera under protected cultivation

29.2 Varieties

A large number of commercially important varieties are grown in different parts of the world. Among them, Regina, Nero, Parade, Mix, Alsmeera, Ibiza, Gold Spot, Sunset, Tara, Lyonella, Ornella, Rosetta, Gloria, Alexias Ginna, Monique, and Anneke are grown in India under protected cultivation for export purpose.

Red	Ruby Red, Sangria
Yellow	Doni, Supernova, Mammut, Talasa
Rose	Rosalin, Salvadore
Pink	Pink Elegance, Marmara, Esmara
Orange	Carrera, Goliath, Marasol
Cream	Farida, Dalma, Snow Flake, Winter Queen

29.3 Diseases

29.3.1 Powdery Mildew, *Golovinomyces (Erysiphe) cichoracearum*

29.3.1.1 Symptoms

The infection causes white powdery coating on the foliage and flowers (Fig. 29.2).

29.3.1.2 Management

The plants should be lightly irrigated by drip irrigation system so that they can be protected by excessive humidity. Destroy the affected plants and the leaves. This disease can be controlled by spraying fenarimol at 0.5 ml/L or Dianocab at 0.3 ml/L.

29.3.2 Collar Rot, *Phytophthora cryptogea*

29.3.2.1 Symptoms

The infection occurs just at the soil surface on the collar portion of the stem. In some cases the leaves turn yellow and the entire plant wilt (Fig. 29.3).

29.3.2.2 Management

Phytophthora cryptogea was eradicated up to 30 cm in soil by solarization as well as by fumigation.

The disease can be controlled by drenching the soil in the root zone of plants either with metalaxyl or with contact fungicide such as captan and copper oxychloride at 0.2 mg/L. Adorn as a drench at 2 fl oz/100 gal prevented *Phytophthora* crown and root rot in two of three trial years. Fenstop as a drench at 14 fl oz/100 gal or Orvego as a drench at rates of 22.5–34 fl oz/100 gal consistently controlled disease in 3 years of trials (Benson and Parker 2011).

29.3.3 Root Rot, *Pythium* sp

29.3.3.1 Symptoms

Several fungi, namely, *Pythium*, *Sclerotium rolfsii*, and *Rhizoctonia solani*, affect the root system of gerbera and cause root rot (Fig. 29.4).

29.3.3.2 Management

Sterilizing the soil before planting and regular application of fungicide like copper oxychloride at 0.2 mg/L to soil may help in controlling the disease.

Fig. 29.2 Powdery mildew on gerbera leaves and flower

Fig. 29.3 Comparison of healthy plant and root system with white roots (*right*) with a diseased plant exhibiting necrotic foliage and complete necrosis of root system (root rot rating = 5, dead) (*left*)

Fig. 29.4 Root rot causing foliage dieback on gerbera

29.3.4 Stem Rot, *Fusarium solani*

29.3.4.1 Symptoms

Petiole of leaves blacken at the base as the plant collapses. When leaf stem is cut, the black vessels can be seen (Fig. 29.5).

29.3.4.2 Management

Plant in pasteurized potting media. Discard infected plants. Spray Topsin/Roko at 2 g/L, Aliette at 2 g/L, Kavach at 1.5 gm/L, and Kocide at 1.5 g/L.

29.3.5 Leaf Spot, *Alternaria* sp

29.3.5.1 Symptoms

Symptoms develop when moisture persists on leaf surface for a longer duration. Brown specks form on florets. Black circular spots appear on the leaves. Centers become white on the leaf spots (Fig. 29.6).

29.3.5.2 Management

Maintain low relative humidity and do not wet leaves when watering. Spray Dithane M 45 at 1.5 g/L, Quintol at 0.5 g/L.

Diseases of gerbera, their symptoms, and management in greenhouse are presented in Table 29.1.

Fig. 29.5 Fusarium stem rot on gerbera

Fig. 29.6 Alternaria leaf spot on gerbera

29.4 Insect and Mite Pests

29.4.1 Whitefly, *Bemisia tabaci*

29.4.1.1 Symptoms

The pest is active during moderately hot and humid conditions (February–July). Small and fragile white adults lay eggs on lower surface of young leaves. Nymphs and adults suck cell sap (Fig. 29.7) resulting in discoloration, deformation, and withering of leaves. In case of severe infestation, sooty mold develops on honeydew secreted by nymphs affecting the plant growth. Severely damaged plants produce small and deformed flowers with crooked stalks.

29.4.1.2 Management

- Excessive dampness and dark conditions should be avoided.
- Removal and burning of infested leaves.
- Spray triazophos 40 EC at 1.5 ml/L or methomyl 40 SP at 2 ml/L or acephate 75 SP at 1.5 g/L followed by pongamia or neem oil 10 ml/L at 15-day interval for nymphs.
- Spray dichlorvos 76 EC at 1 ml/L followed by lambda-cyhalothrin 5 EC at 1 ml/L or deltamethrin 2.8 EC at 1 ml/L at 10-day interval for adults.
- Install yellow sticky traps for attracting adults.

29.4.2 Aphid, *Myzus persicae*

29.4.2.1 Symptoms

Mainly growing buds and flowers are attacked during cool months (October–January). Both nymphs and adults colonize on lower side of petals and suck the sap (Fig. 29.8) leading to discoloration, withering, and deformation of flowers. Development of sooty mold on honeydew secreted by aphids makes flowers look sticky with black dots.

29.4.2.2 Management

- Spray pongamia or neem oil at 10 ml/L at early infestation levels.
- Spray dimethoate 30 EC at 2 ml/L or oxydemeton methyl 25 EC at 2 ml/L or imidacloprid 200 SL at 0.4 ml/L or cartap hydrochloride 50 SP at 1.5 g/L or acephate 75 SP at 1.0 g/L.
- Application of *Verticillium lecanii* at 3 g/L also reduces aphid population.

Table 29.1 Diseases of gerbera and their management under greenhouse

Disease	Symptoms	Control measures
Root rot, *Pythium*	Root skin is easily removed and finally the plants wilt and die as roots rot	Avoid excess watering. Plant in pasteurized potting media. Drench with Benlate or Bavistin or Topsin/Roko all at 2 g /lit
Crown rot, *Phytophthora cryptogea*	Plants wilt suddenly. Leaves turn brown. Roots are rotted and a crown rot develops. Crown of the plants become black	Plant in pasteurized potting media. Avoid overhead and excess watering. Drench with Benlate or Bavistin or Topsin/Roko or Aliette all at 2 g /lit
Fusarium stem rot, *Fusarium solani*	Petiole of leaves blacken at the base as the plant collapses. If leaf stem is cut, you can see the vessels are black	Plant in pasteurized potting media. Discard infected plants. Spray Topsin/Roko or Aliette both at 2 g/lit, Kavach or Kocide both at 1.5 g/lit
Leaf spot, *Alternaria* sp	Develops when moisture persists on leaf surface for a longer duration. Brown specks form on florets. Black circular spots appear on the leaves. Centers become white on the leaf spots	Maintain low relative humidity and do not wet leaves when watering. Spray Dithane M 45 at 1.5 g/lit, Quintol at 0.5 g/lit
Botrytis blight, *Botrytis cinerea*	Occurs especially when relative humidity of the air is above 90 % for 2 hr in the morning. Petioles have long brown spots. Leaves turn yellow and die. Gray spots on the flower petals and rot in the heart of flowers. Stems at soil level are killed. Infected tissues become covered with gray fungal growth	Space plants to ensure good air circulation in the polyhouse. Maintain low humidity. Avoid watering late in the day. Remove crop debris. Spray Dithane M 45 at 1.5 g/lit or Quintol at 0.5 g/lit
Bacterial blight, *Pseudomonas cichorii*	Small to large spots are circular at first and then become irregular and dark brown to black. May have a concentric ring pattern. Brown discoloration along the main vein	Maintain low relative humidity. Avoid overhead watering. Spray streptocycline at 0.2 g/lit or kasugamycin at 1.25 ml/lit
Powdery mildew, *Golovinomyces (Erysiphe) cichoracearum*	White powdery fungal growth develops on the surface of leaves. In case of severe attack, leaves start curling	Avoid excess humidity and maintain proper air circulation in the polyhouse. Spray sulfur at 1 g/lit or Dithane M 45 at 1.5 g/lit or Quintol at 0.5 g/lit
Rhizoctonia crown rot, *Rhizoctonia solani*	Stems at the soil level have a brown lesion. Plants wilt and die	Plant in pasteurized media. Apply a fungicide to protect plants
Thielaviopsis root rot, *Thielaviopsis basicola*	Plants yellow, wilt, and die. Roots are dark brown to black	Plant in pasteurized media. Apply a fungicide to protect plants
Viruses (tobacco rattle, gerbera mosaic, impatiens necrotic spot, and cucumber mosaic viruses)	Leaves may be mottled, have yellow spotting, dead flecks, line or ring spot patterns. Plants may be distorted or have flower color breaking	Maintain good insect and mite control. Plant in pasteurized media. Destroy infected plants

Fig. 29.7 Whiteflies on gerbera leaf

Fig. 29.8 Green peach aphids, *Myzus persicae* on gerbera leaf

29.4.3 Leaf Miner, *Liriomyza trifolii*

29.4.3.1 Symptoms

Incidence starts on newly planted crop at 3–4-leaf stage. Small black flies lay eggs by punching the young leaves. The larvae feed inside leaves on the palisade mesophyll by leaving characteristic serpentine mines (Fig. 29.9) (Parrella et al. 1985) and decrease photosynthesis and yield, directly affecting the marketable produce. Damaged leaves bear extensive mining marks and become brittle and dry in case of severe attack affecting the growth of the plants.

Fig. 29.9 Leaf miner infestation on gerbera

29.4.3.2 Management

- Remove and burn mined leaves.
- Spray triazophos 40 EC at 1.5 ml/L or methomyl 40 SP at 2 ml/L or profenofos 50 EC at 1 ml/L followed by spray of pongamia or neem oil 1 % at 15-day interval.
- Drench the beds after raking up with chlorpyrifos 20 EC at 4 ml/L for killing pupae.
- Spray of deltamethrin 2.8 EC at 1 ml/L for managing adults.
- Install yellow sticky traps.

29.4.4 Thrips, *Thrips palmi*

29.4.4.1 Symptoms

Thrips cause damage by sucking sap from leaves, buds, and flowers from July to February. Damaged leaves develop brown streaks, become brittle, and curl up. Affected buds and flowers get discolored and distorted in shape affecting the quality and quantity of flowers (Fig. 29.10).

29.4.4.2 Management

- Spray dimethoate 30 EC at 2 ml/L or oxydemeton methyl 25 EC at 2 ml/L at 15-day interval if the incidence is low.

Fig. 29.10 Thrips damage on gerbera flower

- Drench the soil with chlorpyrifos 20 EC at 4 ml/L to reduce resting stages.
- Spray imidacloprid 200 SL at 0.4 ml/L or cartap hydrochloride 50 SP at 1.5 g/L or fipronil 5 SC at 1 ml/L or acephate 75 SP at 1.5 g/L followed by pongamia or neem oil 10 ml/l in case of severe incidence.

29.4.5 Mites, *Polyphagotarsonemus latus*

29.4.5.1 Symptoms

These are small brown-colored insects which are normally not seen by naked eyes. They are located on the lower surface of the leaves. They suck the sap and brown spots appear on the leaves (Fig. 29.11). In case of higher incidence of mites, weblike structure is formed. The incidence is more on hot and dry areas.

29.4.5.2 Management

Mites can be controlled by spray of endosulfan (2.0 ml/L) or dicofol (1.5 ml/L). Spray abamectin 1.9 EC at 0.4 ml/L or propargite at 1 ml/L of water.

29.5 Nematodes

29.5.1 Root-Knot Nematode, *Meloidogyne incognita*

29.5.1.1 Symptoms

Symptoms include stunting of plants, yellowing of leaves (Fig. 29.12), wilting of plants, and heavy galling on roots.

29.5.1.2 Management

Preplant treatment of beds with dazomet followed by the application of neem cake (1 kg/m^2, 15 days later) along with *Paecilomyces lilacinus* or *Pochonia chlamydosporia* significantly reduced populations of *M. incognita* and the mortality of plants and suppressed the nematode infection for nearly 2 years. The antagonistic fungi established themselves better in the beds treated with dazomet than in untreated beds. Chlorpyrifos and carbofuran (each applied twice in 6 months) significantly reduced nematode populations in roots and soil. However, there was a buildup of nematode populations in beds treated with these two chemicals after 1 year. On a long-term basis, soil management with preplant treatment of dazomet, followed by the application of oil cakes plus antagonistic fungi, was more effective against *M. incognita* than post-plant treatment with carbofuran, carbosulfan, and chlorpyrifos on gerbera grown in polyhouses.

Soil application of *Bacillus subtilis* (BbV 57) or *Pseudomonas fluorescens* at 2.5 kg/ha at the time of planting is effective for the management of root-knot nematode.

29.5.2 Spiral Nematode, *Helicotylenchus dihystera*

29.5.2.1 Symptoms

Helicotylenchus dihystera was found to be the most predominant nematode as revealed by the community analysis of gerbera rhizosphere samples.

Fig. 29.11 *Left*, healthy and mite-infested gerbera plants. *Right*, mite damage on gerbera leaves

Fig. 29.12 Symptoms of root-knot nematode infection on gerbera plants

29.5.2.2 Management

Soil application of *Pseudomonas fluorescens* and *Paecilomyces lilacinus* (at 2.5 kg/ha and 10 g/m^2, respectively) effectively reduced nematodes number of *R. reniformis* and *H. dihystera*, respectively. These treatments were followed by the application of farm yard manure, neem powder, and carbofuran.

References

Benson DM, Parker KC (2011) Efficacy of fungicides and biopesticides for management of *Phytophthora* crown and root rot of gerbera daisy. Plant Manag Network. doi: 10.1094/PHP-2011-0512-01-RS

Parrella MP, Jones VP, Youngman RR, LeBeck LM (1985) Effect of leaf mining and leaf stippling of *Liriomyza* spp on photosynthetic rates of chrysanthemum. Ann Entomol Soc Am 78:90–93

30 Chrysanthemum

Abstract

Chrysanthemum, also known as "Queen of East," occupies a place of pride as a commercial flower crop and popular exhibition flower. It is also the leading commercial flower, important both as cut flower and as potted plant in the international market. Chrysanthemums are attacked by many insect pests and diseases. The major pests (aphids, thrips, leaf miner, leaf folder, spider mite), diseases (root rot, leaf spot, wilt, white rust, powdery mildew, chrysanthemum stunt virus), foliar nematodes, and their symptoms, biology, spread, and management are discussed.

Keywords

Chrysanthemum • Pests • Diseases • Nematodes • Symptoms • Biology • Spread • Management

30.1 Introduction

Chrysanthemum (*Dendranthema grandiflora*), also known as "Queen of East," is a flower crop next to rose in importance in the international florists' trade and holds third position in India after jasmine and rose. It is also the leading commercial flower, important both as cut flower and as potted plant in the international market. Protected cultivation is beneficial for quality production of flowers for export (Fig. 30.1).

30.2 Varieties

30.2.1 Large Flowered (Exhibits)

30.2.1.1 White Color

Snow Ball, Kasturba Gandhi, Beauty, and Innocence

30.2.1.2 Yellow Color

Chandrama, Sonar Bangla, Super Giant, and Evening Star

P.P. Reddy, *Sustainable Crop Protection under Protected Cultivation*,
DOI 10.1007/978-981-287-952-3_30

Fig. 30.1 Chrysanthemum under protected cultivation

30.2.1.3 Mauve Color
Mahatma Gandhi, Peacock, Raja, and Pink Giant

30.2.1.4 Red Color
Diamond Jubilee, Alfred Wilson, and Distinction

30.2.2 Small Flowered (Cut Flowers)

30.2.2.1 White Color
Birbal Sahni, Apsara, Himani, and Baggi

30.2.2.2 Yellow Color
Nanako, Jayanti, and Kundan

30.2.2.3 Mauve Color
Sharad Prabha, Nilima, and Ajay

30.2.2.4 Red Color
Jaya, Flirt, and Jubilee

30.2.3 Small Flowered (Pot Culture)

30.2.3.1 White Color
Sharad Shoba, Shweta Singar, Jyotsana, and Niharika

30.2.3.2 Yellow Color
Indira, Archana, Sonali Tara, and Sharad Kanti

30.2.3.3 Mauve Color
Sharad Prabha, Mohini, Hemant Sagar, and Fantasy

30.2.3.4 Red Color
Rakhee, Gem, Jaya, Flirt, Arun Singar, and Suhag Singar

30.3 Diseases

30.3.1 Root Rot, *Pythium* spp., *Phytophthora* spp.

30.3.1.1 Symptoms
In this disease, stems turn dark brown to black at the soil line. Plants are stunted, wilt, and die (Fig. 30.2).

30.3.1.2 Management

- Providing good drainage conditions to prevent water logging.
- Soil drenching with thiram or captan or mixture of both at the rate of 2.5 g/m^2 to prevent the infection.
- Mancozeb, metalaxyl, and fosetyl are also effective.

30.3.2 Leaf Spot, *Septoria chrysanthemella*

30.3.2.1 Symptoms
Grayish brown spots appear on leaves which turn surroundings yellow (Fig. 30.3). When flowering starts, the infection occurs on flower buds, which results in complete rotting. The disease spreads from down upward.

30.3.2.2 Management

- Spraying with mancozeb at fortnightly interval helps in controlling the disease.

Fig. 30.2 Pythium root rot on chrysanthemum

- Burning and destroying of infected leaves.
- Spraying with copper oxychloride (0.2 %).

30.3.3 Wilt, *Verticillium dahliae*

30.3.3.1 Symptoms

The leaves turn yellow to gray and the branch or whole plant wilts gradually.

30.3.3.2 Management

- Soil solarization by using black polythene mulch during summer months
- Soil treatment with Dithane M-45 (0.2 %)
- Dipping of rooted cuttings in benomyl suspension before planting
- Use of resistant varieties

30.3.4 White Rust, *Puccinia horiana*

30.3.4.1 Symptoms

Chrysanthemum white rust is a new and destructive disease of chrysanthemum in Europe and the Mediterranean. Initially, circular white or yellow cushions develop on the lower leaf surface and then soon turn brown (Fig. 30.4). The disease is favored by high RH and moderate temperatures (Strider 1985).

30.3.4.2 Management

Healthy plant material and resistant cultivars should be used. If plants are infected or seem to be infected, they should be removed. Overhead irrigation should be avoided because high humidity stimulates the disease. Fungicides should be applied as sprays, as soon as symptoms are seen. Spraying of plants with sulfur and other fungicides such as zineb or captan is effective in reducing the disease incidence.

Regular applications of protective fungicides, such as dithiocarbamates and chlorothalonil, or systemic fungicides, such as bitertanol, oxycarboxin, tolylfluanid, and triforine, are mostly recommended for rust control. Prevention of water condensation is also very effective (Horst 1989).

30.3.5 Powdery Mildew, *Oidium chrysanthemi*

30.3.5.1 Symptoms

Powdery mildew is characterized by a white to ash-gray powdery growth on leaves and occasionally stems. Foliage may become puckered or distorted; severely infected leaves will shrivel and die which may lead to defoliation (Fig. 30.5). The disease is most serious during hot, humid weather. Unlike most fungal diseases, free water is not required for powdery mildew infection; high humidity encourages disease development.

30.3.5.2 Management

Powdery mildew can be avoided by proper plant spacing, good air circulation, low relative humidity, and adequate light levels. Apply preventive fungicides at the first sign of disease with the active ingredients such as copper, azoxystrobin, pyraclostrobin, fludioxonil, triflumizole, myclobutanil, triadimefon, propiconazole, sulfur, potassium bicarbonate, or thiophanate-methyl according to label instructions.

30.3.6 Chrysanthemum Stunt Virus

30.3.6.1 Symptoms

There is an overall reduction in plant size, foliage becomes pale in color, and flower may open pre-

Fig. 30.3 Septoria leaf spot on chrysanthemum leaves

Fig. 30.4 White rust on chrysanthemum leaf

Fig. 30.5 Powdery mildew on chrysanthemum leaves

maturely (Fig. 30.6). This disease occurs during pinching.

30.3.6.2 Management

Use of cuttings from virus-free plants.

Chrysanthemum diseases, their symptoms, and their management in greenhouse are presented in Table 30.1.

30.4 Insect and Mite Pests

30.4.1 Aphid, *Macrosiphoniella sanborni*

30.4.1.1 Symptoms

Greenish-black nymphs and chocolate-brown adults suck the cell sap from tender growing tender parts like stem tips, flower buds, and lower surface of leaves (Fig. 30.7) during cool and cloudy months. The affected flower buds fail to open and dry up before opening. Damage results in loss of vigor, yellowing, premature leaf fall, and stunted growth. Heavy infestations affect bud and flower development considerably. Damage begins in December and is in peak during February–March.

30.4.1.2 Management

- Spray pongamia or neem oil at 10 ml/L at early infestation.

Fig. 30.6 Chrysanthemum stunt viroid infected and healthy plants (*left*), healthy and infected flowers (*right*)

- Spray dimethoate 30 EC at 2 ml/L or oxy-demeton methyl 25 EC at 2 ml/L or imidacloprid 200 SL at 0.4 ml/L or acephate 75 SP at 1.5 g/L at 15-day interval, if the aphid incidence is severe.
- Spray *Verticillium lecanii* at 3 g/L during evening

30.4.2 Thrips, *Microcephalothrips abdominalis*

30.4.2.1 Symptoms

Slender white nymphs and black adults feed on tender leaves causing bronzing and mottling of leaves. Damaged flowers get discolored, distorted, and dry due to feeding by thrips resulting in reduction in quality and quantity of flowers (Fig. 30.8). Thrips cause damage to summer blooming varieties.

30.4.2.2 Management

- Spray dimethoate 30 EC at 2 ml/L or oxy-demeton methyl 25 EC at 2 ml/L at 15-day interval.
- Drench the soil with chlorpyrifos 20 EC at 4 ml/L to reduce resting populations.
- Spray imidacloprid 200 SL at 0.4 ml/L or fipronil 5 SC at 1.5 ml/L or acephate 75 SP at 1.5 g/L, if the incidence is severe.

30.4.3 Leaf Miner, *Phytomyza syngenesiae*

30.4.3.1 Symptoms

The larvae feed inside the leaves and create long sinuous tunnels that show on the upper leaf surface as white or brown meandering lines (Fig. 30.9). The females scrape leaf surface to feed on exuded sap, which causes small pale spots on the upper leaf surface. The leaf mines created by the fly's larvae disfigure the leaves, and heavy infestations can cause a significant loss of the foliage. Incidence of leaf miner is maximum during March–June. The infestation is more severe in polyhouse.

30.4.3.2 Management

30.4.3.2.1 Nonchemical Control

Light infestations can be dealt with by removing the affected leaves or crushing the larvae or pupae at the ends of the tunnels.

30.4.3.2.2 Biological Control

Parasitic wasps (*Dacnusa* and *Diglyphus* spp.) are available for chrysanthemum leaf miner control in glasshouses.

30.4.3.2.3 Chemical Control

If plants become extensively infested, it may be necessary to use a systemic insecticide that can kill the larvae inside the leaves. Suitable insecti-

Table 30.1 Chrysanthemum diseases and their management

Disease (causal agent)	Symptoms	Survival of pathogen and favorable conditions	Management
Gray mold *(Botrytis cinerea)*	Light brown, water-soaked spots on lower petals. Woolly gray fungal spores form on decayed tissues. Rotting of lower leaves. Fungus may enter and girdle stem	In plant debris. Favored by high humidity, low temperatures (50°–60 °F), and water on plant	Keep humidity low (below 98 %); avoid overhead irrigation. Protect foliage with a fungicide, especially lower-dense foliage. Mist blooms with chlorothalonil, iprodione, or fenhexamid
Cottony rot *(Sclerotinia sclerotiorum)*	Stems rotted; flower rot is similar to gray mold. Cottony, white fungal mass may occur on rotted tissues. Black sclerotia may form inside or outside stems	Sclerotia in soil. Spores produced from sclerotia are airborne but infect only through flowers and dead tissues. Favored by high humidity	Same as for gray mold. Also, treat soil with PCNB before planting and spray foliage with iprodione or thiophanate-methyl
Verticillium wilt *(Verticillium dahliae)*	Yellowing and wilting of foliage, may be one-sided. Leaves die and dry upward from the base of the plant. Vascular tissue may be discolored	Soilborne for many years. Carried in cuttings and root divisions. Favored by cool weather followed by hot weather during flowering	Plant in pasteurized soil or soilless mix free of the pathogen. Use resistant cultivars and pathogen-free plants. Fumigate with chloropicrin-methyl bromide combination. In sunny climates, soil solarization might be considered
Fusarium wilt *(Fusarium oxysporum* f sp. *chrysanthemi* and F. *oxysporum* f sp. *tracheiphilum)*	Unilateral chlorosis of one or more leaves near the apex, followed by curvature of the stem toward the affected side. As the disease progresses, there is a general chlorosis and wilt and stunting of leaves. The vascular system becomes a reddish brown. Yellowing of leaves, wilting, and discoloration of the vascular tissue develops on one side of the plant	Soilborne and carried in cuttings. Favored by high soil temperatures (80 °F).	Use disease-free cuttings in clean soil. Treat soil (see *Verticillium* wilt). Adjust pH of soil to 6.5–7.0 and use nitrate nitrogen. Apply a fungicide to protect healthy plants. Avoid planting highly susceptible cultivars ('Bravo', 'Cirbronze', 'Illini Trophy', 'Orange Bowl', 'Royal Trophy', 'Yellow Delaware')
Powdery mildew [*Golovinomyces (Erysiphe) cichoracearum*]	White powdery growth on leaves and stems. Found mainly on older leaves	Airborne spores produced only on living plants. Favored by high humidity, crowding of plants, and cool weather	Spray with piperalin to eradicate existing infections. Protect foliage with myclobutanil, fenarimol, or triadimefon

(continued)

Table 30.1 (continued)

Disease (causal agent)	Symptoms	Survival of pathogen and favorable conditions	Management
Pythium root rot and basal stem rot (*Pythium* spp.)	Girdling black lesions occur on stems near soil line. Plants stunted as a result of reduced root system caused by root rot. Plants wilt and may die	Soilborne pathogen. Spores spread in water or in soil. Favored by excess soil moisture and poor drainage	Treat soil as for *Verticillium* wilt. Drench plants with mefenoxam
Ray blight (*Phoma chrysanthemi*)	Basal leaf and stem rot. Belowground stem infection may cause a one-sided distortion and necrosis of foliage. Blackish rot of petals (ray blight) may extend into flower stalk	In chrysanthemum refuse. Waterborne spores. Favored by rainy weather	Use disease-free cuttings. Avoid wetting foliage and keep humidity low. Protect foliage with chlorothalonil
Rhizoctonia stem rot (*Rhizoctonia solani*)	Stem rotted at soil line and girdle the plant. Reddish-brown dead areas develop at the soil line. Plants stunted. Dark fungal strands may be visible with hand lens. May kill plants. Young infected plants wilt during the day and recover at night	Soilborne fungus. Favored by warm, moist conditions	Plant in pasteurized soil or a soilless mix free of the pathogen. Avoid deep planting. Spray base of transplants with thiophanate-methyl, chlorothalonil, or iprodione after planting, or treat soil with PCNB before planting
Septoria leaf spot (*Septoria obesa, S. chrysanthemi*)	Irregular or circular, brown or black dead spots develop progressively upward from base of plant. Tiny black fungal fruiting bodies in centers of spots	In plant debris and in soil debris for 2 years. Spores spread in splashing water. Favored by wet weather	Protect foliage with a fungicide at first sign of disease. Greenhouse: avoid wetting foliage. Keep humidity low
White rust (*Puccinia horiana*)	Whitish pustules on the lower leaf surface. On the upper surface, the infection is evident as pale-green to yellow spots up to an inch in diameter. Raised, waxy, pink-colored pustules are formed on lower leaf surface. As they mature and produce spores, they turn whitish in color	This microcyclic rust survives on living chrysanthemum foliage. Teliospores in the leaf pustules germinate in place producing basidiospores that are airborne and infect by direct penetration. Favored by the cool, damp weather, except that direct sunlight destroys airborne basidiospores. Principally a greenhouse disease	Requires 3–5 applications of myclobutanil (Eagle or Hoist). Repeated treatments with myclobutanil can cause stunting of growth. Growers in nonregulated areas can rotate with mancozeb (Dithane) and chlorothalonil (Daconil) as a protectant treatment program. Combine fungicide treatments with destruction of infected plants and crop-free periods of 8 weeks. Cleanup of infestations in quarantined areas must be done

(continued)

Table 30.1 (continued)

Disease (causal agent)	Symptoms	Survival of pathogen and favorable conditions	Management
Ray blight [*Ascochyta (Mycosphaerella)*]	Flower development is retarded on one side of the bud. Petals exhibit a brown discoloration. Browning and blackening extends down the stem, causing the flower to droop. Brown to black irregularly shaped spots develop on leaves		Avoid overhead irrigation. Apply a fungicide to protect healthy plants
Ray speck (*Alternaria* sp. or *Stemphylium* sp.)	Pinpoint dead spots develop on petals. These spots may not enlarge. If enough spots are present, the entire flower dies		Avoid overhead irrigation. Maintain greenhouse humidity below 98 %. Apply a fungicide to protect healthy plants
Bacterial blight (*Erwinia chrysanthemi*)	Water-soaked lesions; pith becomes jellylike; tops turn black and exude drops of liquid. Stem may break or split. Cuttings turn dark brown and collapse. Established plants wilt during the day when infected and recover at night	In plant debris. Favored by high temperatures (80°–90 °F), absence of free water, and high humidity	Use disease-free cuttings. Reduce humidity in growing areas. Dip cuttings in streptomycin. Disinfect propagation beds between crops. Destroy infected cuttings
Hollow stem (*Erwinia carotovora*)	Pith of rooted cuttings deteriorates and collapses. Affected tissues are brown. Surviving plants do not grow satisfactorily and pith collapse may extend upward involving several internodes. 'Red Torch', 'Tempo', and 'Tempter' are very susceptible	Bacteria may be present in vascular bundles of symptomless chrysanthemums. Bacteria also present in undecomposed debris. Favored by high temperatures and high moisture	Use disease-free cuttings. Reduce humidity in growing areas. Streptomycin dips may be helpful
Crown gall (*Agrobacterium tumefaciens*)	Irregular or round galls on stems and sometimes leaves	Soil and galls. Infection favored by moist conditions	Destroy infected plants

(continued)

Table 30.1 (continued)

Disease (causal agent)	Symptoms	Survival of pathogen and favorable conditions	Management
Bacterial leaf spot (*Pseudomonas cichorii*)	Small dark brown to black spots on lower leaves enlarge and become irregular in shape. When infected leaves dry, the spots become brittle and crack. The disease often spreads in one side of the plant, eventually to the flowers		Do not plant infected cuttings. Avoid overhead irrigation. Water in a manner that keeps leaf surfaces dry at all times. Protect plants grown outdoors from splashing
Foliar nematode (*Aphelenchoides ritzemabosi*)	Dark green, angular spots in leaves develop progressively upward from base of plant. Leaves turn yellow	Adults can survive up to 3 years in dead leaves. Spread by splashing water. Leaves need to be wet for infection to take place	Use disease-free plants. Discard infected plants. Avoid overhead irrigation. Control weeds. Submerge infected plants in hot water (115 °F) for 10 min
Virus or virus like diseases	**Symptoms**	**Host range and natural spread**	**Management**
Aspermy (*Tomato aspermy virus*)	Flower distortion and reduction in flower size. Color break in florets of red, bronze, and pink cultivars. Foliar symptoms not apparent. Some cultivars are symptomless	Transmitted by handling, cutting tools, and vegetative propagation. Also transmitted by aphids	Use disease-free cuttings. Control aphids; remove infected plants
Chlorotic mottle (*Chrysanthemum chlorotic mottle viroid*)	Mottling followed by complete chlorosis. May be confused with nutritional problems. Symptoms somewhat masked under low light and cool temperature conditions below 70 °F	Spread by handling, cutting tools, and vegetative propagation	Use disease-free cuttings. Remove infected plants and disinfest tools used to handle them. Do not handle healthy chrysanthemums after handling infected plants
Spotted wilt (*tomato spotted wilt virus, impatiens necrotic spot virus*)	Frequently one-sided in plant. Ring patterns on leaves of some cultivars. Leaf distortion and necrosis. Dark necrotic streaks on stems. Flowers may be distorted and with some necrosis	Thrips-transmitted; not spread by cutting knife, but can be spread by vegetative propagation. Many weeds and perennial ornamental plants act as reservoirs of virus (dahlia, calla, nasturtium, mallow, knotweed, plantain, and others)	Eliminate nearby susceptible ornamental plants and weeds. Control thrips inside and outside growing areas. Use virus-free cuttings
Stunt (*Chrysanthemum stunt viroid*)	General stunting of plants. Foliage may be pale with upright, young leaves. Flowers are smaller than normal and some cultivars may flower 7–10 days early	Viroid is easily transmitted by handling, cutting knives, vegetative propagation, etc. Viroid has a wide host range. Symptomless in some plants	Purchase virus-indexed plants that are free of the pathogen. Destroy infected plants and disinfest tools used to handle them. Do not handle healthy chrysanthemums after handling infected plants

Fig. 30.7 Chrysanthemum aphid infestation on leaf (*left*) and adult aphid (*right*)

Fig. 30.8 Thrips damage on chrysanthemum leaf, flower, and adult thrips

Fig. 30.9 Leaf miner on chrysanthemum leaf (*left*) and adult leaf miner (*right*)

Fig. 30.10 Leaf folder larva and adult

cides include thiacloprid or acetamiprid or 0.05 % monocrotophos or 0.05 % triazophos.

30.4.4 Leaf Folder, *Omiodes indicata*

30.4.4.1 Symptoms

All the larval stages attack the plant with the help of silky threads (Fig. 30.10); the larva folds the leaves and starts feeding on leaves from inside. Young larvae live between two leaves spun together. Later instars may spin several leaves together. They feed on the mesophyll of the leaves, but the final instar larvae completely skeletonize the leaves.

30.4.4.2 Management

Spray 0.02 % cypermethrin or 0.02 % decamethrin or 0.05 % quinalphos at fortnightly interval.

30.4.5 Spider Mite, *Tetranychus urticae*

30.4.5.1 Symptoms

Mites mainly feed on the undersurface of leaves (Fig. 30.11) and spread to entire plant in case of heavy infestations. Damage starts in March and reaches a peak during April–June (in hot dry season). Affected leaves become discolored and dry and fall off, while attacked flowers get discolored and dry. The affected flower buds fail to open and dry up even before opening.

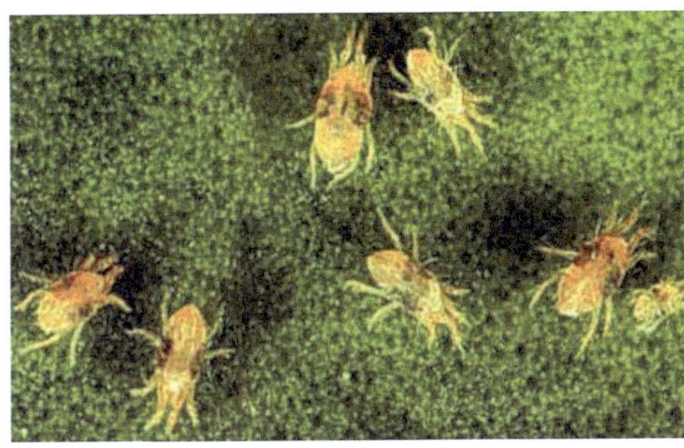

Fig. 30.11 Two-spotted spider mites

30.4.5.2 Management

- Spray of dicofol 18.5 EC at 2.5 ml/L or triazophos 40 EC at 1.5 ml/L or profenofos at 50 EC 1 ml/L or ethion 50 EC at 1 ml/L at 15-day interval followed by pongamia or neem oil at 10 ml/L.
- Spray *Verticillium lecanii* formulation at 3 g/L.

30.5 Nematodes

30.5.1 Foliar Nematode, *Aphelenchoides ritzemabosi*

30.5.1.1 Symptoms

The symptoms on chrysanthemum include characteristic brown spots limited to the veins and a

Fig. 30.12 Foliar nematode symptoms on chrysanthemum leaves

progressive yellowing of the whole leaf. These symptoms are due to combined action of the nematode and other organisms (Cayrol and Combettes 1972). Leaf symptoms on infested chrysanthemum include reddish-yellow lesions on the lower leaves of young plants; in older plants these leaves are markedly chlorotic and a large area of the leaf surface becomes necrotic (Fig. 30.12). The foliage is scanty and the flowers are few and deformed. Leaves in the upper part of plants have shown slightly higher resistance than those in the lower part (Cid del Prado and Sosa Moss 1978). Direct effects are mechanical damage caused by the stylet and damage due to hormones of growth and division (Cayrol and Combettes 1972).

30.5.1.2 Management

Suggested control measures include cleaning and burning infested leaves, submerging infected cuttings in hot water, and spraying of foliage with chlorpyrifos (Gill 1981).

Treatment of chrysanthemum nursery soil with an organophosphorus nematicide was very effective in control of this nematode (Fukazawa and Kobayashi 1971).

References

Cayrol JC, Combettes S (1972) Histopathological study of the chrysanthemum leaf eelworm disease in monoxenic cultures. Ann de Zool-Ecol Animale 4(2):119–128

Cid del Prado I, Sosa Moss C (1978) Occurrence of *Aphelenchoides ritzemabosi* on the foliage of *Chrysanthemum maximum* in Mexico. Nematropica 8:6

Fukazawa N, Kobayashi Y (1971) Chemical control of abnormalities of summer flowering chrysanthemum caused by *Aphelenchoides ritzemabosi* Schwartz. Proc Kanto Toxan Plant Prot Soc 18:126

Gill JS (1981) Fighting the foliar nematode of chrysanthemum. Indian Hortic 25(4):21–22

Horst RK (ed) (1989) Compendium of rose diseases. APS Press, St Paul

Strider DL (ed) (1985) Diseases of floral crops. Praeger Special Studies, New York

31 Gladiolus

Abstract

Gladiolus is becoming a leading cut flower crop and is gaining popularity due to its beautiful and attractive colored flowers. It is cultivated in polyhouses/polytunnel in order to get quality flowers throughout the year. The major pests (aphids, thrips, cutworms, leaf eating caterpillar, mealybug, borer), diseases (Fusarium yellows and corm rot, gray mold, corm rot, Curvularia blight), and their symptoms, biology, spread, and management are discussed.

Keywords

Gladiolus • Pests • Diseases • Symptoms • Biology • Spread • Management

31.1 Introduction

Off-season flowering using portable polythene tunnels of size (3 × 1.5 × 1.5 m) placed over the crop planted in October–November produced quality flowers in February–March. Off-season flowers of gladiolus can be produced under mid-hill conditions successfully (Fig. 31.1).

31.2 Varieties

Cultivars, viz., 'White Friendship' and 'Suchitra', produced quality flowers from February onward under polyhouse conditions (Table 31.1).

31.3 Diseases

31.3.1 Fusarium Yellows and Corm Rot, *Fusarium oxysporum* f sp *gladioli*

31.3.1.1 Symptoms

Wilt is a major fungal disease in gladiolus. It is soilborne and spread through corms from season to season. The fungus development is favored by high temperature, high level of nitrogen, anaerobic condition, and accumulation of carbon dioxide (CO_2). Yellowing of older leaves, browning of vascular tissue, and wilting of whole plant are the symptoms of the disease (Fig. 31.2). Roots

P.P. Reddy, *Sustainable Crop Protection under Protected Cultivation*,
DOI 10.1007/978-981-287-952-3_31

Fig. 31.1 Gladiolus under protected cultivation

cannot absorb nutrients and water due to damage of vascular tissues.

Roots may show discrete brown lesions or a general rotting, and leaf infection is basal and associated with corm rot (Fig. 31.2). Leaf yellowing, starting with oldest leaves, is seen occasionally. Corm rot in the field is soft, invaded by other organisms and is associated with brown wiry roots. In storage, rotting may begin anywhere on corm surface, but most spots are found at base next to the core; by scrapping, browning can be seen.

31.3.1.2 Management

- Use disease-free corm stock.
- The best way to control the wilt is to follow crop rotation.
- Treat corms in hot water (38–40 °C) containing 2.5 g each of Benlate and captan for 30 min for effective management of wilt.
- Farmers should avoid the use of infected/sick soils for cultivation.
- Drench the soil around the infected plant base with carbendazim (1 g/L water) when the disease appears.

Table 31.1 Cultivars of gladiolus for protected cultivation

Color	Cultivars
Yellow	Aldebaron, Jester, Nova Lux, Top Brass, Topaz, Vink's Glory
White	American White Friendship, White Prosperity
Pink	Friendship Pink, Suchitra, Spic, and Span
Red	American Beauty, Oscar, Red Majesty
Purple	Blue Lilac, Red Beauty, Interpid, Trader Horn, Mayur, Marvellous

- Spray zineb (Indofil M-45) at 2 g/L in combination with carbendazim (Bavistin) at1g/L at fortnight interval against corm rot throughout the cropping season after plants are 15 cm tall.
- Soil solarization is reported to give good results, but it is laborious and not practical in large areas.

31.3.2 Gray Mold, *Botrytis cinerea*

31.3.2.1 Symptoms

On leaves, large oval to round, brown spots appear (Fig. 31.3). Spots occur on flowers and stems also. At first the spots are pale brown and then become dark. A soft, brown rotting at the base of florets may develop after rainfall (Fig. 31.3). In a moist, cool atmosphere, the whole flower becomes covered with brownish gray growth of the fungal spores.

31.3.2.2 Management

- Remove old flower spikes and destroy.
- Hot water treatment (52 °C) of corms with carbendazim (1 g/L water) added to the water is effective in eradicating the pathogen from planting stock.
- Control of the disease on flowers depends on controlling leaf infection. Spray of carbendazim (1 g/L of water) and mancozeb (2 g/L of water) during all stages of plant growth prevents disease infection.

Fig. 31.2 *Left*, symptoms of Fusarium wilt on gladiolus plants; *Right*, healthy (*top*) and infected gladiolus corms (*bottom*)

Fig. 31.3 Gray mold symptoms on gladiolus leaves and flowers

31.3.3 Corm Rot, *Fusarium* sp

31.3.3.1 Symptoms

Corm rot is a common problem caused by group of fungi, namely, *Fusarium*, *Curvularia*, *Stromatinia*, and *Penicillium* spp. It causes heavy loss particularly during storage. Anaerobic conditions, storage in airtight rooms, and more humidity favor the infection. Black, brown, greenish, or yellowish moldy growth on corms is seen during the storage (Fig. 31.4). Under poor air circulation, the corms may rot and emit foul smell.

31.3.3.2 Management

Hot water treatment (38–40 °C) containing 2.5 g each of Benlate and captan for 30 min has been quite effective. Damage to corms at the time of lifting, improper curing, and dampness in storage should be avoided.

31.3.4 Curvularia Blight

31.3.4.1 Symptoms

It is caused by many fungi but the most important are *Curvularia trifolii* and *C. eragrostidis*. Water-soaked oval to elongated brownish spots appear on leaves, sheaths, and petals, and later they coalesce in advanced stage. The patches turn to brown and finally black (Fig. 31.5). Moist and warm conditions spread the blight very fast.

31.3.4.2 Management

It is effectively controlled by mancozeb 0.2 % spray at 10-day interval.

Fig. 31.4 Corm rot due to *Fusarium* sp

Fig. 31.5 Curvularia blight on gladiolus leaf

31.4 Insect and Mite Pests

31.4.1 Aphids, *Aphis gossypii*, and *A. craccivora*

31.4.1.1 Symptoms

Aphids suck the sap from tender parts of the plant (Fig. 31.6), and the infected plants become weak and get deformed. They excrete honeydew which attracts sooty mold and ants. Incidence of aphids is commonly seen during February and March.

Fig. 31.6 Aphid infestation on gladiolus leaf

31.4.1.2 Management

It can be controlled by spraying of malathion at 0.1 % or Rogor at 0.1 % at fortnightly intervals.

31.4.2 Thrips, *Taeniothrips simplex*

31.4.2.1 Symptoms

This is a major pest in gladiolus and causes serious damage to the crop. Yellow nymphs and black adults damage leaves and spikes by rasping tissue and sucking the sap. Affected leaves and spikes develop silver-colored streaks (Fig. 31.7) which later turn brown, get deformed, and dry if the damage is severe. It also attacks corms in storage, and infected corms become sticky, shrivel, and produce weak plants when planted.

31.4.2.2 Management

It can be controlled by weekly spraying of dimethoate (Rogor) at 0.2 % or malathion at 0.1 %.

31.4.3 Cutworms, *Agrotis segetum*

31.4.3.1 Symptoms

Incidence of cutworms is normally observed in the first month of crop. Eggs of moth are seen

Fig. 31.7 Thrips damage to gladiolus leaf and flower

near the ground level on plant parts. Larvae feed on emerging shoots and cut the plants at the ground level during night. Sometimes they damage underground corms and developing spikes.

31.4.3.2 Management

Spraying of methyl parathion at 0.05 % or quinalphos at 0.05 % at fortnight interval provides protection.

31.4.4 Leaf Eating Caterpillar, *Spodoptera litura*

31.4.4.1 Symptoms

Egg masses covered with hair are seen on under surface of the leaves. Early instar larvae feed on lower surface of leaves by scraping, while mature larvae eat leaves voraciously (Fig. 31.8) during nighttime. Skeletonization of leaves is the main symptom.

31.4.4.2 Management

Spraying of quinalphos at 0.05 % or chlorpyrifos at 0.05 % or carbaryl at 0.1 % at 10-day interval controls the incidence effectively. Spraying with neem oil at 1 % or neem kernel extract 4 % checks the damage caused by early instar larvae.

31.4.5 Mealybug

31.4.5.1 Symptoms

Both nymphs and adults attack corms by sucking sap causing shriveling and drying of affected corms in severe cases. Incidence of mealybugs starts in the field during dry conditions, and then it is carried to the storage. Movement of ants on plants is the sign of mealybug infestation.

31.4.5.2 Management

Methyl parathion at 0.04 % or dimethoate at 0.04 % or acephate 0.1 % at fortnight interval can be applied to effectively control mealybugs.

31.4.6 Borer, *Helicoverpa armigera*

31.4.6.1 Symptoms

Larvae feed on leaves and unopened buds.

31.4.6.2 Management

Spraying of Thiodan at 0.5–0.8 % or endosulfan at 0.07 % or methyl parathion at 0.05 % or Ekalux at 0.5–0.8 % at fortnight interval can control the pest. Neem kernel extract 4 % or neem oil at 1 % can also control the pest.

31.4.7 Mites, *Tetranychus equatorius*

31.4.7.1 Symptoms

Attack by the mite begins at an early stage when plants are young. Green-colored nymphs with lateral specks and reddish-brown-colored adults colonize on leaves and suck the sap causing discoloration and wilting. The affected leaves fall off.

31.4.7.2 Management

Methyl parathion 0.05, fluvalinate 0.012 %, and diazinon at 0.4 % offer significant control of mites.

Fig. 31.8 Caterpillar damage on gladiolus leaves

Fig. 31.9 Yellowing of leaves due to root-knot nematode infection on gladiolus plant

31.5 Nematodes

31.5.1 Root-Knot Nematodes, *Meloidogyne* spp

31.5.1.1 Symptoms

Severe galling on roots results in yellowing of leaves (Fig. 31.9) which subsequently leads to stunted growth. The nematode invades roots, daughter corms, and cormels which develop after flowering.

31.5.1.2 Survival

The nematodes would survive in corm tissue in the soil as a source of inoculum for next season.

31.5.1.3 Management

31.5.1.3.1 Physical Methods

Hot water treatment of corms at 58 °C for 30 min eliminates root-knot nematode infection.

31.5.1.3.2 Chemical Methods

Dipping of gladiolus corms in thionazin or fensulfothion solution (0.5 g a.i. per liter) gave reduced root-knot nematode infestation (Overman 1970). Application of Vorlex (35 gal/acre) by broadcast method also resulted in better flower yield.

31.5.1.3.3 Integrated Methods

Soil application of neem cake enriched with *Trichoderma harzianum* at 1 t/ha is effective.

Application of 5 t of FYM enriched with *Pseudomonas fluorescens* (with 1×10^9 cfu/g) per ha significantly reduced *M. incognita* by 61–76 % and *Rotylenchulus reniformis* by 65–70 % in the roots of gladiolus and increased the flower yield by 19–22 %.

References

Overman AJ (1970) Gladiolus corm dips for root-knot nematode control. Proc Fla State Hortic Soc 82:362–366

32 Lilies

Abstract
Lilies, both Asiatic and Oriental, along with their hybrids, are more productive and lucrative than any other commercially grown cut flowers. Lilies are prone to attack by several insect pests and diseases. The major pests (aphids, bulb mites, beetle, thrips, weevil, slugs and snails, symphylids), diseases (basal rot, botrytis blight, root rot, southern wilt, damping-off, viral diseases), nematodes (root-knot, lesion, and foliar), and their symptoms, biology, spread, and management are discussed.

Keywords
Lilies • Pests • Diseases • Nematodes • Symptoms • Biology • Spread • Management

32.1 Introduction

Lilies, especially Asiatic and Oriental types grown in greenhouse (Fig. 32.1), are most fascinating in international floriculture trade.

32.2 Varieties

Lily varieties Connecticut King, Gran Paradiso, Elite, Pollyanna, Prato and Solemio (yellow flowers), Navona (white flowers), and Prunello (orange flowers) among Asiatics, while Star Gazer, Marco Polo, Casablanca, Siberia (white flowers), Samour, and Pergamo in case of Orientals are highly promising. Easter lily (*Lilium longiflorum*) var. Osant (white) is also grown under polyhouses.

32.3 Insect and Mite Pests

32.3.1 Aphids, *Neomyzus circumflexus*

32.3.1.1 Symptoms

The most important pests are aphids. The crescent-marked lily aphid is one of the common aphids to attack lily plants in the field and green-

P.P. Reddy, *Sustainable Crop Protection under Protected Cultivation*,
DOI 10.1007/978-981-287-952-3_32

Fig. 32.1 Lily flowers

Fig. 32.2 Lily aphids with crescent markings

house (Fig. 32.2). Several species of aphids that colonize lilies not only spread virus disease, but they also debilitate the plants physically, causing twisted leaves and distortion of flower buds.

32.3.1.2 Management

Insecticidal soaps offer some control against aphids. Applications should be made at regular intervals for maximum efficacy. Growers with aphid problems can apply foliar or soil drenches of a labeled neonicotinoid.

Systemic insecticide sprays should be applied to plantings on a regular basis. They can be combined with a fungicide. The insecticide acephate (e.g., Orthene) has been used quite successfully for aphid control, and malathion, which is sprinkled on the soil and watered in, can protect plants from aphids for an entire growing season.

Oil spraying should be just before sundown. Light summer oil, used at about 1 % dilution, is usually effective. Oils are quite safe to use and can be combined with most insecticides and fungicides.

Aphids can be prevented from migrating into the greenhouse by covering vents and doors with a 300-hole-per-inch mesh screening.

It is important to destroy any dead plant material that may harbor overwintering eggs.

32.3.2 Bulb Mites, *Rhizoglyphus echinopus*

32.3.2.1 Symptoms

Bulb mites (Fig. 32.3) are troublesome but usually secondary pests. In warm climates they are usually present in large numbers, particularly just above the basal plate and between the scales. They attack the roots and basal plate and eventually enter the center of the bulb.

32.3.2.2 Management

The following methods of control may be used against bulb mites:

- Treat bulbs with hot water at 44 °C for 1 h.
- Fumigate dry bulbs with paradichlorobenzene (the active substance in mothballs) in an airtight container. Spread the fumigant over the bottom of the container, using 4 g per liter of air space and exposing the bulbs for 12 h.
- Dust the bulbs with flowers of sulfur.
- Destroy badly infested bulbs by burning them.

Fig. 32.3 Lily bulb mite adult

32.3.3 Beetle, *Lilioceris lilii*

32.3.3.1 Symptoms

Young grubs graze away the underside of leaves, resulting in white or brown dried up patches (Fig. 32.4). The older grubs eat entire leaves, starting at the tips and working back to the stem; they will also feed on the petals, stem, and seed pods. Adult beetles make rounded holes in the leaves and will also feed on petals and seed pods.

32.3.3.2 Biology

Red lily beetle overwinters as adult beetles in soil, leaf litter, and other sheltered places. This could be anywhere, not necessarily in the vicinity of lilies and fritillaries (consequently, there is no advantage in attempting to treat the soil below lily plants). The beetles begin emerging on sunny days in late March and April when they seek out the foliage of host plants. Eggs are laid in small batches on the underside of leaves during April to midsummer. The eggs hatch and the larvae feed on the foliage. When fully fed, the larvae go into the soil to pupate. The next generation of adult beetles emerges from midsummer onward. These beetles add to the feeding damage, but there is only one generation a year, and these late summer adults will not mate and lay eggs until the following year.

32.3.3.3 Management

32.3.3.3.1 Nonchemical

- Where only a few lilies and fritillaries are being grown, the plants should be regularly inspected from late March onward so that adult beetles, larvae, and eggs can be removed by hand.

32.3.3.3.2 Chemical

- For extensive infestations, it may be necessary to use an insecticide, such as thiacloprid (e.g., Bayer Provado Ultimate Bug Killer), acetamiprid (e.g., Bug Clear Ultra), deltamethrin (e.g., Bayer Sprayday Greenfly Killer, Bayer Provado Ultimate Fruit & Vegetable Bug Killer), or lambda-cyhalothrin (e.g., Westland Resolva Bug Killer). Organic gardeners can use pyrethrum (e.g., Py Spray Garden Insect Killer, Doff All in One Bug Spray, Bug Clear Gun for Fruit & Veg, Pyrol Bug & Larvae Killer, or Growing Success Fruit & Veg Bug Killer). Pyrethrum should control newly hatched larvae but will be less successful against the adult beetles.
- Drench soil with an insecticide such as acephate (e.g., Orthene) or neonicotinoid to kill the mature larvae that live just under the soil surface in winter. Also, avoid transporting infested soil to other sites.
- Use a neem-based insecticide such as Azatin, Aza-Direct, or BioNeem. Neem works best on early instar (young) larvae.
- Plants in flower should not be sprayed to avoid harming bees and other pollinating insects.

Fig. 32.4 The lily beetle damage on leaf (*left*) and beetle adult (*right*)

32.3.4 Thrips, *Liothrips vaneeckii*

32.3.4.1 Symptoms

Feeding seems to be localized at the bases of the scales, where thrips (Fig. 32.5) seriously weakens the bulb, rendering it flabby. This allows the entry of bacteria and fungi, frequently resulting in the bulb rotting away.

32.3.4.2 Management

The following controls can be used:

- Treat bulbs with hot water treatment at 44 °C for 1 h to eradicate the pest.
- Dust bulbs with benzene hexachloride (BHC).
- Wash or dip bulbs in a solution of an insecticide.

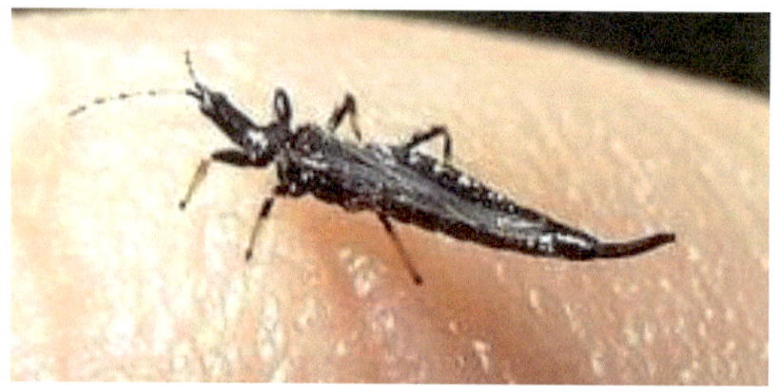

Fig. 32.5 Adult lily thrips

32.3.5 Weevil, *Agasphaerops nigra*

32.3.5.1 Symptoms

The larvae burrow into the lily stem and bulb. Adult weevils (Fig. 32.6) emerge in March and April, feeding on the leaves of plants.

32.3.5.2 Management

Systemic insecticides are highly effective in controlling weevils.

32.3.6 Slugs and Snails

32.3.6.1 Symptoms

Slugs and snails (Fig. 32.7) feed voraciously on lilies when the shoots emerge. Later in the season they can climb the stems, stripping the leaves completely.

32.3.6.2 Management

- Limit habitat around lily plantings by controlling weeds and using mulch instead of ground cover planting.
- These can be destroyed by frequent cultivation of the top few inches of soil, bringing the eggs to light and exposing them to frost and birds.
- Place bait containing metaldehyde among plantings in the early evening hours. This is crucial at the time when lily shoots are emerging and during damp weather. Bait should be renewed after heavy rain. Liquid bait such as Deadline may be less attractive to pets than pelletized bait. Beer traps have been used successfully by some enthusiasts: simply pour the beer into a shallow dish and place it near the lily plantings in the early evening.

32.3.7 Symphylids

32.3.7.1 Symptoms

Symphylids (Fig. 32.8) can damage lily crops severely if their populations get out of control.

Fig. 32.6 Adult lily weevil

Fig. 32.7 Baby slug and eggs

Fig. 32.8 Symphylid maggots

32.3.7.2 Management

The usual control is soil fumigation.

32.4 Diseases

32.4.1 Basal Rot, *Fusarium oxysporum* var *lilii*

Fusarium, however, is the most serious soilborne disease in North America and should be considered a primary pathogen.

32.4.1.1 Symptoms

The usual symptoms of basal rot in growing plants are premature yellowing of the foliage, stunting, and premature senescence. All are typical reactions to ethylene, a gas produced by the decaying bulb tissues. Infected bulbs tend to produce many new scale bulblets, usually on the severed scales (Fig. 32.9); however, such bulblets form at the infected end of the scale and are thus readily infected in turn. The main bulb is frequently destroyed, but masses of stem bulblets often form.

32.4.1.2 Spread and Survival

The pathogen is readily disseminated by spores, which can be carried in the soil or on the surfaces of bulbs, tools, agricultural equipment, or packing crates. Released from the debris of the decayed bulbs, the fungus can remain viable in garden soil for at least 3 years without a host.

32.4.1.3 Management

Avoid fertilizers high in nitrogen, such as ammonium salts; these promote soft, rapid bulb growth and make bulbs very vulnerable to infection. Use well-rotted manures and compost as a mulch; an added benefit is that mulch helps keep the soil cool, which discourages Fusarium.

Control soil moisture. Fusarium is always more active in low, wet sites, so lilies should be planted in well-drained positions. Avoid overwatering during the warm summer months. Since acid soils may also aggravate the disease, an application of lime to increase pH may be advisable in some soils. Avoid mechanical damage

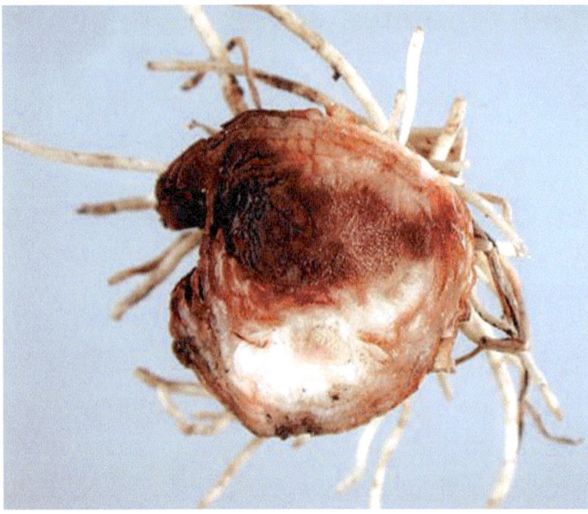

Fig. 32.9 Basal rot on lily bulb

Fig. 32.10 Botrytis damage to lily flower and foliage

during weeding, cleaning up, or transplanting the lilies, and control biting insects such as grubs and nematodes. Any lesion provides easy entry for basal rot organisms.

The high temperatures necessary for scale propagation favor Fusarium infection, and bulbs can perish in infancy. It is therefore critical to scale only the healthiest bulbs, to wash them thoroughly before scaling, and to dip the scales in a fungicide before incubation. Lily hybridizers in the commercial field must make greater efforts to breed varieties specifically for Fusarium resistance. This is even more crucial because key chemical controls are becoming unavailable.

32.4.2 Botrytis Blight, *Botrytis elliptica*, and *B. cinerea*

Botrytis blight is caused by two species of the fungus *Botrytis* that attack the aboveground parts of the plant. Both *B. elliptica* and *B. cinerea* can be present on the same plant, but the former is the more destructive. *Botrytis cinerea* flourishes in cool temperatures and is more apt to infect leaves, open flowers, and seedpods in cool summer weather and late fall. Botrytis is often considered the most important disease of lilies, especially in "Botrytis climates" such as the warm, moist coastal areas.

32.4.2.1 Symptoms

The first signs of Botrytis can be white spots on the leaves; these become teardrop-shaped marks on the upper surface. They are lighter on the margin and darker in the center, so they are often called "pheasant eye" marks. In severe attacks during wet weather and warm, muggy conditions, the spots eventually coalesce, and the whole leaf collapses and decays (Fig. 32.10).

Botrytis does not spread internally through a plant; instead, it continues to appear on new surfaces on the same and neighboring plants as more spores are produced and distributed. In severe cases it enters the stem and the plant collapses. The destruction of foliage can be very rapid. Brown spots appear on open flowers when Botrytis and moisture are present. These are believed to be caused by *B. cinerea* rather than *B. elliptica*, but this has not been conclusively demonstrated.

32.4.2.2 Survival and Spread

The fungus overwinters as small black sclerotia that formed on the leaves in the previous season. These produce spores, which are blown about by the wind and splashed by rain onto the newly developing foliage in spring.

Moisture is essential for the spread of Botrytis. The formation of spores, liberation, and germination all take place within 12 h; thus, 24 h of wet, moderately warm weather may lead to a considerable outbreak. Prolonged rains, frequent showers, fog, and heavy dew accompanied by warm temperatures, with moisture persisting on the foliage, produce perfect conditions for the disease

32.4.2.3 Management

With Botrytis disease, an ounce of prevention is worth a pound of cure. Therefore, early spraying is critical where environmental conditions tend to favor the disease. At the end of the growing season, remove all plant debris, pulling the old stems when they come away easily. Following a severe infection, remove the debris as soon as possible. Pull or cut the stems and rake off as many leaves as possible; all carry the resting sclerotia, which will rise up to bring new infection next spring. Destroy the debris if possible, but do not incorporate it in a compost pile. An application of fresh mulch is very beneficial at this time. Many growers spray the ground with a copper fungicide during the dormant season, but there is no proof that this kills resting or germinating spores.

Infection can occur at temperatures between 2 °C and 24 °C and is most likely in mild, moist, or foggy weather. It is advisable to remove spotted leaves when they are still wet in the morning; this can stop or at least inhibit further spread.

Finally, spraying is only effective when the leaves are dry. Pay particular attention to covering the undersides of the leaves, for this is where infection takes place.

Many lilies are highly resistant to Botrytis. Orientals and Aurelians are much less susceptible than Asiatics. *Lilium lankongense* and its hybrids have shown remarkable resistance; however, *L. davidii* and some of its hybrids are particularly susceptible. The new tetraploid hybrids, with their thicker epidermis, seem more resistant to both Botrytis and leaf scorch.

32.4.3 Root Rot, *Rhizoctonia solani*

Root rot is associated with poor drainage, lack of soil aeration, and planting in soils that are too finely textured, such as heavy clays. Injury by root lesion nematodes (*Pratylenchus penetrans*) opens up roots to infection by rots.

32.4.3.1 Symptoms

Rhizoctonia is a soilborne fungus and a mild parasite of lily bulbs. The symptoms are dark yellow discolorations around the minute lesions caused by the fungal mycelia. The lesions are numerous and give the scales a yellow tint (Fig. 32.11). Damage to the scales is slight, but the lesions can become entry points for other pathogens. *Rhizoctonia* is favored by warm temperatures, especially in greenhouses. There are many strains that can become pathogens in lilies.

32.4.3.2 Management

The best control for root rots is to improve cultural practices, primarily drainage. Overwatering must be avoided at all costs.

If the bulbs are treated annually in a dip containing quintozene, the *Rhizoctonia* is eliminated and the bulbs become white.

32.4.4 Southern Wilt, *Sclerotium rolfsii* var *delphinii*

32.4.4.1 Symptoms

Southern wilt disease can be very serious in Dutch iris. The fungus produces characteristic round brown or reddish resting bodies, the sclerotia, which resemble turnip or cabbage seeds. The disease attacks the bulbs of actively growing plants during the warm summer months. The warm temperatures cause the sclerotia to germinate and attack plant tissues. Affected lily bulbs develop a white chalky or light brown rot (Fig. 32.12), accompanied by conspicuous white strands (mycelia) of the invading fungus. The white, fanlike patches of coarse mycelia are a sure sign of the disease. Affected rows of commercial plantings display telltale patches of brown plants.

32.4.4.2 Survival and Spread

The disease spreads quickly under ideal conditions, and at harvest time all the bulbs will have turned to mush. The sclerotia are abundant in the

Fig. 32.11 Rhizoctonia stem rot on lily

Fig. 32.12 Symptoms of southern blight on lily

soil around plants dying from the disease and can survive without a host for as much as 10 years.

32.4.4.3 Management

If *Sclerotium* is detected, lift and wash the healthy bulbs surrounding the infection; remove and destroy infected bulbs and all surrounding soil to a depth of at least 30 cm. Quintozene has been used as an effective control. A solution of this fungicide can be watered around infected areas to stop further spread. This is done when small pockets of infection occur in commercial plantings.

32.4.5 Damping-Off, *Pythium*, *Phytophthora*, and *Rhizoctonia* spp

32.4.5.1 Symptoms

The cotyledon or first true leaf simply topples over and decays because the fungus has attacked the tissues near the soil line, and the seedlings quickly die.

32.4.5.2 Management

Growers should use pasteurized soil mixes or soilless mixes when growing seedlings indoors. Since damping-off fungi are present in all natural soils, they must be destroyed by heating or chemical treatment before the soil is safe to use. Growers should use only clean, sterilized pots or other containers for growing seedlings. This can be accomplished by dipping containers in a 10 % solution of household bleach (sodium hypochlorite). Growers should use freely draining soil mixes and water very sparingly in the early stages of seedling growth. The surface of the soil should be allowed to become dry. Because high humidity and wet soil provide ideal conditions for damping-off, plants should be watered early in the day so that the soil can dry somewhat before nightfall. Growers should provide seedlings with good air circulation and as much light as possible. Damping-off is seldom a problem under conditions of cool temperature and ample light. Thiophanate-methyl is widely available and combines several fungicides.

32.4.6 Viral Diseases

Three of the most common viruses found in lilies are often transmitted from plant to plant by aphids that have bitten into the plant and ingested the virus. When they land on a noninfected plant, they pass the virus to it when they bite into it. These viruses are *Lily symptomless virus* (*LSV*), *Tulip breaking virus* (*TBV*), and *Cucumber mosaic virus* (*CMV*).

32.4.6.1 Symptoms

32.4.6.1.1 Cucumber Mosaic Virus

This can cause very streaked foliage color and distorted, brittle leaves and flowers. Leaves may become very pale and plants stunted. It is probably the most obvious and widespread virus, one that also infects many other garden plants such as tulips, dahlias, and delphiniums.

32.4.6.1.2 Tulip Breaking Virus

This virus causes the broken color which excited Dutch tulip growers centuries ago. In lilies it causes mottling of the foliage, the lessening in intensity of darker flowers, and sometimes the breaking of the color.

32.4.6.1.3 Lily Symptomless Virus

This difficult-to-spot, creeping disorder must be suspected if the plants look less lively, lacking the joie de vivre of their neighbors (Fig. 32.13). They are on the slippery slope to extinction, and if you feel sure that you have discovered cases, they should be helped on their way without delay. Be careful to differentiate between plants that may not be flourishing as well as they might because of other reasons, such as overcrowding, competition from neighboring plants, damage from rabbits or slugs, or distress from drought conditions.

Fig. 32.13 Symptoms of viral disease on lily flower

32.4.6.2 Management of Viral Diseases

- **Insect control:** Try to minimize the number of aphids feeding on your lilies.
- **Sanitation:** Remove plants that have been diagnosed with virus. They will never get better. Destroy them away from the garden and do not use them in a compost pile.
- **Disinfect:** If you have a virus problem, be sure to disinfect tools used to prune or cut lilies or their bulbs. You can use 1 part bleach to 9 parts water solution.
- **Propagation:** During the early part of the twentieth century, a serious virus infection threatened the entire Easter lily crop. Eventually, they figured out that the virus moved up the plant at a slower rate than the tip of the stem grew. So, they were able to take cuttings from the uninfected tips to produce new, virus-free plants.

Easter lily diseases, their symptoms, and management in greenhouse are presented in Table 32.1.

32.5 Nematodes

32.5.1 Root-Knot, Lesion, and Foliar Nematodes

32.5.1.1 Symptoms

The root-knot nematode can infect lilies. This nematode (several species in the genus *Meloidogyne*) dwells in the soil and infects roots. Infected roots have swellings (knots), which house the female root-knot nematode.

The most harmful nematodes to lilies are the root lesion or meadow nematode (*Pratylenchus* spp) and the leaf-lesion nematode (*Aphelenchoides* spp). These microscopic pests cause serious damage to lily crops in some regions if their populations are not under control.

Nematodes penetrate root tissues, killing cells as they move. They move inside the root, feeding, laying eggs, and destroying additional cells. The roots become soft and flabby, eventually suc-

Table 32.1 Easter lily diseases and their management

Disease/pathogen	Symptoms	Management
Bulb basal rot, *Fusarium* and *Cylindrocarpon*	The basal plate is dark brown and dead. Scales fall off	Use pasteurized potting soil. Do not plant bulbs that have a rotted basal plate. Dip bulbs in a fungicide before planting
Bulb storage rot, *Penicillium* and *Rhizopus*	Bulb scales have brown spots and a severe rot. Bulbs may be soft and covered with a blue-green mold or white fungal growth	Do not plant affected bulbs
Botrytis blight, *Botrytis cinerea*	Oval to circular reddish-brown spots form on leaves. Brown spots develop on flowers	Maintain humidity below 85 %. Apply a fungicide to protect healthy plants
Pythium root rot, *Pythium* sp	Lower leaves yellow and die. Roots are limp and dark brown. Flowers abort	Use pasteurized potting soil. Drench potted bulbs with a fungicide. Or apply a fungicide as a soil drench
Rhizoctonia stem rot, *Rhizoctonia* sp	Bulbs are yellow rather than the normal white. Stems and scales where the stem emerges develop a soft rot, causing the plant to topple	Use pasteurized potting soil. Drench potted bulbs with a fungicide. Repeat treatment as stated on the fungicide label
Scale tip rot, *Pseudomonas, Fusarium* and *Cylindrocarpon*	The tips of scales turn tan or dark brown to black in color	Do not plant affected bulbs
Lily symptomless virus (LSV)	Plants emerge later than normal. Soon after emerging, leaves twist and have white stripes. Tops are crooked. Mildly affected plants grow on and appear normal except for the lower leaves	Maintain greenhouse aphid-free since LSV is carried by aphids from plant to plant
Cucumber mosaic virus (CMV) in combination with LSV	Short white lines develop on the leaves. Entire plant and flowers are stunted	Maintain greenhouses aphid-free since CMV and LSV are carried by aphids from plant to plant

cumbing to infection that moves into the basal plate, turning it into mush. River water often carries nematodes, which can then enter croplands through irrigation. These pests also host bacteria; some species even carry virus diseases. Nematode infestation causes stunting of growth and can severely reduce commercial production. Crops parasitized by nematodes are seldom uniformly affected.

Foliar nematodes move through the stem in a surface film of moisture to invade the leaves and flowers. The foliar nematode (*Aphelenchoides* spp) produces discolored streaks in lily foliage (Fig. 32.14) and will cause the bulbs to decline.

The stem and bulb nematode (*Ditylenchus dipsaci*) causes individual bulb scales to rot and eventually can kill the bulb.

Fig. 32.14 Foliar nematode damage on Easter lily leaves

32.5.1.2 Management

The following controls are used for nematodes:

- Keep the foliage as dry as possible to control foliar nematodes by preventing movement of the organisms. Systemic insecticides are very effective.
- Treat bulbs with hot water at 44 °C for 1 h.
- Fumigate soil with methyl bromide, chloropicrin, and metham sodium (e.g., Vapam). This is a highly successful technique in commercial plantings. Steam sterilization of greenhouse soils is very important.
- Avoid planting lilies continuously in the same site; this prevents harmful nematode populations from building up in the soil.
- Apply a granular nematicide such as fenamiphos (e.g., Nemacur) when planting bulbs.
- Try adding carnivorous nematodes to the soil.

32.5.1.2.1 Soil Sterilization

- In heavily contaminated soils where lilies have been grown for years, soil fumigation or solarization may be the only recourse to eliminate pathogens. Soils are solarized mainly to kill fungi, nematodes, perennial weeds, and weed seeds. The treatment is applied during the summer when soil temperatures are high and moisture levels relatively low; however, the soil must be moist for optimal results.
- The chemicals can be either injected or watered into the soil. A tarp or plastic cover is usually necessary to prevent the chemicals from evaporating out of the soil.
- The soil must be well tilled before solarization, free from clods and coarse organic matter. The tools used should also be clean.

Orchids

33

Abstract

Orchids are one of the world's most fascinating and beautiful creations of God which are valued for their exquisite flowers and long keeping quality, exhibiting a wide range of diversity in form, size, color, and texture of flowers. Commercial production of cut flowers in orchids has expanded tremendously and has become a very profitable occupation. Orchids are relatively sturdy and have few serious problems of pests and diseases. Pest problems include mealybugs, spider mites, scales, and thrips. Fungal (root rot, leaf spots, leaf blights, and spots on flowers) and bacterial diseases of orchids are common because of the high levels of humidity they need to survive. Their symptoms, biology, spread, and management are discussed. The best defense against pests and diseases is growing healthy plants through sound culture and proper sanitation.

Keywords

Orchids • Pests • Diseases • Symptoms • Biology • Spread • Management

33.1 Introduction

Orchidaceae, commonly known as the orchid family, is a diverse and widespread family of flowering plants grown in greenhouses with blooms that are often colorful and fragrant (Fig. 33.1).

33.2 Species and Hybrids

The following orchid species and hybrids are recommended because of their adaptability, ease of growing, and beautiful flowers: *Cattleya*, *Phalaenopsis*, *Dendrobium*, *Oncidiums*, *Vanda*, *Epidendrum*, and *Cymbidium* (Fig. 33.2).

P.P. Reddy, *Sustainable Crop Protection under Protected Cultivation*,
DOI 10.1007/978-981-287-952-3_33

Fig. 33.1 Orchids under protected cultivation

33.2.1 Vandas

Blue Magic, Manuvadee, Miss Joaquim, Nelson Mandela, Pakchong Blue, and Rothschildiana

33.2.2 Dendrobiums

Mini Brown, Margaret Thatcher, Lucky Seven, and Noble Dendrobium

33.2.3 Phalaenopsis

Abendrot, Aphrodite, Bokeh Cream, Everspring King Lee, Gaucho, New Stripes, and Taiwan Glory

33.2.4 Cattleyas

Aquinii, Suzie, Snow Queen, Barbara Belle, New City, Chloris, Helan, and Hermine

33.2.5 Cymbidiums

Best Pink, Golden Leaf-edge Orchid, Yellow Margin Orchid, and Miscanthus Orchid

33.2.6 Oncidiums

Golden Shower, Waimanola Sunset Claire, Golden Anniversary, Passionata Red Galaxy, and Lalita Pia

33.2.7 Epidendrums

Meridith Ann, Messenger, Memory, Spring Fuji, Anna Bibus, Lady Hamilton, and King

33.3 Insect and Mite Pests

33.3.1 Scale, *Diaspis boisduvalii*

33.3.1.1 Symptoms

Scales are sucking insects that attach to and feed on the underside of leaves, in leaf axils, on pseudo bulbs, and on rhizomes. They often are hidden under old leaves and pseudo bulb sheaths (Fig. 33.3). Severe infestations cause chlorotic areas to appear on the leaves and plant surfaces which will turn yellow and may darken and can cause the leaf to drop prematurely. The scale is round flat and yellow in color. The young and mature scales suck the juice from the leaves, petioles, and shoots resulting in stunted growth and drying of shoots in case of severe attack.

Fig. 33.2 Different groups of orchids

33.3.1.2 Management

- Selection of scale-free planting material.
- Cutting and burning of infested parts.
- Application of carbofuran 5–7 g/pot.
- Spray malathion 50 EC at 1 ml + dimethoate 30 EC at 1 ml/L or chlorpyrifos 20 EC at 2 ml/L or ethion 50 EC at 1 ml/L followed by Pongamia or neem oil 10 ml/L. Be sure to spray all plant surfaces, particularly the undersides of leaves and leaf axils.
- Remove old leaves and flower sheaths to eliminate scale hiding places and allow easy inspection.

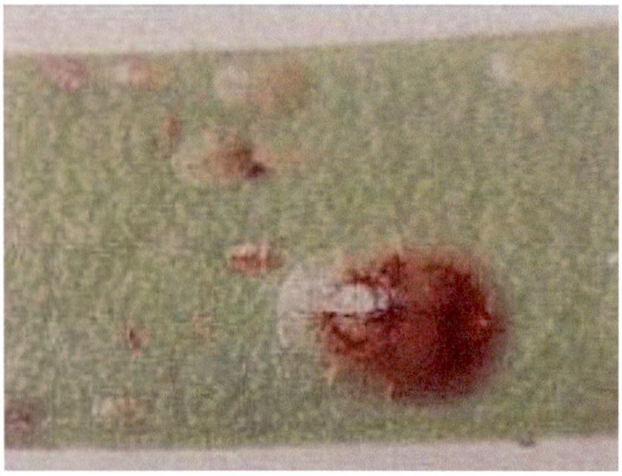

Fig. 33.3 Scale infestation on orchid leaf

33.3.2 Mealybug, *Pseudococcus maritimus*

33.3.2.1 Symptoms

Mealybugs are sucking insects that attack any part of the plant but tend to stay tucked away at the junction of leaf and stem (Fig. 33.4). Severe infestations cause chlorotic areas to appear on the leaves, which may darken, causing the leaf to turn yellow and drop prematurely. Both nymphs and adult mealybugs suck the sap from the plant parts and secrete honeydew leading to sooty mold development.

33.3.2.2 Management

- Prompt collection and destruction of infested parts.
- Crawling of ants on the plants is a sign of mealybug infestation when spraying should be taken up.
- Spray dichlorvos 76 EC at 1 ml/L followed by methyl parathion 50 EC at 1 ml/L or methomyl 40 SP at 2 ml/L or acephate 75 SP at 1.5 g/L at 15 days interval. Be sure to spray all plant surfaces, particularly the undersides of leaves and leaf axils.
- Pongamia or neem oil sprays at 10 ml/L are also effective.
- Remove old leaves and flower sheaths to eliminate hiding places and allow easy inspection.

Fig. 33.4 Mealybug infestation on orchid

Check new plants carefully before adding to the growing area.

33.3.3 Thrips, *Anaphorathrips orchidii*

33.3.3.1 Symptoms

Thrips are very small sucking insects that feed on flowers and occasionally leaves and can transmit virus diseases from plant to plant. Infested buds may not open and flowers may be deformed exhibiting water-soaked spots. Leaves may appear pitted, stippled, silvery, or bleached (Fig. 33.5). Both adults and nymphs suck the cell sap from growing leaves and buds. Affected leaves curl with brown streaks. Damage to developing buds results in discoloration, distortion, and drying.

Fig. 33.5 Thrips infestation on orchid

33.3.3.2 Management

- Spray oxydemeton methyl 25 EC or dimethoate 30 EC each at 2 ml/L at 10 days interval when the infestation is noticed.
- If the incidence is severe, spray cartap hydrochloride 50 SP at 1.5 g/L or fipronil 5 SC at 1.5 m/L or imidacloprid 200 SL at 0.4 ml/L followed by Pongamia or neem oil at 10 ml/L.
- Plants and flowers can be sprayed with pesticides like Orthene, malathion or safer soap, applied in accordance with label instructions. Repeat applications will be required because thrips remain hidden on the plant or can be reintroduced to the plant from other flowers in the landscape.
- Good sanitation will help prevent infestation as well as keeping plant hosts (flowers, citrus, gardenias, eucalyptus, etc.) separate from orchids.

33.3.4 Aphids, *Macrosiphum luteum*

33.3.4.1 Symptoms

Aphids are sucking insects that attack buds, flowers, and new growths and transmit virus diseases from plant to plant. Buds and flower may fail to open and leaves may have a sticky deposit. The honeydew excreted by aphids and other sucking insects is attractive to ants and is an ideal medium for sooty mold (Fig. 33.6). When sooty mold is present, inspect the plants for aphids, mealybugs, scale, and mites.

33.3.4.2 Management

Wash aphids away from the plant with a jet of water. Plants can be sprayed with pesticides like malathion, Orthene, or safer soap using the product in accordance with label instructions.

33.3.5 Whiteflies, *Aleurothrix* sp.

33.3.5.1 Symptoms

Whiteflies are small, moth-like insects that attack buds, flowers, and new growth. The tell-tale sign of whiteflies is a cloud of tiny white insects arising from an affected plant when it is moved or disturbed (Fig. 33.7).

33.3.5.2 Management

Good sanitation and elimination of weeds will help prevent infestation as well as keeping plant hosts separate from orchids.

Plants can be sprayed with malathion, Orthene, or safer soap following label instructions. Repeat applications at 4 days intervals until whiteflies are no longer present.

33.3.6 Caterpillars

33.3.6.1 Symptoms

Caterpillars are the immature stage of moths and butterflies. While not common, they are voracious feeders that can do a great deal of damage to flowers and leaves in a short period of time (Fig. 33.8).

33.3.6.2 Management

Caterpillars can be physically picked off the plant and destroyed; check the underside of leaves for their presence. *Bacillus thuringiensis* or Bt is a naturally occurring bacterium of insects; it is a safe and natural product that can be sprayed in the growing area following label instructions.

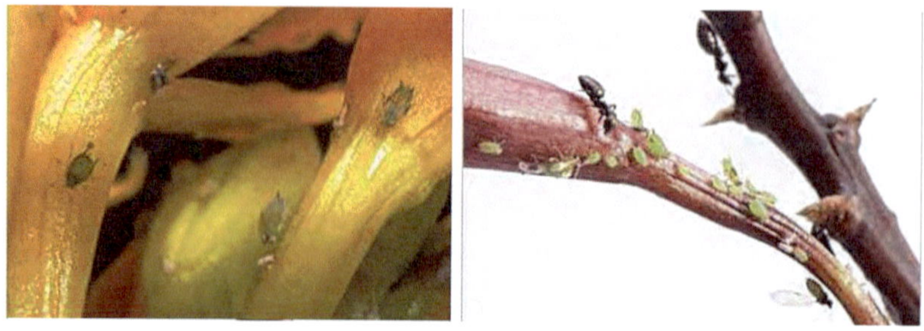

Fig. 33.6 Aphid infestation on orchid

Fig. 33.7 Whitefly infestation on orchid

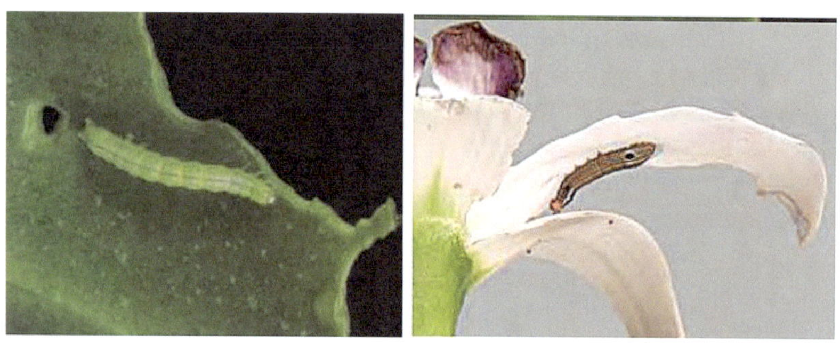

Fig. 33.8 Caterpillar infestation on orchid leaf and flower

Keep the growing area clean of fallen leaves and debris in which insect pests and their eggs can hide. Keep the landscape free of caterpillars.

33.3.7 Mite, *Brevipalpus* sp.

33.3.7.1 Symptoms

Mites typically feed on the underside of the leaves and can be found under the leaves as small, red to brown pests. A hand lens may be needed to see them. Leaf undersides may have webbing and brown splotches from the mite excrement. The upper surface of a damaged leaf may have a silvery sheen that eventually becomes sunken and turns brown. Leaves may be streaked, stippled, or spotted due to lack of chlorophyll (Fig. 33.9). When the incidence is severe, the affected leaves turn brown and dry.

33.3.7.2 Management

- Spray of dicofol 18.5 EC at 2.5 ml/L or triazophos 40 EC at 1.5 ml/L or ethion at 50 EC at 1 ml/L followed by Pongamia or neem oil at 10 ml/L at 15 days interval.

Fig. 33.9 Mite infestation on orchid leaf

- When infestations are severe, spray abamectin 1.9 EC at 0.5 ml/L or fenazaquine 10 EC at 1 ml/L.
- Plants can be sprayed with a miticide like Kelthane being particularly careful to contact all the undersides of the leaves. During warm weather, new generations mature every 6 days, so repeat applications will be required, perhaps three applications at 4 days intervals.
- Mites appear during warm, dry weather. Increasing humidity and leaf wetness and, if possible, decreasing temperature help prevent infestations.

33.3.8 Snails and Slugs, *Achatina fulica*

33.3.8.1 Symptoms

These molluscs will leave holes and notches in the leaves, flowers (Fig. 33.10), and roots and may chew off the growing tips. Adult and young ones devour plants during nights. Snail's attack is seen mainly during monsoon period when high moist conditions prevail.

33.3.8.2 Management

- Handpicking and killing of snails by dropping them in 5 % salt solution.

Fig. 33.10 Slugs and snails infestation on orchid leaves and flower

- Sprinkling of salt crystals in the paths or around pots.
- Spread Snail Kill (3 % metaldehyde pellets) at 50 kg/ha.
- Spray neem oil at 10 ml/L or soap nut extract at 60 g/L on foliage.
- Chemical baits may be placed in the growing area. Ash and diatomaceous earth can be spread on horizontal surfaces to create a barrier though water will deactivate it.
- Beer in shallow tins can be spread in the growing area and the drowned pests removed the next day.

Common orchid pests and their management are presented in Table 33.1.

33.4 Diseases

33.4.1 Black Rot, *Pythium* and *Phytophthora* spp.

33.4.1.1 Symptoms

The infection usually starts on the leaves, new leads or roots, though all plant parts are susceptible. The disease spreads rapidly and will kill the plant unless treated promptly.

Leaf symptoms first appear on the underside as small, irregular, watery, brown spots which rapidly become purplish-brown or purplish-black. The spots may have a yellowish advancing margin. The lesions enlarge with age and may ooze water if pressed. Old lesions sometimes become dry and black (Fig. 33.11), often allow-

Table 33.1 Common orchid pests and their management

Pest	First line of defense	Second line of defense	Comments
Aphids	Wash off with warm water	Insecticidal soap, Orange Guard (orange oil), horticultural oil, isopropyl alcohol	If aphids are on the flower buds of orchids, try repeatedly washing them off with warm water
Mealybugs	Use a cotton swab drenched with isopropyl alcohol	Insecticidal soap, horticultural oil, neem products	For orchids with mealybugs on their roots, remove the orchid from the pot, soak the roots in a solution of insecticidal soap for a few hours, then repot in a clean new pot with new potting material
Thrips	Neem, horticultural oil, insecticidal soap	Malathion, Orthene	Malathion and Orthene are effective on a broad range of insect problems but they both reek; apply outside of living area
Scale	Orange Guard; or dip cotton swab in isopropyl alcohol and wipe across the armored shell of this insect. Make sure to penetrate this shell	Insecticidal soap, neem, horticultural oils	Difficult to eradicate. Apply controls repeatedly to get rid of it
Spider mites	Wash off with a strong stream of warm water	Insecticidal soap, horticultural oils, Orthene	Prevent spider mites with proper watering; avoiding excessively hot growing temperatures
Snails and slugs	Put out a shallow platter of old beer, and wait for these creatures to belly up to the bar at nightfall		

Fig. 33.11 Black rot infection on orchids

ing other diseases to attack the plant. The disease may spread rapidly to the rhizome and roots, particularly when the temperature and humidity are high.

New leads show a purple or purple-brown area with a yellowish advancing margin and may be pulled off easily.

Pseudo bulbs, roots, or rhizomes show infections as purplish-black, often sharply delineated, discolored area in the center of the plant. The infection often starts in the roots and may spread upward to the base of the pseudo bulb or leaf, causing the plant to wilt.

Cattleyas may show a creamy yellow discoloration on one or both sides of the pseudo bulbs.

The discoloration eventually turns black or brown and softens, and the bulb rots.

33.4.1.2 Management

Unless the plant is valuable, the best approach is to discard it, as the disease is highly contagious and will spread from plant to plant from splashing water. If the plant is valuable, isolate it from other plants, remove infected tissue with a sterile tool, and drench with a suitable fungicide like Subdue or Banrot following label instructions.

High temperatures and humidity contribute to the spread of the disease. Consider using a preventative fungicide spray, particularly during hot humid periods.

33.4.2 Fusarium Wilt, *Fusarium* sp.

33.4.2.1 Symptoms

Fusarium blocks the flow of moisture through the plant's vascular system plugging the phloem. Infected leaves are yellow, thin, shriveled, wrinkled, or wilted and eventually die.

The diagnostic symptom in the plant is a circle or band of purple or pinkish-purple discoloration on the outer layers of the rhizome evident when the rhizome is cut (Fig. 33.12). If the disease is extensive, the entire rhizome may turn purple, and the discoloration may extend to the pseudo bulbs. The pathogen is spread through improper hygiene, generally as a result of using non-sterile cutting tools, which transfers the fungus from plant to plant. Severely infected plants may die in 3–9 weeks, while mildly infected plants gradually decline over a year or so.

Fig. 33.12 Fusarium wilt symptoms on orchid

33.4.2.2 Management

Prevention is a simple matter of following proper hygiene. Discard infected part of rhizome and pseudo bulb if the purple band is evident. Repot only the part of plant showing no purple discoloration. Drench sanitized plants in a thiophanate methyl following label instructions. Be diligent in disinfecting growing area and cutting tools. Each time the cutting tool contacts infected tissue, it should be sterilized before making a second cut, preferably through flame sterilization.

33.4.3 Root Rot, *Rhizoctonia* sp.

33.4.3.1 Symptoms

Root rot occurs when the medium breaks down, drainage is poor, and/or plants are overwatered. Rot sets in quickly when roots are damaged by injury or salt buildup from hard water or over fertilizing. *Rhizoctonia* is very contagious and if the disease is not controlled immediately, infected plants develop brown root rot and die.

Rhizoctonia is primarily a root disease, but the symptoms can be noticed on aerial parts of the plant. Leaves and pseudo bulbs become yellow, shriveled, thin, and twisted, and new growths become progressively smaller. The roots usually show a brown rot (Fig. 33.13) with white or brown fungal growth. In severe infections, the fungus girdles and kills the plant. The infection quickly invades the lower leaves and rhizomes of small seedlings.

33.4.3.2 Management

Remove infected part of roots and leaves using a sterile cutting tool; drench the remaining plant in a protectant fungicide like thiophanate methyl or systemic fungicide (such as Subdue) following label instructions. Disinfect growing area with 10 % bleach solution.

Make sure that potting media is fresh and plants are not overwatered. When disease is suspected in other plants or when repotting is overdue, depot the plants, check their roots, and repot

Fig. 33.13 Root rot symptoms on orchid

as necessary. In hard-water areas, pots should be flushed at least monthly to prevent root damage by watering heavily to solubilize the salts and then watering heavily an hour later to flush the salts from the pot.

33.4.4 Anthracnose, *Colletotrichum* and *Glomerella* spp.

33.4.4.1 Symptoms

This fungal disease infects the aerial portion of the plant. The leaves are most often attacked. Leaf tips turn brown beginning at the apex and proceeding toward the base (Fig. 33.14). Dark brown or light gray patches develop, sometimes as concentric rings or as numerous dark bands across the leaf. The affected area is usually sharply defined and somewhat sunken, while the remainder of the leaf appears normal. Spore bodies develop in the infected area.

Flowers develop watery, black or brown pustules which are usually raised and occur on the underside of older sepals and petals. The spots may merge and cover the entire flower.

33.4.4.2 Management

Normal sanitation, good air movement, lower temperatures (if possible), and increased light may help to reduce the spread of this disease. The pathogen is most active in warm weather when light is low and moisture is high.

Systemic fungicide like thiophanate methyl or protectant fungicides like mancozeb should be sprayed following label instructions. Alternate systemic and protectant fungicide use.

33.4.5 Leaf Spots, *Cercospora, Septoria, Guignardia/Phyllosticta* spp.

33.4.5.1 Symptoms

33.4.5.1.1 *Cercospora* Symptoms

Infection shows first as a yellow spot on the underside of the leaf. Soon after the infection occurs, the yellow-green area may be noted on the top surface of the leaf. The spots continue to enlarge in a circular or irregular pattern and may eventually cover the entire leaf. With age, the spots become slightly sunken and necrotic and change to purple-brown or purple-black (Fig. 33.15). The advancing margin remains yellow. Heavily infected leaves usually fall from the plant prematurely, especially if the infection started near the base of the leaf. As the spots enlarge in irregular patterns, they become sunken and turn purplish-brown to purplish-black. The top surface of the leaf first becomes chlorotic and finally necrotic.

33.4.5.1.2 *Guignardia/Phyllosticta* Symptoms

Spotting from *Guignardia/Phyllosticta* may start anywhere on the leaf or pseudo bulb. The lesions are tiny, yellow, and slightly sunken. As they enlarge, they become round to oval and more sunken, especially if the infection is on the leaves. With age, they turn tan to dark brown and develop a slightly raised, red to purple-black margin (Fig. 33.16). Eventually, tiny black, raised spore structures develop in the center of the spots. Individual spots are about 0.35 cm across. Severely infected leaves may drop prematurely.

Fig. 33.14 Symptoms of anthracnose on orchid leaves

Fig. 33.15 Cercospora leaf spot on orchid

Its presence may indicate insufficient light. This blight is also known as *Guignardia*; the names apply to two different sexual stages of the same fungus.

33.4.5.1.3 Septoria Symptoms

The tiny spots may start on either leaf surface as sunken, yellow lesions. They continue to enlarge, becoming dark brown to black, circular or irregular lesions. Spots may merge to form large, irregular patches on the leaf. Heavily infected leaves fall prematurely.

33.4.5.2 Management of Leaf Spot Diseases

Remove infected leaves with a sterile instrument and reduce leaf wetness. Spray with a systemic fungicide such as thiophanate methyl or a protectant fungicide like mancozeb, following label instructions. Alternate the use of systemic and protectant fungicides.

Provide good sanitation with good air movement. Reduce leaf wetness; water on the leaves may lead to infection. If the fungus is a continuing problem, monthly fungicide sprays may offer effective prevention.

33.4.6 Petal Blight, *Botrytis* sp.

33.4.6.1 Symptoms

Very small, black or light brown, spots appear on the flowers. The spots may enlarge and cover the entire flower (Fig. 33.17). If conditions are moist, a gray fungal growth may appear on severely infected or decaying flowers.

33.4.6.2 Management

This fungus is common in the environment and cannot be eradicated. Remove infected flowers since these are reservoirs of infection, then spray with a protectant fungicide like Daconil in enclosed areas. Always follow label instructions.

Infection may be reduced through careful sanitation, increased air circulation, reduced humidity, and warmer night temperatures (>68 °F).

33.4.7 Bacterial Soft and Brown Rot, *Erwinia* spp.

33.4.7.1 Symptoms

Small water-soaked spots appear on the leaves and often are surrounded by yellow halos (Fig. 33.18). If unchecked, the infection will rapidly rot the leaves and roots and spread more slowly into the rhizomes or pseudo bulbs. This wet rot may have a foul odor and has water-soaked appearance.

In Phalaenopsis, the disease spreads so rapidly that plants may be completely rotted in 2–3 days. The bacteria are opportunistic organisms that can enter through wounds.

Fig. 33.16 *Guignardia/Phyllosticta* leaf spot on orchids

Fig. 33.17 Botrytis petal blight on orchid

Dendrobium leaves appear yellow and water-soaked and become black and sunken.

Vanda leaves develop translucent patches which become black and sunken.

Paphiopedilum leaves develop small, round spots that are initially yellow and water-soaked but eventually become reddish brown and sunken. The spot enlarges in all directions and may reach the growing crown before the leaf tip is affected. If untreated, the disease quickly spreads throughout the plant, leaving it a dark, shriveled mass.

Grammatophyllum leaves have water-soaked, browning spots which become black and sunken.

33.4.7.2 Management

Immediately remove infected tissue using a sterile instrument, spray bactericides like physan or copper compounds on infected and adjacent plants following label instructions (copper should not be used on dendrobiums or blooming plants), disinfect growing area with 10 % bleach solution. Treat nearby plants as well as those that are diseased.

The disease is spread by splashing water, so avoid overhead watering if the disease is present. The pathogen is favored by hot and moist conditions, so if infection occurs, keep leaves dry, increase air circulation, and reduce temperature and humidity (if possible). Periodic preventive sprays with copper compounds help to prevent infection, particularly during hot and humid weather.

33.4.8 Bacterial Brown Spot, *Acidovorax* sp. (syn *Pseudomonas* sp.)

33.4.8.1 Symptoms

The symptoms may appear anywhere on the leaf as a small, soft, water-soaked blister. Initially dirty green in color, the infected spot enlarges, coalesces, and eventually becomes brown or black, dried up, and sunken (Fig. 33.19). It oozes bacteria-laden liquid, particularly when the disease reaches the tip of the leaf. It is most prevalent during the warmer weather.

In Cattleya, the infection enters through wounds on older plants and usually affects only older leaves. It advances slowly and is rarely fatal.

Fig. 33.18 Symptoms of bacterial soft and brown rot on orchid

Fig. 33.19 Bacterial *brown spot* symptoms on orchid leaf and stem

In Phalaenopsis, the blister-like spots may be surrounded with a yellowish or pale green halo. Spots coalesce, and the infection spreads rapidly. If the diseased area invades the crown, the plant will die.

33.4.8.2 Management

Immediately remove infected tissue and spray the plant with a bactericide like physan or copper compounds following label instructions. Disinfect growing area with 10 % bleach solution.

Pseudomonas cattleya is a water-borne pathogen that prefers warm, moist conditions. Reduce humidity and temperature (if possible), eliminate overhead watering, and increase air circulation.

33.4.9 Viral Diseases

33.4.9.1 Symptoms

Symptoms include chlorotic and necrotic spots, streaks, lines, and rings in the leaves. Flowers may show necrotic spots and streaks as well as color break (Fig. 33.20). The virus is present in all parts of the plant.

33.4.9.2 Management

There is no treatment for a virus infected plant. Destroy the plant to prevent it from infecting other plants. If the plant is valuable, isolate it completely from other plants and follow precautions to prevent infecting other plants.

33.4.9.2.1 Use Sterile Cutting Tools

Viruses are spread by transmitting the plant sap from one plant to another via mechanical means. The primary means by which viruses are spread from plant to plant is by improperly sterilized cutting tools. There are two ways to keep cutting tools sterilized, either use sterile single-edged razor blades that are discarded after each and every use or use a hot flame to sterilize cutting tools after using the tools on a given plant. The cutting tool should be sterilized for 15–20 s with a hot flame on each side.

33.4.9.2.2 Control During Repotting

Viruses can be spread whenever there is mechanical transmission of sap from an infected plant to another plant, even by leaves rubbing against one another. Observe these additional controls:

Fig. 33.20 *Top row* – Cymbidium mosaic virus (*left*), tobacco mosaic virus (*right*), *bottom row* – tobacco spotted wilt virus (*left*), orchid fleck virus (*middle*), and tospovirus (*right*)

(a) *Latex Gloves*: Wear latex gloves when handling a given plant and discard those gloves when you are done handling the plant. Your bare hands can come into contact with plant sap containing the virus and infect the next plant.

(b) *Newspaper on Potting Surface*: Keep the potting surface sterile. Keep a stack of newspapers handy and when repotting, place newspaper under the potting area. Upon completion, wrap up the newspaper, gloves and other detritus and discard them before touching the next plant.

(c) *Disinfect Pots Prior to Use*: Make sure your pots are sterile. Plastic pots can be disinfected by first washing them with soap to remove residual organic matter, then soaking them for an hour in a 20 % bleach solution, then soaking them for an hour in physan mixed as per label instructions. Clay pots are porous and cannot be sterilized against viruses by using bleach and physan alone. Follow the normal disinfection routine for plastic pots above and then bake them in the oven at 350 °F for an hour to kill any residual virus.

33.4.9.2.3 Control During Routine Activities

Viruses can be spread whenever there is mechanical transmission of sap from an infected plant to another plant, even by leaves rubbing against one another. Observe these additional controls:

(a) *Cutting Inflorescences*: Use a sterile tool to cut each inflorescence from the plant. The easiest way to do this is to use a sterile, single-edged razor blade to remove the inflorescence and discard it after each use (or bake in a 35 °F oven for an hour). If you use shears, you should flame sterilize them between each inflorescence.
(b) *Do Not Remove Inflorescences by Hand*: A virus, if present, can be unknowingly transmitted to your hand and you can infect the next plant when you touch it or remove the next dead flower bud by hand. Instead, use a sterile razor blade and discard it after each use.

34 Anthuriums

Abstract

Anthurium is one the most valued cut flowers which has a great export potential. Successful production of anthuriums requires the management of numerous pests and diseases. The major diseases (bacterial blight, anthracnose, bacterial wilt, root rot, and black nose), pests (scales, mealybugs, thrips, whiteflies, caterpillars, mites, and snails), and the burrowing nematode and their symptoms, biology, spread, and management are discussed.

Keywords

Anthurium • Pests • Diseases • Nematodes • Symptoms • Biology • Spread • Management

34.1 Introduction

Anthurium is one of the beautiful cut flowers cultivated mostly in tropical humid climate (Fig. 34.1). It is the largest and probably the most complex genus of arum family. The word anthurium is derived from the Greek words “anthos” and “oura,” which means “bloom” and “tair”, respectively. The genus *Anthurium* belongs to the complex family of Araceae. This family has eight subfamilies, among them anthurium belongs to Pothoideae. Anthurium occupies important position in cut flower trade on account of its beauty. It is grown for its attractive flowers as well as foliage. It is an evergreen tropical herbaceous plant. The most popular and economically important species are *Anthurium andraeanum* and *A. scherzerianum* which posses attractive long-lasting inflorescence.

34.2 Varieties

There are many varieties of anthurium with different colors like dark red, orange, white, pink, and bicolor (two colors on the spathe). Varieties of anthurium in the Netherlands are well known as private companies from this country have developed many varieties. Tissue-cultured plants of these varieties are available in India. Some of

P.P. Reddy, *Sustainable Crop Protection under Protected Cultivation*,
DOI 10.1007/978-981-287-952-3_34

Fig. 34.1 Anthuriums under protected cultivation

the varieties, viz., Ivory (white), Chichas (Maroon), Aymara (pink), and Jewel (red with green), can perform well in India.

The popular cultivars grown throughout the world both for cut flowers and in pots are Abe (bright pink), Avo-Jose (white), Haga White, Nova Aurora (red), Ozaki (red), Red Elf, Rico (rose), Sun Burst (bright orange), Sarina (white and rose), and Uniwai (white).

34.3 Insect and Mite Pests

34.3.1 Whitefly, *Aleurotulus anthuricola*

34.3.1.1 Symptoms

White and waxy-coated adults and nymphs colonize under surface of the leaves (Fig. 34.2) and petiole sheath. The pest sucks the sap from infested parts causing yellowing, wilting, and drying of affected plant parts.

34.3.1.2 Management

- Remove and burn infested leaves.
- Spray Methomyl 40 SP at 2 ml/L or acephate 75 SP at 1.5 g/L alternating with triazophos 40 EC at 1.5 ml/L at 15 days interval for managing nymphs.

Fig. 34.2 Anthurium spiraling whitefly

- Dichlorvos 76 EC at 1 ml/L or malathion 50 EC at 1 ml/L spray checks adult flies.
- Install yellow sticky traps for attracting and killing adults.

34.3.2 Thrips, *Chaetanaphothrips orchidii*

34.3.2.1 Symptoms

Damage starts in March and reaches to a peak in August to September. Thrips produce white- or brown-colored streaks and distortion of spathe by sucking cell sap (Fig. 34.3). Severe damage occurs especially on young leaves.

Fig. 34.3 Damage to anthurium spathe by thrips (*left*) and adult thrips

34.3.2.2 Management

- Spray oxydemeton methyl 25 EC at 2 ml/L or dimethoate 30 EC at 2 ml/L or Profenofos 50 EC at 1 ml/L or acephate 75 SP at 1.5 g/L.
- Spray Vertimec at 0.4 ml/L at least 1–2 sprays early in the morning.
- After 15 days give a follow-up spray of Pongamia or Jatropha oil at 10 ml/L.
- Spray Imidacloprid 200 SL at 0.4 ml/L or Fipronil 5 SC at 1.5 ml/L, if the incidence is severe.

Malathion and Topcide, a pyrethroid, provided good control of thrips, and flowers had very little or no injury (Fig. 34.4).

34.3.3 Mites, *Brevipalpus* spp

34.3.3.1 Symptoms

Mite (Fig. 34.5) colonies are seen on lower surface of mature leaves. They suck sap and cause severe specking, mottling, and wilting of affected leaves in severe cases. Bronzing occurs on upper and lower surfaces of leaf.

34.3.3.2 Management

- Spray dicofol 18.5 EC at 2.5 ml/L or triazophos 40 EC at 1.5 ml/L or ethion 50 EC at 1 ml/L at 15 days interval followed by Pongamia or Neem oil at 10 ml/L.
- If the incidence is severe, spray abamectin 1.9 EC at 0.5 ml/L or flufenoxuron 10 DC at 1 ml/L.
- Spray Kelthane at 1 ml/L or Vertimec at 0.4 ml/L at least 1–2 sprays early in the morning.

34.3.4 Citrus Red Mite, *Panonychus citri*

34.3.4.1 Symptoms

On leaves, citrus red mite (Fig. 34.6) feeding results in a pale stippling visible primarily on the upper surface of the leaf. Attacked tissue becomes yellow or white or takes on a bronzy appearance. In severe infestations, the stippling enlarges to dry necrotic areas (commonly called mesophyll collapse). Eventually, leaves may drop and twigs dieback.

34.3.4.2 Management

Avid and Floramite gave good initial and residual control of citrus red mite (CRM) up to 28 days after treatment (DAT) with lower mite counts per leaf than Conserve and untreated controls (Fig. 34.7) (Yogi et al. 2000).

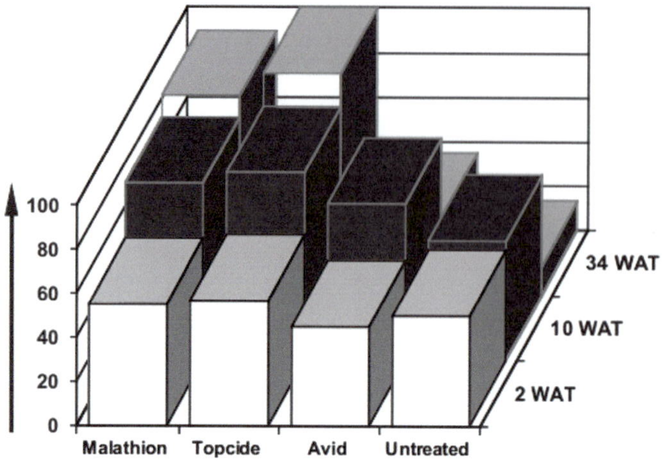

Fig. 34.4 Effect of chemicals on anthurium flower quality against thrips

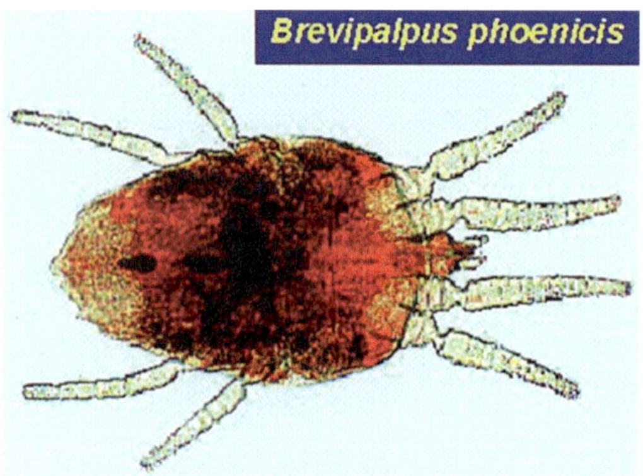

Fig. 34.5 False spider mite

Fig. 34.6 Citrus red mite

34.3.5 Caterpillars

34.3.5.1 Symptoms

Makes large holes on the tender leaves (Fig. 34.8).

34.3.5.2 Management

Spray with Lannate at 1 ml/L

34.3.6 Snails

34.3.6.1 Symptoms

They eat root tips and damage flowers and buds (Fig. 34.9).

34.3.6.2 Management

Spray metaldehyde with mixture of chick feed by which snails are attracted which can be collected in the morning and killed.

34.4 Diseases

34.4.1 Bacterial Blight, *Xanthomonas axonopodis* pv *dieffenbachiae* (Syn *Xanthomonas campestris* pv *dieffenbachiae*)

34.4.1.1 Symptoms

Early foliar symptoms start as water-soaked spots visible near the margins where hydathodes, filled with guttation fluid, serve as the most common port of entry. Tissues surrounding the infected areas turn yellow. Water-soaked spots coalesce, eventually forming large necrotic zones at leaf margins (Fig. 34.10). The pathogen quickly moves into vascular tissues of petioles and stems, preventing the translocation of nutrients and water and producing symptoms of water stress that may resemble natural senescence. The main

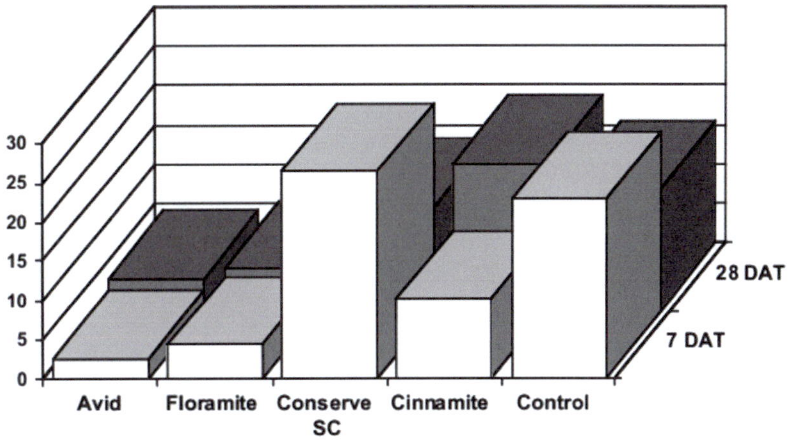

Fig. 34.7 Efficacy of chemicals against citrus red mite

Fig. 34.8 Large holes made by caterpillar pests on anthurium leaf

stem of systemically infected plants turns dark brown, and the growing point deteriorates, eventually leading to death of the plant. When the spathe is infected, the disease is often called "flower blight" (Fig. 34.10). Less frequently, bacteria enter stomates, forming circular water-soaked lesions surrounded by chlorotic zones. Stomatal invasion often results in limited colonization of mesophyll tissues and does not necessarily lead to systemic infection.

34.4.1.2 Management

34.4.1.2.1 Cultural Controls

Field sanitation, which involves removal of leaves showing early infections and elimination of systemically infected plants, was the principal method of reducing anthurium blight in early years (Nishijima 1988). Disinfection of cutting tools is important to prevent the spread of blight, since plant materials which show no symptoms have the potential for latent infection and the pathogen is spread during harvest. These two approaches, while useful, are insufficient for stopping disease spread.

Chase (1988) suggested that lower fertilizer rates for potted anthuriums could result in reduced disease and greater flower production. Sakai (1990) reported that higher levels of ammonium fertilizer led to higher amounts of amino compounds in guttation fluid when compared to nitrate fertilizers. He suggested that this promoted bacterial multiplication in the guttation fluid. The use of nitrate or inorganic fertilizers for plant growth was expected to reduce the amount of amino compounds in guttation fluid and thereby reduce blight incidence (Sakai 1991, Sakai et al. 1992). Greenhouse trials indicated a relationship between fertilizer treatments and blight susceptibility (Higaki et al. 1990, 1992).

Growing plants under plastic or glasshouses coupled with drip irrigation rather than overhead or sprinkler irrigation reduced the spread of the bacteria through aerosols and water splash and significantly reduced the incidence of blight in anthurium-seedling culture (Kamemoto and Kuehnle 1989). Growing anthuriums under cool and shaded conditions slows the progression of the disease. Inoculated plants exposed to temperatures greater than 31 °C were more susceptible to disease than inoculated plants exposed to 26 °C or lower temperatures (Alvarez et al. 1990). This difference in temperature can be regulated

Fig. 34.9 Snails on anthurium spathe (*left*) and leaf (*right*)

Fig. 34.10 Bacterial blight symptoms on leaves and spathe

in commercial shade houses by strategically increasing airflow.

34.4.1.2.2 Chemical Methods

Products containing copper (CuPro, Phyton 27, Camelot), mancozeb (Protect T/O, Dithane), and *Bacillus subtilis* (Cease, Companion) are effective against *Xanthomonas*.

34.4.1.2.3 Tissue-Cultured Anthuriums

The best single means for blight disease management entails the use of plant material which is guaranteed to be pathogen-free. Tissue-cultured plants, although highly regarded and recommended to growers, have the potential of latent infection with *X. axonopodis* pv *dieffenbachiae* (Norman and Alvarez 1994, Norman et al. 1993). Triple indexing protocol developed for anthuriums ensures that tissue-cultured plants do not serve as sources of inoculums (Tanabe et al. 1992).

34.4.1.2.4 Development of Resistant Cultivars

Genetic engineering serves as a means of introducing resistance genes from non-plant origins into anthurium plants. *Agrobacterium*-mediated gene transfer has been used to successfully transform anthuriums (Kuehnle et al. 1991). Genes that code the antibacterial peptides attacin and cecropin have been isolated from the cecropia moth (*Hyalophora cecropia*) and genetically engineered into anthuriums (Kuehnle et al. 1992, 1993, 2004). Transgenic anthurium plants expressing attacin were less susceptible to *X. axonopodis* pv *dieffenbachiae* and had fewer numbers of bacteria present when compared to non-transgenic plants (Kamemoto and Kuehnle 1996).

Kuehnle et al. (2004) reported that two cultivars transformed to express the Shiva-1 lytic peptide (a synthetic analog of cecropin B) significantly resisted anthurium blight. Disease progression in a transgenic line of 'Paradise Pink' was significantly reduced compared to the non-transgenic control, indicating increased tolerance.

34.4.1.2.5 Biological Control

Foliar applications of the bacterial community reduced infection of anthurium leaves through hydathodes and wounds (Fukui et al. 1999). The strains were identified as *Sphingomonas chlorophenolica, Microbacterium testaceum, Brevundimonas vesicularis,* and *Herbaspirillum rubrisulbalbicans*. All four species survived on the surfaces of microplants up to 2 months and were effective in protecting microplants against infection (Alvarez and Mizumoto 2001). Striking differences were observed when susceptible cultivar 'Rudolph' was inoculated with *X. axonopodis* pv *dieffenbachiae* and either treated (sprayed weekly) or not treated with the biocontrol agents. Non-treated plants developed typical blight symptoms after inoculation with the pathogen and died. In contrast, 75 % of the treated plants survived (Fig. 34.11) and produced flowers after 22 weeks (Fig. 34.12). In repeated trials disease was reduced by 25–40 % on treated plants, whereas all non-treated plants died within 28–35 days after inoculation.

Biostimulation was observed as an unexpected outcome in studies involving anthurium treatment for biological control (Alvarez and Mizumoto 2001). Stage 4 tissue cultured microplants treated with the bacterial consortium discussed above had stronger root systems and greater survival rates during transplant to individual pots than non-treated plants. Treated plants were more vigorous, flowered sooner, and were larger in plant height, leaf area, leaf number, and shoot and root dry weights. Growth effects were most striking on *A. andraeanum* cultivars 'Rudolph', 'Marian Seefurth', and several numbered cultivars.

34.4.1.2.6 Integrated Methods

Components of an integrated management program for anthurium blight include sanitation, disinfection of harvesting implements and containers, chemical sprays, modification of cultural practices, production of pathogen-free planting stocks in vitro, use of resistant cultivars, and biological control.

34.4.2 Bacterial Wilt, *Ralstonia solanacearum*

34.4.2.1 Symptoms

Leaf yellowing (chlorosis) is usually the first symptom observed. The disease spreads rapidly throughout the vascular system of the plant, turning veins in the leaves and stems a brown, bronze color (Fig. 34.13). Bacterial ooze (brown slime) will be present if cuts are made into the stems of highly infected plants. Plants will exhibit wilt symptoms even though adequate soil moisture is available. Wilt symptoms are the byproduct of *Ralstonia* bacteria clogging the vascular system of the plant.

34.4.2.2 Factors Favoring the Disease

Cool greenhouse temperatures may temporarily mask symptoms and give bacteria time to spread. Symptoms appear rapidly during hot weather.

34.4.2.3 Management

A strict sanitation program is the most successful way to stop the spread of this pathogen and eventually eradicate it from a production facility. Fungicides that contain phosphorous acid have also been shown to be effective in preventing infection; however, they do not cure systemically infected plants (Norman et al. 2006).

Bacterial wilt is spread via contaminated soil, water, tools, or worker contact. Use disease-free propagation material. The bacterial wilt pathogen is easily spread via infected cuttings. Because the bacteria survive well in soil, both contaminated plant material and the supporting soil should be discarded. If pots and trays from contaminated infected plants are to be reused, they should be scrubbed free of adhering soil and then soaked in a disinfectant to kill the remaining bacteria. Knives and clippers should be sterilized between plants with a disinfectant containing a quaternary ammonium compound (Physan 20, Green-Shield) or diluted solution of bleach to prohibit spread.

Fig. 34.11 Disease reduction on 'Marian Seefurth' plants treated with beneficial bacteria (four plants on *right*) compared to four non-treated plants (*left*) 28 days after inoculation with *X. axonopodis* pv *dieffenbachiae* (Photo by A Alvarez)

Fig. 34.12 'Marian Seefurth' plants after continued weekly post-inoculation treatments. Diseased plants eventually recovered and produced flowers (Photo by A Alvarez)

Fig. 34.13 *Ralstonia* bacterial wilt causes yellowing (chlorosis) of anthurium leaves (Credit: D Norman, UF/IFAS)

34.4.3 Rhizoctonia Root Rot, *Rhizoctonia solani*

34.4.3.1 Symptoms

The term "damping-off" is used to describe these classical symptoms. Young, tender stems are girdled, become water soaked, and are unable to support the weight of the plant. *Rhizoctonia* attacks the roots and lower stems of plants (Fig. 34.14), but under wet conditions it can also attack and spread in the upper leaf canopy. Discolored brown roots are one of the symptoms observed with *Rhizoctonia* infections

34.4.3.2 Spread and Survival

Rhizoctonia can survive within soil for years without a host plant. The fungus produces small mats of tightly woven mycelia called sclerotia. Sclerotia are irregular in shape and brown in color and resemble particles of soil. Sclerotia provide a seedlike mechanism for the fungus to survive unfavorable conditions, such as drought or cold weather. These small sclerotia stick to trays and pot surfaces and are one of the ways the fungus spreads through nurseries.

34.4.3.3 Factors Favoring the Disease

Water-saturated soils are conducive for disease development.

34.4.3.4 Management

Never incorporate native soils into media mixes without steam sterilizing. Use well-drained soil

Fig. 34.14 *Rhizoctonia* root rot symptoms on *Anthurium* (Credit: D Norman, UF/IFAS)

mixes. Never store peat moss, sphagnum moss, chips, or potting media mixes directly on soil surfaces where they can be colonized by the fungus. Plants should be cultivated on raised benches to limit root contact with soil. *Rhizoctonia* frequently gains access to production facilities via infected propagation material.

Many fungicides are effective against outbreaks of *Rhizoctonia*. Examples include Clearys 3336, Fungo/Allban (Thiophanate methyl), Medallion (Fludioxonil), and Contrast (flutolanil).

34.4.4 Root Rot, *Phytophthora/Pythium*

Phytophthora and *Pythium* species cause substantial damage to *Anthurium* and numerous tropical foliage crops.

34.4.4.1 Symptoms

Plants infected with either of these pathogens (*Phytophthora* and *Pythium*) exhibit wilting, leaf yellowing (chlorosis), and root dieback.

Phytophthora and *Pythium* infections primarily attack root systems. Plants will wilt even though adequate soil moisture is available. Root sloughing is the primary diagnostic tool. Under severe conditions, the foliage may exhibit black to brown leaf lesions (Fig. 34.15).

These symptoms are similar to symptoms caused by *Rhizoctonia*; however, fungal strands (mycelial growth) are rarely observed with *Phytophthora* and *Pythium* infections.

34.4.4.2 Factors Favoring the Disease

Water-saturated soils are conducive to disease development. These diseases can usually be avoided by using light, well-drained soil mixes.

34.4.4.3 Management

Use well-drained, synthetic soil mixes. Use disease-free stock plants. Plants with symptoms of disease should be discarded, and the rest of the production facility should be treated with a fungicide drench. If potting containers are reused, they should be scrubbed and sterilized. Cutting shears, knives, and tools should be dipped in an appropriate disinfectant between plants.

Fungicides such as Mefenoxam (Subdue Maxx) and Aluminum tris/Fosetyl-AL (Aliette WDG), Dimethomorph (Stature), fluopicolide (Adorn), and phosphorous acid (Alude, K-Phite, Vital) may be used to control *Phytophthora* and *Pythium*.

34.4.5 Black Nose, *Colletotrichum gloeosporioides*

In *Anthurium* the pathogen is highly specific, only attacking the spadix portion of the flower (the nose).

34.4.5.1 Symptoms

Black nose can cause havoc in cut flower and potted-plant production. Flowers and flowering potted plants cannot be sold with this condition. If left uncontrolled, this disease could cause significant cut flower loss.

Symptoms start as tiny brownish spots on the flower spadix. During high humidity these spots enlarge, appear water-soaked, and turn necrotic (Fig. 34.16). Sometimes the entire spadix will turn black as lesions coalesce. The shape of most lesions is, however, angular due to the shape of the spadix tissue. As the disease becomes more severe, masses of orange colored spores form on

Fig. 34.15 Anthurium wilt caused by *Phytophthora* (*left*). *Pythium* and *Phytophthora* cause root sloughing (*right*) (Credit: D Norman, UF/IFAS)

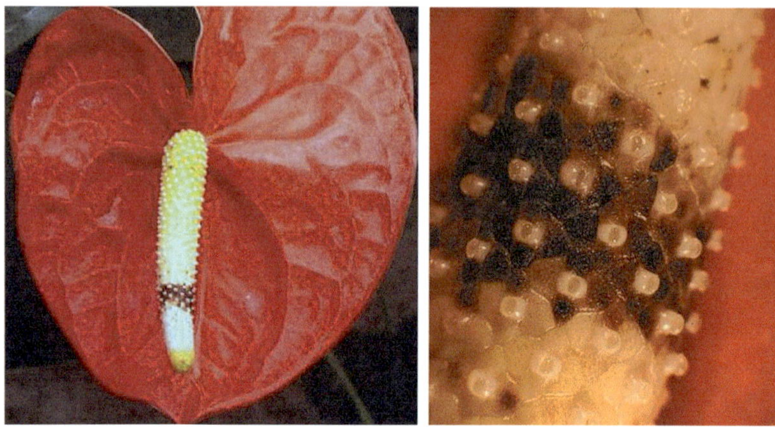

Fig. 34.16 Black nose disease on *Anthurium* causes spadix darkening (Credit: D Norman, UF/IFAS)

necrotic areas. Leaves and spathes are rarely if ever infected. The spadix may eventually fall off. Growers may also observe black, spore-containing structures (acervuli) on dead leaves and stems.

34.4.5.2 Factors Favoring the Disease

This disease is most severe during humid, warm conditions. *Colletotrichum* readily invades plant tissues previously damaged by pesticides, fertilizer, or bacterial blight (*Xanthomonas*).

34.4.5.3 Management

The *Colletotrichum* fungus produces thousands of small hot-dog-shaped spores that can readily be moved by splashing water, air movement, and workers. A strict sanitation program is crucial to control the spread of this pathogen in a production facility.

Mancozeb (Manzate 200) and Benomyl (Benlate) at the recommended dosages can easily control the disease at an economic level. Some other fungicides that can be used include chemicals such as Tilt, Bayleton, Baycor, and Prochloraz, but they are a bit expensive. Fungicide applications are usually discouraged because chemical residues diminish the marketability of flowers and plants.

Anthurium plant-breeding programs both in Hawaii and Florida have incorporated disease resistance into many of the current cultivars. Newer cultivars are highly resistant to this pathogen and rarely exhibit black nose. It is worthwhile to note that several anthurium varieties grown by Caribbean farmers are indeed resistant to anthracnose disease.

34.5 Nematodes

34.5.1 Burrowing Nematode, *Radopholus similis*

34.5.1.1 Symptoms

Root rots and plant decline are well-known symptoms of burrowing nematode infection. Anthurium root rots caused by *R. similis* are brown or dark brown to black. The rots develop relatively slowly. Initially, although older roots are infected and rot, new roots are produced and the plant often continues to grow well. But compared to healthy plants, the amount of functional roots in diseased plants is greatly reduced. With time, fewer new roots are produced, and gradually the entire root mass is destroyed (Fig. 34.17). The steady and progressive destruction of the root system usually causes plant decline in the second to fourth years. The leaves turn yellow and may have other symptoms of nutrient deficiency. The plants become smaller and lose vigor, producing fewer and smaller flowers.

In moist environments, the nematodes are able to migrate short distances outside the plant, above the soil line. They maybe splashed onto or migrate up stems to aerial roots, where they penetrate the soft, fleshy root tips. Invasion of the root tips frequently cause cessation of root growth. Infected tips are brown (Fig. 34.18) and may be missing. When the environment is dry, nematodes either migrate back into the soil or desiccate, and many rotted tips are devoid of nematodes. The tips produce a callus, stop growing, and have a rounded form that suggests the description "stubby root." When the environment becomes favorable for root growth, a new lateral root is produced at or near the tip of the original root. This new tip can be infected and become stubby also. Repetition of this cycle results in a short aerial root with several stubby tips (Fig. 34.18). Stubby roots and branched aerial roots are not normal in anthurium. While this condition is not always caused by nematodes, these abnormal roots should be a clue to the possible presence of burrowing nematodes.

Fig. 34.17 Burrowing nematode infected (*left*) and healthy roots (*right*) of anthurium

The first symptom of rot on roots growing in cinder potting medium are small, pink, orange-brown or light brown streaks or elongated rots. These develop into brownish streaks, with or without yellow borders, and then become distinctly long, blackish-brown rots. With time, these rots expand, and large sections of the root are destroyed. At times, if the nematodes feed and grow deep into the root, only a gray streak is visible on the external surface of a white or light colored root.

34.5.1.2 Biology

Burrowing nematodes emerge from eggs and mature in 18–20 days (Shurtleff and Averre 2000). A typical female may produce as many as 100 eggs in her lifetime.

34.5.1.3 Management

Drenching potted anthuriums at 49–50 °C for 15–20 min eliminated 95–100 % of burrowing nematodes, infecting the roots and stems of plants to ≤ 1 g^{-1} in infested roots of potted anthuriums.

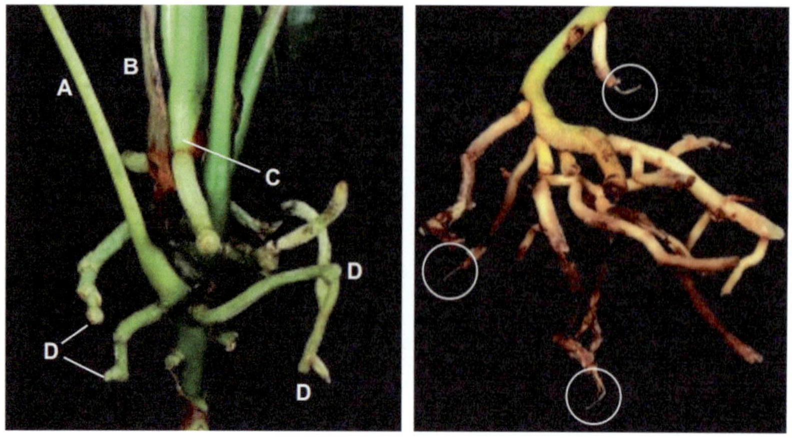

Fig. 34.18 (*Left*) Typical stubby roots on an anthurium plant: A, petiole; B, sheath; C, node; D, stubby roots. (*Right*) Early stages of root rot on anthurium. Note the missing root tips with exposed vascular strands

Table 34.1 Pest management strategies (non-chemical and chemical controls) for anthuriums

Pest	Damage or problem	Nonchemical control	Chemical controls
Aphids	Sooty mold caused by honeydew secretions; spotting caused by sucking mouth parts	Highly parasitized by wasps and predated on by lady beetles and lacewings. Remove severely infested flowers and leaves	Discus, Marathon, Neem products, Orthene, pyrethrins, pyrethroids, oils, soaps, TriStar
Thrips (Anthurium and banana rust thrips)	White streaking of flower spathe; damage inflicted early in bud stage. Thrips may be present on mature, open flowers. Banana rust thrips may be more prevalent than anthurium thrips; gross characteristics of damage may be indistinguishable between species	Remove severely infested buds. Thrips pupate in media below plant; apply control measures to media and emerging bud area	Avid, Conserve, Discus, Malathion, Marathon, TriStar, pyrethroids
Whitefly, anthurium	Waxy secretions in leaf sheath area; quarantine problem	Remove severely infested plants	Discus, Distance, Marathon, Orthene, Talus, Tristar
Beetle, black twig borer	Tiny hole in petiole associated with death of petiole and leaf distant to hole	Strict sanitation; fallen petioles contain live beetles. Dispose of all infested petioles prior to insecticide application	Discus, Dursban/DuraGuard, Marathon, pyrethroids
Mite, false spider and citrus red	Bronzing of upper and lower surfaces of leaf and flower spathe. Citrus red mite cause bronzing primarily on the upper leaf surface	Monitor plants for mite damage on upper and lower surfaces of flower spathe. Initiate control measures if damage is detected. If unsprayed, fast-moving brown predatory mites and lady beetles may be present especially at high pest population	Avid, Floramite, Hexygon, Ovation, Sanmite, Tetrasan

(continued)

Table 34.1 (continued)

Pest	Damage or problem	Nonchemical control	Chemical controls
Nematode, burrowing	Stunting of plants due to root damage and rot	A 3–6 month bare, weed-free fallow period with well-decomposed plant parts will eliminate nematodes. Plant only nematode-free stocks. Hot water treatment at 50 °C for 15 min will disinfest stems. Tests show that hot air at 50 °C, 60 % R.H. for 15 min may also disinfest stem	Avid, DiTera

The following Table 34.1 gives important insect, mite, and nematode pests of anthuriums and their management.

References

Alvarez A, Mizumoto C (2001) Bioprotection and stimulation of aroids with phylloplane bacteria. Phytopathology 91:S3

Alvarez A, Lipp R, Norman D, Gladstone L (1990) Epidemiology and control of anthurium blight. In: Alvarez AM (ed) Proceedings of the 3rd anthurium blight conference, Hawaii Inst Trop Agric Human Res, University of Hawaii, Honolulu, pp 27–30

Chase AR (1988) Chemical and nutritional aspects of controlling *Xanthomonas* diseases on Florida ornamentals. In: Alvarez AM (ed) Proceedings of the 1st anthurium blight conference, Hawaii Inst Trop Agric Human Res, University of Hawaii, Honolulu, pp 32–34

Fukui R, Fukui H, Alvarez AM (1999) Suppression of bacterial blight by a community isolated from the guttation fluids of anthuriums. Appl Environ Microbiol 65:1020–1028

Higaki T, Imamura J, Tanabe M, Nishijima W, Hara A, Deardoff D, Sewake K (1990) Nutritional and cultural effects on anthurium bacterial blight. In: Alvarez AM (ed) Proceedings of the 3rd anthurium blight conference, Hawaii Inst Trop Agric Human Res, University of Hawaii, Honolulu, pp 7–11

Higaki T, Imamura JS, Moniz D (1992) Nutritional and cultural effects on bacterial blight of anthurium. In: Delate KM, Tome CHM (eds) Proceedings of the 5th anthurium blight conference, Hawaii Inst Trop Agric Human Res, University of Hawaii, Honolulu, pp 44–45

Kamemoto H, Kuehnle A (1989) Breeding for blight resistance: a progress report. In: Fernandez JA, Nishijima WT (eds) Proceedings of the 2nd anthurium blight conference, Hawaii Inst Trop Agric Human Res, University of Hawaii, Honolulu, p 10

Kamemoto H, Kuehnle A (1996) Breeding anthuriums in Hawaii. University of Hawaii Press, Honolulu

Kuehnle AR, Chen FC, Sugii N, Jaynes JM (1991) Engineering of blight resistance in anthurium. In: Alvarez AM, Deardorff DC, Wadsworth KB (eds) Proceedings of the 4th anthurium blight conference, Hawaii Inst Trop Agric Human Res, University of Hawaii, Honolulu, pp 42–43

Kuehnle AR, Chen FC, Jaynes JM, Norman D, Alvarez A (1992) Engineering blight resistant anthurium: a progress report. In: Delate KM, Tome CHM (eds) Proceedings of the 5th anthurium blight conference, Hawaii Inst Trop Agric Human Res, University of Hawaii, Honolulu, pp17–18

Kuehnle AR, Chen FC, Jaynes JM (1993) Status of genetically engineered anthuriums. In: Delate KM, Yoshimura ER (eds) Proceedings of the 6th Hawaii anthurium international conference, Hawaii Inst Trop Agric Human Res, University of Hawaii, Honolulu, pp 7–8

Kuehnle AR, Fujii T, Mudalige R, Alvarez A (2004) Gene and genome mélange in breeding of *Anthurium* and *Dendrobium* Orchid. Acta Hortic (ISHS) 651:115–122

Nishijima WT (1988) Anthurium blight: an overview. In: Alvarez AM (ed) Proceedings of the 1st anthurium blight conference, Hawaii Inst Trop Agric Human Res, University of Hawaii, Honolulu, pp 6–8

Norman DJ, Alvarez AM (1994) Latent infections of in vitro anthurium caused by *Xanthomonas campestris* pv *dieffenbachiae*. Plant Cell Tiss Organ Cult 39:55–61

Norman D, Alvarez A, Lipp R (1993) Latent infections of *Xanthomonas campestris* pv *dieffenbachiae* in tissue-cultured anthurium. In: Delate KM, Yoshimura ER (eds) Proceedings of the 6th Hawaii anthurium international conference, Hawaii Inst Trop Agric Human Res, University of Hawaii, Honolulu, pp 12–16

Norman DJ, Chen J, Mangravita-Nova A, Yuen JMF (2006) Control of bacterial wilt of geranium with phosphorus acid. Plant Dis 90(6):798–802

Sakai DS (1990) The effect of nitrate and ammonium fertilizer on the contents of anthurium guttation fluid. In: Alvarez AM (ed) Proceedings of the 3rd anthurium blight conference, Hawaii Inst Trop Agric Human Res, University of Hawaii, Honolulu, pp 18–19

Sakai DS (1991) The effect of nitrogen fertilizer levels on amino compounds in guttation fluid of anthurium and incidence of bacterial blight. In: Alvarez AM, Deardorff DC, Wadsworth KB (eds) Proceedings of the 4th anthurium blight conference, Hawaii Inst Trop Agric Human Res, University of Hawaii, Honolulu, pp 51–52

Sakai WS, Okimura S, Hanohano T, Furutani SC, Sakai DS (1992) A detailed study of nitrogen fertilization, glutamine production, and systemic blight on anthurium cultivars Ellison, Onizuka and Calypso. In: Delate KM, Tome CHM (eds) Proceedings of the 5th anthurium blight conference, Hawaii Inst Trop Agric Human Res, University of Hawaii, Honolulu, pp 47–50

Shurtleff M, Averre C (2000) Diagnosing plant diseases caused by nematodes. APS, St Paul

Tanabe M, Fernandez J, Moriyasu P, Crane S, Wolff W, Liu RW (1992) Anthurium in vitro triple indexing. In: Delate KM, Tome CHM (eds) Proceedings 5th anthurium blight conference, Hawaii Inst Trop Agric Human Res, University of Hawaii, Honolulu, pp 8–11

Yogi JT, Jacobsen CM, Hara AH (2000) Efficacy of cinnamite, conserve, avid and floramite against citrus red mite on anthuriums, 1999. Arthropod Manage Tests 25:345–346

35 The Way Forward

Abstract

Among the productivity-enhancing technologies, protected cultivation has a tremendous potential to increase the yield of vegetables and flower crops by several folds. Greenhouse crop production is now a growing reality throughout the world with an estimated 405,000 ha of greenhouses spread over in all the continents. The purpose of growing crops under greenhouse conditions is to extend their cropping season and to protect them from adverse environmental conditions, such as extreme temperatures and precipitation, and from diseases and pests. Quality is a high priority for greenhouse crops, requiring much care in pest and disease management, not only to secure yields but also to obtain a high cosmetic standard.

Historically, not enough attention has been paid to exploiting and amending production technology for the control of pests and diseases. This makes the control of pests and diseases in protected crops even more challenging, with many important problems being unresolved and new ones arising as the industry undergoes more changes in production systems. Integrated pest management (IPM) is a systematic approach to manage pests that involves the integration of cultural, physical, biological, host resistance including transgenic plants, and chemical practices to grow crops with minimal use of pesticides. With a long-term perspective, it is easier to see that an investment in IPM can pay for itself in a higher-quality crop and a cleaner environment.

Keywords

Productivity • Protected cultivation • Biocontrol • Pests • Diseases • Integrated pest management

P.P. Reddy, *Sustainable Crop Protection under Protected Cultivation*,
DOI 10.1007/978-981-287-952-3_35

35.1 Introduction

The complex control systems of modern greenhouses are designed to maintain an ideal environment for the crop, both economically and physiologically. These conditions also provide a protected, favorable environment for pests and pathogens: optimal humidity and temperature, no rain and no wind. The warm, humid conditions and abundant food under protected conditions provide an excellent, stable environment for pest development. Often, the natural enemies that keep pests under control outside are not present under protected environment. The same crop or crops are grown year after year; particular care must therefore be exercised, especially regarding pests which persist in the soil or on the structure of the greenhouse itself. For these reasons, pest situations often develop in the indoor environment more rapidly and with greater severity than outdoors. Producers of vegetable crops generally can accept a higher level of damage than those of ornamental crops that are produced for their aesthetic value.

Maintaining a healthy greenhouse crop is more than essential for an economic and sustainable greenhouse vegetable and ornamental production. This requires first consistent monitoring, which involves systematically checking the greenhouse crop at regular intervals and at critical times to gather information about the crop, pests and their natural enemies, diseases, and their antagonists. Visual observation of symptoms, laboratory analyses of soils or plant parts, weather data, sticky color traps, and pheromone traps are some of the available tools which should be used to collect the maximum information necessary for an informed decision to make sure there is a clear distinction between non-parasitic afflictions (plant stress and other physiological disorders) and those caused by pests and pathogens. The more often a crop is monitored, the more information a grower has about what is happening in the greenhouse.

A multifaceted approach to greenhouse crop protection, with integration of chemical, cultural, biological, mechanical, and physical control of pests and diseases, will be more successful and make adaptive changes in pests and pathogens less likely. IPM has become the general crop protection policy in greenhouse crops and is the wise answer to the overuse of chemical pesticides.

Modern protection technologies will play an important role to meet the challenge of productivity enhancement in vegetable and ornamental crops under greenhouse conditions. In the context of ensuring environmental safety and economic input use pattern, crop pest and disease management techniques under protected conditions need to be reoriented to achieve the sustainable crop production.

35.2 Biological Control

Natural enemies may be scarce or entirely absent in greenhouses, unless accidentally or purposefully introduced; on the other hand, the stability of the greenhouse environment allows natural enemies of pests to be used as an effective means of control.

Greenhouses offer a privileged environment for disease biocontrol, but implementation is still very limited. However, insect biocontrol has grown to a standardized approach throughout the greenhouse market. Crop protection scientists and companies investing in biocontrol products should likewise view the future of biological control of pests in greenhouse systems with optimism. A few products have already been registered and several more should be commercialized within the next few years. Success stories against a number of pests will be important both to validate biocontrol of pests and, most important, to gain acceptance by growers. The biological control of insects and mites has resulted in a significant reduction (over 60 %) in pesticide use.

Biological control of pests will probably become more widespread in the next few years. However, it is unlikely that biological control will be able to completely control plant diseases and therefore it is important to research the possibilities for integrating biological control with other control measures. Integrated production and protection (IPP) means integration of biological control with chemical control but also with alternative control measures, such as application of plant extracts and use of good agricultural practices (GAP). Partially resistant cultivars used in combination with biological control offer better prospects than very susceptible cultivars.

It is generally accepted that insect resistance in plants is compatible with biological control. New activities in plant breeding research should aim at (i) combining aspects of host-plant resistance with biological control, (ii) selecting plant cultivars that are able to attract natural enemies after they have been attacked by pest organisms, and (iii) selecting plant cultivars that produce a better "working environment" for biological control agents. Because of specific advantages of biological control for the growers, greenhouse crops will be produced without the need to use conventional pesticides in the very near future. At the same time, this results in a cleaner environment, in satisfying the demand of consumers for pesticide-free food, and in sustainable crop protection.

More scientific efficacy trials with proper replication and statistical analysis are needed under commercial or near-commercial conditions. Biocontrol registrations now request data on the interaction of pesticides with beneficial insects. Growers need to know whether new products are compatible with their current pest management strategies. Knowledge of epidemiology and ecology of pathogens in the greenhouse, which may be different from the field, is required. Finally, the challenge of production and formulation of biocontrol agents remains, with each organism bringing its own set of problems. Effective production and formulation protocols are usually proprietary, involving substantial investment to develop economic production and a formulation with adequate shelf life, stability, and titer.

35.3 Biorational Pesticides

Biorational pest control, referring to pesticides of natural origin that have limited or no adverse effects on the environment or beneficial organisms, is an important tool to control pests under protected cultivation. The main objective of this approach is to optimize pest control in an economically and ecologically sound way. Biorational pesticides which include botanicals (neem products, insecticidal soap, rosemary oil, soybean oil, garlic extracts), minerals (sulfur, potassium bicarbonate, phosphites, and copper compounds), and insect growth regulators (Azatin, Dimilin, Neemazad, Trigard) are becoming popular owing to environmental awareness and consumer concern.

A combination of biorational pest control products will be a better option for growers of protected crops and also help organic growers to manage their crops in a sustainable and effective way.

Biorational pesticides currently represent only a small fraction, 1.3 % of the world pesticide market. However, the growth rate of the biorational pesticide sector over the next 10 years has been forecasted at 10–15 % annually in comparison to 2 % for chemical pesticides. Biorational crop protection materials give an opportunity to manage pest populations without toxic chemicals.

35.4 Management of Soilborne Pathogens

Soilborne pathogens can be a problem in greenhouses, because of limited possibilities for crop rotation. Earlier, methyl bromide (MBr) was widely used for chemical soil disinfestation, but due to its negative effects on the ozone layer, it is being phased out. Other alternatives include soil solarization, biofumigation, use of antagonists, and use of grafted seedlings. Incorporation of antagonist *Verticillium*, *Pochonia*, *Paecilomyces*, and *Pseudomonas* in the soil in the greenhouses was found effective to reduce the disease pressure. The use of grafted seedlings in combination with chemical disinfestation (but not with MBr) is more effective.

35.4.1 Solarization

Solarization to control of soilborne pathogens is simple and economic in its application and leaves no toxic residues for humans, animals, and plants. These attributes make solarization suitable for application in a wide variety of production systems. Soil solarization should be done during the hottest period of the year. It offers a satisfactory and environmentally friendly solution for the control of soilborne pathogens. This method is easily used into organic, conventional, and integrated control growing system, and it is more economical than chemical treatments.

Solarization is easily combined with other control methods to reduce the need for chemical control. Together with reduced dosages of chemical fumigants, it allows better management of soilborne pathogens, which are otherwise difficult to control. Solarization followed by use of biocontrol agents has good potential, facilitating the introduction of antagonists, especially in warm regions.

35.4.2 Grafting Technology

Grafting technology has evolved into a unique component in the production of several Solanaceous and Cucurbitaceous vegetables for pest management and for improvement of crop productivity. The ready availability of disease-resistant rootstocks and development of highly efficient grafting techniques have led to expansion in the use of grafted plants worldwide. Previous studies have provided an enhanced understanding of grafting-conferred defense mechanisms from inherent resistance within rootstocks to induced systemic resistance. However, there is still a lack of in-depth research on rootstock–scion interactions and long-distance signaling as related to improved resistance in grafted plants. In addition to resistance to soilborne diseases and nematodes, foliar and virus disease resistances of grafted plants warrant more studies. To differentiate the grafting effect per se from rootstock–scion interactions, well-designed studies should include a complete set of non-grafted and self-grafted rootstock and scion treatments in addition to the scion–rootstock grafts. Because crosstalk may exist among pathways that cope with biotic and abiotic stresses, a comprehensive understanding of the defense mechanisms in grafted plants will greatly assist with the effective development of desirable rootstocks to serve sustainable agriculture.

Even though there are many problems associated with cultivating grafted vegetable seedlings, the demand for successfully grafted seedlings is growing rapidly. Breeding multipurpose rootstock and developing efficient grafting machines and techniques will undoubtedly encourage increased use of grafted seedlings in many countries. Large-scale commercial production of vegetable seedlings is expanding rapidly in many developed countries, and this will lead to an increased commercial supply and use of grafted vegetable seedlings throughout the world.

35.5 Host Resistance

Host-plant resistance to disease is widely used; plant resistance to insects and nematodes is increasingly the focus of research and development. A more logical approach, albeit more difficult and time-consuming, is to combine as many genes as possible into each cultivar. Resistance of this type shall break down only if the pathogen evolves a complex race capable of overcoming all the resistant genes at once. Some plant breeders have followed this approach by combining all the known sources of TMV resistance into tomato cultivars. These cultivars have been grown commercially for a number of years and so far no TMV strain capable of overcoming their resistance has been found. Resistance as an IPM component remains the most effective way of combating viral diseases.

35.6 Guided Chemical Control

For control of pest and diseases, the pesticides still act as first line of defense. The use of pesticides will remain, at least in the short to medium term, an important strategy, allowing the grower to continue to produce economically a quality crop. Guided chemical control aims to reduce pesticide use by determining whether a control for a certain pest is necessary and, if so, when it should be applied (for optimum effect). In guided control, chemical control is only deemed necessary when the economic benefits counterbalance the costs; it implies use of curative rather than preventive pesticides. It is essential that the use of chemicals in IPM be based on economic threshold and only after considering the impact of the factors regulating the populations of pests and pathogens, making sure that there are no other effective management tools. IPM is foreseen as a way to keep pesticide applications to the minimum required and at the lowest effective dose with the most selective products, while observing the necessary safety intervals before harvest.

Select pesticides which least interfere with natural enemy activity. If selective pesticides are not available, alternatives, such as selective spraying, exist. Attempts have been made to improve the compatibility of beneficial organisms with pesticide application by selecting beneficials with resistance to chemical pesticides.

The increasing awareness on the harmful effects of pesticides on human health and environment are warring to the fore and there is a need to popularize eco-friendly crop protection technologies.

35.7 Genetic Engineering

Genetic engineering (GE) approaches to crop breeding can greatly facilitate the introgression of novel resistance traits into elite crop cultivars. This could increase the genetic diversity of pest resistance traits and reduce the negative impact of pests on crop yields. Crucially, it would provide a means of introducing resistance traits at a faster rate than that by which the pests evolve counter-resistance. This is why GE could provide a route to sustainable crop protection. However, the strength of this approach relies on discovery of novel resistance genes to incorporate using GE techniques. Another new direction for insect pest control is to introduce traits with a nontoxic mode of action for emission of semiochemicals that repel pests or attract their natural enemies.

Genetically modified (GM) crops are already widely used to control Lepidopteran pests and virus-resistant GM crops. The coat protein (CP) gene of cucumber mosaic virus (CMV) was cloned from a Chinese CMV isolate, the CaMV promoter and NOS terminator added, and the gene construct was transformed into sweet pepper plants to confer resistance to CMV.

35.8 Integrated Pest Management

Integrated pest management (IPM) offers a practical method for the effective management of pests under greenhouses. Through the adoption of sound cultural practices and monitoring techniques, accurate problem identification, and timely implementation and evaluation of appropriate management strategies, growers can improve crop production while minimizing their reliance on routine pesticide applications. IPM makes use of many different management options: cultural, physical, mechanical, biological, and chemical. Routine crop inspection alerts growers to developing pest and cultural problems while they are still minor and easily manageable. Early detection and intervention is the foundation of an IPM program.

In the greenhouses, efforts should be made to introduce and enhance biological control. IPM provides a doorway toward sustainable crop production intensification – the "Save and Grow" approach promoted by FAO for greenhouse crops. Integrated management of pests and diseases without conventional chemical pesticides is a goal that should be realized for most of the important vegetables and ornamentals within a decade.

To adopt the IPM package, a combined interdisciplinary action must be implemented by a consortium of competent institutions. The institutions that could be involved include State Agricultural Universities, ICAR Research Institutes, and State Agriculture and Horticulture Departments (for transfer of technology). The consortium shall carry out identification of potential area of IPM implementation, monitoring of key pests, controlling the soilborne pathogens through soil solarization, controlling the key pests through integrated pest management programs, and IPM package designing, based on the previous steps.

35.9 Conclusions

Greenhouse production is rapidly growing throughout the world providing plentiful food and ornamental crops. At the same time, the public is demanding greenhouse-grown food to be produced with minimal pesticide applications. IPM provides an avenue to meet these demands by heavily relying on alternatives to pesticides such as biological control and good horticultural practices and turning to pesticides only when absolutely necessary.

Successful crop production requires that crop pests and diseases be managed so that their effects on the plants are minimized. The management of crop pests and diseases is directed at preventing their establishment and minimizing their development and spread to prevent their establishment in the crop. The presence of pests and diseases is a facet of crop production, and growers must use all available options and strategies to avoid serious pest and disease problems.

Research efforts are needed for developing disease management technology under protected environment with emphasis on avoidance and selective use of pesticides. Safe waiting intervals based on harvest time pesticide residues needed to be established for the crops under protected environment as this information is lacking completely. Apart from this, emphasis to improve the awareness level of the growers for timely diagnosis and judicious use of pesticides needs to be taken up on priority.

The challenge ahead for researchers is to identify non-pesticide methods and integrate sustainable pest management technologies that can contribute to increased productivity, such as breeding for durable resistance, biological control, and cultural methods and which will have minimal adverse environmental and social impacts.

Index

P.P. Reddy, *Sustainable Crop Protection under Protected Cultivation*,
DOI 10.1007/978-981-287-952-3

MIX
Papier aus verantwortungsvollen Quellen
Paper from responsible sources
FSC® C105338

If you have any concerns about our products,
you can contact us on
ProductSafety@springernature.com

In case Publisher is established outside the EU,
the EU authorized representative is:
Springer Nature Customer Service Center GmbH
Europaplatz 3, 69115 Heidelberg, Germany

Printed by Libri Plureos GmbH
in Hamburg, Germany